www.wadsworth.com

wadsworth.com is the World Wide Web site for Wadsworth and is your direct source to dozens of online resources.

At *wadsworth.com* you can find out about supplements, demonstration software, and student resources. You can also send email to many of our authors and preview new publications and exciting new technologies.

wadsworth.com
Changing the way the world learns®

Constructions of Deviance

Constructions of Deviance
Social Power, Context, and Interaction

4th Edition

PATRICIA A. ADLER
University of Colorado

PETER ADLER
University of Denver

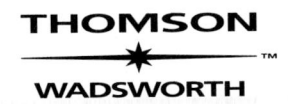

THOMSON
WADSWORTH

Australia • Canada • Mexico • Singapore • Spain
United Kingdom • United States

THOMSON
WADSWORTH

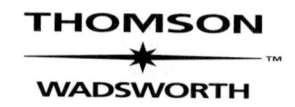

Sociology Editor: Robert Jucha
Assistant Editor: Stephanie Monzon
Editorial Assistant: Matthew Goldsmith
Technology Project Manager: Dee Dee Zobian
Marketing Manager: Matthew Wright
Marketing Assistant: Michael Silverstein
Project Manager, Editorial Production:
    Erica Silverstein
Print/Media Buyer: Robert King

Permissions Editor: Elizabeth Zuber
Production Service: Peggy Francomb,
    Shepherd, Inc.
Copy Editor: Colleen Yonda
Cover Designer: Andrew Ogus ■ Book Design
Cover Image: © Malcom Tarlofsky
Text and Cover Printer: The Maple-Vail Book
    Manufacturing Group
Compositor: Shepherd, Inc.

COPYRIGHT © 2003 Wadsworth, a division of Thomson Learning, Inc. Thomson Learning™ is a trademark used herein under license.

ALL RIGHTS RESERVED. No part of this work covered by the copyright hereon may be reproduced or used in any form or by any means—graphic, electronic, or mechanical, including but not limited to photocopying, recording, taping, Web distribution, information networks, or information storage and retrieval systems—without the written permission of the publisher.

Printed in the United States of America

1 2 3 4 5 6 7 06 05 04 03 02

For more information about our products, contact us at:
Thomson Learning Academic Resource Center
1-800-423-0563

For permission to use material from this text,
contact us by: **Phone:** 1-800-730-2214
    Fax: 1-800-730-2215
    Web: http://www.thomsonrights.com

Library of Congress Cataloging-in-Publication Data
Constructions of deviance : social power, context, and interaction / [edited by] Patricia A. Adler, Peter Adler.—4th ed.
    p. cm.
Includes index.
ISBN 0-534-55379-6
    1. Deviant behavior. 2. Social interaction. I. Adler, Patricia A. II. Adler, Peter

HM291 .C642 2002
302.5'42—dc21
                        2002024965

Wadsworth/Thomson Learning
10 Davis Drive
Belmont, CA 94002-3098
USA

Asia
Thomson Learning
5 Shenton Way
#01–01
UIC Building
Singapore 068808

Australia
Nelson Thomson Learning
102 Dodds Street
South Melbourne, Victoria 3205
Australia

Canada
Nelson Thomson Learning
1120 Birchmount Road
Toronto, Ontario M1K 5G4
Canada

Europe/Middle East/Africa
Thomson Learning
High Holborn House
50151 Bedford Row
London WC1R 4LR

Latin America
Thomson Learning
Seneca, 53
Colonia Polanco
11560 Mexico D.F.
Mexico

Spain
Paraninfo Thomson Learning
Calle/Magallanes, 25
28015 Madrid, Spain

To Diane and Dana
Who remind us that the zest of life lies near the margins

and

To Chuck
Who brings out the deviance in everyone

and

To John
Who showed us the miracle of life and rebirth

and

To Jane
Who lives the ordinary as deviant

Contents

PREFACE   XV

GENERAL INTRODUCTION   1

PART I
Defining Deviance   7

1   On the Sociology of Deviance   11
    Kai T. Erikson

2   A Typology of Deviance Based on Middle Class Norms   19
    Charles R. Tittle and Raymond Paternoster

3   Positive Deviance   30
    Druann Maria Heckert

PART II
Theories of Deviance   43

4   The Normal and the Pathological   55
    Emile Durkheim

5	*Social Structure and Anomie* Robert K. Merton	*60*
6	*Differential Association* Edwin H. Sutherland and Donald R. Cressey	*67*
7	*Labeling Theory* Howard S. Becker	*70*
8	*Control Theory of Delinquency* Travis Hirschi	*75*
9	*Conflict Theory of Crime* Richard Quinney	*84*
10	*Deviance: The Constructionist Stance* Joel Best	*90*

PART III
Studying Deviance — 95

11	*Child Abuse Reporting* Douglas J. Besharov with Lisa A. Laumann	*99*
12	*Survey of Sexual Behavior of Americans* Edward O. Laumann, John H. Gagnon, Robert T. Michael, and Stuart Michaels	*106*
13	*Researching Dealers and Smugglers* Patricia A. Adler	*116*

PART IV
Constructing Deviance — 133

Moral Entrepreneurs

14	*The Social Construction of Drug Scares* Craig Reinarman	*137*
15	*Blowing Smoke: Status Politics and the Smoking Ban* Justin L. Tuggle and Malcolm D. Holmes	*149*
16	*Moral Panics: The Case of Satanic Day Care Centers* Mary deYoung	*160*

Differential Social Power

17	*The Saints and the Roughnecks* William J. Chambliss	*169*

18	*The Police and the Black Male* Elijah Anderson	183
19	*Homophobia and Women's Sport* Elaine M. Blinde and Diane E. Taub	195

PART V
Deviant Identity   207

20	*The Adoption and Management of a "Fat" Identity* Douglas Degher and Gerald Hughes	211
21	*Becoming Bisexual* Martin S. Weinberg, Colin J. Williams, and Douglas W. Pryor	222
22	*Anorexia Nervosa and Bulimia* Penelope A. McLorg and Diane E. Taub	233

PART VI
Accounts   245

23	*Convicted Rapists' Vocabulary of Motive* Diana Scully and Joseph Marolla	247
24	*The Influence of Situational Ethics on Cheating* Donald L. McCabe	263

PART VII
Stigma Management   273

25	*Identity and Stigma of Women with STDs* Adina Nack	277
26	*Stigma Management and Collective Action Among the Homeless* Leon Anderson, David A. Snow, and Daniel Cress	286
27	*Collective Stigma Management and Shame: Avowal, Management, and Contestation* Daniel D. Martin	305

PART VIII
The Social Organization of Deviance   325

Individuals
28 *Sexual Asphyxia*   327
    Shearon A. Lowery and Charles V. Wetli

Subcultures
29 *Real Punks and Pretenders: The Social Organization of a Counterculture*   337
    Kathryn J. Fox

Gangs
30 *Gender and Victimization Risk Among Young Women in Gangs*   353
    Jody Miller

Formal Organizations
31 *International Organized Crime*   367
    Roy Godson and William J. Olson

Corporations
32 *The Crash of ValuJet Flight 592: A Case Study in State-Corporate Crime*   380
    Rick A. Matthews and David Kauzlarich

PART IX
Structure of the Deviant Act   393

Cooperation
33 *Cruising for Sex in Public Places*   395
    Richard Tewksbury

34 *The Manufacture of Fantasy*   408
    Amy Flowers

Conflict
35 *Fraternities and Rape on Campus*   416
    Patricia Yancey Martin and Robert A. Hummer

36 *Opportunity and Crime in the Medical Professions*   431
    John Liederbach

PART X
Phases of the Deviant Career 441

Entering Deviance
37 *Joining a Gang* 445
Martín Sánchez Jankowski

Managing Deviance
38 *Gay Male Christian Couples and Sexual Exclusivity* 467
Andrew K. T. Yip

Leaving Deviance
39 *Shifts and Oscillations in Deviant Careers: The Case of Upper-Level Drug Dealers and Smugglers* 481
Patricia A. Adler and Peter Adler

Post-Deviance
40 *The Professional Ex-: An Alternative for Exiting the Deviant Career* 494
J. David Brown

REFERENCES FOR THE GENERAL AND PART INTRODUCTIONS 507

Preface

In the wake of the tragedies that occurred on September 11, 2001 at the World Trade Center in New York City and at the Pentagon in Washington, DC, the term *deviance* has, perhaps, taken on new meaning that none of us could have ever imagined before these atrocities took place. These events and the incumbent problems that have ensued cannot be done justice with any sort of sociological, psychological, economic, or political theories that we have had at our disposal. Our global society, no doubt, has witnessed numerous examples of groups that have transgressed society's norms, morals, and laws, but no one was prepared for these attacks, and the explanations for them, though far-reaching, will continue to be advanced for the centuries to come. The sociology of deviance is a starting point for looking at how this type of evil can be perpetuated on a civilization, but it would be foolhardy to say that, within the confines of previous research and extant theories, a full explanation can be offered that can account for how these types of acts can happen. As confused as this has left us, contained within the pages of this book are at least some examples of how social scientists have approached the topic of deviance and tried to grapple with the conundrum of how, when, and why people deviate. You will find a wide swath of subjects that will pique your curiosity, make you question your beliefs, and provide some preliminary answers to why people violate the codes of society. In addition you will discover that, in some cases, the so-called deviants themselves are just like you and us, no different in their outlooks, appearances, and beliefs, and that the causes of their violations of society's expectations do not lie within themselves, but in the society that so labels them. Ranging from structural to individualistic explanations, the sociology of deviant behavior has added to our knowledge of how social power, social context, and social interaction all interact, symbiotically, to create the realm of deviance in society.

As the shifting sands of morality in American society continue to transform our culture, deviance and the changing definitions are at the fore. As Gusfield (1967) showed us more than three decades ago, with these social changes, activities that were considered nondeviant a generation ago have taken on deviant characteristics (the near prohibition on cigarettes in public settings is a case in point), whereas activities that were once deviant are now acknowledged as commonplace, albeit still stigmatized (the fad of tattooing and piercing, especially among teenagers, being perhaps the most remarkable). For better or worse, we are living at a time when rapid social change is occurring virtually right in front of our eyes. With this type of backdrop, there can be no better time to study the altering definitions of deviant behavior and the consequences that follow for society. It is our hope that, in this fourth edition of *Constructions of Deviance,* we are able to highlight the excitement that this field holds. We want to show students the liveliness of the debates, the graphic images that sociologists paint, and the wide array of activities that fall under the domain of deviance.

NEW TO THIS EDITION

In all the editions of this book we have tried to keep pace with the nuances that have happened in the research on deviant behavior. It has been our hope that this book represents a collection of both the classic pieces that have stood the test of time and the latest developments in the field that reflect the swiftly occurring changes in American culture. We are pleased to offer to you, then, many of the same selections that students have enjoyed in the past editions, while presenting twelve new pieces that we think will be around for the next set of students who read future editions of this book. New to this volume are pieces on obesity and obesity control groups, women and sexually-transmitted diseases, stigma work among the homeless, gay male cruising, women in gangs, state-corporate crime and airplane safety (which has taken on a particularly new meaning in light of the events of September 11), homophobia and women athletes, the telephone sex industry, and medical fraud. In addition we have added a stronger emphasis on middle-class (mainstream) definitions and middle-class forms of crime and a new original piece on the social constructionist perspective on deviance written by Joel Best, who may not have invented the term, but who by now has become practically synonymous with it.

Organizationally we have changed the book slightly by providing greater separation and clarification of the different parts of the book. We hope that this accentuates for students the differences between the concepts around which the book is focused. Further we have significantly expanded some of the part introductions, particularly in the theory section, so that this book can more fruitfully be used as a text, as well as a reader. Finally, in an effort to show students the relevance of each article, we have added introductory synopses that help to interpret and relate the specific researches to the overall theme of the book.

The book still leans toward the social constructionist approach, remaining a text that builds on our intellectual backgrounds in symbolic interactionism and ethnographic research. As such it retains its vibrant appeal to students by offering the most up-to-date empirical readings that are drawn from participant-observation studies rich in experiential description of deviance from the everyday lived perspective. At the same time, the book has continued to move toward the mean of the field, incorporating classical theoretical and innovative methodological elements that satisfy the needs of professors.

We offer this collection as a testimonial to the continuing vibrancy of deviance research. Our sense is that there has never been a better time to study this phenomenon and to look at the changes that they represent about society. Above all else, we hope that you find the readings enjoyable, enlightening, and thought-provoking.

ACKNOWLEDGMENTS

Merely eight years after this project was initially conceptualized we are now putting together it fourth iteration. This almost seems impossible, but once again, speaks to the vitality of this subject matter. By now literally thousands of students have been exposed to the readings we have presented and have been christened to the new sociology of deviance. Many people have provided critical feedback that has helped us in fashioning this fourth edition. First and foremost are our many students, particularly in Patti's class, "Deviance in U.S. Society," at the University of Colorado, where over 500 bodies cram into Chem 140 each semester, and Peter's students at the University of Denver in his class, "Deviance and Society," who continue to brave the elements, despite its reputation as one of the toughest courses on campus. They have provided us with a template for what the contemporary collegian desires, does, and dreams about. These people remind us of the diversity of sentiments—moral and immoral, normative and deviant, radical and conservative—that exist.

There have been some special people, such as Katherine Coroso, Marci Eads, Abby Fagan, Molly George, Tamera Gugelmeyer, Joanna Gregson Higginson, Tom Hoffman, Katy Irwin, Jennifer Lois, Adina Nack, Joe Settle, and Sarah Sutherland, who not only served as TAs for this course but have provided much of the impetus for the changes and amendments we have made throughout. Our friends in the discipline continue to suggest studies, supply encouragement, and lend support for our endeavors. Whether a quick conversation in the hallway or a convention hotel, an e-mail message, a lengthy letter, or a harangue over the telephone, they remind us that we should keep the edge and continue to search for the latest examples to hold their students' interest. During a visit to Australia, we were pleased to see that this book has an audience Down Under, and we gained a great deal of insight into the Australian brand of deviance from discussions with, especially, Alec Pemberton and Jill Hooley, both at the University of Sydney.

We have been extremely fortunate to get a set of constructive comments from the reviewers Wadsworth commissioned: Kim Davies, Augusta State University; Jackie Eller, Middle Tennessee State University; E. Ernest Wood, Edinboro University of Pennsylvania; Barbara Perry, Northern Arizona University; Phyllis Kitzerow, Westminster College; Thomas C. Wilson, Florida Atlantic University; Paul Price, Pasadena City College; Kathryn Fox, University of Vermont. The previous edition was reviewed by Kerry Ferris, University of California, Los Angeles; Scott Hunt, University of Kentucky; Teresa Lagrange, Cleveland State University; Matt Lee, University of Delaware; Jay Livingstone, Montclair State University; and Angus Vail, University of Connecticut. The original project received helpful comments from Tom Cook, Wayne State College; William M. Hall, Alfred University; John D. Hewlitt, Northern Arizona University; Ruth Horowitz, University of Delaware; Phyllis G. Kitzerow, Westminster College; and Michael Nusbaumer, Indiana University-Purdue University at Fort Wayne. The stalwart staff at Wadsworth has provided unending support during the process of revisions. We are fortunate to have worked with such diligent professionals as Paula Begley-Jenkins, Halee Dinsey, Jerilyn Emori, Peggy Francomb, Jennifer Jones, Bob Jucha, Bob Kauser, Ari Levenfeld, Lin Marshall, Reilly O'Neal, Jane Pohlenz, Erica Silverstein, Denise Simon, Steve Spangler, Staci Wolfram, Matthew Wright, Dee Dee Zobian and Beth Zuber. Special commendation must go to Eve Howard, our special editor who has worked hand-in-hand with us since the second edition, and Serina Beauparlant, the editor who originally conceived this project. At the University of Denver, the irrepressible Dorene Miller continues to do her stalwart job, assuring that the project stayed on track and was finished on time. Although you can't always tell a book by its cover, we would like to take this opportunity to thank Malcolm Tarlofsky, artist of the covers of all four editions, who added to the book's success by producing, in our opinion, the finest covers in the field.

One of the pleasures of editing this book has been sharing it with our friends. As we respectfully dedicated the first edition to our partners in crime Diane Duffy and Dana Larsen, the second edition to our dear friend Chuck Gallmeier, and the third edition to the implacable John Irwin, we sincerely and warmly dedicate this edition to Jane Horowitz, whose friendship now spans over thirty years and who has stuck with us through all of the trials and tribulations of life. Finally, our children, Jori and Brye, remind us that deviance is often a youthful endeavor and that we need to be respectful of their desires to test the boundaries. To all of the readers of previous editions, thanks for the support; to the new readers of this fourth edition, welcome to the journey.

About the Editors

Patricia A. Adler (Ph.D., University of California, San Diego) is Professor of Sociology at the University of Colorado. She has written and taught in the areas of deviance, drugs in society, and the sociology of children. A second edition of her book *Wheeling and Dealing* (Columbia University Press), a study of upper-level drug traffickers, was released in 1993.

Peter Adler (Ph.D., University of California, San Diego) is Professor of Sociology at the University of Denver. His research interests include social psychology, qualitative methods, and the sociology of work, sport, and leisure.

Together the Adlers have edited the *Journal of Contemporary Ethnography* and were the founding editors of *Sociological Studies of Child Development*. In 2001, their anthology, *Sociological Odyssey,* was published by Wadsworth, as well as *Encyclopedia of Criminology and Deviant Behavior,* Volume 1, edited with Jay Corzine. Among their many books are *Backboards and Blackboards,* a participant-observation study of college athletes that was published by Columbia University Press in 1991, and *Peer Power,* a study of the culture of elementary schoolchildren that was published by Rutgers University Press in 1998. Currently they are studying the subculture of resort workers, and have just begun research on people who cut themselves (self-mutilation). Patti and Peter have two children, a son, Brye, who attends Emory University, and a daughter, Jori, who graduated from Emory in 2000 and now works for the Turner Broadcasting System.

About the Contributors

Elijah Anderson is currently Charles and William L. Day Professor of the Social Sciences and Professor of Sociology at the University of Pennsylvania. An expert on the sociology of Black America, he is the author of *A Place on the Corner, Streetwise,* and *Code of the Street.* He is interested in the social psychology of organizations, field methods of social research, social interaction, and social organization.

Leon Anderson is Professor and Chair in the Department of Sociology and Anthropology at Ohio University. He is the coauthor with David Snow of *Down on Their Luck: A Study of Homeless Street People* (University of California Press, 1993). His current research interests include social service organizations and the commercial blood products industry.

Howard S. Becker lives and works in San Francisco. He is the author of *Outsiders, Arts Worlds, Writing for Social Scientists,* and *Tricks of the Trade.* He has taught at Northwestern University and the University of Washington. In 1998 the American Sociological Association bestowed upon him the Career of Distinguished Scholarship Award, the association's highest honor.

Douglas J. Besharov, a lawyer, is the Joseph J. and Violet Jacobs Scholar in Social Welfare Studies at the American Enterprise Institute for Public Policy Research and Professor at the University of Maryland School of Public Affairs. He was the first director of the U.S. National Center on Child Abuse and Neglect. Among his publications is *Recognizing Child Abuse: A Guide for the Concerned* (Free Press).

Joel Best is Professor and Chair of the University of Delaware's Department of Sociology and Criminal Justice. His books include *Threatened Children* (1990), *Random Violence* (1999), and *Damned Lies and Statistics* (2001).

Elaine M. Blinde is Professor and Graduate Coordinator of Physical Education at Southern Illinois University at Carbondale. Her research relates to sociological analysis of sport, with a particular focus on gender issues. Recent work and publications relate to disability and sport, gender beliefs of young girls, and women's relationship to baseball.

J. David Brown earned his Ph.D. in sociology at the University of Denver. He is currently the Director of Educational Assessment and Institutional Research at Red Rocks Community College. His most recent scholarly activities have focused on access, mobility, and higher education partnerships in the New Independent States of Eastern Europe.

William J. Chambliss is Professor of Sociology at George Washington University. He is the author and editor of over twenty books, including *Law, Order, and Power* (with Robert Seidman); *On the Take: Petty Crooks to Presidents; Organizing Crime; Exploring Criminology; Boxman: A Professional Thief's Journey* (with Harry King); *Crime and the Legal Process; Making Law* (with Marjorie S. Zatz); and most recently an introductory text, *Sociology* (with Richard P. Applebaum). He is the Past President of the Society for the Study of Social Problems and the American Society of Criminology. He is currently writing a book on the political economy of piracy and smuggling.

Daniel M. Cress is Associate Professor of Sociology at Western State College in Gunnison, Colorado. His teaching and research interests encompass social movements, social inequality, environmental sociology, and research methods. He has published work on protest activity by homeless people and is currently engaged in a project on wildland fire fighters.

Donald R. Cressey was Professor of Sociology at the University of California, Santa Barbara, when he died in 1987. His most well-known publications include *Principles of Criminology* (with Edwin H. Sutherland); *Other People's Money;* and *Theft of the Nation.* Cressey received many honors for his research and teaching, and he cherished none more than the Edwin H. Sutherland Award presented by the American Society of Criminology in 1967.

Douglas Degher is Professor of Sociology at Northern Arizona University. His major interests are in social theory, identity, and the sociology of sport. His most recent work is an analysis of the "ESPNization of contemporary sport."

Mary deYoung is Professor and Chair of the Department of Sociology at Grand Valley State University. Her work focuses on the nexus of trauma and culture and memory. She has two books forthcoming on ritual abuse moral panic, *The Ritual Abuse Controversy: An Annotated Bibliography* (McFarland, 2002) and *The Day Care Ritual Abuse Moral Panic* (McFarland, 2003).

Emile Durkheim (1858–1917) was a French sociologist who is generally considered to be the father of sociology. His major works are *Suicide; The Rules of Sociological Method; The Division of Labor in Society;* and *The Elementary Forms of Religious Life.*

ABOUT THE CONTRIBUTORS

Kai T. Erikson has been Professor of Sociology and American Studies at Yale University for more than thirty years. He is the author of several books, including *Wayward Puritans; Everything in Its Path;* and *A New Species of Trouble.* He served as President of the American Sociological Association 1984–1985.

Amy Feldstein Flowers earned her Bachelor's degree in philosophy from the University of California at Berkeley and her Ph.D. in sociology from the University of Southern California. Her book, *The Fantasy Factor: An Insider's View of the Phone Sex Industry,* was published by the University of Pennsylvania Press in 1998, and was recently translated into Korean. She is an independent scholar and practicing sociologist in private industry. She lives in the metropolitan New York area.

Kathryn J. Fox is an Associate Professor of Sociology at the University of Vermont. She teaches and studies in the area of deviance and social control. Her ethnographic studies include punk subcultures, an AIDS prevention project with injection drug users, and prison therapy for violent offenders. She is currently beginning a new project focusing on delinquency in schools.

John H. Gagnon was Professor of Sociology at the State University of New York, Stony Brook until 1998. He is the author or coauthor of such books as *Sex Offenders; Sexual Conduct; Human Sexualities;* and *The Social Organization of Sexuality.* In addition, he has been the coeditor of a number of books, most recently *Conceiving Sexuality* and *Encounter with AIDS: Gay Men and Lesbians Confront the AIDS Epidemic,* as well as the author of many scientific articles.

Roy Godson is Professor of Government at Georgetown University and President of the National Strategy Information Center. He has served as a consultant to the president's Foreign Intelligence Advisory Board, the National Security Council, and related agencies of the U.S. government. He has written, coauthored, or edited over sixteen books on intelligence and national security.

Druann Maria Heckert received her B.A. from Frostburg State, M.A. from the University of Delaware, and Ph.D. from the University of New Hampshire. She currently teaches at Fayetteville (North Carolina) State University. Her major research interests/publications are in the areas of positive deviance, lethal violence, and appearance norms (hair color).

Travis Hirschi received his Ph.D. in sociology from the University of California, Berkeley. He is currently Regents Professor Emeritus at the University of Arizona. He served as President of the American Society of Criminology and has received that society's Edwin H. Sutherland award. His books include *Delinquency Research* (with Hanan C. Selvin); *Causes of Delinquency; Measuring Crime* (with Michael Hindelang and Joseph Weis); and *A General Theory of Crime* (with Michael R. Gottfredson). His most recent book is a coedited volume with Michael Gottfredson, *The Generality of Deviance.*

Malcolm D. Holmes is Professor of Criminal Justice at the University of Wyoming. He has published a number of articles on the effects of ethnicity on criminal justice decision making. His current research examines the myth of "predatory" crime.

Gerald Hughes is Professor of Sociology at the University of Northern Arizona. His primary interests are in identity theory, with his most recent work focusing on motorcycle riders and identity transformation.

Robert A. Hummer is Director of the Population Research Center and Associate Professor of Sociology at the University of Texas at Austin. His current research focuses on the health patterns in the United States, particularly as they relate to race/ethnicity, migration, and religious involvement. He is coauthor (with Richard Rogers and Charles Nam) of *Living and Dying in the USA: Behavioral, Health and Social Differentials of Adult Mortality* (Academic Press, 2000).

David Kauzlarich is Assistant Professor of Sociology at Southern Illinois University, Edwardsville. He has published widely on the topics of state crime and nuclear weapons and is coauthor of two books: *Introduction to Criminology*, 8th edition (2002, with Hugh D. Barlow) and *Crimes of the American Nuclear State: At Home and Abroad* (1998, with Ronald C. Kramer).

Edward O. Laumann is the George Herbert Mead Distinguished Service Professor in the Department of Sociology and the College at the University of Chicago. Previously he has been Editor of the *American Journal of Sociology* and Dean of the Social Sciences Division and Provost at the University of Chicago. Two volumes on sexuality were published in 1994: *The Social Organization of Sexuality* and *Sex in America*, and along with Robert Michael, he has recently published *Sex, Love, and Health: Private Choices and Public Policy* (University of Chicago Press, 2001).

Lisa Laumann-Billings completed her doctorate work in developmental and clinical psychology at the University of Virginia. She has worked and published in the areas of divorce, family conflict, and child maltreatment for the past 10 years. She is currently working with Dr. David Olds at the Prevention Research Center in Denver, Colorado on preventative strategies for reducing child maltreatment and family violence in high-risk families.

John Liederbach is Assistant Professor of Criminal Justice at the University of North Texas. His primary research interests are in the areas of white-collar and professional crime, as well as police behavior in small-town, rural, and suburban jurisdictions.

Shearon A. Lowery is a member of the Department of Sociology and Anthropology at Florida International University.

Joseph Marolla is Director of the Center for Teaching Excellence and Chairperson in the Department of Sociology and Anthropology at Virginia Commonwealth University. He is doing research in the sociology of sport and continues to remain interested in the social construction of deviance.

Daniel D. Martin is Assistant Professor in the Department of Sociology, Gerontology, and Anthropology at Miami University of Ohio. He teaches in the areas of inequality, organizations, and urban sociology. Since 1998 he has been conducting community action research with low-income families of murdered children and neighborhood organizers in Cincinnati and Columbus, Ohio. Dr. Martin is a member of Oxford Citizens for Peace and Justice, and Professors for Peace, a newly formed group on the campus of Miami University in Oxford, Ohio.

Patricia Yancey Martin is Daisy Flory Professor of Sociology at Florida State University. Martin's teaching and research on gender and organizations had led her to focus on rape crisis centers and men's social fraternities. She co-edited a book with Myre Marx Ferree, *Feminist Organizations: Harvest of the New Women's Movement* (1995), and has recent articles on group rape and gender dynamics in police departments, prosecutors' offices, hospitals, and rape crisis centers. Martin tries to make the invisible dynamics of gender visible so people can be assessed based on competence rather than stereotypes, so service recipients (like rape victims)can be treated responsively, and so organizations can avoid hurting people unintentionally. She has papers forthcoming on gender dynamics in large corporations (in *ORGANIZATIONS*)and in the legal profession (*SIGNS*).

Rick A. Matthews is Assistant Professor of Sociology at Ohio University. His research interests include corporate and government crime as well as the political economy of street crime. He is currently conducting research on corporate crime in Nazi Germany during the Holocaust.

Donald L. McCabe is Professor of Organizational Management at Rutgers University. Over the last decade he has done extensive research on college cheating and has surveyed over 16,000 students at more than 60 colleges and universities around the country. He also recently completed a nationwide survey of 4500 high school students. He has published this research widely in business, education, and sociology journals and is founding president of the Center for Academic Integrity, a consortium of more than 200 colleges and universities from around the country who are joined in a united effort to promote academic integrity among college students. He received his Ph.D. in management from New York University.

Penelope A. McLorg is Adjunct Assistant Professor of Anthropology at Southern Illinois University, Carbondale. Her specialization is biological anthropology, with concentrations in human biology, biomedical anthropology, aging, and women's health. She recently received Southern Illinois University, Carbondale's Outstanding Dissertation Award for her research on aging patterns in customary blood glucose among Maya women in rural Yucatan, Mexico. Her published works concern sociocultural factors in eating disorders, infant feeding attitudes and behaviors, and bone anatomy and measurement.

Robert K. Merton is University Professor Emeritus at Columbia University and a member of the Adjunct Faculty of the Rockefeller University. His books include *Social Theory and Social Structure; Sociology of Science; Sociological Ambivalence;* and *On the Shoulders of Giants.* He is the only sociologist awarded the National Medal of Science, the nation's highest scientific honor.

Robert T. Michael is the Eliakim Hastings Moore Distinguished Professor and Dean of the Harris Graduate School of Public Policy at the University of Chicago. From 1984 to 1989 he served as Director of the National Opinion Research Center (NORC).

Stuart Michaels served as Project Manager of the National Health and Social Life Survey (NHSLS).

Jody Miller is Assistant Professor of Criminology and Criminal Justice at the University of Missouri, St. Louis. Dr. Miller specializes in feminist theory and qualitative methods. Her research focuses on gender, crime and victimization, particularly in the contexts of youth gangs, urban communities, and the commercial sex industry. Her monograph, *One of the Guys: Girls, Gangs, and Gender,* was published by Oxford University Press in 2001.

Adina Nack received her Ph.D. in sociology at the University of Colorado in 2001 and is an Assistant Professor in the Department of Sociology at the University of Maine. Her research focuses on examining issues of deviance, social psychology, and inequality within areas of medical sociology such as sexual health. Currently she is working on a study of the interactional and structural factors that may explain substantial recent decreases in Maine's teen pregnancy rates.

William J. Olson is Staff Director for the Caucus on International Narcotics Control for the U.S. Senate. Previously he was Deputy Assistant Secretary of State in the Bureau of International Narcotics Matters. He was also Director and served as Acting Deputy Assistant Secretary of Defense for Low Intensity Conflict. His published works include books and articles on U.S. strategic interests in the Persian Gulf, guerrilla warfare, counterinsurgency, and the war on drugs.

Raymond Paternoster is Professor in the Department of Criminology and Criminal Justice, University of Maryland. He is interested in the testing of theories of crime and the unfolding of crime over the life course. He currently is conducting a detailed study of the imposition of the death penalty in the state of Maryland.

Douglas W. Pryor is Associate Professor in Sociology and Acting Chair of the Department of Sociology, Anthropology, and Criminal Justice at Towson University in Maryland. He is coauthor of *Dual Attractions: Understanding Bisexuality* and author of *Unspeakable Acts: Why Men Sexually Abuse Children.* His current research and writing includes papers on bisexuality, community-based treatment of familial sex offenders, and fear of rape among women.

Richard Quinney is Professor of Sociology Emeritus at Northern Illinois University. His books include *The Social Reality of Crime; Critique of Legal Order; Class, State, and Crime; Criminology as Peacemaking* (with Harold E. Pepinsky); and *Criminal Behavior Systems* (with Marshall B. Clinard and John Wildeman). His autobiographical reflections are contained in *Journey to a Far Place* and *For the Time Being.*

Craig Reinarman is Professor and Chair of Sociology at the University of California, Santa Cruz, and Visiting Scholar at the Center for Drug Research at the University of Amsterdam. He has served on the Board of Directors of the College on Problems of Drug Dependence, as a consultant to the World Health Organization's Programme on Substance Abuse, and a principal investigator on research grants from the National Institute on Drug Abuse. Dr. Reinarman is the author of *American States of Mind* (Yale University Press,

1987), coauthor of *Cocaine Changes* (Temple University Press, 1991), and coeditor of *Crack in America* (University of California Press, 1997).

Martín Sánchez-Jankowski is Professor of Sociology and Director of the Center for Urban Ethnography at the University of California, Berkeley. His specialties are in the areas of poverty, interpersonal violence, and racial/ethnic stratification. Currently he is finishing a book on social change within poverty neighborhoods, which is based on eight years of field work in five New York and Los Angeles neighborhoods.

Diana Scully is Professor of Sociology and Women's Studies at Virginia Commonwealth University in Richmond, Virginia where she is also Director of the Women's Studies Program. Her books include *Men Who Control Women's Health* (Teachers College Press, 1994) and *Understanding Sexual Violence* (Routledge, 1994).

David A. Snow is Professor of Sociology at the University of California, Irvine. He has published widely on various aspects of social movements, framing process in the context of movements, conversion processes, self and identity, ethnographic field methods, and homelessness. He is the author of *Shakubuku: A Study of Nicheren Shoshu Buddhist Movements in America, 1960–1975,* and coauthor of *Down on Their Luck: A Study of Homeless Street People* (with Leon Anderson), winner of numerous scholarly awards, and of *Social Movements: Readings on Their Emergence, Mobilization, and Dynamics* (with Dough McAdam).

Edwin H. Sutherland (1883–1950) received his Ph.D. from the University of Chicago. His major works include his enduring textbook, *Criminology,* which was first published in 1924, and *The Professional Thief* and *White Collar Crime.* He is generally regarded as the founder of differential association theory.

Diane E. Taub is Associate Dean of the College of Liberal Arts and Professor of Sociology at Southern Illinois University, Carbondale. Her research primarily involves the sociology of deviance, with a focus on the experiences of women. Recent publications concern eating disorders in women and the lived experiences of individuals with physical disabilities.

Richard Tewksbury is Professor of Justice Administration at the University of Louisville. He received his Ph.D. in sociology from the Ohio State University. His research focuses on men's studies, constructions of sex, gender, and sexuality identities, criminal victimization risks, and institutional corrections.

Charles R. Tittle received his Ph.D. in sociology in 1965 from the University of Texas, Austin. He has served on the faculties of Indiana University, Florida Atlanta University, Washington State University, and is currently Professor of Sociology at North Carolina State University where he holds the Goodnight/Glaxo-Wellcome Endowed Chair of Social Science. His interests are in social theory and urban sociology.

Justin L. Tuggle received his B.A. from Humboldt State University and his M.A. from the University of Wyoming. He teaches fifth grade at Grant Elementary School in Redding, California, and teaches sociology part-time at Shasta Community College. Currently he is interested in patterns of drug use with organized sports.

Martin S. Weinberg received his Ph.D. at Northwestern University and is Professor of Sociology at Indiana University. He is coauthor of *Deviance: The Interactionist Perspective; Sexual Preference; Homosexualities; Male Homosexuals;* and *Dual Attraction,* and has contributed articles to such journals as the *American Sociological Review, Social Problems, Journal of Contemporary Ethnography, Archives of Sexual Behavior,* and *Journal of Sex Research.*

Charles V. Wetli is affiliated with the Medical Examiner Office, Dade County, Florida, and the University of Miami School of Medicine.

Colin J. Williams is Professor of Sociology at Indiana University-Purdue University, Indianapolis. He is coauthor of *Male Homosexuals; Sex and Morality in the U.S.;* and *Dual Attraction.*

Andrew K. T. Yip has been researching the interplay of sexuality and spirituality, as well as same-sex relationships. He is the author of *Gay Male Christian Couples: Life Stories* (Praeger, 1997). He has also published widely in edited books and journals such at the *British Journal of Sociology* and *Sociology of Religion.* He is currently involved in two research projects: British non-heterosexual Muslims and non-heterosexual aging.

Constructions of Deviance

General Introduction

The topic of deviance has held an enduring fascination for students of sociology, gripping their interest for several reasons. Some people hold career plans that include law or law enforcement and want to expand their base of practical knowledge. Others feel a special closeness for the subject of deviance based on personal experience or inclination. A third group is drawn to deviance merely because it is different, offering the promise of excitement or the exotic. The sociological study of deviance can fulfill all these goals, taking us deep into the criminal underworld, inward to the familiar, and outward to the fascinating and bizarre. In the following pages we peer into the deviant realm, looking at both deviants and those who define them as such. In so doing, we look at a range of deviant behaviors, discuss why people engage in these, and analyze how they are sociologically organized. We begin in Part I by defining deviance, in an effort to lay down the parameters of its scope.

STUDYING DEVIANCE

Reasonable theories and social policies pertaining to deviance must be based on a firm foundation of accurate knowledge. Social scientists have an array of different methodologies and sources of data at their disposal, including survey research, experimental design, historical methods, official statistics, and field research. All of these methods have obvious strengths and weaknesses and have been used by sociologists in studying deviance. While some generate statistical portraits about the extensiveness of deviant behaviors, we undertake in this book to offer a richer, more experiential understanding of what goes on in deviant worlds, showing *how* things happen and what they *mean* to participants. It is the reports of field researchers, individuals who immerse themselves personally in deviant settings, that yield such depth and descriptive accounts. It is also our belief that, due to the often secretive nature of deviance, methods that objectify or distance researchers from the people being researched will be less likely to accurately portray deviant worlds. Thus, the works in this book are tied together by their experiential richness and by the belief that researchers must study deviance as it naturally occurs in the real world. The most appropriate methodology for this, field research, advocates that sociologists should get as close as possible to the people they are studying in order to understand their worlds (Adler and Adler 1987). Despite the problems that arise from the secretive and hidden nature of deviant acts, sociologists have devised techniques to penetrate secluded deviant worlds. These methodological ploys often come complete with perils, so it is wise for people who are considering studying deviance to be aware of these issues. Part III discusses field research methods and two of the other common sources of information about deviant behavior.

CONSTRUCTING DEVIANCE

In Parts I, II, and IV we delve into the origins and definitions of deviant behavior. Some have traditionally considered defining deviance as a simple task, implying that a widespread agreement exists about what is deviant and what is not. This view corresponds to the **absolutist position** on deviance, the perspective that something obvious within an act, belief, or condition makes it different from the norm in everybody's eyes. In its fundamental core it embodies the unambiguous, objective essence of true or real deviance. This view has its roots in both religious and naturalistic assumptions, with people arguing that certain acts are contrary to the strictures of God or the laws of nature. Contemporary religious leaders, especially those of the charismatic and evangelical persuasions, often use these arguments in advancing their moral beliefs in written and verbal oratory. Absolutist views of deviance are eternal and global. If something is judged to have been intrinsically morally wrong in the past (for example, adultery or divorce), it should be recognized as wrong now and always in the future. Similarly, if something is considered to be morally wrong in one place, it should be judged wrong everywhere (for example, the practice of female genital mutilation, which is viewed with horror in America but considered essential to defining a proper, chaste, marriageable woman in parts of Africa).

Deviance, then, is not viewed as something that is determined by social norms, customs, or rules, but that is located in an intrinsic essence that stands apart from and exists before the creation of these socially created codes. According to this perspective, deviance is an objective fact, and by any other name it would appear the same. People have backed up this belief system by pointing to the existence of universal taboos surrounding such acts as murder, incest, and lying. These acts, they claim, are deviant in their very nature, according to an absolute moral order.

The contrasting view is the **social constructionist position,** which holds that there are no absolute, unchanging, or universal features that define deviance. This position suggests that definitions are forged by people, and reforged by them anew in different eras and locations, leading to significant constructions and reconstructions of deviance. When different social contexts frame and give meaning to the perception and interpretation of acts, the same act may be alternately perceived as deviant or normative. Constructionists argue that deviance is thus lodged in the eyes of the beholder rather than in the act itself. This creates the possibility of multiple definitions of acts simultaneously existing, both deviant and nondeviant, among different groups. We see this clearly in cases of controversial, morally debated acts such as getting an abortion, praying in school, and consuming illicit drugs. Major campaigns have been assembled to sway public opinion over the morality and appropriateness of these behaviors, with large segments of the country falling into oppositional camps. Groups that are fairly alike in their composition may forge shared agreements about norms that would be differently received among a broader segment of the general public. For example, more widely acceptable behavior, such as dancing, is condemned in some thoroughly conservative environments, such as Bible Belt religious colleges. At the same time, language that is widely used throughout the country may be condemned as politically incorrect and morally offensive on extremely liberal campuses. Even these examples of acts considered intrinsically, essentially deviant can be seen to have flexibility in their definitions as there are conditions under which these acts would be considered nondeviant. To kill somebody for personal gain, vengeance, freedom, or through negligence might be considered deviant (and, more likely, criminal), but when these acts are committed by the state (for executions, war, or covert intelligence), in self defense, or on one's own property (in states where they have "make my day" laws), they are considered not only nondeviant but heroic. Similarly, anthropologists have found cultures that condone certain forms of incest to quiet or soothe infants (Henry 1964; Weatherford 1986). Often, too, people differentiate between normal lies and white lies, those designed to spare the feelings of someone other than the speaker or differentiated by circumstances that are deemed to be acceptable. Social constructionists argue that we should regard definitions of deviance as *relative* to their situations and surroundings.

Assuming shared agreement, the absolutist position on deviance guides researchers to study the structural conditions fostering deviance, leading them to generate causal correlations that might predict and ultimately modify or control deviance through deterrence and social policy. Yet the strength of its simplistic approach is also its weakness, as general agreement on societal norms and values cannot always be found. Interpretations may be complex and varied, especially in a society as broad and diverse as the United States, because people belong to many pluralistic subcultural groups. Yet even if people shared broad agreement, definitions of deviance would not necessarily be absolute. The social constructionist perspective, in contrast, lodges the definition, and hence focus of attention, on the audience that frames and reacts to people's actions. Deviance, it argues, is not located in the act but in the societal reaction. As Becker (1963:9) stated in advancing labeling theory, one of the main components of a social constructionist approach to deviance,

> . . . *social groups create deviance by making the rules whose infraction constitutes deviance*, and by applying these rules to particular people and labeling them as outsiders. From this point of view, deviance is *not* a quality of the act the person commits, but rather a consequence of the application by others of rules and sanctions to an "offender." (*Italics in original.*)

Deviance is defined through the social meanings collectively applied to people's behavior or conditions, which is rooted in the interaction between individuals and social groups. Those who have the power to make and apply rules onto others control the normative order. The politically, socially, and economically dominant groups enforce their definitions onto the downtrodden and powerless. Deviance is thus a representation of unequal power in society.

THE SOCIAL ORGANIZATION OF DEVIANCE

We conclude this volume, in Parts VIII, IX, and X, with a discussion of how deviants and their deviance are socially organized. Earlier sections of the book concentrate on macro and micro levels of addressing deviance by considering the movements and powers that shape deviant definitions at the societal level and by looking at how people's identities are shaped at the interpersonal level. Here, we take a midlevel focus by looking at how deviants organize their social organization and relationships, activities and acts, and careers, in connection with others. We begin with the study of deviant organization, examining the various ways members of deviant scenes organize their relations with each other. These range from individuals acting on their own, outside of relationships with other deviants, to subcultures, to more tightly connected gangs, to highly committed international cartels, and finally to corporations. We then consider the structure of deviant acts. Some forms of deviance involve cooperation between the participants, where people mutually exchange illicit goods or services. Others are characterized by conflict, where some parties to the act take advantage of others, often against their will. Finally, we look at the phases and contours of deviant careers, beginning with people's entry into the world of deviant behavior and associates, continuing with the way they fashion their involvement in deviance, and concluding with their often problematic, and occasionally inconclusive, retirements from the compelling world of deviance.

PART I

Defining Deviance

In order to study the topic of deviance we must first clarify what we mean by the term. What behaviors or conditions fall into this category, and what is the relation between deviance and other categories, such as crime? When we speak of deviance, we refer to violations of social norms. Norms are behavioral codes or prescriptions that guide people into actions and self-presentations conforming to social acceptability. Norms need not be agreed upon by every member of the group doing the defining, but a clear or vocal majority must agree.

One of the founding sociologists, William Sumner (1906), conceptualized norms into three categories: *folkways, mores,* and *laws.* He defined folkways as simple everyday norms based on custom, tradition, or etiquette. Violations of folkway norms do not generate serious outrage, but might cause people to think of the violator as odd. Common folkway norms include standards of dress, demeanor, physical closeness to or distance from others, and eating behavior. People who come to class dressed in bathing suits, who never seem to be paying attention when they are spoken to, who sit or stand too close to others, or who eat with their hands instead of silverware would be violating a folkway norm. We would not arrest them nor would we impugn their moral character, but we might think that there was something odd about them.

Mores are norms based on broad societal morals whose infraction would generate more serious social condemnation. Interracial marriage, illegitimate childbearing, and drug addiction all constitute moral violations. Upholding these norms is seen as critical to the fabric of society, so that their violation threatens the social order. Interracial marriage threatens racial purity and the stratification hierarchy based on race; illegitimate childbearing threatens the institution of marriage and the transference of money, status, and family responsibility from one generation to the next; and drug addiction represents the triumph of hedonism over rationality, threatening the responsible behavior necessary to hold society together and accomplish its necessary tasks. People who violate mores may be considered wicked and potentially harmful to society.

Laws are the strongest norms because they are supported by codified social sanctions. People who violate them are subject to arrest and punishment ranging from fines to imprisonment. Many laws are directed toward behaviors that used to be folkway violations or, especially, more violations, but became encoded into laws. Others are regarded as necessary for maintaining social order. Although violating a law by acts such as traffic violations will bring the stigma associated with arrest, it will not usually brand the violator as deviant.

This discussion returns us to the question about the relationship between *deviance* and *crime*. Are they identical terms, is one a subset of the other, or are they overlapping categories? To answer this question we must consider one facet of it at a time. First, do some things fall into both categories, crime and deviance? The overlap between these two is extensive, with crimes of violence, crimes of harm, and theft of personal property considered both deviant and illegal. Second, are there types of deviance that are not crimes? Actually, much deviance is noncriminal, such as obesity, stuttering, physical handicaps, racial intermarriage, and unwed pregnancy. Deviance is not a subset of crime, then. Finally, is there crime that is nondeviant? Although much crime is considered deviant, and derives from various lesser deviant categories, some criminal violations do not violate norms or bring moral censure. Examples of this include some white collar crimes commonly regarded as merely aggressive business practices, such as income tax evasion, and forms of civil disobedience, where people break laws to protest them. Thus, crime is not a subset of deviance. Crime and deviance, then, are overlapping categories with independent dimensions.

People can be labeled deviant as the result of the ABCs of deviance: their *attitudes, behaviors,* or *conditions*. First, they may be branded deviant for alternative sets of attitudes or belief systems. These beliefs can fall into the religious or political category, with people who hold radical or unusual views of the supernatural (cult members, Satanists, fundamentalists) or who hold extreme political attitudes (far leftists or rightists, terrorists) considered deviant. Mental illness also falls into the deviant attitudinal category, as people with deviant worldviews are often considered mentally ill, and people with chemical, emotional, or psychological problems may be considered deviant.

The behavioral category is the most familiar one, with people coming to be regarded as deviant for their outward actions. Deviant behaviors may be intentional or inadvertent, and include such activities as violating dress or speech conventions, engaging in kinky sexual behavior, or committing murder. People cast into the deviant realm for their behaviors have an *achieved deviant status;* they have earned the deviant label through something they have done.

Other people regarded as deviant may have an *ascribed deviant status,* based on something they acquire from birth. This would include having: a deviant socioeconomic status such as being poor; a deviant racial status such as being a person of color (in a dominantly Caucasian society); or a congenital physical handicap. Here, there is nothing that such people have done to become deviant and little or nothing they can do to repair the deviant status. Moreover, there may be nothing necessarily inherent in these statuses that make them deviant: they may become deviant through the result of a social definitional process that gives unequal weight to powerful and dominant groups in society. Conditional deviant status may be ascribed or achieved. On the one hand, people may be born with conditional deviance due to their personal and/or racial/ethnic characteristics (height, weight, color). On the other hand, a conditional deviant status can also be achieved, such as when people burn or disfigure themselves severely, when they become too fat or too thin, or when they cover their bodies with adornments such as tattoos, piercings, or scarification. Some of these may also be changed, moving the deviant back within the norm.

1

On the Sociology of Deviance

KAI T. ERIKSON

This classic selection examines the functions of deviance for society. Deviance and the social reactions it evokes, one of the key focal concerns of every community, are scrutinized by the mass media, law enforcement, and ordinary citizens. Although we no longer attend public hangings or observe people held in stockades, we are aware of the way society punishes (or fails to punish) deviant acts; in so doing we continually redraw the social boundaries of acceptability. Rather than being a fixed property, norms are subject to shift and evolution, and the interactions between deviants and agents of social control locate the margins between deviance and respectability. Erikson notes, ironically, that the very institutions and agencies mandated to manage deviance tend to reinforce it, gathering offenders together, socializing them to the skills and attitudes they need to know, and alienating them from mainstream society. Once individuals have been identified as deviant, they undergo "commitment ceremonies," where they are negatively labeled, a status change that is hard to reverse. Society's expectations that deviants will not reform foster the self-fulfilling prophesy by which norm violators reproduce their deviance, living up to the negative images society holds of them. Erikson notes several valuable functions that deviance performs in a society: it fosters boundary maintenance so that people know what is acceptable and unacceptable; it bolsters cohesion and solidarity, thus preserving the stability of social life; and it promotes full employment, guaranteeing jobs for the people working in the deviance and crime management sector.

Human actors are sorted into various kinds of collectivity, ranging from relatively small units such as the nuclear family to relatively large ones such as a nation or culture. One of the most stubborn difficulties in the study of deviation is that the problem is defined differently at each one of these levels: behavior that is considered unseemly within the context of a single family may be entirely acceptable to the community in general, while behavior that attracts severe censure from the members of the community may go altogether unnoticed elsewhere in the culture. People in society, then, must learn to deal separately with deviance at each one of these levels and to distinguish among them in his own daily activity. A man may disinherit his son for conduct that violates old family traditions or ostracize a neighbor for conduct

From Kai T. Erikson, *Wayward Puritans* Copyright © 1966 Macmillan Publishing Company. Reprinted by permission of Allyn & Bacon.

that violates some local custom, but he is not expected to employ either of these standards when he serves as a juror in a court of law. In each of the three situations he is required to use a different set of criteria to decide whether or not the behavior in question exceeds tolerable limits.

In the next few pages we shall be talking about deviant behavior in social units called "communities," but the use of this term does not mean that the argument applies only at that level of organization. In theory, at least, the argument being made here should fit all kinds of human collectivity—families as well as whole cultures, small groups as well as nations—and the term "community" is only being used in this context because it seems particularly convenient.[1]

The people of a community spend most of their lives in close contact with one another, sharing a common sphere of experience which makes them feel that they belong to a special "kind" and live in a special "place." In the formal language of sociology, this means that communities are boundary maintaining: each has a specific territory in the world as a whole, not only in the sense that it occupies a defined region of geographical space but also in the sense that it takes over a particular niche in what might be called cultural space and develops its own "ethos" or "way" within that compass. Both of these dimensions of group space, the geographical and the cultural, set the community apart as a special place and provide an important point of reference for its members.

When one describes any system as boundary maintaining, one is saying that it controls the fluctuation of its consistent parts so that the whole retains a limited range of activity, a given pattern of constancy and stability, within the larger environment. A human community can be said to maintain boundaries, then, in the sense that its members tend to confine themselves to a particular radius of activity and to regard any conduct which drifts outside that radius as somehow inappropriate or immoral. Thus the group retains a kind of cultural integrity, a voluntary restriction on its own potential for expansion, beyond that which is strictly required for accommodation to the environment. Human behavior can vary over an enormous range, but each community draws a symbolic set of parentheses around a certain segment of that range and limits its own activities within that narrower zone. These parentheses, so to speak, are the community's boundaries.

Now people who live together in communities cannot relate to one another in any coherent way or even acquire a sense of their own stature as group members unless they learn something about the boundaries of the territory they occupy in social space, if only because they need to sense what lies beyond the margins of the group before they can appreciate the special quality of the experience which takes place within it. Yet how do people learn about the boundaries of their community? And how do they convey this information to the generations which replace them?

To begin with, the only material found in a society for marking boundaries is the behavior of its members—or rather, the networks of interaction which link these members together in regular social relations. And the interactions which do the most effective job of locating and publicizing the group's outer edges would seem to be those which take place between deviant persons on the one side and official agents of the community on the other. The deviant is a person whose activities have moved outside the margins of the group, and when the community calls him to account for that vagrancy it is making a statement about the nature and placement of its boundaries. It is declaring how much variability and diversity can be tolerated within the group before it begins to lose its distinctive shape, its unique identity. Now there may be other moments in the life of the group which perform a similar service: wars, for instance, can publicize a group's boundaries by drawing attention to the line separating the group from an adversary, and certain kinds of religious ritual, dance ceremony, and other traditional pageantry can dramatize the difference between "we" and "they" by portraying a symbolic encounter between the two. But on the whole, members of a community inform one another about the placement of their boundaries by participating in the confrontations which occur when persons who venture out to the edges of the group are met by policing agents whose special business it is to guard the cultural integrity of the community. Whether these confrontations take the form of criminal trials, excommunication hearings, courts-martial, or even psychiatric case conferences, they act as boundary-maintaining devices in the sense that they demonstrate to whatever audience is concerned where the line is drawn between behavior that belongs in the special universe of the group and behavior that does not. In general, this kind of information is not easily relayed by the straightforward use of language. Most readers of this paragraph, for instance, have a fairly clear idea of the line separating theft from more legitimate forms of commerce, but few of them have ever seen a published statute describing these differences. More likely than not, our information on the subject has been drawn from publicized instances in which the relevant laws were applied—and for that matter, the law itself is largely a collection of past cases and decisions, a synthesis of the various confrontations which have occurred in the life of the legal order.

It may be important to note in this connection that confrontations between deviant offenders and the agents of control have always attracted a good deal of public attention. In our own past, the trial and punishment of offenders were staged in the market place and afforded the crowd a chance to participate in a direct, active way. Today, of course, we no longer parade deviants in the town square or expose them to the carnival atmosphere of a Tyburn, but it is interesting that the "reform" which brought about this change in penal practice coincided almost exactly with the development of newspapers as a medium of mass information. Perhaps this is no more than an accident of history, but it is nonetheless true that newspapers (and now radio and television) offer much the same kind of entertainment as public hangings or a Sunday visit to the local gaol. A considerable portion of what we call "news" is devoted to reports about deviant behavior and its consequences, and it is no simple matter to explain why these items should be considered newsworthy or why they should command the extraordinary attention they do. Perhaps they appeal to a number of psychological perversities among the mass audience, as commentators have suggested, but at the same time they constitute one of our main sources of information about the normative outlines of society. In a figurative sense, at least, morality and immorality meet at the public scaffold, and it is during this meeting that the line between them is drawn.

Boundaries are never a fixed property of any community. They are always shifting as the people of the group find new ways to define the outer limits of their universe, new ways to position themselves on the larger cultural map. Sometimes changes occur within the structure of the group which require its members to make a new survey of their territory—a change of leadership, a shift of mood. Sometimes changes occur in the surrounding environment, altering the background against which the people of the group have measured their own uniqueness. And always, new generations are moving in to take their turn guarding old institutions and need to be informed about the contours of the world they are inheriting. Thus single encounters between the deviant and his community are only fragments of an ongoing social process. Like an article of common law, boundaries remain a meaningful point of reference only so long as they are repeatedly tested by persons on the fringes of the group and repeatedly defended by persons chosen to represent the group's inner morality. Each time the community moves to censure some act of deviation, then, and convenes a formal ceremony to deal with the responsible offender, it sharpens the authority of the violated norm and restates where the boundaries of the group are located.

For these reasons, deviant behavior is not a simple kind of leakage which occurs when the machinery of society is in poor working order, but may be, in controlled quantities, an important condition for preserving the stability of social life. Deviant forms of behavior, by marking the outer edges of group life, give the inner structure its special character and thus supply the framework within which the people of the group develop an orderly sense of their own cultural identity. Perhaps this is what Aldous Huxley had in mind when he wrote:

> Now tidiness is undeniably good—but a good of which it is easily possible to have too much and at too high a price. . . . The good life can only be lived in a society in which tidiness is preached and practised, but not too fanatically, and where efficiency is always haloed, as it were, by a tolerated margin of mess.[2]

This raises a delicate theoretical issue. If we grant that human groups often derive benefit from deviant behavior, can we then assume that they are organized in such a way as to promote this resource? Can we assume, in other words, that forces operate in the social structure to recruit offenders and to commit them to long periods of service in the deviant ranks? This is not a question which can be answered with our present store of empirical data, but one observation can be made which gives the question an interesting perspective—namely, that deviant forms of conduct often seem to derive nourishment from the very agencies devised to inhibit them. Indeed, the agencies built by society for preventing deviance are often so poorly equipped for the task that we might well ask why this is regarded as their "real" function in the first place.

It is by now a thoroughly familiar argument that many of the institutions designed to discourage deviant behavior actually operate in such a way as to perpetuate it. For one thing, prisons, hospitals, and other similar agencies provide aid and shelter to large numbers of deviant persons, sometimes giving them a certain advantage in the competition for social resources. But beyond this, such institutions gather marginal people into tightly segregated groups, give them an opportunity to teach one another the skills and attitudes of a deviant career, and even provoke them into using these skills by reinforcing their sense of alienation from the rest of society.[3] Nor is this observation a modern one:

> The misery suffered in gaols is not half their evil; they are filled with every sort of corruption that poverty and wickedness can generate; with all the shameless and profligate enormities that can be produced by the impudence of ignominy, the range of want, and the malignity of dispair. In a prison the check of the public eye is removed; and the power of the law is spent. There are few fears, there are no blushes. The lewd inflame the more modest; the audacious harden the timid. Everyone fortifies himself as he can against his own remaining sensibility; endeavoring to practise on others the arts that are practised on himself; and to gain the applause of his worst associates by imitating their manners.[4]

These lines, written almost two centuries ago, are a harsh indictment of prisons, but many of the conditions they describe continue to be reported in even the most modern studies of prison life. Looking at the matter from a long-range historical perspective, it is fair to conclude that prisons have done a conspicuously poor job of reforming the convicts placed in their custody; but the very consistency of this failure may have a peculiar logic of its own. Perhaps we find it difficult to change the worst of our penal practices because we *expect* the prison to harden the inmate's commitment to deviant forms of behavior and draw him more deeply into the deviant ranks. On the whole, we are a people who do not really expect deviants to change very much as they are processed through the control agencies we provide for them, and we are often reluctant to devote much of the community's resources to the job of rehabilitation. In this sense, the prison which graduates long rows of accomplished criminals (or, for that matter, the state asylum which stores its most severe cases away in some back ward) may do serious violence to the aims of its founders; but it does very little violence to the expectations of the population it serves.

These expectations, moreover, are found in every corner of society and constitute an important part of the climate in which we deal with deviant forms of behavior.

To begin with, the community's decision to bring deviant sanctions against one of its members is not a simple act of censure. It is an intricate rite of transition, at once moving the individual out of his ordinary place in society and transferring him into a special deviant position.[5] The ceremonies which mark this change of status, generally, have a number of related phases. They supply a formal stage on which the deviant and his community can confront one another (as in the criminal trial); they make an announcement about the nature of his deviancy (a verdict or diagnosis, for example); and they place him in a particular role which is thought to neutralize the harmful effects of his misconduct (like the role of prisoner or patient). These commitment ceremonies tend to be occasions of wide public interest and ordinarily take place in a highly dramatic setting.[6] Perhaps the most obvious example of a commitment ceremony is the criminal trial, with its elaborate formality and exaggerated ritual, but more modest equivalents can be found wherever procedures are set up to judge whether or not someone is legitimately deviant.

Now an important feature of these ceremonies in our own culture is that they are almost irreversible. Most provisional roles conferred by society—those of the student or conscripted soldier, for example—include some kind of terminal ceremony to mark the individual's movement back out of the role once its temporary advantages have been exhausted. But the roles allotted the deviant seldom make allowance for this type of passage. He is ushered into the deviant position by a decisive and often dramatic ceremony, yet is retired from it with scarcely a word of public notice. And as a result, the deviant often returns home with no proper license to resume a normal life in the community. Nothing has happened to cancel out the stigmas imposed upon him by earlier commitment ceremonies; nothing has happened to revoke the verdict or diagnosis pronounced upon him at that time. It should not be surprising, then, that the people of the community are apt to greet the returning deviant with a considerable degree of apprehension and distrust, for in a very real sense they are not at all sure who he is.

A circularity is thus set into motion which has all the earmarks of a "self-fulfilling prophesy," to use Merton's fine phrase. On the one hand, it seems quite obvious that the community's apprehensions help reduce whatever chances the deviant might otherwise have had for a successful return home. Yet at the same time, everyday experience seems to show that these suspicions are wholly reasonable, for it is a well-known and highly publicized fact that many if not most ex-convicts return to crime after leaving prison and that large numbers of mental patients require further treatment after an initial hospitalization. The common feeling that deviant persons never really change, then, may derive from a faulty premise; but the feeling is expressed so frequently and with such conviction that it eventually creates the facts which later "prove" it to be correct. If the returning deviant encounters this circularity often enough, it is quite understandable that he, too, may begin to wonder whether he has fully graduated from the deviant role, and he may respond to the uncertainty by resuming some kind of deviant activity. In many respects, this may be the only way for the individual and his community to agree what kind of person he is.

Moreover this prophesy is found in the official policies of even the most responsible agencies of control. Police departments could not operate with any real effectiveness if they did not regard ex-convicts as a ready pool of suspects to be tapped in the event of trouble, and psychiatric clinics could not do a successful job in the community if they were not always alert to the possibility of former patients suffering relapses. Thus the prophesy gains currency at many levels within the social order, not only in the poorly informed attitudes of the community at large, but in the best informed theories of most control agencies as well.

In one form or another this problem has been recognized in the West for many hundreds of years, and this simple fact has a curious implication. For if our culture has supported a steady flow of deviation throughout long periods of historical change, the rules which apply to any kind of evolutionary thinking would suggest that strong forces must be at work to keep the flow intact—and this because it contributes in some important way to the survival of the culture as a whole. This does not furnish us with sufficient warrant to declare that deviance is "functional" (in any of the many senses of that term), but it should certainly make us wary of the assumption so often made in sociological circles that any well-structured society is somehow designed to prevent deviant behavior from occurring.[7]

It might be then argued that we need new metaphors to carry our thinking about deviance onto a different plane. On the whole, American sociologists have devoted most of their attention to those forces in society which seem to assert a centralizing influence on human behavior, gathering people together into tight clusters called "groups" and bringing them under the jurisdiction of governing principles called "norms" or "standards." The questions which sociologists have traditionally asked of their data, then, are addressed to the uniformities rather than the divergencies of social life: how is it that people learn to think in similar ways, to accept the same group moralities, to move by the same rhythms of behavior, to see life with the same eyes? How is it, in short, that cultures accomplish the incredible alchemy of making unity out of diversity, harmony out of conflict, order out of confusion? Somehow we often act as if the differences between people can be taken for granted, being too natural to require comment, but that the symmetry which human groups manage to achieve must be explained by referring to the molding influence of the social structure.

But variety, too, is a product of the social structure. It is certainly remarkable that members of a culture come to look so much alike; but it is also remarkable that out of all this sameness a people can develop a complex division of labor, move off into diverging career lines, scatter across the surface of the territory they share in common, and create so many differences of temper, ideology, fashion, and mood. Perhaps we can conclude, then, that two separate yet often competing currents are found in any society: those forces which promote a high degree of conformity among the people of the community so that they know what to expect from one another, and those forces which encourage a certain degree of diversity so that people can be deployed across the range of group space to survey its potential, measure its capacity, and, in the case of those we call deviants, patrol its boundaries. In such a scheme, the deviant would appear as a natural product of group differentiation. He is not a bit of debris spun out by faulty social machinery, but a relevant figure in the community's overall division of labor.

NOTES

1. In fact, the first statement of the general notion presented here was concerned with the study of small groups. See Robert A. Dentler and Kai T. Erikson, "The Functions of Deviance in Groups," *Social Problems,* VII (Fall 1959), pp. 98–107.

2. Aldous Huxley, *Prisons: The "Carceri" Etchings by Piranesi* (London: The Trianon Press, 1949), p. 13.

3. For a good description of this process in the modern prison, see Gresham Sykes, *The Society of Captives* (Princeton, N.J.: Princeton University Press, 1958). For discussions of similar problems in two different kinds of mental hospital, see Erving Goffman, *Asylums* (New York: Bobbs-Merrill, 1962) and Kai T. Erikson, "Patient Role and Social Uncertainty: A Dilemma of the Mentally Ill," *Psychiatry,* XX (August 1957), pp. 263–274.

4. Written by "a celebrated" but not otherwise identified author (perhaps Henry Fielding) and quoted in John Howard, *The State of the Prisons,* London, 1777 (London: J. M. Dent and Sons, 1929), p. 10.

5. The classic description of this process as it applies to the medical patient is found in Talcott Parsons, *The Social System* (Glencoe, Ill.: The Free Press, 1951).

6. See Harold Garfinkel, "Successful Degradation Ceremonies," *American Journal of Sociology,* LXI (January 1956), pp. 420–424.

7. Albert K. Cohen, for example, speaking for a dominant strain in sociological thinking, takes the question quite for granted: "It would seem that the control of deviant behavior is, by definition, a culture goal." See "The Study of Social Disorganization and Deviant Behavior" in Merton, et al., *Sociology Today* (New York: Basic Books, 1959), p. 465.

2

A Typology of Deviance Based on Middle Class Norms

CHARLES R. TITTLE AND RAYMOND PATERNOSTER

This selection discusses the difficulty of defining deviance, as definitions are relative, rooted in particular situations and contexts that vary across time, place, and group. Tittle and Paternoster, though, offer a typology that identifies behavior that would be regarded as deviant by the middle class, the group that represents the largest population in American society. Dealing with issues such as lying, disloyalty, hedonism, irresponsibility, and invasion of privacy, the authors define ten middle class norms, describe what forms their violation might take, and discuss the kinds of sanctions violators are likely to incur. While these norms do not represent the sentiment of all groups, it is likely that they are widely shared, providing a foundational basis for collective sentiment and solidarity in American culture.

It would be easier to study deviance if we could classify deviant behavior into categories that are homogeneous with respect to crucial characteristics. Classification brings order to seemingly disparate phenomena by grouping the individual instances into similar abstract types. Moreover, a typology might suggest underlying principles that could simplify the obvious complexities of this subject matter.

DIFFICULTY OF CLASSIFICATION

Unfortunately, classification by homogeneous traits is difficult because deviance has no inherent characteristics such as harmfulness, badness, or seriousness that would enable us to identify types. Whether some behavior is deviant or not depends upon social definitions that vary from place to place and from time to time. A given behavior may at one time be thought by a social group to be harmful or bad but in other historical eras not be so feared. Or behavior that one group or society thinks is important or serious may be considered quite

---

Charles R. Tittle and Raymond Paternoster, *Social Deviance and Crime*, Los Angeles: Roxbury, 2000.

innocuous by another. "Harmfulness" or "badness" cannot be objectively determined, and if it cannot be objectively determined then the placing of particular deviant acts into categories will necessarily be somewhat arbitrary.

Deviance is relative rather than absolute; its important features are not intrinsic but instead are products of the social contexts from which judgments of social acceptability or unacceptability emanate. Deviance makes sense only as a reflection of a particular normative system, and its classification must be intimately linked with a classification of the norms of that system at that historical moment, which are also emergent and relative. Thus, any typology of deviant behavior is necessarily bounded by time, place, status, and group. Moreover, because norms are seldom completely clear, any typology of deviation will incorporate ambiguities and redundancies, thereby inevitably violating one or another of the rules of good classification, particularly those requiring all instances to fit in the scheme and to fit in one place.

Not only does the relative, socially arbitrary nature of deviance make classification difficult, but the efforts of social groups themselves to classify the deviance they recognize often complicate attempts by scholars to develop workable typologies. When social groups try to classify behaviors, they do so unsystematically and piecemeal. As a result, indigenous typologies are usually even less coherent, comprehensive, and systematic than those constructed by scholars. Therefore, when scholars try to incorporate indigenous categories into their own typologies, it usually does not work. For example, modern societies collectively recognize a category of behaviors called crime, subject to management by official functionaries. At first, one might think that a classification scheme of deviance should include a distinct category for crime. We will now see that this is not so easily done.

The trouble with using crime as a distinct category in classifying deviance is that some criminal behavior is not deviant, and criminal acts are themselves extremely diverse. Indeed, since law making is political and arbitrary, any behavior whatsoever may be designated as crime, regardless of its characteristics. Contemporary legal codes prohibit such dissimilar acts as marijuana smoking, murder, and oral sex; in Hammurabi's ancient legal code, the diluting of drinks by tavern keepers, adultery, and blasphemy were all similarly designated as criminal (Hammurabi, 1904). Furthermore, any state legislature in the United States could at any time declare coffee drinking (or any other behavior) to be criminal so that it automatically would have to be included in the general category of crime, along with such acts as tax evasion, theft, murder, and rape. There is simply nothing about criminal behavior that makes it distinguishable as a category—other than the fact that all such acts have been designated as subject to official management by the lawmaking body of the particular society in question.

To some extent the law itself even recognizes this heterogeneity, and it attempts to subclassify criminal acts into at least two types: serious (felonies) and less serious (misdemeanors). But this distinction does not suffice because it fails to capture even a fraction of the diversity inherent in any legal code. In addition, the felony/misdemeanor distinction is arbitrary, inconsistent, and often contrary to collective perceptions about the behaviors in question. The point is this: a completely workable classification or taxonomy of deviant behavior is impossible, especially one that attempts to build upon "folk" or indigenous categories. Despite the seeming impossibility of the task, the attempt to develop typologies is nevertheless useful, and it is enlightening to study the classification schemes developed by others.

A TYPOLOGY BASED ON MIDDLE-CLASS NORMS

The scheme to be presented takes middle-class American (U.S.) norms as its reference base. It demonstrates the variety of deviant behaviors that can be recognized from just one normative perspective. After studying this typology you should be able to appreciate the volume and diversity of deviance that is possible when numerous normative contexts, over time and both within and across modern societies, are taken into account. . . .

The classification scheme is summarized in Table 1. Ten middle-class norms are identified and listed in Column 1. Associated with each norm is a category of deviant behavior listed in Column 2. Column 3 lists specific examples of the various deviant acts within each category. The scheme ranks norms from most to least importance for middle-class people. . . .

Loyalty/Apostasy

A primary, and possibly the most important, norm among middle-class Americans concerns the ultimate right of the group or collectivity to sustain itself through subordinating individual interests to group survival. Recognition of this right is expressed in the norm of loyalty. By middle-class standards, all people must commit themselves to the group or society as a whole and maintain that commitment against all challenges. Any behavior that seems to express disloyalty, weak commitment, or disrespect for the group is, therefore, unacceptable. Such behavior can be referred to as *apostasy*. Behaviors within the category of apostasy include revolutionary actions, betrayal of government secrets, cooperation with an enemy nation (treason), draft dodging, defiling the flag, surrendering one's citizenship, and advocating contrary government philosophies.

Revolutionary actions such as participation in a conspiracy to overthrow the government obviously display disloyalty, as do selling of military secrets to an enemy and gunrunning during war or helping an invader to establish control over an area. In a less obvious way, attempts to avoid the draft are taken to be indicative of disloyalty to the group, another way of saying that the person lacks the commitment to sacrifice for the interests of the group. And defiling the flag or advocating another governmental or economic philosophy (such as communism) is assumed to prove that the person holds the group in such low regard that he would be disloyal in a critical situation.

Table 1 A Classification of U.S. Middle-Class Deviance

Norm	Deviance	Examples
Group Loyalty	Apostasy	Revolution, Betraying national secrets, Treason, Draft dodging, Flag defilement, Giving up citizenship, Advocating contrary government philosophy
Privacy	Intrusion	Theft, Burglary, Rape, Homicide, Voyeurism, Forgery, Record spying
Prudence	Indiscretion	Prostitution, Homosexual behavior, Incest, Bestiality, Adultery, Swinging, Gambling, Substance abuse
Conventionality	Bizarreness	"Mentally ill behavior" (handling excrement, nonsense talk, eating human flesh, fetishes), Separatist life styles
Responsibility	Irresponsibility	Family desertion, Reneging on debts, Unprofessional conduct, Improper role performance, Violations of trust, Pollution, Fraudulent business
Participation	Alienation	Non-participatory life styles (hermitry, street living), Perpetual unemployment, Receiving public assistance, Suicide
Moderation	Hedonism—Asceticism	Chiseling—Rate busting, Atheism—Fanaticism, Teetotaling—Alcoholism, Total honesty—Total deceit, Hoarding—Wasting, Ignoring children—Smothering them
Honesty	Deceitfulness	Selfish lying, Price-fixing, Exploitation of the weak and helpless, Bigamy, Welfare cheating
Peacefulness	Disruption	Noisy disorganizing behavior, Boisterous reveling, Quarreling, Fighting, Contentiousness
Courtesy	Uncouthness	Private behavior in public places (picking nose, burping), Rudeness (smoking in prohibited places, breaking in a line), Uncleanliness

Clearly, most middle-class Americans endorse the norm of loyalty and consider it exceptionally important. No personal or other group obligation can excuse apostate behavior; and middle-class people usually display disgust and loathing for those who are guilty. Moreover, violations of the norm of loyalty almost always result in sanctions—often quite severe—and they usually evoke lifetime stigmatization whether or not there is official sanction.

Privacy/Intrusion

A second major middle-class American norm revolves around the concept of privacy. It holds that every person has the right to exclusive control over some things, especially private places and personal items. Sometimes there are disputes about the limits of exclusive control; for example, do parents or pet owners have a right to abuse their charges? Moreover, exclusive control over property is limited in view of potential consequences for others (you cannot burn your own house if it poses a fire hazard for other homes), and exclusivity is never absolute. In addition, the fact of ownership may be disputed. But, when exclusive control is recognized, the principle of privacy prevails; that is, only the owner may invade that domain. Hence the associated deviance is called *intrusion,* and it consists of acts that deny the controller or owner of some domain the exclusivity implied by ownership. Examples of intrusion include theft, burglary, rape, homicide, voyeurism, forgery, and record spying (unauthorized examination of bank accounts, hospital records, or other confidential information).

Of course, most acts of intrusion are subject to sanction and, indeed, typically are sanctioned. In fact, most acts of intrusion are among the most feared of crimes, the so-called "index crimes" for which the FBI collects and publishes statistics. Yet the stigma and sanction are not as severe or long lasting as in the case of apostate behavior. It appears that recognition of the ultimate superiority of the group, reflected in the norm of loyalty, supersedes the rights of the individual, as it must for society to continue. In the final analysis, the only persons who may intrude into private domains with impunity are official representatives of the collectivity who act on authority of the group as a whole.

Prudence/Indiscretion

The third most important normative area among middle-class Americans appears to be that concerning *prudence.* All people are expected to exercise selectivity in the practice of activities that are pleasurable. They are to use pleasure as a means to an end, not as an end in itself, and the expression of pleasure must be within specified limits or boundaries. The prescribed end is expected to be some contribution to the economic or social maintenance of society. A person is supposed to avoid activities that are frivolous or primarily oriented around self-gratification as well as activities that may involve nonproductive emotional involvements or that may disrupt productive emotional ties. Violation of the norm of prudence can be characterized as *indiscretion.* Indiscrete behavior would include prostitution, homosexual behavior, bestiality, adultery, incest, swinging, gambling, and abuse of drugs.

Prostitution (sexual activities as a commercial exchange), swinging (consensual exchange of sexual partners among married people), and bestiality (sexual activities with animals) all involve sexual interaction without emotion, thus challenging the usual linkage between emotional involvement and sexual expression that motivates the establishment and maintenance of stable, productive, and socially useful unions. Homosexual behavior (sexual relations with someone of one's own gender), incest (sexual relations with a relative), and adultery (sexual relations of a married person with someone other than his or her spouse) may involve emotional attachments that cannot be productive and that may provoke disruptive conflicts. Gambling and abusive recreational drug use convey images of obsession with self gratification and frivolity. All of these indiscrete behaviors are sanctionable, but most of them usually do not receive official sanctions. They all do, however, ordinarily provoke substantial stigma from others, which expresses fairly long-lasting group disapproval.

Conventionality/Bizarreness

A fourth norm of middle-class America mandates that all must practice personal habits and lead lives that are similar to the conventions followed by most middle-class people. Violations of the norm of conventionality are called *bizarre* behavior; that is, relative to most middle-class people, the behavior in question is unusual or statistically atypical to such an extent that the sanity or "normality" of the individual is questioned. But not all statistically unusual behavior is deviant by middle-class standards. To qualify as bizarre, the behavior must be incomprehensible to the typical person. The average middle-class individual must be unable to imagine herself or anyone like her committing such an act; that is, the behavior must not have a good reason behind it. Some atypical behaviors are readily understandable to "normals." Clearly, religious celibacy is acceptable since it reflects devotion to an established faith and is believed to be good and useful. Likewise, skydiving, while unusual, is nevertheless comprehensible as a recreational search for adventure—fulfillment of the individualistic ideal of bravery—and is therefore not deviant. On the other hand, total sexual abstinence for no apparent good reason (such as among wedded people who fail to consummate the marriage) would be bizarre.

Some of the activities described here as bizarre behavior are subject to some form of official sanction; usually, however, sanctions are informal—ridicule, harassment, and rejection. Many of the more bizarre acts are formally sanctioned by incarceration in mental institutions, where therapists are employed to help their practitioners "recover," which means to begin to act in conventional ways. In all instances, however, there is a strong tendency toward stigmatization, which can be erased only with great difficulty.

Responsibility/Irresponsibility

Another norm of middle-class Americans prescribes responsible conduct. People, especially those on whom others depend, must be reliable. Violations of this norm constitute instances of *irresponsible* behavior. Examples include family desertion, refusal or failure to meet financial obligations, negligence in maintaining property or performing occupational roles, failure to fulfill professional standards, violations of trust, pollution of the environment, and selling defective, harmful, or useless products or services.

For instance, parents are expected to provide financial and emotional support for children, and spouses are required to fulfill marital obligations toward each other. Children and spouses, as well as members of the larger society, have a stake in faithful parental and marital role performance. When husbands or wives desert families, for whatever reason, they leave their dependents open to potential harm, and they may be helping to create burdens for the community should the children or deserved spouse need public assistance. This is thought by middle-class people to be irresponsible. Similarly, all individuals who participate in the world of commerce must be able to rely upon each other for basic financial dependability that justifies the trust on which contracts are based. Individuals, therefore, commit deviance when they renege on debts, delay payments, or evade contractual obligations. . . .

Professionals such as physicians, lawyers, accountants, and police bear a special middle-class responsibility. Norms prescribe that they perform competent services for those who entrust themselves as clients, including in the case of the police, the entire community, and that the interests of the clients must take precedence over the self-interests of the practitioners. A physician who performs unnecessary surgery simply for the purpose of earning the fee for service is engaging in irresponsible behavior, as is one who prescribes medicine to be bought in a pharmacy the physician owns, or who reports to unrelated third parties about the illnesses of another. Similarly, a male physician who takes advantage of female patients who trust him to examine their unclothed bodies behaves irresponsibly and deviantly in violation of the role obligations of a physician. And it is deviant for a physician to keep people sick to guarantee income rather than to help them recover. . . .

Some of the newest, and to some extent still controversial, forms of deviance concern behavior that jeopardizes others through damage to the environment—even small incremental damages. Just as individuals may endanger public health and safety by spitting on the sidewalk, dumping garbage on the street, or putting carbon residues into the air through their automobile exhausts, now corporations are being held responsible for polluting the environment through discharge of toxic chemicals into rivers or the air, or by destruction of natural protective systems such as forests or sand dunes. Similarly, although swindlers have always been considered to be doing deviance by selling defective home siding (so-called "Tin Men") or roof painting jobs, now "respectable" businesses are being castigated for marketing defective or dangerous products, such as automobiles with improperly designed gasoline tanks that explode on rear-end impacts or children's pajamas that are excessively flammable (see Simon and Eitzen, 1993). The major forms of deviance involving useless products or services are medical frauds. Charlatans often sell ineffective or dangerous machines or drugs to desperate people who suffer continuous pain or incurable diseases. . . .

The ambivalence of these norms is reflected in the relative leniency of official sanctions. Even though the listed behaviors may be disapproved, usually they go unpunished or, if violations are punished, they are given only a "slap on the wrist." More often, these transgressions provoke private civil suits and some stigma, although usually not great or long lasting.

Participation/Alienation

Another middle-class norm specifies that everybody will take an active part in the social and economic life of the community or society and in the institutions it spawns. Those who fail to participate are guilty of *alienation*, unless they can provide an acceptable excuse. Deliberate defiance of the participation norm is illustrated by hermits, tramps, and bums, as well as by suicides. The perpetually unemployed, those receiving public assistance, the aged, and the handicapped all exhibit forms of involuntary alienation, but they are nevertheless held in contempt unless their inactivity is justified. It may be excused by visible physical defects but only if the symptoms are sufficiently debilitating to make participation impossible, or by an individual's reputation for perseverance that allows the middle-class audience to conclude that the nonparticipator has done all he or she could to fulfill expectations. . . .

The importance of the norm of participation is shown by the suspiciousness with which the middle class regards any deviation. People receiving public assistance are assumed to be chiselers or "welfare cheats" unless they can individually redeem themselves, and middle-class people are ever alert to detect any evidence of immorality, deceit, or laziness by welfare mothers, the infirm, or the aged that would confirm their suspicions. If a person is inactive because of age, he or she must not be seen walking without a crutch or cane. An unemployed man must not laugh, drink beer, or be seen in recreation, no matter how long he has been unemployed or how hard he has searched for a job; nor must he refuse any kind of employment, no matter how degrading or dangerous it might be. And nonparticipating handicapped individuals must show their handicap so observers will know they have a right to be excused from the alienated status. Otherwise, such persons will be regarded as fakes and cheats, false claimants to special dispensation.

Although alienation is rarely dealt official punishment, it is the object of constant informal sanctions, and its practice carries a stigma that may endure into future generations, as when the police suspiciously keep a close watch on children from welfare families. . . The behavior of tramps, bums, street people, and welfare recipients is also sometimes a matter of official review but more often an object of informal scorn. The nonparticipation of the aged, handicapped, infirm, and suicides is almost completely a matter of informal criticism.

Moderation/Hedonism and Asceticism

Middle-class Americans subscribe to Emerson's golden mean. All things should be in moderation, and extremes of any kind—even for desirable activities—are unacceptable. This concern with moderation produces judgment of acts as deviant that fall on one end or another of a continuum from hedonism (too much of something) to asceticism (not enough of something). These rules of moderation apply to all realms of activity from work to play. Thus it is good to nurture a child but it is bad to practice "smother love" (hedonism) or to reject a child; one is expected to perform an honest day's work, but it is wrong to be a "workaholic" or "rate-buster" (Roethlisberger and Dickson, 1964 [1939]); it is good to drink socially but unacceptable to abuse alcohol or to teetotal. . . .

The golden mean rules the middle-class psyche especially in matters of politics. Extreme conservatives as well as extreme liberals are anathema. The ideal politician is middle of the road, moderate, balanced. And while middle-class Americans ostensibly love freedom and hate totalitarianism, they tolerate "extremes in the defense of freedom" no better than extremes in the service of dictatorship. The history of American government has shown a repeated cyclical pattern as the pendulum swings from the "liberalism" of one generation of governmental policy to the "conservatism" of the next, each in its turn laying claim to having brought policy back to the middle ground.

Here, as with most deviance, violations are sometimes managed by formal sanctions but more often by informal reactions. While an administrator who cuts too many corners may be arrested, most simply find themselves on more and more precarious ground with their peers and coworkers. And administrators who slavishly adhere to the rules may find themselves formally dismissed because of an inability to get the job done, but more likely they will simply go nowhere in the organization, a type of informal reaction. Moreover, in most areas there are no formal sanctions to deal with hedonism or asceticism. Those who extravagantly waste their money are ridiculed and pitied as are those whose penuriousness deprives them of important comforts. But neither is formally dealt with unless the behavior is also so bizarre as to invoke the intervention of mental health personnel. And the "greasy grind," curve busting student along with the "space cadet" are subject to no more punishment than occasional scorn, ridicule, or social isolation.

Honesty/Deceitfulness

The middle-class normative system requires a certain degree of honesty, at least in important things. Everybody is expected to avoid lying, fraud, or misrepresentation, but this prohibition is subject to the restraint of the golden mean mentioned before. Persons who exceed the limit of tolerance for mendacity are exhibiting the deviance of *deceitfulness*. Examples include selfish lying, price-fixing by business people, fraudulent business activities, exploitation of the weak and helpless, bigamy, and welfare cheating.

Unfortunately, the limits within which honesty is to be practiced are not consistent across social contexts; the norm is situationally variable. For example, if you visit a sick male friend and are struck by his poor appearance, it is not good manners for you to tell him how bad he looks. Even if the man proclaims it himself, you are supposed to assure him that he will soon be well. Or if you are a male and your girlfriend buys a new dress and seeks your opinion, you mustn't tell her that you think it's ugly; middle-class norms call for diplomatic concealment in cases like this where the truth serves only to hurt someone. But when the sick person gets well and asks if you covered for him at work during the illness, you can't, according to the norm of honesty, say you did if, in fact, you didn't. Or if the girlfriend with the new dress asks her boyfriend what he did that day she was shopping, the fellow must not deny that he was with friends she does not like. There seems to be an understanding that selfish lying, which promotes one's personal interest, is unacceptable but that unselfish lying, which serves an altruistic purpose, is all right. Yet even this guideline is not infallible. An individual who lies in a court of law to protect a friend is no less guilty of perjury than one who lies to save his own skin; both will be condemned as disreputable, although the latter perhaps more thoroughly than the former.

Peacefulness/Disruption

Quiet, tranquility, and order are the hallmarks of American middle-class communities and social life. Middle-class people dislike contentiousness and conflict; they seek agreement, cooperation, and harmony. Violations of these norms are called *disruption*. Examples include noisy or disruptive protests; boisterous revelry; quarrels, fights, or brawls, particularly in open or public places; and disagreeableness or contentiousness.

Contentiousness is disruptive because it disturbs the air of tolerance that middle-class people like to display. In conversation, one is supposed to avoid conflict-generating topics, and if they are broached, every point of view is to be granted respect if not equal validity. And if someone believes something passionately, he or she is to show restraint in its presentation and defense. In all things a person is expected to be reasonable. Those who have a "negative" attitude, who like to argue, who are loud and boisterous, or who don't want to go along with the group are regarded as disruptive and may be excluded from further group interaction, gossiped about, or ridiculed (although probably not face to face). The ultimate expression of this norm is evident in hospitals when people are dying. No matter how much pain the person is suffering or how dreadful the prospect of death, a dying individual is cajoled to restrain any groans, screams, or yells on the grounds that the noise disturbs others and is in bad taste as well (Sudnow, 1967). Hence one must even bear the agony of death in conformity with the peacefulness norm of the middle class, whose influence permeates hospitals and other social institutions dealing with sickness and death. . . .

In short, any behavior that interferes with the orderly, peaceful flow of life must be avoided. Consideration for others is paramount, so disruption is regarded as unacceptable, to be managed mainly by avoidance or informal means, although sometimes by official sanctions, as when the police break up noisy parties, arrest participants in a fight, or disperse demonstrators who block traffic.

Courtesy/Uncouthness

Finally, there are a large number of middle-class rules concerning interpersonal interaction. The point of these middle-class norms is to ensure that an individual's behavior does not make the ordinary business of social intercourse too unpleasant and that necessary accommodation to other people be elevated to a civilized plane. Therefore, in almost every realm of activity there are middle-class norms requiring that the presence of others be taken into account so that they are not offended. Violations of those norms constitute a form of deviance called *uncouthness*.

Uncouth behavior includes any private behavior that is done in public—passing gas, scratching one's genitals, picking one's teeth or nose, spitting, performing bodily functions, vomiting, sleeping, burping, or having sex. Such behavior tends to offend others by introducing unpleasant smells, sounds, and sights. In the middle-class mind, it is enough that one endure these things when they are personally generated without having to share others' experience of them as well. Other rude acts are smoking in an elevator or a nonsmoking section of a restaurant, rushing ahead of someone to go through a door, interrupting a person who is speaking, breaking into a ticket line, or coughing and sneezing without covering one's mouth. Failing to keep one's body clean or to use deodorant is a form of uncouthness that offends others, as are crude table manners like making slurping noises while eating soup or drinking coffee.

The usual method of dealing with these transgressions is informal—a reaction of disgust or sometimes deliberate silence, followed by ostracising the guilty party. Very rarely are there official sanctions for uncouthness, although sometimes arrests are made for sleeping or performing bodily functions in public places.

SUMMARY

This typology of middle-class norms and deviance shows that there are enormous numbers and kinds of deviance—far more than usually come to mind when the subject of deviance is discussed. It is easy to imagine the large number and variety of deviant acts that would be evident if we built typologies based on the norms of all social classes, regions, ethnic groups, or subcultures. Yet if we were to construct such typologies, we would find substantial overlap with this one. That is because the middle class dominates U.S. society, both by imposing its standards through the schools, the mass media, and the law and by enjoying a degree of natural hegemony in behavior styles and thinking that flows from the admiration and emulation of those with higher status. For example, lower-class norms differ a little bit in a few of the categories, but, by and large, the lower class shares most norms with the middle class. And so it is for almost any normative reference we might choose. The fact is, there is considerable evidence of an overarching American culture exhibiting remarkable degrees of normative agreement. Nevertheless, we must be attuned to the many variations that do exist.

REFERENCES

Hammurabi, King of Babylon. 1904. *The Code of Hammurabi, King of Babylon, About 2250 B.C.,* Tr. R. F. Harper. Chicago: University of Chicago Press.

Roethlisberger, Fritz, and William J. Dickson. 1964 [1939]. *Management and the Worker.* NY: Wiley.

Simon, David R., and D. Stanley Eitzen. 1993. *Elite Deviance.* 4th edition. Boston: Allyn and Bacon.

Sudnow, David. *Passing On: The Social Organization of Dying.* Englewood Cliffs, NJ: Prentice-Hall.

3

Positive Deviance

DRUANN MARIA HECKERT

Differences of opinion exist concerning the range of deviance definitions; must they necessarily occupy the realm of the undesirable, or can they point to acts or conditions that depart from the norm in a positive way? Can some acts or conditions be regarded as deviant because they are too far above the bar instead of below it? While some regard positive deviance as an oxymoron, internally contradictory by definition, Heckert not only lays a functional and relativistic groundwork for the notion, but articulates its parameters as well. Illustrating the idea with a variety of examples, Heckert categorizes assorted types of positive deviance, discussing how they differ, yet all fall within this framework. In so doing, she points out that positive deviance is not necessarily associated with positive deeds (Hitler might represent someone with an excess of charisma), but merely marks people possessing unusually high levels of these socially desirable attributes. She concludes that people accord stigma to and often distance themselves from those with this halo effect, indicating a distinct social ambivalence towards this type of deviance and anything that falls too far outside of the comfortable normative range.

Positive deviance has been variously defined in the literature. Additionally, divergent examples, ranging from extreme intelligence to accomplished athletes have been advanced as pertinent examples of positive deviance. While the actors and/or actions that have been mentioned in the literature do have in common that there has been a deviation in a positive direction, the diversity of the examples is great. Consequently, a classificatory model, developed from examples that have been cited in the literature on positive deviance is presented. The types (i.e., ideal types) include the following: altruism, charisma, innovation, supra-conforming behavior, and innate characteristics. The category of ex-deviants is also advanced.

POSITIVE DEVIANCE

Not specifically utilizing the term, Sorokin (1950) had by 1950 recognized the validity of the concept. Convinced that Western culture had entered a "de-

---

Reprinted by permission of the publisher from *Free Inquiry in Creative Sociology*, Vol. 26, No. 1, May 1998.

clining sensate phase," Sorokin felt that a negative orientation permeated these societies. This stance also dominated the social sciences. According to Sorokin:

> For decades Western social science has been cultivating . . . an ever-increasing study of crime and criminals; of insanity and the insane; of sex perversion and perverts; of hypocrisy and hypocrites. . . . In contrast to this, Western social science has paid scant attention to positive types of human beings, their positive achievements, their heroic actions, and their positive relationships. The criminal has been "researched" incomparably more thoroughly than the saint or the altruist; the idiot has been studied much more carefully than the genius; perverts and failures have been investigated much more intensely than integrated persons or heroes. (1950)

Sorokin (1950) suggested that a more thorough understanding of positive types of individuals was essential, especially in terms of the ability of humans to understand the negative.

Various conceptualizations of positive deviance have emerged during the last several decades. One important point is that, unlike the scholarly theorizing regarding deviance (negative deviance), in general, certain analysts (Best, Luckenbill 1982; Goode 1991; Sagarin 1985) contend that positive deviance does not exist. For example, in an acerbic denunciation, Sagarin (1985) contended that positive deviance is an oxymoron and should occupy no place in the study of deviance and Goode (1991) also proclaimed that the concept was not viable. Nevertheless, this opinion is not universally accepted.

Currently, existing literature in positive deviance is scant in comparison to the voluminous literature in negative deviance. However, social scientists have advanced the point of view that the concept of positive deviance is important, and furthermore, pertinent to the study of deviance, in general. According to Ben-Yehuda (1990), ". . . it will open new and exciting theoretical and empirical windows for research." Considering the multitude and the divergency of definitions and definitional approaches to the concept of deviance, it should not be surprising that there has also been a variety of definitions/definitional approaches offered for positive deviance. As such, these can be separated into the following categories: discussions of positive deviance that do not specifically use the terminology, definitions postulating a norm-violation perspective, definitions that utilize a labeling or societal reaction approach, and definitions that advocate a single or unique form of behavior only.

Certain theorists (Katz 1972; Lemert 1951; Liazos 1975; Sorokin 1950; Wilkins 1965) have recognized the validity of analyzing positive forms of behaviors within the general context of the study of deviance. Nevertheless, they did not employ the term, positive deviance. For example, Wilkins (1965) wrote that some types of deviance are functional to society. Geniuses, reformers, and religious leaders are all examples of deviants, in addition, to those examples more often thought about, such as criminals. Wilkins (1965) suggested that deviance could be examined by utilizing the analogy of a continuous distribution which ranged from bad to good. Normal behaviors constitute the major portion of the continuum; at the negative end are acts such as serious crimes and at the good end are behaviors, such as those performed by saints. For example, regarding intelligence, most people fall into the middle part of the continuum, while there are a small number of those of very low intelligence (negative deviants) as well as a very small number of geniuses (positive deviants).

Perhaps, not always explicitly stating a preference of a specific paradigm, some theorists (Sorokin 1950; Wilkins 1965; Winslow 1970) have offered a view of positive deviance as that which violates norms, in that norms are exceeded. Similar to Wilkins (1965), Winslow (1970) noted that deviance can be constructed as a concept which is "relative to statistical norms." When deviance is conceptualized as approximating a normal curve, normative acts are in the middle of this curve. At one extreme end of the curve, beyond tolerance limits, are disapproved behaviors, such as mental illness and suicide. Positive deviance refers to approved deviation, beyond the tolerance limits, such as wealth, health, wisdom, virtue, and patriotism.

On the other hand, while not always stating their adoption of the paradigm, various theorists (Freedman, Doob 1968; Hawkins, Tiedeman 1975; Norland, Hepburn, Monette 1976; Scarpitti, McFarlane 1975; Steffensmeier, Terry 1975) have explained positive deviance from a labeling or societal reaction paradigmatic stance and in synthesis with a non-Marxist Conflict approach, so does Ben-Yehuda (1990). As an example, Freedman and Doob (1968) analyzed positive deviance from a psychological frame of reference, while for all intents and purposes proffering a labeling approach. From their point of view, deviance is an ephemeral characteristic which varies by situation. Differences are important. Various characteristics can be labeled deviant if others involved in a situation in which the individual is enmeshed do not share the same trait. As Freedman and Doob wrote:

> Gulliver was as deviant among the Brobdingnags when he was unimaginably small and weak then when he lived in Lilliput where he was fantastically big and powerful. The genius is as deviant as the idiot. . . . It is perhaps remarkable that the term "exceptional" children is used to refer not only to the unusually intelligent, but also the mentally retarded, the physically handicapped, the emotionally disturbed and so on. (1968)

The reaction of others is significant since certain acts will require a major difference from the norm to be judged deviant while with other acts, only a small variation from the norm will result in a designation of deviance. Simply put, as Steffensmeier and Terry (1975) noted, "Deviance consists of differentially valued phenomenon." Optimally desirable phenomenon include great beauty or heroism as examples of positively valued behaviors.

As a final approach, some theorists (Ewald 1981; Buffalo, Rodgers 1971) have suggested that positive deviance refers to only a very specific type of action. Ewald advanced the idea of positive deviance as excessive conformity when he wrote:

> Positive deviance is where the relationship to societal norms is not one of blatant violation but rather extension, intensification, or enhancement of

social rules. In this case, the zealous pursuit or overcommitment to normative prescriptions is what earns the individual or group the label of deviant. The individual or group is essentially true to normative standards but simply goes "too far" in that plausible or actual results are judged inappropriate by the general culture. (1981)

In a nutshell, positive deviance has been conceptualized as follows: from a norm-violation stance, from a labeling perspective, and from the reference of describing only one type of act. Some integration can be achieved with the norm-violation and reactionist approaches. Therefore, positive deviance is defined as behavior that people label (publicly evaluate) in a superior sense. That labeling will typically occur because the behavior departs from that which is considered normative in the particular case.

EXAMPLES OF POSITIVE DEVIANCE

A myriad of behaviors and/or actions have been advanced as examples of positive deviance. Specifically, the following have been referred to as examples of positive deviance: Nobel Prize winners (Szasz 1970), the gifted (Huryn 1986), motion picture stars (Lemert 1951), superstar athletes (Scarpitti, McFarlane 1975), pro quarterbacks (Steffensmeier, Terry 1975), geniuses (Hawkins, Tiedeman 1975), exceptionally beautiful women (Lemert 1951), reformers (Wilkins 1965), altruists (Sorokin 1950), Congressional Medal of Honor winners (Steffensmeier, Terry 1975), religious leaders (Wilkins 1965), straight-A students (Hawkins, Tiedeman 1975), zealous weight lifters and runners (Ewald 1981), innovative/creative people, such as Freud or Darwin (Palmer 1990), and social idealists (Scarpitti, McFarlane 1975).

These behaviors and/or actions are similar to the extent that they are all examples of positive deviance. Consequently, people will label (publicly evaluate) the behaviors and/or actors in a superior manner. In essence, there is a departure from that which is deemed to be normative in a society. As a result of the behavior being non-normative, several potential consequences ensure the similarity of the divergent types of positive deviance. For example, positive deviants due to the fact that in essence they are as different from "normal" as negative deviants and perhaps threatening to the dominant social order, can at times, be originally labeled negative deviants (e.g., the French Impressionists, Galileo, civil rights leaders) by the powers that be. Also, even many types of positive deviance, that are for the most part viewed positively, often concomitantly, are subject in some respects to negative treatment. For example, inordinately intelligent individuals are considered positive deviants, according to Scarpitti and McFarlane (1975). Nevertheless, derogatory traits are often imputed to them. This process is intuitively obvious to the gifted child who is simultaneously termed gifted, yet perniciously assumed to be "geeky" or socially unacceptable to peers. In essence, various types of positive deviance share many attributes in common. Perhaps, positive deviants even have similarities to negative deviants that they do not share with non-deviants.

Nevertheless, a problem emerges due to the diversity of behaviors and/or actors that have been posited to be examples of positive deviants. In reality, a Congressional Medal of Honor winner, a charismatic religious leader, and a beauty queen winner are actually quite disparate. Comparatively, the mentally ill, criminals, and the physically handicapped are also different. Consequently, to delve further into the nature of positive deviance, a typology of positive deviance would assist in the elucidation of positive deviance.

POSITIVE D EVIANCE: A CLASSIFICATORY MODEL

The following types of positive deviance are advanced: altruism, charisma, innovation, supra-conformity, and innate characteristics. This classificatory scheme was developed by examining and categorizing the examples provided in the existing literature on positive deviance. The typology may not yet be exhausted at this point; indeed, another potential type of positive deviant, the ex-deviant, is suggested. Additionally, other types of positive deviance could also be postulated at some further point. This model is composed of ideal types.

Altruism

The first form of positive deviance postulated is altruism. Sorokin (1950) specifically discussed altruists in general (including saints and good neighbors as examples), Scarpitti and McFarlane (1975) mentioned self-sacrificing heroes, and in a variation on that particular theme, Steffensmeier and Terry (1975) referred to Congressional Medal of Honor winners. Interestingly, while altruism has been primarily researched by psychologists in the modern era, Auguste Comte (1966) was the first social scientist to use and analyze the concept. Altruism involves an act undertaken voluntarily to assist another person or other people without any expectation of reward (Leeds 1963; Cialdini, Kerrick, Bauman 1982; Grusec 1981; Macaulay, Berkowitz 1970). As Sorokin so eloquently noted,

> Genuine altruism is pure also in its motivation: altruistic actions are performed for their own sake, quite apart from any consideration of pleasures of utility. (1948)

Rosenhan (1970) has dichotomized altruism into normal altruism which includes acts such as donating small amounts of money and does not require much effort and autonomous altruism, which refers to actors, such as abolitionists who did exert themselves and sacrifice themselves to a much greater degree. Autonomous altruism is more descriptive of positive deviance.

Charisma

Charisma is the second type of positive deviance. Sorokin (1950) discussed the historical examples of Gandhi and Jesus as examples, and Wilkins (1965) cited religious leaders in general as positive deviants. According to the seminal work of Weber (1947), the charismatic claim to legitimate authority (as opposed to rational-legal or traditional authority) is rooted in the devotion of followers to the believed (not necessarily tangible) extraordinary qualities of their leader and the authority is based on the willingness of the followers to obey their leader. More comprehensively, Weber wrote:

> The term "charisma" will be applied to a certain quality of an individual personality by virtue of which he is set apart from ordinary men and treated as endowed with supernatural, superhuman, or at least specifically exceptional powers of qualities. These are such as are not accessible to the ordinary person, but are regarded as of divine origin or as exemplary, and on the basis of them the individual concerned is treated as a leader. . . . How the quality in question would be ultimately judged from any ethical, esthetic, or their such point of view is naturally entirely indifferent for purposes of definition. What is alone important is how the individual is actually regarded by those subject to charismatic authority by his "followers" or "disciples." (1947)

One important point that Weber (1947) made was that this quality can be attributed by followers to people perceived as having gifts in different areas, including, for example, intellectuals, shamans, (magicians), war leaders, heroes, and prophets. Essentially, the charismatic relationship is composed of two important elements: a situation in which there is a following that wants to be led and a leader who has the capability to catalyze their needs and/or desires.

Innovation

Innovation is another form of positive deviance. As examples, Szasz (1970) discussed Nobel Prize winners, Palmer (1990) analyzed innovative/creative figures including Freud and Darwin, and Wilkins (1965) suggested reformers. In essence, innovation (or invention) has been basically defined as the combining of already existing cultural elements in a novel manner, or the modifying of already existing cultural elements to produce a new one (Lenski, Lenski 1982; Linton 1936; Ogburn 1964; Rogers, Shoemaker 1971). Innovations cover a myriad of areas, as they range from the abstract to the pragmatic, and from art to technology. As Kallen (1964) notes, innovations are a fundamental factor of a society as innovations can occur in these crucial areas of culture: food, clothing, shelter, defense, disease prevention, production, recreation, religion, science, thought, literature, and art. Innovators, as positive deviants, profoundly impact the life of a culture. The willingness of a society to foster change, which is a condition present to a greater extent in modern societies, will relate to the acceptance of the innovator.

Supra-Conformity

A fourth kind of positive deviance is supra-conformity. Hawkins and Tiedemen (1975) pointed to straight-A students, Ewald (1981) analyzed zealous weight lifters and runners, and Scarpitti and McFarlane (1975) mentioned extreme moralists. Additionally, Buffalo and Rodgers (1971) and Ewald (1981) have utilized the concept of positive deviance to suggest only supra-conforming behavior. Supra-conformity is behavior that is at the level of the idealized within a culture. That is, as Gibbs noted,

> Collective evaluations refer to what behavior *ought to be* in a society, whereas collective expectations denote what behavior actually *will be*. (1965)

In relation to the normative structure of a society, there is a tendency for the idealized version of the norms to be attained less often than the realized versions of the norm (Homans 1950; Johnson 1978; White 1961). In other words, norms operate at two levels—the ideal, which most people believe is better but few achieve and the realistic version, which most people can achieve. The negative deviant fails to abide by either level; the "normal" person operates at the realistic level, but does not achieve the idealized level; and the positive deviant is able to attain or behave at the idealized level. Cohen expressed this idea in the following manner when he noted that only a small percentage of people can reach that which is idealized in a society:

> The ideal is one thing, the practice another. In other words, persons may be variously socialized into the ideological traditions of their society so that the two—ideology and its achievement—are not simply the same thing from different perspectives, but are quite independently variable entities. (1966)

Thus, a supra-conformist demonstrates desire and ability to pursue, perhaps, even in quixotic style if necessary that which is idealized for a particular norm.

Innate Characteristics

Finally, innate characteristics constitute a fifth kind of positive deviance. Certain actions/actors, that are positive deviance, are at least partially rooted in innate characteristics. Examples that have been referred to as positive deviance include beautiful women (Hawkins, Tiedeman 1975; Lemert 1951), superstar athletes (Scarpitti, McFarlane 1975), and movie stars (Lemert 1951). The use of the terminology, innate characteristics, is actually not the best choice to describe this type of positive deviance. These traits (e.g., beauty, intelligence, talent) are innate to a certain, as to yet, unspecifiable extent, and to a certain, as to yet, unspecifiable extent, are modified by environmental conditions. In addition, these characteristics are culturally defined. For example, Rebelsky and Daniel (1976) clearly note that intelligence is culturally defined and according to Morse, Reis, Gruzen, and Wolff (1974), attractiveness is culturally defined, as individuals from the same cultural background do tend to coincide in their assessment of what is physically attractive. As such, innate characteristics can be considered a fifth type of positive deviance. As Scarpitti and McFarlance noted,

> Deviant attributes often are the products of one's biological inheritance, which accounts for such conditions as rare beauty, extraordinary intelligence, or dwarfism. (1975)

Another Potential Type: The Ex-Deviant

The potential for new types of positive deviance, not previously cited in the literature, certainly exists. For example, the ex-deviant might possibly be deemed a positive deviant. The previously stigmatized person, labeled in a negative fashion, that manages to convert to a status of normative person is essentially a novel way to think of a positive deviant. According to Pfuhl and Henry, destigmatization

> . . . refers to the processes used to negate or expunge a deviant identity and replace it with one that is essentially non-deviant or normal. (1993)

Subsumed as types of destigmatization are purification ". . . whereby one's defective self is replaced by a moral or 'normal' self, either by sacred or secular norms," and transcendence whereby the deviant manages ". . . to display a 'better' self rather than to eliminate the former self." An example of destigmatization is an ex-convict; an example of transcendence is an accomplished person with a physical disability. Purification essentially involves a destigmatization by which the person exits a stigmatized role. While the previous stigmatization might still taint the individual, society tends to positively evaluate the purification. As such, an ex-deviant might potentially be considered in relation to the concept of positive deviance.

More specifically, Ebaugh has defined the ex-role as,

> The process of disengagement from a role that is central to one's self-identity and the reestablishment of an identity in a new role that takes into account one's ex-role. (1988)

As Ebaugh (1988) notes, certain ex-roles are in fact potentially stigmatizing as they are not generally culturally construed as positive role changes (e.g., ex-spouse, ex-nun). On the other hand, other role changes are societally constructed as positive in that the deviant has been rehabilitated to a more positive status (e.g., ex-alcoholic, ex-prostitute, ex-convict). Additionally, those role changes viewed as socially positive are deemed to be more within the control of individuals.

Crucially, the exiting process is a fairly difficult one, mediated by various factors. Ebaugh (1988) hypothesizes that the role exit is a fairly long-term process generally consisting of the following stages: first doubts, or doubting the previous role; seeking and evaluating alternatives to the role (including "conscious cuing, anticipatory socialization, role rehearsal, and shifting reference groups"); turning points; and establishing the ex-role. Even while the role exit from deviant to non-deviant is positively evaluated and labeled, the person still often experiences the remnants of the stigmatization that typically accompanies the previous role. Thus, there is a tenuousness to exiting a role. Perhaps, the fragile and difficult path from deviant to ex-deviant produces the positive evaluation of the category of ex-deviant.

One of the most dramatic role changes is that of ex-convict. Irwin and Austin have outlined the extraordinary difficulty in the transformation of an incarcerated individual to an ex-convict who does not relapse, as follows:

> During this period of supervision, many released inmates experience tremendous difficulties in adjusting to the outside world without being rearrested and returning to prison and jail. In general, most inmates are rearrested at least once after being released from prison. (1997)

Among the most critical factors in facilitating recidivism, or impeding rehabilitation, according to Irwin and Austin (1997) are the following: the trauma of reentering the world after being incarcerated in a total institution; the difficulty of attaining employment for the all too often undereducated and underskilled ex-convict; the intensive supervision and law enforcement mandate of parole agents; drug testing; intensive supervision programs; and electronic monitoring. While Irwin and Austin (1997) conclude that the majority of incarcerated individuals do intend to lead a conforming life after their release from prison, these difficult obstacles result in most inmates ending up dependent, drifting between conventionality and criminality, and dereliction. Some do make it.

How do the formerly incarcerated achieve the positively evaluated status of returning to conformity? Irwin and Austin conclude:

> They usually do so only because of the random chance of securing a good job and a niche in some conventional social world by virtue of their own individual efforts to "straighten up" often with the help of their family, friends, or primary assistance organization. But even members of this group are likely to face periodic obstacles in being accepted fully as a citizen. (1997)

Shover (1983) has most extensively analyzed the successful passage, from the perspective of ex-convicts (in this case, ordinary property offenders). According to ex-convicts, two types of changes assisted the transition from the negative status of convict to ex-convict: temporal changes and interpersonal changes. The temporal changes, perhaps congruent to a certain extent with the processes of adult maturation, included the following: an identity shift, in the confrontation with a past of unsuccessful criminality; a perception that their time had not been well spent and that time was not infinite; a lessening of youthful material goals; and a sense of tiredness at the thought of dealing with a criminal justice system that while not omnipotent, is certainly potent. Additionally, interpersonal contingencies primarily revolved around involvement with a significant relationship and secure employment. Essentially, these were the factors identified by one group of ex-deviants as the most pertinent in their advancing beyond their formerly negative status.

Thus, the ex-deviant transcends the stigmatization, that Goffman (1963) deemed so critical in shaping the individual. While the ex-deviant still may be tinged with a previous status, this particularly unique category is another potential type of positive deviance. Perhaps, other types will be outlined in the future.

CONCLUSION

One point should be noted. Various actions or actors probably transcend more than one category. As an example, Mother Teresa lived a life of altruism (rather than just having engaged in one dramatic altruistic incident), yet was also a supra-conformist, as she abided by the idealized norms of religious adherents, rather than just the expected behavioral norms. Additionally, while Martin Luther King was primarily a charismatic leader, he was innovative in that he combined cultural elements in a new way, by applying the techniques of nonviolent civil disobedience to the civil rights movement. At the same time, with his specifically exquisite oratorical skills, he also fits into the category of having been a possessor of innate characteristics. All in all, many actions and/or actors can be explained by more than one type. Nevertheless, the present typology seems the best way to begin the categorization of positive deviance, since as previously noted, each type can be considered an ideal type.

Hopefully, this typology will help to clarify the concept of positive deviance and facilitate the emergence of other questions and other issues, especially those issues that have been suggested in relationship to deviance (negative deviance). For example, various theorists contrast major forms of deviance, with minor ones. Curra (1994) and Raybeck (1991) differentiate between soft deviance, or unique behaviors not consistent with social norms but not threatening to the social system, and hard deviance, or more serious and ominous forms of behavior. Along these same lines, Thio (1988) contrasts higher-consensus deviance with lower-consensus deviance, depending on the seriousness of the act and the degree of societal consensus in relationship to the perception of the act. In reference to examples of positive deviance that have been cited in the literature, most are probably soft deviance, in that the acts do not generally harm others and are not reacted to as serious. For example, altruists and straight-A students do not potentially harm others. In some cases, the predominant paradigms ensconced in the social order are potentially challenged by such examples of positive deviants as innovators in any realm of the social order, from science to art to politics to religion, or by reformers. This phenomenon might address the issue of why certain positive deviants are not generally easily accepted in their time and place. Deviance is relative; many positive deviants also experience this relativity in that the initial reception to their actions is negative. In this sense, these types of positive deviants can potentially, at least, be deemed hard deviance, in the sense that the social order is challenged.

Another interesting issue is the following and also relates to the relativity of deviance. Are certain actions and/or actors (or categories of positive deviance) more likely to be positively labeled, negatively labeled, or neutrally labeled at first? Perhaps, since innovation can be more psychologically threatening to a culture, innovation is more often negatively labeled in the beginning. On the other hand, altruism, because it involves self-sacrifice, and is not usually potentially threatening to society, is more often positively evaluated at first. In addition, physical attractiveness as a form of innate characteristic, seems usually to result in an initial positive label and minimal negative treatment. According to

Dion, Berscheid, and Walster (1972) "what is beautiful is good" since the attractive are the recipients of ubiquitous advantageous treatment, extending to various parts of their life.

Additionally, the notion of stigma, outlined by Goffman (1963), has been central to the examination of deviance. The ambivalence toward positive deviance does raise the possibility that stigma is applicable in this case, also. The central reason is that positive deviants are also different. For example, as previously suggested, the entire social construction of the "geek" with its accompanying stereotypes, would suggest that straight-A students and/or the gifted are not entirely positively received. The sword is dual-edged in that while there is a positive treatment, the stigmatization is also profound and quite potentially has a negative impact on individuals so categorized. The clownish, or not completely human construction of the geek (or nerd or dweeb or dork) is perhaps similar to the village idiot or the in-group deviant, as presented by Goffman (1963), a "mascot" not fully rejected and partially admired for academic acumen, yet not fully accepted. Positive deviants are different; due to their difference, the possibility of stigmatization is great.

Another useful way to think about deviance, that might also be pertinent to positive deviance, is the manner in which deviance is functional to society. Cohen (1966) has maintained that deviance can contribute to society in the following manner: opposing red tape and dealing with anomalies, serving as a safety valve, clarifying the rules, uniting a group in opposition to the deviant, uniting a group in support of the deviant, accentuating conformity, and performing as a warning sign to society. Along these same lines, perhaps, positive deviance also provides some of the same opportunities to benefit the social order. For example, reformers clearly provide a warning signal to society that the social order is in dire need of change. As another example, positive deviants accent conformity. As Cohen (1966) describes, "The good deed, as Shakespeare noted, shines brightest in a naughty world." Thus, as deviants (negative deviants) are a reference for the contrast between bad and good, so can positive deviants, ranging from straight-A students to altruists serve as a reference. Altruists also contrast with conforming behavior, serving as a guide for human potentiality. As such, positive deviance, like deviance also contributes to the social order of society.

The concept of positive deviance needs to be further expanded. Yet, it does appear that critical ideas related to deviance could also be applied to positive deviance. Additionally, this typology posits that there is more than one type of positive deviance. Each type needs to be examined further—within a framework and within the parameters of positive deviance and of deviance theory—as a unique and as an important entity.

REFERENCES

Ben-Yehuda, N. 1990. "Positive and negative deviance: More fuel for a controversy." *Deviant Behavior* 11: 221–243.

Best, J., and D. F. Luckenbill. 1982. *Organizing Deviance*. Englewood Cliffs, NJ: Prentice-Hall.

Buffalo, M. D., and J. W. Rodgers. 1971. "Behavioral norms, moral norms, and attachment: Problems of deviance and conformity." *Social Problems* 19: 101–113.

Cialdini, R. J., D. T. Kerrick, and D. J. Bauman. 1982. "Effects of mood on prosocial behavior in children and adults." Pp. 339–359 in N. Eisenberg, ed., *The Development of Prosocial Behavior*. New York: Academic Press.

Cohen, A. K. 1966. *Deviance and Control*. Englewood Cliffs, NJ: Prentice-Hall.

Comte, A., 1966. *System of Positive Polity, Vol. 1. General Views of Positivism and Introductory Principles*. Tr. J. H. Bridges. New York: Burt Franklin.

Curra, J. 1994. *Understanding Social Deviance: From the Near Side to the Outer Limits*. New York: Harper Collins.

Dion K., E. Berscheid, and E. Walster. 1972. "What is beautiful is good." *J Personality Social Psychol* 24: 285–290.

Ebaugh, H. R. F. 1988. *Becoming an Ex: The Process of Role Exit*. Chicago: University of Chicago Press.

Ewald, K. 1981. *The Extension of the Becker Model of Socialization to Positive Deviance: The Cases of Weight Lifting and Running*. Unpublished Ph.D. Dissertation: Ohio State University.

Freedman, J. L., and A. N. Doob. 1968. *Deviancy: The Psychology of Being Different*. New York: Academic Press.

Gibbs, J. P. 1965. "Norms: The problem of definition and classification." *Amer J Sociology* 70: 586–594.

Goffman, E. 1963. *Stigma: Notes on the Management of Spoiled Identity*. Englewood Cliffs, NJ: Prentice-Hall.

Goode, E. 1991. "Positive deviance: A viable concept." *Deviant Behavior* 12: 289–309.

Grusec, J. E. 1981. "Socialization processes and the development of altruism." Pp. 65–90 in P. Rushton, R. M. Sorrentino, eds., *Altruism and Helping Behavior*. Hillsdale, NJ: Lawrence Erlbaum.

Hawkins, R., and G. Tiedeman. 1975. *The Creation of Deviance*. Columbus, OH: Charles E. Merrill.

Homans, G. C. 1950. *The Human Group*. New York: Harcourt, Brace and Company.

Huryn, J. S. 1986. "Giftedness as deviance: A test of interaction theories." *Deviant Behavior* 7: 175–186.

Irwin, J., and J. Austin. 1997. *It's About Time: America's Imprisonment Binge*, 2nd ed., Belmont, CA: Wadsworth.

Johnson, E. H. 1978. *Crime, Correction, and Society*. New York: Harcourt, Brace, and Company.

Kallen, H. M. 1964. "Innovation." Pp. 427–430 in A. Etzioni and E. Etzioni, eds., *Social Change*. New York: Basic Books.

Katz, J. 1972. "Deviance, charisma, and rule-defined behavior." *Social Problems* 20: 186–202.

Leeds, R. 1963. "Altruism and the norm of giving." *Merrill-Palmer Qrtly* 9: 229–240.

Lemert, E. M. 1951. *Social Pathology*. New York: McGraw-Hill.

Lenski, G., and J. Lenski. 1982. *Human Societies*. New York: McGraw-Hill.

Liazos, A. 1975. "The poverty of the sociology of deviance: Nuts, sluts, and perverts." Pp. 9–29 in D. J. Steffensmeier and R. M. Terry, eds., *Examining Deviance Experimentally*. New York: Alfred Publishing.

Linton, R. 1936. *The Study of Man*. New York: C Appleton-Century.

Macauley, J. R., and L. Berkowitz. 1970. "Overview." Pp. 1–9 in J. Macauley and L. Berkowitz, eds., *Altruism and Helping Behavior*. New York: Academic Press.

Morse, S. T., H. T. Reis, J. Gruzen, and E. Wolff. 1974. "The eye of the beholder: Determinants of physical attractiveness judgments in the United States and South Africa." *J Personality* 42: 528–541.

Norland, S., J. R. Hepburn, and D. Monette. 1976. "The effects of labeling and consistent differentiation in the construction of positive deviance." *Sociology Soc Res* 61: 83–95.

Ogburn, W. F. 1964. *On Culture and Social Change: Selected Papers.* O. D. Duncan, ed. Chicago: University of Chicago Press.

Palmer, S. 1990. *Deviant Behavior.* New York: Plenum Press.

Pfuhl, E., and S. Henry. 1993. *The Deviance Process,* 3rd ed. New York: Aldine de Gruyter.

Raybeck, D. 1991. "Deviance: Anthropological perspectives." Pp. 51–72 in M. Freilich, D. Raybeck, and J. Savishinsky, eds. *Deviance: Anthropological Perspectives.* New York: Bergin and Garvey.

Rebelsky, F. A., and P. A. Daniel. 1976. "Cross-cultural studies of infant intelligence." Pp. 279–297 in M. Lewis, ed., *Origins of Intelligence.* New York: Plenum Press.

Rogers, E. M., and F. F. Shoemaker. 1971. *Communication of Innovations.* New York: Free Press of Glencoe.

Rosenhan, D. 1970. "The naturalistic socialization of altruistic autonomy." Pp. 251–268 in J. Macauley and L. Berkowitz, eds., *Altruism and Helping Behavior.* New York: Academic Press.

Sagarin, E. 1985. "Positive deviance: An oxymoron." *Deviant Behavior* 6: 169–181.

Scarpitti, F. R., and P. T. McFarlane. 1975. *Deviance: Action, Reaction, Interaction.* Reading, MA: Addison-Wesley.

Shover, N. 1983. "The later stages of ordinary property offender careers." *Social Problems* 31: 208–218.

Sorokin, P. A. 1948. *The Reconstruction of Humanity.* Boston: Beacon Press.

———, 1950. *Altruistic Love.* Boston: Beacon Press.

Steffensmeier, D. J., and R. M. Terry. 1975. *Examining Deviance Experimentally.* Port Washington, NY: Alfred Publishing.

Szasz, T. S. 1970. *The Manufacture of Madness.* New York: Harper and Row.

Thio, A. 1988. *Deviant Behavior,* 3rd ed. New York: Harper Collins.

Weber, M. 1947. *The Theory of Social and Economic Organization.* Trs. A. M. Henderson and T. Parsons. New York: Free Press.

White, W. 1961. *Beyond Conformity.* Westport, CT: Greenwood Press.

Wilkins, L. T. 1965. *Social Deviance.* Englewood Cliffs, NJ: Prentice-Hall.

Winslow, R. W. 1970. *Society in Transition: A Social Approach to Deviance.* NY: Free Press.

PART II

Theories of Deviance

Deviance holds a special intrigue for scholars of theory. Given its pervasive nature in society, its enigmatic conditions, and its generic appeal, even the earliest sociologists attempted to explain how and why deviance occurs. Especially considering people's inclination to conformity, the pressing question for scholars has dealt with *why* individuals engage in norm-violating behavior. Explanations for deviant behavior are as divergent as the acts they explain, ranging from acts of delinquency to professional theft, acts of integrity to a search for kicks, acts of desperation to those of bravado and daring. Below we outline some of the major attempts at understanding deviant behavior.

THE STRUCTURAL PERSPECTIVE

The dominant theory in sociology for the first half of the twentieth century, structural functionalism, also commanded the greatest amount of sway in explaining deviant behavior. Emile Durkheim, the French scholar generally considered the founding father of sociology, advanced the theory that society is a moral phenomenon, that at its root the morals (norms, values, and laws) that individuals are taught constrain their behavior. Durkheim believed that youngsters are taught the "rights" and "wrongs" of society early in life, with most people conforming to these expectations throughout adulthood. In large measure these moral beliefs determine how people behave, what they want, and who they are. Durkheim suggested that societies with high degrees of social integration (bonding, community involvement) would increase the conformity of its members. What concerned Durkheim, however, was that in the modern French society in which he was living, more and more people were becoming distanced from each other, people were partially losing their sense of belonging to their communities, and the norms and expectations of their groups were becoming less clearly defined. He believed that this condition, which he referred to as *anomie,* was producing a concomitant amount of social disintegration, leading to greater degrees of deviance. Thus for Durkheim, while norms still existed on the societal level, the lack of social integration created a situation where they were no longer becoming as significant a part of each individual.

Despite his concerns about the increasing rates of deviance that society would produce, Durkheim also subscribed to the idea that deviance was functional for society. Curiously, despite its obvious negative effects, Durkheim felt that deviance produces some positive benefits as well. At a time when people are worrying about the moral breakdown and social disintegration of society, deviance serves to remind us of the moral boundaries in society. Each time a deviant act is committed and publicly announced, society is united in indignation against the perpetrator. This serves to bring people together, rather than tear them apart. At the same time society is reminded about what is "right" and "wrong" and, for those who conform, a greater social integration ensues. These ideas were perhaps best illustrated by Yale sociologist Kai Erikson who, in his 1966 book *Wayward Puritans* (excerpted earlier), demonstrated the role of deviance in defining morality and bringing people together. Erikson examined Puritan patterns of isolating and treating offenders. He believed that deviance serves as a means to promote a *contrast* with the rest of the community, thus giving members of the larger society more strength in their moral convictions. Erikson's analysis focused on the transformation of the seventeenth century Bay Colony, as a group of revolutionaries tried to establish a new community in New England. These deviants, the revolutionaries, played an important role in the transformation of norms and values—their behavior elicited societal reaction, which served to *clearly define* the new community's norms and values. In addition, punishing some people for norm violations reminded others of the rewards for conformity. Selection 4, "The Normal and the Pathological," by Durkheim, lays out his theory of the inevitability of deviance in all societies. In his ironic twist, Durkheim argues that deviance is normal rather than pathological, serving a positive function in society. To achieve the maximum benefit, however, a society needs a manageable amount of deviance. When the numbers of people declared deviant by current moral standards rises or falls too much, society alters its moral criteria to maintain the level of deviance in the optimal range. At different times it may "define deviancy down," as Moynihan (1993) has suggested in looking at the way the bar defining acceptable behavior has been lowered by normalizing the high levels of violence, divorce, and deinstitutionalized mental patients inhabiting the streets. Or it may "define deviancy up," as Krauthammer (1993) has suggested, raising the bar defining normality so that behavior formerly considered innocent, such as the way parents discipline their children, the sex people may have on dates, and the prejudice people feel towards members of other groups has been redefined as child abuse, date rape, and thought crime. What deviance does for society is to define the moral boundaries for everyone. Violation of norms serves to remind the masses what is acceptable and what is not; in Durkheim's words, it enforces the "collective conscience" of the group. The public outcry after the revelation of the Starr Report on the Clinton sex scandal offers us a good illustration of that. Perhaps it is difficult to imagine that behavior that disgusts, reviles, or even nauseates you are not the acts of immoral, sick, or evil people, but are normal and even beneficial parts of all societies. Durkheim's theory suggests that structural needs of the society as a whole, beyond the scope of its individual members, foster the continuing recurrence of deviance.

The structural perspective locates the root cause of crime and deviance outside of individuals, in the invisible social structures that make up our society. Structural explanations for deviance look at features of society, or of some societies compared to other societies, that seem to generate higher rates of crime or deviance among some societies or groups within them. In looking for explanations for why some societies are likely to have higher crime rates than others, sociologists have suggested that those with greater degrees of inequality are likely to show more crime than those where people have roughly similar amounts of the good things available. Looking within each society, structuralists locate the cause of crime in two main factors: the differential opportunity structure, and prejudice and discrimination towards certain groups. In a society with inequality, some groups will clearly have greater structural access to certain opportunities than others. Groups with access to greater power, political, and economic opportunity may use these to define their acts as legitimate and the acts of others as deviant, at the same time as they corruptly use their power to their advantage. Not everyone has equal opportunity to dispense political favors, to manipulate stock prices, or to conduct covert operations. At the same time groups with less access to the legitimate opportunity structure due to reduced educational opportunity, diminished access to health care, lower-class background, and disadvantaged legitimate networks and connections, do not have the same opportunity to succeed normatively. Members of these groups may be propelled by their position in the social structure into alternate pathways.

It was Robert Merton, a mid-twentieth century sociologist from Columbia University, who actually extended Durkheim's ideas and built them into a specific structural theory of deviant behavior. In a wide-sweeping and influential article, selection 5, "Social Structure and Anomie," he claimed that contradictions are implicit in a stratified system in which the culture dictates success goals for all citizens, while institutional access is limited to just the middle and upper strata. In other words, despite the American Dream of rags-to-riches, some people, most often lower-class individuals, are systematically excluded from the competition. Instead of merely going through the motions while knowing that their legitimate path to success (measured in American society by financial wealth) is blocked, some members of the lower class retaliate by choosing a deviant alternative. Merton believed that these people have accepted society's goals (to be comfortable, to get rich), but they have insufficient access to the approved means of attaining these goals (deferred gratification, education, hard work). The problem lies in the social structure in society, where even if people follow the approved means, there are roadblocks prohibiting them from rising through the stratification system. Deviant behavior occurs when socially sanctioned means are not available for the realization of highly desirable goals. The only way to achieve these goals is to detour around them, to bypass the approved means in order to get at the approved goals. For example, for young men raised in urban ghettos with poor housing facilities, dilapidated schools, and inconsistent family lives, the road to success is more likely to be through dealing drugs, pimping, or robbing, than it is through the normative route of school and hard work. According to Merton, then, *anomie* results from the lack of access to culturally prescribed goals and the lack of availability of legitimate means for attaining those goals. Deviance (or, more specifically, crime) is the obvious alternative. Once again structural opportunities or their lack are seen as the root cause of deviance, rather than some psychological or individual pathology.

Richard Cloward and Lloyd Ohlin, in *Delinquency and Opportunity* (1960), thought that Merton was correct in directing us toward the notion that members of disadvantaged socioeconomic groups have less opportunity for achieving success in a legitimate manner, but they thought that Merton wrongly assumed that those groups, when confronted with the problem of differential opportunity, could automatically choose deviance and crime. Cloward and Ohlin suggested that all disadvantaged people have some lack of opportunity for legitimate pursuits, but they do not have the same opportunity for participating in illegitimate practices. What Cloward and Ohlin believed was that deviant behavior depends on people's access to illegitimate opportunities. They found that three types of deviant opportunities are present: (1) criminal (similar to the type Merton described), which arise from access to deviant subcultures, though not all disadvantaged youth enjoy these avenues; (2) conflict, which attract people who have a propensity for violence and fighting; and (3) retreatist, which attract people who are not inclined toward illegitimate means or violent actions, but who want to withdraw from society, such as drug users. According to Cloward and Ohlin, groups of people may have greater or lesser opportunity to climb the illicit opportunity ladder by virtue of several factors: (1) some neighborhoods are rife with more criminal opportunities, networks, and enterprises than others, and people reared in these grow up amidst these better opportunities, (2) some forms of illicit enterprise are dominated by people of particular racial or ethnic groups, so that members of these groups have an easier time rising to the top of these businesses or organizations, and (3) the upper echelons of crime display a distinct glass ceiling for women, with men dominating the positions of decision-making, earning, and power. Thus Cloward and Ohlin extended Merton's theory by specifying the existence of differential illegitimate opportunities available to members of disadvantaged groups. It is in this opportunity structure, rather than individual motivation, they argue, that the explanation for deviance can be found.

Selection 9, "Conflict Theory of Crime," by Richard Quinney, offers an explanation that is not functionalist but still structural. Conflict theorists see society differently from functionalists in their view of society as pluralistic, heterogeneous, and conflictual rather than unified and consensual. Social conflict arises out of the incompatible interests of diverse groups in society, such as businesses versus their workers, conservatives versus liberals, whites versus people of color, and the rich versus the poor. These groups have a structural conflict of interest with each other that stands above and beyond the individual members, framing the way they come to recognize their interests and act in the world. Not only is conflict a natural outcome of this arrangement, but crime is as well. In a succinct summation of conflict theory's major tenets, Quinney describes how crime exists because some behaviors conflict with the interests of the dominant class. These powerful members of society create legal definitions of human conduct, casting those behaviors that threaten its interests as criminal. Then the dominant class enforces those laws onto the less powerful groups in society, through the police, the legal, and the criminal justice systems ensuring that their interests are protected. Members of subordinate classes are compelled to commit those actions that have been defined as crimes because their poverty presses them to do so. The dominant group can then create and disseminate their ideology of crime, which is that the most dangerous criminal elements in society can be found in the subordinate classes and that these groups deserve arrest, prosecution, and imprisonment. Through class struggle and class conflict, crime is constructed, formulated, and applied so that less powerful groups are subdued and more powerful groups are strengthened. These processes are illustrated by the diagram in Quinney's selection. This approach shows how larger social forces, such as group and class interests, shape the behavior of individual members, leading some to use their advantage to dominate over others, while the others react to their structural subordination by engaging in those behaviors already defined as deviant and deserving of punishment. All of these structural theories place the cause for the incidence of deviance on the structures of society, rather than on individuals and their problems.

THE CULTURAL PERSPECTIVE

While structuralist theories had enormous impact on sociologists' thinking about deviance, other authors arose who felt that these were not all-encompassing explanations. These theorists believed that deviance was a collective act, driven and carried out by groups of people. Building on conflict theory's view that multiple groups with different interests exist in society, the subcultural theorists examined the implications of membership in these groups. Groups with conflicting interests include not only dominant and subordinate groups, but a variety of social, religious, political, ethnic, and economic factions. Membership in each of these groups places people in distinct subcultures, each of which contain their own set of distinct norms and values. A pluralistic nation that was once thought of as the world's "melting pot," we have become, in part, a nation of many different groups, each with its own distinct subculture. Thorsten Sellin, writing about "The Conflict of Conduct Norms" (1938) suggested that to some extent the norms and values of these subcultures incorporated and meshed with the norms and values of the overarching American culture, but to some extent they were different and in conflict. The disparities and different cultural codes between subcultural groups may become apparent in three situations. First, when people from one culture cross over into the territory of another culture, such as when people from a rural area move to the city, they find that their country ways do not mesh with the city ways. In these cases, they may find themselves subject to the urban norms and values. Second, cultural conflict may occur when the laws of one cultural group are extended to apply to another, such as when one group moves into and takes over the territory of another. This may happen when people regentrify a downtown area, and the latitude that used to be enjoyed by the former occupants to congregate outdoors, be homeless, do or deal drugs, or solicit prostitution becomes lost. In these cases the norms of the group that has taken over may apply. Third, cultural codes may clash on the border of contiguous cultural areas, such as when people from different cultural groups find themselves in contact. They may be on a national border, a neighborhood border, or they may simply be individuals encountering individual members of another subculture. In these cases no clear set of norms and values necessarily dominates, but individuals have to negotiate their cultural understandings delicately, trying to understand each others' norms and values. This could happen when new first-year students find themselves rooming with someone from a distinctly different background, where neither of their ways necessarily predominates. Sometimes this even happens when college students go home for the holidays, only to find that the norms under which they were living at college are rather different from those in their parents' houses. In each of these three cases, people may find themselves torn between the norms and values of different group memberships. Following the norms and values of their subcultural group may produce behavior that becomes defined as deviant by the standards of the broader culture. Yet from their subcultural perspective, their behavior may be viewed as representing the acts of good people working to uphold the behaviors they honor. In his writing, Sellin was particularly thinking about the deviance of children belonging to immigrant ethnic or racial groups moving into the United States, caught in the struggle between two cultures, but his theory applies equally well to the large number of diverse subcultural groups in our country. He extended his model to apply to all conflicts between cultural groups that share a close geographic area, especially where there is normative and value domination by one culture over another.

Building on this idea, Albert Cohen, in *Delinquent Boys* (1955), argued that working-class adolescent males comprise a subculture with a different value system from the dominant American culture. These boys, Cohen asserted, have the greatest degree of difficulty in achieving success, since the establishment's standards are so different from their own. They may try, at first, to fit in with the cultural expectations, but find that they are unsuccessful. Exposed to middle-class aspirations and judgments they cannot reasonably fulfill, they develop a blockage (or strain) that leads them to experience "status frustration." What results from this frustration is the formation of an oppositionally reactive subculture that allows them to achieve status based on nonutilitarian, malicious, and negativistic behavior. These boys, in reaction against society's perceived unfairness towards them, substitute norms that reverse those of the larger society. Cohen claimed that delinquent boys, due to their societal rejection, turn the society's norms "upside down," rejecting middle-class standards and adopting values in direct opposition to those of the majority.

Walter Miller (1958), writing just after Cohen, further delineated the importance of subcultural values for the development of deviant behavior. He believed that the values of the lower-class culture produce deviance for its members because these are "naturally" in discord with middle-class values. Young people who conform to the lower-class culture in which they were born almost automatically become deviant. The culture of the lower-class, by which members attain status in the eyes of their peers, is characterized by several "focal concerns:" getting into trouble, showing toughness, maintaining autonomy, demonstrating street-smartness, searching for excitement, and being tied in their lot to the capricious whims of fate. When these individuals follow the norms of their subculture, they become deviant according to the predominantly middle-class societal norms and values.

The lasting impact of subcultural theories has been to suggest that conflicting values may exist in society. When one part of society can impose its definitions on other parts, the dominant group has the ability to label the minority group's norms and values and the behavior that results from these as deviant. Thus any act can be considered deviant if it is so defined. These theories are suited to illustrating the motivations of people from minority, youth, alternative, or disadvantaged subcultures that are not well-aligned with the dominant culture. They locate the explanation for deviance not in the structures that shape society, but in the flesh of the norms and values that compose different subcultural groups. Through cultural transmission, groups pass their norms and values down from one generation to the next, ensuring their survival and social placement as well as the continuance of cultural conflict.

THE INTERACTIONIST PERSPECTIVE

While these previous theories produced insight into some explanations for deviance, there are interactional forces that inevitably intervene between the larger causes these sociologists propose and the way deviant behavior takes shape. Many people are exposed to the same structural conditions and the same cultural conflicts and pressures that have been theorized as accounting for deviance, but still resist engaging in deviant behaviors. Left unaddressed is how people from the same structural groups and same subcultures can turn out so differently, how members of some families turn to deviance while others do not, and how members of the same family turn out so differently from each other. Interactionist theories fill this void by looking in a more micro fashion at people's everyday life behavior to try to understand why some people engage in deviance and become so labeled, while others do not. Interactionist theories deal with real flesh and blood people in specific times and places. They look at how people actually encounter specific others, and they look at the influence of these others. They seek to understand not only why deviance occurs, but how it happens. Many of these theories look at specific social-psychological and interactional dynamics, such as family dynamics, the influence of role models, and the role of peer groups. When people confront the problems, pressures, excitements, and allures of the world, they most often do so in conjunction with their peer groups. It is within peer groups that people make decisions about what they will do and how they will do it. Their core feelings about themselves develop and become rooted in such groups. People's actions and reactions are thus guided by the collective perceptions, interpretations, and actions of their peer groups.

Edwin Sutherland and Donald Cressey recognized this point when they proposed their "Differential Association" theory of deviance, selection 6. The key feature of this view is the belief that deviant behavior is socially learned, and not from just anyone, but from people's most intimate friends and family members. People may be exposed to a variety of deviant and nondeviant ideas and contacts without that necessarily leading them to engage in deviance. But as their circle of contacts shifts from being primarily composed of people who hold nondeviant ideas to having greater numbers holding deviant ideas and favorable definitions of deviant acts, they become more likely to engage in deviance. The more their friends hold deviant attitudes and engage in deviant behavior, the more likely they are to follow suit.

Sutherland and Cressey further suggested that people learn a variety of elements critical to deviance from their associates: the norms and values of the deviant subculture, the rationalizations for legitimizing deviant behavior, the techniques necessary to commit the deviant acts, and the status system of the subculture by which members evaluate themselves and others. People, thus, do not decide, at a fixed point in time, to become deviant, but move toward these attitudes and behavior as they shift their circle of associates from more normative friends to more deviant friends. Sutherland argued that people rarely stumble onto deviance through their own devising or by seeing acts of deviance in the mass media (as many would suggest), but rather by having the knowledge, skills, attitudes, values, traditions, and motives passed down to them through interpersonal (not impersonal) means. Influencing this is the age at which they encounter this deviance (earlier in life is likely to be more significant) and the intimacy of the deviant relationships (closer friends will have greater sway).

Also looking at the interactional level, David Matza (1964) noted that this movement into deviant subcultures occurs through a process of "drift," as people gradually leave their old crowd and become enmeshed in a circle of deviant associates. In so doing, Matza suggested that rather than just jumping immediately into deviance, they may drift between deviance and legitimacy, keeping one foot in each world. By simultaneously participating in both deviant and legitimate worlds, people can learn about and experience the nuances of the deviant world without having to abandon the advantages of their status within the legitimate world. They may drift indefinitely, without having to make a commitment to either for quite some time. For example college students may experiment with a different sexual orientation without revealing this to everyone and without necessarily giving up their claim to heterosexuality. At some point they may decide to align more firmly with one side, or this may be forced on them by outside events (getting caught, moving away, becoming sick). Being confronted by someone who discovers the deviance may force people to make a choice and get off the fence, or perhaps just leaving college and having to choose which lifestyle and social network to align themselves with may force a choice. An alternate is that individuals choose one path, after a time, and decide to follow it. But Matza suggests that the dual membership condition may precede such a decision, where individuals try out both alternatives for a time without making a commitment. Thus Matza suggests that it is rare for people to turn to deviance overnight, but more common for them to take smaller steps, gradually moving through making deviant acquaintances, becoming familiar with deviant ideology, thinking about engaging in deviant acts, trying some out, and then expanding their frequency and range of deviance. Quitting deviance may be a similarly gradual and difficult process, requiring the abandonment of the group of deviant friends and reintegration in conventional circles, perhaps one foot in both worlds again, before normative behavior becomes the mode that is finally chosen.

A third theory under the interactionist perspective is "Labeling Theory," the subject of selection 7. This approach suggests that many people dabble to greater or lesser degrees in various forms of deviance. Studies of juvenile delinquency suggest that rates of youthful participation are extremely widespread, nearly universal. How many people can claim to have reached adulthood without experimenting in illicit drinking, drug use, stealing, or vandalism? Yet do all of these people consider themselves deviants? Most do not. Many people retire out of deviance as they mature, avoiding developing the deviant identity altogether. Others go on to engage in what Becker (1963) has called "secret deviance," conducting their acts of norm violation without ever seriously encountering the deviant label. Yet others, many of them no more experienced in the ways of deviance than the youthful delinquent or the secret deviant, become identified and identify themselves as deviants. What causes this difference? One critical difference, labeling theory suggests, lies in who gets caught. Getting caught sets off a chain reaction of events that leads to profound social and self-conceptual consequences. Frank Tannenbaum (1938) has described how individuals are publicly identified as norm violators and branded with that tag. They may go through official or unofficial social sanctioning where people identify and treat them as deviant. Becker (1963:9) noted that "the deviant is one to whom that label has successfully been applied; deviant behavior is behavior that people so label." Deviance exists at the macro societal level of social norms and definitions through the collective attitudes we assign to certain acts and conditions. But it also comes into being at the micro everyday life level when the deviant label is applied onto someone. The thrust of labeling theory is twofold, focusing on diverse levels and forces. As Edwin Schur (1979:160) summarized its complexity:

> The twin emphases in such an approach are on *definition* and *process* at all the levels that are involved in the production of deviant situations and outcomes. Thus, the perspective is concerned not only with what happens to specific individuals when they are branded with deviantness ("labeling," in the narrow sense) but also with the wider domains and processes of social definitions and collective rule-making that frequently lie behind such concrete applications of negative labels. *(Italics in original.)*

In our labeling theory selection, Becker emphasizes that deviance lies in the eye of the beholder. There is nothing inherently deviant in any particular act, he claims, until some powerful group defines the act as deviant. Taking the onus off of the individual, Becker emphasizes the importance of looking at the process by which people are labeled deviant and understanding that deviance is a consequence of others' reactions. This approach forces us to look, then, at how people are defined as deviant, why some acts are labeled and others ignored, and the circumstances that surround the commission of the act. Thus deviance only exists when it is created by society. The key emphasis of the labeling theory approach to deviance, then, lies in the importance of human peer interaction in understanding the root cause of human behavior.

Rooted at the micro level, but looking less at the specific dynamics of interaction and more at the relationship between individuals and society, is Travis Hirschi's "Control Theory of Delinquency," the subject of selection 8. Like labeling theory, which took as given that people readily engage in acts of deviance but focused its explanation on the process of identity change that occurs when individuals are caught and labeled, control theory finds it unnecessary to seek for the causes of deviant behavior. These are obvious, Hirschi asserts, as deviance and crime are not only fun, but they offer shortcuts and yield immediate, tangible benefits (albeit, with a risk). What we should be seeking to understand, instead, is what holds people back from committing these acts, what forces constrain and control deviance. Hirschi's answer is that social control lies in the extent to which people develop a stake in conformity, a *bond to society.* People who have a greater investment in society will be less likely to risk losing this through norm or law violations and will follow the rules more willingly. They may have such a stake through their job, relationships to friends and family, or their reputation in the community. Their stake may be fostered by any of the four components discussed in this selection: attachment to conventional others; commitment to conventional institutions; involvement in conventional activities, and deep beliefs in conventional norms. The extent to which a society is able to foster greater bonds between itself and its potentially deviant members by giving them a greater stake in achieving success will affect the constraint or spread of its deviance, particularly Hirschi notes, of the delinquent variety. It is these ties, interactionally forged and maintained, that influence individuals in their choice between deviant or nondeviant pathways.

Although these overarching perspectives and the theories nested within them differ in the level at which they place their explanations, they all locate them squarely in the social domain. In this they renounce the prevailing tendency towards unidimensional psychological explanation that roots causation in pathology, compulsion, neurosis, or maladjustment—explanations whose oversimplicity and inadequacy in a modern, complex world cannot be overstated.

The most recent perspective on deviance theory has been advanced by social constructionists. Joel Best, one of its leading proponents, describes its history and views in selection 10, "Deviance: The Social Constructionist Stance." Initially coming from a microinteractionist approach, constructionist theorists sought to revitalize labeling theory by bridging the gap between the way labels are applied to individuals, who then internalize them in their everyday life context, and a larger, more macro awareness of the power structure in society that influences the way these labels are defined and enforced. Having access to greater degrees of social power enables the dominant groups in society to rationalize their ideologies and behavior as legitimate, while defining those of less powerful groups as deviant. In so doing they use the vehicle of deviance and its enforcement to boost their own power while disempowering those they construct as illegitimate. The definitions they forge are then applied to individuals who lack the power to repel them, and the agents of social control act to carry out their moral edicts. The powerless segments of society become the recipients of the activities of moral entrepreneurs (rule creators and rule enforcers), who lay claims that certain behaviors are menacing and dangerous. Individuals connected to these activities may be defined as deviant, isolated, sought out, labeled, and stigmatized. Social constructionism thus builds on the basic labeling theory foundation of identity construction by integrating conflict theory's sensitivity to inequality, looking at how the power struggle between dominant and subordinate groups is directly tied to interactional and identity consequences.

4

The Normal and the Pathological

EMILE DURKHEIM

One of the founders of the functionalist approach to sociology, Durkheim integrates deviance into his overarching view of society. If, as he believes, society can be compared to a living organism, then all of its social institutions, like the parts of a human body, must contribute to its continuing existence. Under this view deviance, because of its pervasive presence cross-culturally and over the entire course of history, is not an illness or pathology of the system, but rather something that contributes to society's positive functioning. In fact like Erikson, he notes several positive functions that it provides, including serving as a means of introducing social change as new behaviors move through criminal and deviant status into respectability. Durkheim argues that deviance is so critical to society that if people stopped engaging in it immediately (if we lived in a "society of saints"), we would have to redefine acts now considered acceptable as deviant. Punishing and curing criminals cannot simply, then, be regarded as the objective of society.

Crime is present not only in the majority of societies of one particular species but in all societies of all types. There is no society that is not confronted with the problem of criminality. Its form changes; the acts thus characterized are not the same everywhere; but, everywhere and always, there have been men who have behaved in such a way as to draw upon themselves penal repression. If, in proportion as societies pass from the lower to the higher types, the rate of criminality, i.e., the relation between the yearly number of crimes and the population, tended to decline, it might be believed that crime, while still normal, is tending to lose this character of normality. But we have no reason to believe that such a regression is substantiated. Many facts would seem rather to indicate a movement in the opposite direction. From the beginning of the [nineteenth] century, statistics enable us to follow the course of criminality. It has everywhere increased. In France the increase is nearly 300 percent. There is, then, no phenomenon that presents more

---

Reprinted with the permission of The Free Press, a Division of Smion & Schuster, Inc., From *The Rules of Sociological Method,* by Emile Durkheim translated by Sarah A. Solovay and John H. Mueller. Edited by George E. G. Catlin. Copyright © 1938 by George E. G. Catlin; copyright renewed 1966 by Sarah A. Solovay, John H. Mueller, George E. G. Catlin.

55

indisputably all the symptoms of normality, since it appears closely connected with the conditions of all collective life. To make of crime a form of social morbidity would be to admit that morbidity is not something accidental, but, on the contrary, that in certain cases it grows out of the fundamental constitution of the living organism; it would result in wiping out all distinction between the physiological and the pathological. No doubt it is possible that crime itself will have abnormal forms, as, for example, when its rate is unusually high. This excess is, indeed, undoubtedly morbid in nature. What is normal, simply, is the existence of criminality, provided that it attains and does not exceed, for each social type, a certain level, which it is perhaps not impossible to fix in conformity with the preceding rules.[1]

Here we are, then, in the presence of a conclusion in appearance quite paradoxical. Let us make no mistake. To classify crime among the phenomena of normal sociology is not to say merely that it is an inevitable, although regrettable phenomenon, due to the incorrigible wickedness of men; it is to affirm that it is a factor in public health, an integral part of all healthy societies. This result is, at first glance, surprising enough to have puzzled even ourselves for a long time. Once this first surprise has been overcome, however, it is not difficult to find reasons explaining this normality and at the same time confirming it.

In the first place crime is normal because a society exempt from it is utterly impossible. Crime, we have shown elsewhere, consists of an act that offends certain very strong collective sentiments. In a society in which criminal acts are no longer committed, the sentiments they offend would have to be found without exception in all individual consciousnesses, and they must be found to exist with the same degree as sentiments contrary to them. Assuming that this condition could actually be realized, crime would not thereby disappear; it would only change its form, for the very cause which would thus dry up the sources of criminality would immediately open up new ones.

Indeed, for the collective sentiments which are protected by the penal law of a people at a specified moment of its history to take possession of the public conscience or for them to acquire a stronger hold where they have an insufficient grip, they must acquire an intensity greater than that which they had hitherto had. The community as a whole must experience them more vividly, for it can acquire from no other source the greater force necessary to control these individuals who formerly were the most refractory. For murderers to disappear, the horror of bloodshed must become greater in those social strata from which murderers are recruited; but, first it must become greater throughout the entire society. Moreover, the very absence of crime would directly contribute to produce this horror; because any sentiment seems much more respectable when it is always and uniformly respected.

One easily overlooks the consideration that these strong states of the common consciousness cannot be thus reinforced without reinforcing at the same time the more feeble states, whose violation previously gave birth to mere infraction of convention—since the weaker ones are only the prolongation, the attenuated form, of the stronger. Thus robbery and simple bad taste injure the same single altruistic sentiment, the respect for that which is another's. However, this same sentiment is less grievously offended by bad taste than by robbery; and since, in addition, the average consciousness has not sufficient intensity to react keenly to the bad taste, it is treated with greater tolerance. That is why the person guilty of bad taste is merely blamed, whereas the thief is punished. But, if this sentiment grows stronger, to the point of silencing in all consciousnesses the inclination which disposes man to steal, he will become more sensitive to the offenses which, until then, touched him but lightly. He will react against them, then, with more energy; they will be the object of greater opprobrium, which will transform certain of them from the simple moral faults that they were and give them the quality of crimes. For example, improper contracts, or contracts improperly executed, which only incur public blame or civil damages, will become offenses in law.

Imagine a society of saints, a perfect cloister of exemplary individuals. Crimes, properly so called, will there be unknown; but faults which appear venial to the layman will create there the same scandal that the ordinary offense does in ordinary consciousness. If, then, this society has the power to judge and punish, it will define these acts as criminal and will treat them as such. For the same reason, the perfect and upright man judges his smallest failings with a severity that the majority reserve for acts more truly in the nature of an offense. Formerly, acts of violence against persons were more frequent than they are today, because respect for individual dignity was less strong. As this has increased, these crimes have become more rare; and also, many acts violating this sentiment have been introduced into the penal law which were not included there in primitive times.[2]

In order to exhaust all the hypotheses logically possible, it will perhaps be asked why this unanimity does not extend to all collective sentiments without exception. Why should not even the most feeble sentiment gather enough energy to prevent all dissent? The moral consciousness of the society would be present in its entirety in all the individuals, with a vitality sufficient to prevent all acts offending it—the purely conventional faults as well as the crimes. But a uniformity so universal and absolute is utterly impossible; for the immediate physical milieu in which each one of us is placed, the hereditary antecedents, and the social influences vary from one individual to the next, and consequently diversify consciousnesses. It is impossible for all to be alike, if only because each one has his own organism and that these organisms occupy different areas in space. That is why, even among the lower peoples, where individual originality is very little developed, it nevertheless does exist.

Thus, since there cannot be a society in which the individuals do not differ more or less from the collective type, it is also inevitable that, among these divergences, there are some with a criminal character. What confers this character upon them is not the intrinsic quality of a given act but that definition which the collective conscience lends them. If the collective conscience is stronger, if it has enough authority practically to suppress these divergences, it will also be more sensitive, more exacting; and, reacting against the slightest deviations with the energy it otherwise displays only against more considerable infractions, it will attribute to them the same gravity as formerly to crimes. In other words, it will designate them as criminal.

Crime is, then, necessary; it is bound up with fundamental conditions of all social life, and by that very fact it is useful, because these conditions of which it is part are themselves indispensable to the normal evolution of morality and law.

Indeed, it is no longer possible today to dispute the fact that law and morality vary from one social type to the next, nor that they change within the same type if the conditions of life are modified. But, in order that these transformations may be possible, the collective sentiments at the basis of morality must not be hostile to change, and consequently must have but moderate energy. If they were too strong, they would no longer be plastic. Every pattern is an obstacle to new patterns, to the extent that the first pattern is inflexible. The better a structure is articulated, the more it offers a healthy resistance to all modification; and this is equally true of functional, as of anatomical, organization. If there were no crimes, this condition could not have been fulfilled; for such a hypothesis presupposes that collective sentiments have arrived at a degree of intensity unexampled in history. Nothing is good indefinitely and to an unlimited extent. The authority which the moral conscience enjoys must not be excessive; otherwise no one would dare criticize it, and it would too easily congeal into an immutable form. To make progress, individual originality must be able to express itself. In order that the originality of the idealist whose dreams transcend his century may find expression, it is necessary that the originality of the criminal, who is below the level of his time, shall also be possible. One does not occur without the other.

Nor is this all. Aside from this indirect utility, it happens that crime itself plays a useful role in this evolution. Crime implies not only that the way remains open to necessary changes but that in certain cases it directly prepares these changes. Where crime exists, collective sentiments are sufficiently flexible to take on a new form, and crime sometimes helps to determine the form they will take. How many times, indeed, it is only an anticipation of future morality—a step toward what will be! According to Athenian law, Socrates was a criminal, and his condemnation was no more than just. However, his crime, namely, the independence of his thought, rendered a service not only to humanity but to his country. It served to prepare a new morality and faith which the Athenians needed, since the traditions by which they had lived until then were no longer in harmony with the current conditions of life. Nor is the case of Socrates unique; it is reproduced periodically in history. It would never have been possible to establish the freedom of thought we now enjoy if the regulations prohibiting it had not been violated before being solemnly abrogated. At that time, however, the violation was a crime, since it was an offense against sentiments still very keen in the average conscience. And yet this crime was useful as a prelude to reforms which daily became more necessary. Liberal philosophy had as its precursors the heretics of all kinds who were justly punished by secular authorities during the entire course of the Middle Ages and until the eve of modern times.

From this point of view the fundamental facts of criminality present themselves to us in an entirely new light. Contrary to current ideas, the criminal no longer seems a totally unsociable being, a sort of parasitic element, a strange and unassimilable body, introduced into the midst of society.[3] On the contrary, he plays a definite role in social life. Crime, for its part, must no longer be conceived as an evil that cannot be too much suppressed. There is no occasion for self-congratulation when the crime rate drops noticeably below the average level, for we may be certain that this apparent progress is associated with some social disorder. Thus, the number of assault cases never falls so low as in times of want.[4] With the drop in the crime rate, and as a reaction to it, comes a revision, or the need of a revision in the theory of punishment. If, indeed, crime is a disease, its punishment is its remedy and cannot be otherwise conceived; thus, all the discussions it arouses bear on the point of determining what the punishment must be in order to fulfill this role of remedy. If crime is not pathological at all, the object of punishment cannot be to cure it, and its true function must be sought elsewhere.

NOTES

1. From the fact that crime is a phenomenon of normal sociology, it does not follow that the criminal is an individual normally constituted from the biological and psychological points of view. The two questions are independent of each other. This independence will be better understood when we have shown, later on, the difference between psychological and sociological facts.

2. Calumny, insults, slander, fraud, etc.

3. We have ourselves committed the error of speaking thus of the criminal, because of a failure to apply our rule (*Division du travail social,* pp. 395–96).

4. Although crime is a fact of normal sociology, it does not follow that we must not abhor it. Pain itself has nothing desirable about it; the individual dislikes it as a society does crime, and yet it is a function of normal physiology. Not only is it necessarily derived from the very constitution of every living organism, but it plays a useful role in life, for which reason it cannot be replaced. It would, then, be a singular distortion of our thought to present it as an apology for crime. We would not even think of protesting against such an interpretation, did we not know to what strange accusations and misunderstandings one exposes oneself when one undertakes to study moral facts objectively and to speak of them in a different language from that of the layman.

5

Social Structure and Anomie

ROBERT K. MERTON

Merton's notable contribution to deviance theory locates the root causes of crime in the structure of society, not in the individuals involved in criminal actions. Broader patterns of crime characterize certain groups of people because of the social "strain" they face between the good things society offers and their inability to legitimately attain them. Merton identifies a disjuncture for people between what he calls cultural goals, *or the ends toward which we are all socialized to strive (largely conceived by him in a material sense), and* institutional norms, *or the acceptable means that we use to reach these goals. He refers to this condition where people cannot attain their goals legitimately as a "blocked opportunity structure." Societies where goals are more strongly emphasized than the legitimate means of achieving them are likely to see people pursue the innovation adaptation, as when schools make grades more critically important than the learning required to attain them, and when some groups lack the resources to legitimately earn good grades. Merton sketches other ways that groups adapt to their balance of means and ends (conformity, ritualism, retreatism, rebellion) as alternate adaptations to their structural conditions in society.*

The framework set out in this essay is designed to provide one systematic approach to the analysis of social and cultural sources of deviant behavior. Our primary aim is to discover how some *social structures exert a definite pressure upon certain persons in the society to engage in nonconforming rather than conforming conduct.* If we can locate groups peculiarly subject to such pressures, we should expect to find fairly high rates of deviant behavior in these groups, not because the human beings comprising them are compounded of distinctive biological tendencies but because they are responding normally to the social situation in which they find themselves. Our perspective is sociological. We look at variations in the *rates* of deviant behavior, not at its incidence. Should our quest be at all successful, some forms of deviant behavior will be found to be as psychologically normal as conformist behavior, and the equation of deviation and psychological abnormality will be put in question.

Reprinted with the permission of The Free Press, a Division of Simon & Schuster, Inc. from *Social Theory and Social Structure,* Revised and Enlarged Edition by Robert K. Merton. Copyright © 1967, 1968 by Robert K. Merton.

PATTERNS OF CULTURAL GOALS AND INSTITUTIONAL NORMS

Among the several elements of social and cultural structures, two are of immediate importance. These are analytically separable although they merge in concrete situations. The first consists of culturally defined goals, purposes and interests, held out as legitimate objectives for all or for diversely located members of the society. The goals are more or less integrated—the degree is a question of empirical fact—and roughly ordered in some hierarchy of value. Involving various degrees of sentiment and significance, the prevailing goals comprise a frame of aspirational reference. They are the things "worth striving for." They are a basic, though not the exclusive, component of what Linton has called "designs for group living." And though some, not all, of these cultural goals are directly related to the biological drives of man, they are not determined by them.

A second element of the cultural structure defines, regulates, and controls the acceptable modes of reaching out for these goals. Every social group invariably couples its cultural objectives with regulations, rooted in the mores or institutions, of allowable procedures for moving toward these objectives. These regulatory norms are not necessarily identical with technical or efficiency norms. Many procedures which from the standpoint of particular individuals would be most efficient in securing desired values—the exercise of force, fraud, power—are ruled out of the institutional area of permitted conduct. At times, the disallowed procedures include some which would be efficient for the group itself—for example, historic taboos on vivisection, on medical experimentation, on the sociological analysis of "sacred" norms—since the criterion of acceptability is not technical efficiency but value-laden sentiments (supported by most members of the group or by those able to promote these sentiments through the composite use of power and propaganda.) In all instances, the choice of expedients for striving toward cultural goals is limited by institutionalized norms.

We shall be primarily concerned with the first—a society in which there is an exceptionally strong emphasis upon specific goals without a corresponding emphasis upon institutional procedures. If it is not to be misunderstood, this statement must be elaborated. No society lacks norms governing conduct. But societies do differ in the degree to which the folkways, mores and institutional controls are effectively integrated with the goals which stand high in the hierarchy of cultural values. The culture may be such as to lead individuals to center their emotional convictions upon the complex of culturally acclaimed ends, with far less emotional support for prescribed methods of reaching out for these ends. With such differential emphases upon goals and institutional procedures, the latter may be so vitiated by the stress on goals as to have the behavior of many individuals limited only by considerations of technical expediency. In this context, the sole significant question becomes: Which of the available procedures is most efficient in netting the culturally approved value? The technically most effective procedure, whether culturally legitimate or not, becomes typically preferred to institutionally prescribed conduct. As this process of attenuation continues, the society becomes unstable and there develops what Durheim called "anomie" (or normlessness).

The working of this process eventuating in anomie can be easily glimpsed in a series of familiar and instructive, though perhaps trivial, episodes. Thus, in competitive athletics, when the aim of victory is shorn of its institutional trappings and success becomes construed as "winning the game" rather than "winning under the rules of the game," a premium is implicitly set upon the use of illegitimate but technically efficient means. The star of the opposing football team is surreptitiously slugged; the wrestler incapacitates his opponent through ingenious but illicit techniques; university alumni covertly subsidize "students" whose talents are confined to the athletic field. The emphasis on the goal has so attenuated the satisfactions deriving from sheer participation in the competitive activity that only a successful outcome provides gratification. Through the same process, tension generated by the desire to win in a poker game is relieved by successfully dealing one's self four aces, or when the cult of success has truly flowered, by sagaciously shuffling the cards in a game of solitaire. The faint twinge of uneasiness in the last instance and the surreptitious nature of public delicts indicate clearly that the institutional rules of the game are *known* to those who evade them. But cultural (or idiosyncratic) exaggeration of the success-goal leads men to withdraw emotional support from the rules.

This process is of course not restricted to the realm of competitive sport, which has simply provided us with microcosmic images of the social macrocosm. The process whereby exaltation of the end generates a literal *demoralization,* that is, a de-institutionalization, of the means occurs in many groups where the two components of the social structure are not highly integrated.

Contemporary American culture appears to approximate the polar type in which great emphasis upon certain success-goals occurs without equivalent emphasis upon institutional means. It would of course be fanciful to assert that accumulated wealth stands alone as a symbol of success just as it would be fanciful to deny that Americans assign it a place high in their scale of values. In some large measure, money has been consecrated as a value in itself, over and above its expenditure for articles of consumption or its use for the enhancement of power. "Money" is peculiarly well adapted to become a symbol of prestige. As Simmel emphasized, money is highly abstract and impersonal. However acquired, fraudulently or institutionally, it can be used to purchase the same goods and services. The anonymity of an urban society, in conjunction with these peculiarities of money, permits wealth, the sources of which may be unknown to the community in which the plutocrat lives or, if known, to become purified in the course of time, to serve as a symbol of high status. Moreover, in the American Dream there is no final stopping point. The measure of "monetary success" is conveniently indefinite and relative. At each income level, as H. F. Clark found, Americans want just about 25 percent more (but of course this "just a bit more" continues to operate once it is obtained).

In this flux of shifting standards, there is no stable resting point, or rather, it is the point which manages always to be "just ahead." An observer of a community in which annual salaries in six figures are not uncommon reports the anguished words of one victim of the American Dream: "In this town, I'm snubbed socially because I only get a thousand a week. That hurts."

To say that the goal of monetary success is entrenched in American culture is only to say that Americans are bombarded on every side by precepts which affirm the right or, often, the duty of retaining the goal even in the face of repeated frustration. Prestigeful representatives of the society reinforce the cultural emphasis. The family, the school and the workplace—the major agencies shaping the personality structure and goal formation of Americans—join to provide the intensive disciplining required if an individual is to retain intact a goal that remains elusively beyond reach, if he is to be motivated by the promise of a gratification which is not redeemed. As we shall presently see, parents serve as a transmission belt for the values and goals of the groups of which they are a part—above all, of their social class or of the class with which they identify themselves. And the schools are of course the official agency for the passing on of the prevailing values, with a large proportion of the textbooks used in city schools implying or stating explicitly "that education leads to intelligence and consequently to job and money success." Central to this process of disciplining people to maintain their unfulfilled aspirations are the cultural prototypes of success, the living documents testifying that the American Dream can be realized if one but has the requisite abilities.

Coupled with this positive emphasis upon the obligation to maintain lofty goals is a correlative emphasis upon the penalizing of those who draw in their ambitions. Americans are admonished "not to be a quitter" for in the dictionary of American culture, as in the lexicon of youth, "there is no such word as 'fail.'" The cultural manifesto is clear: one must not quit, must not cease striving, must not lessen his goals, for "not failure, but low aim, is crime."

Thus the culture enjoins the acceptance of three cultural axioms: First, all should strive for the same lofty goals since these are open to all; second, present seeming failure is but a way-station to ultimate success; and third, genuine failure consists only in the lessening or withdrawal of ambition.

In rough psychological paraphrase, these axioms represent, first a symbolic secondary reinforcement of incentive; second, curbing the threatened extinction of a response through an associated stimulus; third, increasing the motive-strength to evoke continued responses despite the continued absence of reward.

In sociological paraphrase, these axioms represent, first, the deflection of criticism of the social structure onto one's self among those so situated in the society that they do not have full and equal access to opportunity; second, the preservation of a structure of social power by having individuals in the lower social strata identify themselves, not with their compeers, but with those at the top (whom they will ultimately join); and third, providing pressures for conformity with the cultural dictates of unslackened ambition by the threat of less than full membership in the society for those who fail to conform.

It is in these terms and through these processes that contemporary American culture continues to be characterized by a heavy emphasis on wealth as a basic symbol of success, without a corresponding emphasis upon the legitimate avenues on which to march toward this goal. How do individuals living in this cultural context respond? And how do our observations bear upon the doctrine that deviant behavior typically derives from biological impulses breaking through the restraints imposed by culture? What, in short, are the consequences for the behavior of people variously situated in a social structure of a culture in which the emphasis on dominant success-goals has become increasingly separated from an equivalent emphasis on institutionalized procedures for seeking these goals?

TYPES OF INDIVIDUAL ADAPTATION

Turning from these culture patterns, we now examine types of adaptation by individuals within the culture-bearing society. Though our focus is still the cultural and social genesis of varying rates and types of deviant behavior, our perspective shifts from the plane of patterns of cultural values to the plane of types of adaptation to these values among those occupying different portions in the social structure.

We here consider five types of adaptation, as these are schematically set out in the following table, where (+) signifies "acceptance," (−) signifies "rejection," and (±) signifies "rejection of prevailing values and substitution of new values."

A Typology of Modes of Individual Adaptation

Modes of Adaptation	Culture Goals	Institutionalized Means
I. Conformity	+	+
II. Innovation	+	−
III. Ritualism	−	+
IV. Retreatism	−	−
V. Rebellion	±	±

I. Conformity

To the extent that a society is stable, adaptation type I—conformity to both cultural goals and institutionalized means—is the most common and widely diffused. Were this not so, the stability and continuity of the society could not be maintained. . . .

II. Innovation

Great cultural emphasis upon the success-goal invites this mode of adaptation through the use of institutionally proscribed but often effective means of attaining at least the simulacrum of success—wealth and power. This response occurs when the individual has assimilated the cultural emphasis upon the goal without equally internalizing the institutional norms governing ways and means for its attainment. . . .

It appears from our analysis that the greatest pressures toward deviation are exerted upon the lower strata. Cases in point permit us to detect the sociological mechanisms involved in producing these pressures. Several researches have shown that specialized areas of vice and crime constitute a "normal" response to a situation where the cultural emphasis upon pecuniary success has been absorbed, but where there is little access to conventional and legitimate means for becoming successful. The occupational opportunities of people in these areas are largely confined to manual labor and the lesser white-collar jobs. Given the American stigmatization of manual labor *which has been found to hold rather uniformly in all social classes,* and the absence of realistic opportunities for advancement beyond this level, the result is a marked tendency toward deviant behavior. The status of unskilled labor and the consequent low income cannot readily compete *in terms of established standards of worth* with the promises of power and high income from organized vice, rackets and crime.

For our purposes, these situations exhibit two salient features. First, incentives for success are provided by the established values of the culture *and* second, the avenues available for moving toward this goal are largely limited by the class structure to those of deviant behavior. It is the *combination* of the cultural emphasis and the social structure which produces intense pressure for deviation. . . .

III. Ritualism

The ritualistic type of adaptation can be readily identified. It involves the abandoning or scaling down of the lofty cultural goals of great pecuniary success and rapid social mobility to the point where one's aspirations can be satisfied. But though one rejects the cultural obligation to attempt "to get ahead in the world," though one draws in one's horizons, one continues to abide almost compulsively by institutional norms. . . .

We should expect this type of adaptation to be fairly frequent in a society which makes one's social status largely dependent upon one's achievements. For, as has so often been observed, this ceaseless competitive struggle produces acute status anxiety. One device for allaying these anxieties is to lower one's level of aspiration—permanently. Fear produces inaction, or, more accurately, routinized action.

The syndrome of the social ritualist is both familiar and instructive. His implicit life-philosophy finds expression in a series of cultural clichés: "I'm not sticking *my* neck out," "I'm playing safe," "I'm satisfied with what I've got," "Don't aim high and you won't be disappointed." The theme threaded through these attitudes is that high ambitions invite frustration and danger whereas lower aspirations produce satisfaction and security. It is the perspective of the frightened employee, the zealously conformist bureaucrat in the teller's cage of the private banking enterprise or in the front office of the public works enterprise.

IV. Retreatism

Just as Adaptation I (conformity) remains the most frequent, Adaptation IV (the rejection of cultural goals and institutional means) is probably the least common. People who adapt (or maladapt) in this fashion are, strictly speaking, *in* the society but not *of* it. Sociologically these constitute the true aliens. Not sharing the common frame of values, they can be included as members of the *society* (in distinction from the *population*) only in a fictional sense.

In this category fall some of the adaptive activities of psychotics, autists, pariahs, outcasts, vagrants, vagabonds, tramps, chronic drunkards and drug addicts. They have relinquished culturally prescribed goals and their behavior does not accord with institutional norms. The competitive order is maintained but the frustrated and handicapped individual who cannot cope with this order drops out. Defeatism, quietism and resignation are manifested in escape mechanisms which ultimately lead him to "escape" from the requirements of the society. It is thus an expedient which arises from continued failure to near the goal by legitimate measures and from an inability to use the illegitimate route because of internalized prohibitions.

V. Rebellion

This adaptation leads men outside the environing social structure to envisage and seek to bring into being a new, that is to say, a greatly modified social structure. It presupposes alienation from reigning goals and standards. These come to be regarded as purely arbitrary. And the arbitrary is precisely that which can neither exact allegiance nor possess legitimacy, for it might as well be otherwise. In our society, organized movements for rebellion apparently aim to introduce a social structure in which the cultural standards of success would be sharply modified and provision would be made for a closer correspondence between merit, effort and reward.

THE STRAIN TOWARD ANOMIE

The social structure we have examined produces a strain toward anomie and deviant behavior. The pressure of such a social order is upon outdoing one's competitors. So long as the sentiments supporting this competitive system are distributed throughout the entire range of activities and are not confined to the final result of "success," the choice of means will remain largely within the ambit of institutional control. When, however, the cultural emphasis shifts from the satisfactions deriving from competition itself to almost exclusive concern with the outcome, the resultant stress makes for the breakdown of the regulatory structure.

6

Differential Association

EDWIN H. SUTHERLAND AND DONALD R. CRESSEY

Like Durkheim, Sutherland and Cressey do not regard crime and deviance as the result of either pathology in society or pathological behavior patterns. Crime, they argue, is learned in much the same way as all ordinary behavior and represents the expression of the same behavioral needs and values as other behavior. Crime is less likely to be learned, in fact, from frightening or suspicious outsiders than from people's own intimate associates. In this way, Sutherland and Cressey cast the learning of crime and deviance as a normal process, and note that it is a likely occurrence as people become surrounded with increasing numbers of deviant friends. Sutherland and Cressey place the learning of deviance within people's most intense and personal relations: families and peer groups. By repeatedly watching others modeling crime, by learning from them how to effectively do it, and by becoming convinced by their rationalizations and neutralizations that such behaviors are acceptable, individuals move into criminal behavior patterns.

The following statements refer to the process by which a particular person comes to engage in criminal behavior.

1. *Criminal behavior is learned.* Negatively, this means that criminal behavior is not inherited, as such; also, the person who is not already trained in crime does not invent criminal behavior, just as a person does not make mechanical inventions unless he has had training in mechanics.

2. *Criminal behavior is learned in interaction with other persons in a process of communication.* This communication is verbal in many respects but includes also "the communication of gestures."

3. *The principal part of the learning of criminal behavior occurs within intimate personal groups.* Negatively this means that the impersonal agencies of communication, such as movies and newspapers, play a relatively unimportant part in the genesis of criminal behavior.

4. *When criminal behavior is learned, the learned includes (a) techniques of committing the crime, which are sometimes very complicated, sometimes very simple; (b) the specific direction of motive, drives, rationalizations, and attitudes.*

From Edwin H. Sutherland, Donald R. Cressey, and David F. Luckenbill, *Principles of Criminology,* 11th Edition (pp. 88–90), 1992. Reprinted by permission of Alta Mira Press.

5. *The specific direction of motives and drives is learned from definitions of the legal codes as favorable or unfavorable.* In some societies an individual is surrounded by persons who invariably define the legal codes as rules to be observed, while in others he is surrounded by persons whose definitions are favorable to the violation of the legal codes. In our American society these definitions are almost always mixed, with the consequence that we have culture conflict in relation to the legal codes.

6. *A person becomes delinquent because of an excess of definitions favorable to violation of law over definitions unfavorable to violation of law.* This is the principle of differential association. It refers to both criminal and anti-criminal associations and has to do with counteracting forces. When persons become criminal, they do so because of contacts with criminal patterns and also because of isolation from anticriminal patterns. Any person inevitably assimilates the surrounding culture unless other patterns are in conflict; a southerner does not pronounce *r* because other southerners do not pronounce *r*. Negatively, this proposition of differential association means that associations which are neutral so far as crime is concerned have little or no effect on the genesis of criminal behavior. Much of the experience of a person is neutral in this sense, for example, learning to brush one's teeth. This behavior has no negative or positive effect on criminal behavior except as it may be related to associations which are concerned with the legal codes. This neutral behavior is important especially as an occupier of the time of a child so that he is not in contact with criminal behavior during the time he is so engaged in the neutral behavior.

7. *Differential associations may vary in frequency, duration, priority, and intensity.* This means that associations with criminal behavior and also associations with anticriminal behavior vary in those respects. "Frequency" and "duration" as modalities of associations are obvious and need no explanation. "Priority" is assumed to be important in the sense that lawful behavior developed in early childhood may persist throughout life, and also that delinquent behavior developed in early childhood may persist throughout life. This tendency, however, has not been adequately demonstrated, and priority seems to be important principally through its selective influence. "Intensity" is not precisely defined, but it has to do with such things as the prestige of the source of a criminal or anticriminal pattern and with emotional reactions related to the associations. In a precise description of the criminal behavior of a person, these modalities would be rated in quantitative form and a mathematical ratio reached. A formula in this sense has not been developed, and the development of such a formula would be extremely difficult.

8. *The process of learning criminal behavior by association with criminal and anti-criminal patterns involves all of the mechanisms that are involved in any other learning.* Negatively, this means that the learning of criminal behavior is not restricted to the process of imitation. A person who is seduced, for instance, learns criminal behavior by association, but this process would not ordinarily be described as imitation.

9. *While criminal behavior is an expression of general needs and values, it is not explained by those general needs and values, since noncriminal behavior is an expression of the same needs and values.* Thieves generally steal in order to secure money, but likewise honest laborers work in order to secure money. The attempts by many scholars to explain criminal behavior by general drives and values, such as the happiness principle, striving for social status, the money motive, or frustration, have been, and must continue to be, futile, since they explain lawful behavior as completely as they explain criminal behavior. They are similar to respiration, which is necessary for any behavior, but which does not differentiate criminal from noncriminal behavior.

It is not necessary, at this level of explanation, to explain why a person has the associations he has; this certainly involves a complex of many things. In an area where the delinquency rate is high, a boy who is sociable, gregarious, active, and athletic is very likely to come in contact with other boys in the neighborhood, learn delinquent behavior patterns from them, and become a criminal; in the same neighborhood the psychopathic boy who is isolated, introverted, and inert may remain at home, not become acquainted with the other boys in the neighborhood, and not become delinquent. In another situation, the sociable, athletic, aggressive boy may become a member of a scout troop and not become involved in delinquent behavior. The person's associations are determined in a general context of social organization. A child is ordinarily reared in a family; the place of residence of the family is determined largely by family income; and the delinquency rate is in many respects related to the rental value of the houses. Many other aspects of social organization affect the kinds of associations a person has.

The preceding explanation of criminal behavior purports to explain the criminal and noncriminal behavior of individual persons. It is possible to state sociological theories of criminal behavior which explain the criminality of a community, nation, or other group. The problem, when thus stated, is to account for variations in crime rates and involves a comparison of the crime rates of various groups or the crime rates of a particular group at different times. The explanation of a crime rate must be consistent with the explanation of the criminal behavior of the person, since the crime rate is a summary statement of the number of persons in the group who commit crimes and the frequency with which they commit crimes. One of the best explanations of crime rates from this point of view is that a high crime rate is due to social disorganization. The term *social disorganization* is not entirely satisfactory, and it seems preferable to substitute for it the term *differential social organization*. The postulate on which this theory is based, regardless of the name, is that crime is rooted in the social organization and is an expression of that social organization. A group may be organized for criminal behavior or organized against criminal behavior. Most communities are organized for both criminal and anticriminal behavior, and, in that sense the crime rate is an expression of the differential group organization. Differential group organization as an explanation of variations in crime rates is consistent with the differential association theory of the processes by which persons become criminals.

7

Labeling Theory

HOWARD S. BECKER

Becker's classic statement of the labeling perspective agrees with Merton that the essence of crime is located outside of the individual person committing the deviant act. But rather than looking at the structural conditions that produce acts that might innately be regarded as criminal, Becker looks at societal reactions to behavior after the fact. Under other circumstances, the same behavior may be viewed very differently. Becker notes that variations may arise due to the diverse temporal contexts framing acts, to the social position and power of those who commit or who have been harmed by acts, and by the consequences that arise from acts. These framing elements, which are sometimes unrelated to the behavior itself, may elevate one act as heinous and relegate another, similar one, to obscurity. Becker thus locates the root of deviance in the response of other people rather than the act itself, and in the chain of events that is unleashed once people have labeled an act and its perpetrator as deviant.

The interactionist perspective . . . defines deviance as the infraction of some agreed-upon rule. It then goes on to ask who breaks rules, and to search for the factors in their personalities and life situations that might account for the infractions. This assumes that those who have broken a rule constitute a homogeneous category, because they have committed the same deviant act.

Such an assumption seems to me to ignore the central fact about deviance: it is created by society. I do not mean this in the way it is ordinarily understood, in which the causes of deviance are located in the social situation of the deviant or in "social factors" which prompt his action. I mean, rather, that *social groups create deviance by making the rules whose infraction constitutes deviance,* and by applying those rules to particular people and labeling them as outsiders. From this point of view, deviance is *not* a quality of the act the person commits, but rather a consequence of the application by others of rules and sanctions to an "offender." The deviant is one to whom the label has successfully been applied; deviant behavior is behavior that people so label.[1]

Since deviance is, among other things, a consequence of the responses of others to a person's act, students of deviance cannot assume that they are

---

Reprinted with the permission of The Free Press, a Division of Simon & Schuster, Inc., from *Outsiders: Studies in the Sociology of Deviance* by Howard S. Becker. Copyright © 1963 by The Free Press; copyright © Renewed 1991 by Howard S. Becker.

dealing with a homogeneous category when they study people who have been labeled deviant. That is, they cannot assume that those people have actually committed a deviant act or broken some rule, because the process of labeling may not be infallible; some people may be labeled deviant who in fact have not broken a rule. Furthermore, they cannot assume that the category of those labeled deviant will contain all those who actually have broken a rule, for many offenders may escape apprehension and thus fail to be included in the population of "deviants" they study. Insofar as the category lacks homogeneity and fails to include all the cases that belong in it, one cannot reasonably expect to find common factors of personality or life situation that will account for the supposed deviance. What, then, do people who have been labeled deviant have in common? At the least, they share the label and the experience of being labeled as outsiders. I will begin my analysis with this basic similarity and view deviance as the product of a transaction that takes place between some social group and one who is viewed by that group as a rule-breaker. I will be less concerned with the personal and social characteristics of deviants than with the process by which they come to be thought of as outsiders and their reactions to that judgement. . . .

The point is that the response of other people has to be regarded as problematic. Just because one has committed an infraction of a rule does not mean that others will respond as though this had happened. (Conversely, just because one has not violated a rule does not mean that he may not be treated, in some circumstances, as though he had.)

The degree to which other people will respond to a given act as deviant varies greatly. Several kinds of variation seem worth noting. First of all, there is variation over time. A person believed to have committed a given "deviant" act may at one time be responded to much more leniently than he would be at some other time. The occurrence of "drives" against various kinds of deviance illustrates this clearly. At various times, enforcement officials may decide to make an all-out attack on some particular kind of deviance, such as gambling, drug addiction, or homosexuality. It is obviously much more dangerous to engage in one of these activities when a drive is on than at any other time. (In a very interesting study of crime news in Colorado newspapers, Davis found that the amount of crime reported in Colorado newspapers showed very little association with actual changes in the amount of crime taking place in Colorado. And, further, that people's estimate of how much increase there had been in crime in Colorado was associated with the increase in the amount of crime news but not with any increase in the amount of crime.)[2]

The degree to which an act will be treated as deviant depends also on who commits the act and who feels he has been harmed by it. Rules tend to be applied more to some persons than others. Studies of juvenile delinquency make the point clearly. Boys from middle-class areas do not get as far in the legal process when they are apprehended as do boys from slum areas. The middle-class boy is less likely, when picked up by the police, to be taken to the station; less likely when taken to the station to be booked; and it is extremely unlikely that he will be convicted and sentenced.[3] This variation occurs even though the original infraction of the rule is the same in the two cases. Similarly, the law is differentially applied to Negroes and whites. It is well known that a Negro believed to have attacked a white woman is much more likely to be punished than a white man who commits the same offense; it is only slightly less well known that a Negro who murders another Negro is much less likely to be punished than a white man who commits murder.[4] This, of course, is one of the main points of Sutherland's analysis of white-collar crime: crimes committed by corporations are almost always prosecuted as civil cases, but the same crime committed by an individual is ordinarily treated as a criminal offense.[5]

Some rules are enforced only when they result in certain consequences. The unmarried mother furnishes a clear example. Vincent[6] points out that illicit sexual relations seldom result in severe punishment or social censure for the offenders. If, however, a girl becomes pregnant as a result of such activities the reaction of others is likely to be severe. (The illicit pregnancy is also an interesting example of the differential enforcement of rules on different categories of people. Vincent notes that unmarried fathers escape the severe censure visited on the mother.)

Why repeat these commonplace observations? Because, taken together, they support the proposition that deviance is not a simple quality, present in some kinds of behavior and absent in others. Rather, it is the product of a process which involves responses of other people to the behavior. The same behavior may be an infraction of the rules at one time and not at another; may be an infraction when committed by one person, but not when committed by another; some rules are broken with impunity, others are not. In short, whether a given act is deviant or not depends in part on the nature of the act (that is, whether or not it violates some rule) and in part on what other people do about it.

Some people may object that this is merely a terminological quibble, that one can, after all, define terms any way he wants to and that if some people want to speak of rule-breaking behavior as deviant without reference to the reactions of others they are free to do so. This, of course, is true. Yet it might be worthwhile to refer to such behavior as *rule-breaking behavior* and reserve the term *deviant* for those labeled as deviant by some segment of society. I do not insist that this usage be followed. But it should be clear that insofar as a scientist uses "deviant" to refer to any rule-breaking behavior and takes as his subject of study only those who have been *labeled* deviant, he will be hampered by the disparities between the two categories.

If we take as the object of our attention behavior which comes to be labeled as deviant, we must recognize that we cannot know whether a given act will be categorized as deviant until the response of others has occurred. Deviance is not a quality that lies in behavior itself, but in the interaction between the person who commits an act and those who respond to it. . . .

In any case, being branded as deviant has important consequences for one's further social participation and self-image. The most important consequence is a drastic change in the individual's public identity. Committing the improper act and being publicly caught at it place him in a new status. He has been revealed as a different kind of person from the kind he was supposed to be. He is labeled a "fairy," "dope fiend," "nut" or "lunatic," and treated accordingly.

In analyzing the consequences of assuming a deviant identity let us make use of Hughes' distinction between master and auxiliary status traits.[7] Hughes notes that most statuses have one key trait which serves to distinguish those who belong from those who do not. Thus the doctor, whatever else he may be, is a person who has a certificate stating that he has fulfilled certain requirements and is licensed to practice medicine; this is the master trait. As Hughes points out, in our society a doctor is also informally expected to have a number of auxiliary traits: most people expect him to be upper middle-class, white, male, and Protestant. When he is not, there is a sense that he has in some way failed to fill the bill. Similarly, though skin color is the master status trait determining who is Negro and who is white, Negroes are informally expected to have certain status traits and not to have others; people are surprised and find it anomalous if a Negro turns out to be a doctor or a college professor. People often have the master status trait but lack some of the auxiliary, informally expected characteristics; for example, one may be a doctor but be a female or a Negro.

Hughes deals with this phenomenon in regard to statuses that are well thought of, desired, and desirable (noting that one may have the formal qualifications for entry into a status but be denied full entry because of lack of the proper auxiliary traits), but the same process occurs in the case of deviant statuses. Possession of one deviant trait may have a generalized symbolic value, so that people automatically assume that its bearer possesses other undesirable traits allegedly associated with it.

To be labeled a criminal one need only commit a single criminal offense, and this is all the term formally refers to. Yet the word carries a number of connotations specifying auxiliary traits characteristic of anyone bearing the label. A man who has been convicted of housebreaking and thereby labeled criminal is presumed to be a person likely to break into other houses; the police, in rounding up known offenders for investigation after a crime has been committed, operate on this premise. Further, he is considered likely to commit other kinds of crimes as well, because he has shown himself to be a person without "respect for the law." Thus, apprehension for one deviant act exposes a person to the likelihood that he will be regarded as deviant or undesirable in other respects.

There is one other element in Hughes' analysis we can borrow with profit: the distinction between master and subordinate statuses.[8] Some statuses, in our society as in others, override all other statuses and have a certain priority. Race is one of these. Membership in the Negro race, as socially defined, will override most other status considerations in most other situations; the fact that one is a physician or middle-class or female will not protect one from being treated as a Negro first and any of these other things second. The status of deviant (depending on the kind of deviance) is this kind of master status. One receives the status as a result of breaking a rule, and the identification proves to be more important than most others. One will be identified as a deviant first, before other identifications are made. . . .

NOTES

1. The most important earlier statements of this view can be found in Frank Tannenbaum, *Crime and the Community* (New York: Columbia University Press, 1938), and E. M. Lemert, *Social Pathology* (New York: McGraw-Hill Book Co., 1951). A recent article stating a position very similar to mine is John Kitsuse, "Societal Reaction to Deviance: Problems of Theory and Method," *Social Problems* 9 (Winter, 1962): 247–256.

2. F. James Davis, "Crime News in Colorado Newspapers," *American Journal of Sociology* LVII (January 1952): 325–330.

3. See Albert K. Cohen and James F. Short, Jr., "Juvenile Delinquency," p. 87 in Robert K. Merton and Robert A. Nisbet, eds., *Contemporary Social Problems* (New York: Harcourt, Brace and World, 1961).

4. See Harold Garfinkel, "Research Notes on Inter- and Intra-Racial Homicides," *Social Forces* 27 (May 1949): 369–381.

5. Edwin Sutherland, "White Collar Criminality," *American Sociological Review* V (February 1940): 1–12.

6. Clark Vincent, *Unmarried Mothers* (New York: The Free Press of Glencoe, 1961), pp. 3–5.

7. Everett C. Hughes, "Dilemmas and Contradictions of Status," *American Journal of Sociology* L (March 1945): 353–359.

8. *Ibid.*

8

Control Theory of Delinquency

TRAVIS HIRSCHI

Hirschi focuses on youthful delinquency, the age where most deviation from the norms generally occurs. In contrast to previous perspectives, Hirschi's model is a more individualistic one, looking at the bond between each person and society. Conforming behavior is reinforced by individuals' attachment to norm-abiding members of society, the commitment and investment they have made in a legitimate life and identity (educational credentials, respectable reputation), their level of involvement in legitimate activities and organizations, and their subscription to the commonly held beliefs and values characterizing normative society. People who violate norms have a flaw in one or more of these bonds to society, and can be brought back into the normative ranks by strengthening and reinforcing these weak bonds. Hirschi's perspective is thus more social psychological than structural.

Control theories assume that delinquent acts result when an individual's bond to society is weak or broken. Since these theories embrace two highly complex concepts, the *bond* of the individual to *society*, it is not surprising that they have at one time or another formed the basis of explanations of most forms of aberrant or unusual behavior. It is also not surprising that control theories have described the elements of the bond to society in many ways, and that they have focused on a variety of units as the point of control....

ELEMENTS OF THE BOND

Attachment

In explaining conforming behavior, sociologists justly emphasize sensitivity to the opinion of others.[1] Unfortunately, . . . they tend to suggest that man *is* sensitive to the opinion of others and thus exclude sensitivity from their explanations of deviant behavior. In explaining deviant behavior, psychologists, in contrast, emphasize insensitivity to the opinion of others.[2] Unfortunately, they too tend to ignore variation, and, in addition, they tend to tie sensitivity

---

From Travis Hirschi, *Causes of Delinquency*, pp. 16–26, Berkeley: University of California Press, 1969. Reprinted by permission of the author.

inextricably to other variables, to make it part of a syndrome or "type," and thus seriously to reduce its value as an explanatory concept. The psychopath is characterized only in part by "deficient attachment to or affection for others, a failure to respond to the ordinary motivations founded in respect or regard for one's fellow";[3] he is also characterized by such things as "excessive aggressiveness," "lack of superego control," and "an infantile level of response."[4] Unfortunately, too, the behavior that psychopathy is used to explain often becomes part of the *definition* of psychopath. As a result, in Barbara Wootton's words: "[The psychopath] is . . . *par excellence,* and without shame or qualification, the model of the circular process by which mental abnormality is inferred from anti-social behavior while anti-social behavior is explained by mental abnormality."[5]

The problems of diagnosis, tautology, and name-calling are avoided if the dimensions of psychopathy are treated as causally and therefore problematically interrelated, rather than as logically and therefore necessarily bound to each other. In fact, it can be argued that all of the characteristics attributed to the psychopath follow from, are effects of, his lack of attachment to others. To say that to lack attachment to others is to be free from moral restraints is to use lack of attachment to explain the guiltlessness of the psychopath, the fact that he apparently has no conscience or superego. In this view, lack of attachment to others is not merely a symptom of psychopathy, it *is* psychopathy; lack of conscience is just another way of saying the same thing; and the violation of norms is (or may be) a consequence.

For that matter, given that man is an animal, "impulsivity" and "aggressiveness" can also be seen as natural consequences of freedom from moral restraints. However, since the view of man as endowed with natural propensities and capacities like other animals is peculiarly unpalatable to sociologists, we need not fall back on such a view to explain the amoral man's aggressiveness.[6] The process of becoming alienated from others often involves or is based on active interpersonal conflict. Such conflict could easily supply a reservoir of *socially derived* hostility sufficient to account for the aggressiveness of those whose attachments to others have been weakened.

Durkheim said it many years ago: "We are moral beings to the extent that we are social beings."[7] This may be interpreted to mean that we are moral beings to the extent that we have "internalized the norms" of society. But what does it mean to say that a person has internalized the norms of society? To violate a norm is, therefore, to act contrary to the wishes and expectations of other people. If a person does not care about the wishes and expectations of other people—that is, if he is insensitive to the opinion of others—then he is to that extent not bound by the norms. He is free to deviate.

The essence of internalization of norms, conscience, or superego thus lies in the attachment of the individual to others.[8] This view has several advantages over the concept of internalization. For one, explanations of deviant behavior based on attachment do not beg the question, since the extent to which a person is attached to others can be measured independently of his deviant behavior. Furthermore, change or variation in behavior is explainable in a way that it is not when notions of internalization or superego are used. For example, the divorced man is more likely after divorce to commit a number of deviant acts, such as suicide or forgery. If we explain these acts by reference to the superego (or internal control), we are forced to say that the man "lost his conscience" when he got a divorce; and, of course, if he remarries, we have to conclude that he gets his conscience back.

This dimension of the bond to conventional society is encountered in most social control-oriented research and theory. F. Ivan Nye's "internal control" and "indirect control" refer to the same element, although we avoid the problem of explaining changes over time by locating the "conscience" in the bond to others rather than making it part of the personality.[9] Attachment to others is just one aspect of Albert J. Reiss's "personal controls"; we avoid his problems of tautological empirical *observations* by making the relationship between attachment and delinquency problematic rather than definitional.[10] Finally, Scott Briar and Irving Piliavin's "commitment" or "stake in conformity" subsumes attachment, as their discussion illustrates, although the terms they use are more closely associated with the next element to be discussed.[11]

Commitment

"Of all passions, that which inclineth men least to break the laws, is fear. Nay, excepting some generous natures, it is the only thing, when there is the appearance of profit or pleasure by breaking the laws, that makes men keep them."[12] Few would deny that men on occasion obey the rules simply from fear of the consequences. This rational component in conformity we label commitment. What does it mean to say that a person is committed to conformity? In Howard S. Becker's formulation it means the following:

> First, the individual is in a position in which his decision with regard to some particular line of action has consequences for other interests and activities not necessarily [directly] related to it. Second, he has placed himself in that position by his own prior actions. A third element is present though so obvious as not to be apparent; the committed person must be aware [of these other interests] and must recognize that his decision in this case will have ramifications beyond it.[13]

The idea, then, is that the person invests time, energy, himself, in a certain line of activity—say, getting an education, building up a business, acquiring a reputation for virtue. When or whenever he considers deviant behavior, he must consider the costs of this deviant behavior, the risk he runs of losing the investment he has made in conventional behavior.

If attachment to others is the sociological counterpart of the superego or conscience, commitment is the counterpart of the ego or common sense. To the person committed to conventional lines of action, risking one to ten years in prison for a ten-dollar holdup is stupidity, because to the committed person the costs and risks obviously exceed ten dollars in value. (To the psychoanalyst, such an act exhibits failure to be governed by the "reality-principle.")

In the sociological control theory, it can be and is generally assumed that the decision to commit a criminal act may well be rationally determined—that the actor's decision was not irrational given the risks and costs he faces. Of course, as Becker points out, if the actor is capable of in some sense calculating the costs of a line of action, he is also capable of calculational errors: ignorance and error return, in the control theory, as possible explanations of deviant behavior.

The concept of commitment assumes that the organization of society is such that the interest of most persons would be endangered if they were to engage in criminal acts. Most people, simply by the process of living in an organized society, acquire goods, reputations, prospects that they do not want to risk losing. These accumulations are society's insurance that they will abide by the rules. Many hypotheses about the antecedents of delinquent behavior are based on this premise. For example, Arthur L. Stinchcombe's hypothesis that "high school rebellion . . . occurs when future status is not clearly related to present performance"[14] suggests that one is committed to conformity not only by what one has but also by what one hoped to obtain. Thus "ambition" and/or "aspiration" play an important role in producing conformity. The person becomes committed to a conventional line of action, and he is therefore committed to conformity.

Most lines of action in a society are of course conventional. The clearest examples are educational and occupational careers. Actions thought to jeopardize one's chances in these areas are presumably avoided. Interestingly enough, even nonconventional commitments may operate to produce conventional conformity. We are told, at least, that boys aspiring to careers in the rackets or professional thievery are judged by their "honesty" and "reliability"—traits traditionally in demand among seekers of office boys.[15]

Involvement

Many persons undoubtedly owe a life of virtue to a lack of opportunity to do otherwise. Time and energy are inherently limited: "Not that I would not, if I could, be both handsome and fat and well dressed, and a great athlete, and make a million a year, be a wit, a bon vivant, and a lady killer, as well as a philosopher, a philanthropist, a statesman, warrior, and African explorer, as well as a 'tone-poet' and saint. But the thing is simply impossible."[16] The things that William James here says he would like to be or do are all, I suppose, within the realm of conventionality, but if he were to include illicit actions he would still have to eliminate some of them as simply impossible.

Involvement or engrossment in conventional activities is thus often part of a control theory. The assumption, widely shared, is that a person may be simply too busy doing conventional things to find time to engage in deviant behavior. The person involved in conventional activities is tied to appointments, deadlines, working hours, plans, and the like, so the opportunity to commit deviant acts rarely arises. To the extent that he is engrossed in conventional activities, he cannot even think about deviant acts, let alone act out his inclinations.[17]

This line of reasoning is responsible for the stress placed on recreational facilities in many programs to reduce delinquency, for much of the concern with the high school dropout, and for the idea that boys should be drafted into the army to keep them out of trouble. So obvious and persuasive is the idea that involvement in conventional activities is a major deterrent to delinquency that it was accepted even by Sutherland: "In the general area of juvenile delinquency it is probable that the most significant difference between juveniles who engage in delinquency and those who do not is that the latter are provided abundant opportunities of a conventional type for satisfying their recreational interests, while the former lack those opportunities or facilities."[18]

The view that "idle hands are the devil's workshop" has received more sophisticated treatment in recent sociological writings on delinquency. David Matza and Gresham M. Sykes, for example, suggest that delinquents have the values of a leisure class, the same values ascribed by Veblen to *the* leisure class: a search for kicks, disdain of work, a desire for the big score, and acceptance of aggressive toughness as proof of masculinity.[19] Matza and Sykes explain delinquency by reference to this system of values, but they note that adolescents at all class levels are "to some extent" members of a leisure class, that they "move in a limbo between earlier parental domination and future integration with the social structure through the bonds of work and marriage."[20] In the end, then, the leisure of the adolescent produces a set of values, which, in turn, leads to delinquency.

Belief

Unlike the cultural deviance theory, the control theory assumes the existence of a common value system within the society or group whose norms are being violated. If the deviant is committed to a value system different from that of conventional society, there is, within the context of the theory, nothing to explain. The question is, "Why does a man violate the rules in which he believes?" It is not, "Why do men differ in their beliefs about what constitutes good and desirable conduct?" The person is assumed to have been socialized (perhaps imperfectly) into the group whose rules he is violating; deviance is not a question of one group imposing its rules on the members of another group. In other words, we not only assume the deviant *has* believed the rules, we assume he believes the rules even as he violates them.

How can a person believe it is wrong to steal at the same time he is stealing? In the strain theory, this is not a difficult problem. (In fact, . . . the strain theory was devised specifically to deal with this question.) The motivation to deviance adduced by the strain theorist is so strong that we can well understand the deviant act even assuming the deviator believes strongly that it is wrong.[21] However, given the control theory's assumptions about motivation, if both the deviant and the nondeviant believe the deviant act is wrong, how do we account for the fact that one commits it and the other does not?

Control theories have taken two approaches to this problem. In one approach, beliefs are treated as mere words that mean little or nothing if the other forms of control are missing. "Semantic dementia," the dissociation between rational faculties and emotional control which is said to be characteristic of the psychopath, illustrates this way of handling the problem.[22] In short, beliefs, at least insofar as they are expressed in words, drop out of the picture; since they do not differentiate between deviants and nondeviants, they are in the same class as "language" or any other characteristic common to all members of the group. Since they represent no real obstacle to the commission of delinquent acts, nothing need be said about how they are handled by those committing such acts. The control theories that do not mention beliefs (or values), and many do not, may be assumed to take this approach to the problem.

The second approach argues that the deviant rationalizes his behavior so that he can at once violate the rule and maintain his belief in it. Donald R. Cressey had advanced this argument with respect to embezzlement,[23] and Sykes and Matza have advanced it with respect to delinquency.[24] In both Cressey's and Sykes and Matza's treatments, these rationalizations (Cressey calls them "verbalizations," Sykes and Matza term them "techniques of neutralization") occur prior to the commission of the deviant act. If the neutralization is successful, the person is free to commit the act(s) in question. Both in Cressey and in Sykes and Matza, the strain that prompts the effort at neutralization also provides the motive force that results in the subsequent deviant act. Their theories are thus, in this sense, strain theories. Neutralization is difficult to handle within the context of a theory that adheres closely to control theory assumptions, because in the control theory there is no special motivational force to account for the neutralization. This difficulty is especially noticeable in Matza's later treatment of this topic, where the motivational component, the "will to delinquency," appears *after* the moral vacuum has been created by the techniques of neutralization.[25] The question thus becomes: Why neutralize?

In attempting to solve a strain-theory problem with control-theory tools, the control theorist is thus led into a trap. He cannot answer the crucial question. The concept of neutralization assumes the existence of moral obstacles to the commission of deviant acts. In order plausibly to account for a deviant act, it is necessary to generate motivation to deviance that is at least equivalent in force to the resistance provided by these moral obstacles. However, if the moral obstacles are removed, neutralization and special motivation are no longer required. We therefore follow the implicit logic of control theory and remove these moral obstacles by hypothesis. Many persons do not have an attitude of respect toward the rules of society; many persons feel no moral obligation to conform regardless of personal advantage. Insofar as the values and beliefs of these persons are consistent with their feelings, and there should be tendency toward consistency, neutralization is unnecessary; it has already occurred.

Does this merely push the question back a step and at the same time produce conflict with the assumption of a common value system? I think not. In the first place, we do not assume, as does Cressey, that neutralization occurs in order to make a specific criminal act possible.[26] We do not assume, as do Sykes and Matza, that neutralization occurs to make many delinquent acts possible.

We do not assume, in other words, that the person constructs a system of rationalizations in order to justify commission of acts he *wants* to commit. We assume, in contrast, that the beliefs that free a man to commit deviant acts are *unmotivated* in the sense that he does not construct or adopt them in order to facilitate the attainment of illicit ends. In the second place, we do not assume, as does Matza, that "delinquents concur in the conventional assessment of delinquency."[27] We assume, in contrast, that there is *variation* in the extent to which people believe they should obey the rules of society, and, furthermore, that the less a person believes he should obey the rules, the more likely he is to violate them.[28]

In chronological order, then, a person's beliefs in the moral validity of norms are, for no teleological reason, weakened. The probability that he will commit delinquent acts is therefore increased. When and if he commits a delinquent act, we may justifiably use the weakness of his beliefs in explaining it, but no special motivation is required to explain either the weakness of his beliefs or, perhaps, his delinquent act.

The keystone of this argument is of course the assumption that there is variation in belief in the moral validity of social rules. This assumption is amenable to direct empirical test and can thus survive at least until its first confrontation with data. For the present, we must return to the idea of a common value system with which this section was begun.

The idea of a common (or perhaps better, a single) value system is consistent with the fact, or presumption, of variation in the strength of moral beliefs. We have not suggested that delinquency is based on beliefs counter to conventional morality; we have not suggested that delinquents do not believe delinquent acts are wrong. They may well believe these acts are wrong, but the meaning and efficacy of such beliefs are contingent on other beliefs and, indeed, on the strength of other ties to the conventional order.[29]

NOTES

1. Books have been written on the increasing importance of interpersonal sensitivity in modern life. According to this view, controls from within have become less important than controls from without in *producing* conformity. Whether or not this observation is true as a description of historical trends, it is true that interpersonal sensitivity has become more important in *explaining* conformity. Although logically it should also have become more important in explaining nonconformity, the opposite has been the case, once again showing that Cohen's observation that an explanation of conformity should be an explanation of deviance cannot be translated as "an explanation of conformity has to be an explanation of deviance." For the view that interpersonal sensitivity currently plays a greater role than formerly in producing conformity, see William J. Goode, "Norm Commitment and Conformity to Role-Status Obligations," *American Journal of Sociology* LXVI (1960): 246–258. And, of course, also see David Riesman, Nathan Glazer, and Rouel Denney, *The Lonely Crowd* (Garden City, New York: Doubleday, 1950), especially Part I.

2. The literature on psychopathy is voluminous. See William McCord and Joan McCord, *The Psychopath* (Princeton: D. Van Nostrand, 1964).

3. John M. Martin and Joseph P. Fitzpatrick, *Delinquent Behavior* (New York: Random House, 1964), p. 130.

4. *Ibid.* For additional properties of the psychopath, see McCord and McCord, *The Psychopath,* pp. 1–22.

5. Barbara Wootton, *Social Science and Social Pathology* (New York: Macmillan, 1959), p. 250.

6. "The logical untenability [of the position that there are forces in man 'resistant to socialization'] was ably demonstrated by Parsons over 30 years ago, and it is widely recognized that the position is empirically unsound because it assumes [!] some universal biological drive system distinctly separate from socialization and social context—a basic and intransigent human nature" (Judith Blake and Kingsley Davis, "Norms, Values, and Sanctions," *Handbook of Modern Sociology,* ed. Robert E. L. Faris [Chicago: Rand McNally, 1964], p. 471).

7. Emile Durkheim, *Moral Education,* trans. Everett K. Wilson and Herman Schnurer (New York: The Free Press, 1961), p. 64.

8. Although attachment alone does not exhaust the meaning of internalization, attachments and beliefs combined would appear to leave only a small residue of "internal control" not susceptible in principle to direct measurement.

9. F. Ivan Nye, *Family Relationships and Delinquent Behavior* (New York: Wiley, 1958), pp. 5–7.

10. Albert J. Reiss, Jr., "Delinquency as the Failure of Personal and Social Controls," *American Sociological Review* XVI (1951): 196–207. For example, "Our observations show . . . that delinquent recidivists are less often persons with mature ego ideals or nondelinquent social roles" (p. 204).

11. Scott Briar and Irving Piliavin, "Delinquency, Situational Inducements, and Commitment to Conformity," *Social Problems* XIII (1965): 41–42. The concept "stake in conformity" was introduced by Jackson Toby in his "Social Disorganization and Stake in Conformity: Complementary Factors in the Predatory Behavior of Hoodlums," *Journal of Criminal Law, Criminology and Police Science* XLVIII (1957): 12–17. See also his "Hoodlum or Business Man: An American Dilemma," *The Jews,* ed. Marshall Sklare (New York: The Free Press, 1958), pp. 542–550. Throughout the text, I occasionally use "stake in conformity" in speaking in general of the strength of the bond to conventional society. So used, the concept is somewhat broader than is true for either Toby or Briar and Piliavin, where the concept is roughly equivalent to what is here called "commitment."

12. Thomas Hobbes, *Leviathan* (Oxford: Basil Blackwell, 1957), p. 195.

13. Howard S. Becker, "Notes on the Concept of Commitment," *American Journal of Sociology* LXVI (1960): 35–36.

14. Arthur L. Stinchcombe, *Rebellion in a High School* (Chicago: Quadrangle, 1964), p. 5.

15. Richard A. Cloward and Lloyd E. Ohlin, *Delinquency and Opportunity* (New York: The Free Press, 1960), p. 147, quoting Edwin H. Sutherland, ed., *The Professional Thief* (Chicago: University of Chicago Press, 1937), pp. 211–213.

16. William James. *Psychology* (Cleveland: World Publishing Co., 1948), p. 186.

17. Few activities appear to be so engrossing that they rule out contemplation of alternative lines of behavior, at least if estimates of the amount of time men spend plotting sexual deviations have any validity.

18. *The Sutherland Papers,* ed. Albert K. Cohen et al. (Bloomington: Indiana University Press, 1956), p. 37.

19. David Matza and Gresham M. Sykes, "Juvenile Delinquency and Subterranean Values," *American Sociological Review* XXVI (1961): 712–719.

20. *Ibid.,* p. 718.

21. The starving man stealing the loaf of bread is the image evoked by most strain theories. In this image, the starving man's belief in the wrongness of his act is clearly not something that must be explained away. It can be assumed to be present without causing embarrassment to the explanation.

22. McCord and McCord, *The Psychopath,* pp. 12–15.

23. Donald R. Cressey, *Other People's Money* (New York: The Free Press, 1953).

24. Gresham M. Sykes and David Matza, "Techniques of Neutralization: A Theory of Delinquency," *American Sociological Review* XXII (1957), 664–670.

25. David Matza, *Delinquency and Drift* (New York: Wiley, 1964), pp. 181–191.

26. In asserting that Cressey's assumption is invalid with respect to delinquency, I do not wish to suggest that it is invalid for the question of embezzlement, where the problem faced by the deviator is fairly specific and he can reasonably be assumed to be an upstanding citizen. (Although even here the fact that the embezzler's non-sharable financial problem often results from some sort of hanky-panky suggests that "verbalizations" may be less necessary than might otherwise be assumed.)

27. *Delinquency and Drift,* p. 43.

28. This assumption is not, I think, contradicted by the evidence presented by Matza against the existence of a delinquent subculture. In comparing the attitudes and actions of delinquents with the picture painted by delinquent subculture theorists, Matza emphasizes—and perhaps exaggerates—the extent to which delinquents are tied to the conventional order. In implicitly comparing delinquents with a supermoral man, I emphasize—and perhaps exaggerate—the extent to which they are not tied to the conventional order.

29. The position taken here is therefore somewhere between the "semantic dementia" and the "neutralization" positions. Assuming variation, the delinquent is, at the extremes, freer than the neutralization argument assumes. Although the possibility of wide discrepancy between what the delinquent professes and what he practices still exists, it is presumably much rarer than is suggested by studies of articulate "psychopaths."

9

Conflict Theory of Crime

RICHARD QUINNEY

Quinney's conflict theory of crime offers a view of society that radically departs from previous perspectives. While Merton has a fairly unified view of society, assuming general agreement among people on the acceptable ends and the means of getting there, Durkheim and Erikson assume a broad social consensus about boundaries and the appropriate means of enforcing them, and Hirschi assumes a broad common value system, Quinney offers a more fragmented view of society. The conflict perspective envisions two groups in society: the dominant class and those they dominate. Defining and enforcing crime becomes a means of reproducing the power and socioeconomic inequalities between these groups, as the dominant class has interests that conflict with their subordinates. Quinney asserts that those dominating society use their power to formulate laws, to apply these laws to less powerful groups, and to rationalize their behavior as legitimate. In turn, those defined as criminal become more likely to engage in further behavior that will be defined as criminal. Quinney's conflict theory suggests that crime is one of the coercive means through which the dominance of the elite is maintained over the masses.

A theory that helps us begin to examine the legal order critically is the one I call the *social reality of crime*. Applying this theory, we think of crime as it is affected by the dynamics that mold the society's social, economic, and political structure. First, we recognize how criminal law fits into capitalist society. The legal order gives reality to the crime problem in the United States. Everything that makes up crime's social reality, including the application of criminal law, the behavior patterns of those who are defined as criminal, and the construction of an ideology of crime, is related to the established legal order. The social reality of crime is constructed on conflict in our society.

The theory of the social reality of crime is formulated as follows.

I. **The Official Definition of Crime:** Crime as a legal definition of human conduct is created by agents of the dominant class in a politically organized society.

The essential starting point is a definition of crime that itself is based on the legal definition. Crime, as *officially* determined, is a *definition* of behavior that is

From Richard Quinney, *Criminology* (Boston: Little, Brown, 1975), pp. 37–41. Reprinted by permission of the author.

conferred on some people by those in power. Agents of the law (such as legislators, police, prosecutors, and judges) are responsible for formulating and administering criminal law. Upon *formulation* and *application* of these definitions of crime, persons and behaviors become criminal.

Crime, according to this first proposition, is not inherent in behavior, but is a judgment made by some about the actions and characteristics of others. This proposition allows us to focus on the formulation and administration of the criminal law as it applies to the behaviors that become defined as criminal. Crime is seen as a result of the class-dynamic process that culminate in defining persons and behaviors as criminal. It follows, then, that the greater the number of definitions of crime that are formulated and applied, the greater the amount of crime.

II. Formulating Definitions of Crime: Definitions of crime are composed of behaviors that conflict with the interests of the dominant class.

Definitions of crime are formulated according to the interests of those who have the power to translate their interests into public policy. Those definitions are ultimately incorporated into the criminal law. Furthermore, definitions of crime in a society change as the interests of the dominant class change. In other words, those who are able to have their interests represented in public policy regulate the formulation of definitions of crime.

The powerful interests are reflected not only in the definitions of crime and the kinds of penal sanctions attached to them, but also in the *legal policies* on handling those defined as criminals. Procedural rules are created for enforcing and administering the criminal law. Policies are also established on programs for treating and punishing the criminally defined and programs for controlling and preventing crime. From the initial definitions of crime to the subsequent procedures, correctional and penal programs, and policies for controlling and preventing crime, those who have the power regulate the behavior of those without power.

III. Applying Definitions of Crime: Definitions of crime are applied by the class that has the power to shape the enforcement and administration of criminal law.

The dominant interests intervene in all the stages at which definitions of crime are created. Because class interests cannot be effectively protected merely by formulating criminal law, the law must be enforced and administered. The interests of the powerful, therefore, also operate where the definitions of crime reach the *application* stage. As Vold has argued, crime is "political behavior and the criminal becomes in fact a member of a 'minority group' without sufficient public support to dominate the control of the police power of the state." Those whose interests conflict with the ones represented in the law must either change their behavior or possibly find it defined as criminal.

The probability that definitions of crime will be applied varies according to how much the behaviors of the powerless conflict with the interests of those in power. Law enforcement efforts and judicial activity are likely to increase when the interests of the dominant class are threatened. Fluctuations and variations in applying definitions of crime reflect shifts in class relations.

Obviously, the criminal law is not applied directly by those in power; its enforcement and administration are delegated to authorized *legal agents*. Because the groups responsible for creating the definitions of crime are physically separated from the groups that have the authority to enforce and administer law, local conditions determine how the definitions will be applied. In particular, communities vary in their expectations of law enforcement and the administration of justice. The application of definitions is also influenced by the visibility of offenses in a community and by the public's norms about reporting possible violations. And especially important in enforcing and administering the criminal law are the legal agents' occupational organization and ideology.

The probability that these definitions will be applied depends on the actions of the legal agents who have the authority to enforce and administer the law. A definition of crime is applied depending on their evaluation. Turk has argued that during "criminalization," a criminal label may be affixed to people because of real or fancied attributes: "Indeed, a person is evaluated, either favorably or unfavorably, not because he *does* something, or even because he *is* something, but because others react to their perceptions of him as offensive or inoffensive." Evaluation by the definers is affected by the way in which the suspect handles the situation, but ultimately the legal agents' evaluations and subsequent decisions are the crucial factors in determining the criminality of human acts. As legal agents evaluate more behaviors and persons as worthy of being defined as crimes, the probability that definitions of crime will be applied grows.

IV. How Behavior Patterns Develop in Relation to Definitions of Crime: Behavior patterns are structured in relation to definitions of crime, and within this context people engage in actions that have relative probabilities of being defined as criminal.

Although behavior varies, all behaviors are similar in that they represent patterns within society. All persons—whether they create definitions of crime or are the objects of these definitions—act in reference to *normative systems* learned in relative social and cultural settings. Because it is not the quality of the behavior but the action taken against the behavior that gives it the character of criminality, that which is defined as criminal is relative to the behavior patterns of the class that formulates and applies definitions. Consequently, people whose behavior patterns are not represented when the definitions of crime are formulated and applied are more likely to act in ways that will be defined as criminal than those who formulate and apply the definitions.

Once behavior patterns become established with some regularity within the segments of society, individuals have a framework for creating *personal action patterns*. These continually develop for each person as he moves from one experience to another. Specific action patterns give behavior an individual substance in relation to the definitions of crime.

People construct their own patterns of action in participating with others. It follows, then, that the probability that persons will develop action patterns with a high potential for being defined as criminal depends on (1) structured opportunities, (2) learning experiences, (3) interpersonal associations and identifications, and (4) self-conceptions. Throughout the experiences, each person creates a conception of self as a human social being. Thus prepared, he behaves according to the anticipated consequences of his actions.

In the experiences shared by the definers of crime and the criminally defined, personal-action patterns develop among the latter because they are so defined. After they have had continued experience in being defined as criminal, they learn to manipulate the application of criminal definitions.

Furthermore, those who have been defined as criminal begin to conceive of themselves as criminal. As they adjust to the definitions imposed on them, they learn to play the criminal role. As a result of others' reactions, therefore, people may develop personal-action patterns that increase the likelihood of their being defined as criminal in the future. That is, increased experience with definitions of crime increases the probability of their developing actions that may be subsequently defined as criminal.

Thus, both the definers of crime and the criminally defined are involved in reciprocal action patterns. The personal-action patterns of both the definers and the defined are shaped by their common, continued, and related experiences. The fate of each is bound to that of the other.

V. Constructing an Ideology of Crime: An ideology of crime is constructed and diffused by the dominant class to secure its hegemony.

This ideology is created in the kinds of ideas people are exposed to, the manner in which they select information to fit the world they are shaping, and their way of interpreting this information. People behave in reference to the *social meanings* they attach to their experiences.

Among the conceptions that develop in a society are those relating to what people regard as crime. The concept of crime must of course be accompanied by ideas about the nature of crime. Images develop about the relevance of crime, the offender's characteristics, the appropriate reaction to crime, and the relation of crime to the social order. These conceptions are constructed by communication, and, in fact, an ideology of crime depends on the portrayal of crime in all personal and mass communication. This ideology is thus diffused throughout the society.

One of the most concrete ways by which an ideology of crime is formed and transmitted is the official investigation of crime. The President's Commission on Law Enforcement and Administration of Justice is the best contemporary example of the state's role in shaping an ideology of crime. Not only are we as citizens more aware of crime today because of the President's Commission, but official policy on crime has been established in a crime bill, the Omnibus Crime Control and Safe Streets Act of 1968. The crime bill, itself a reaction to the growing fears of class conflict in American society, creates an image of a severe crime problem and, in so doing, threatens to negate some of our basic constitutional guarantees in the name of controlling crime.

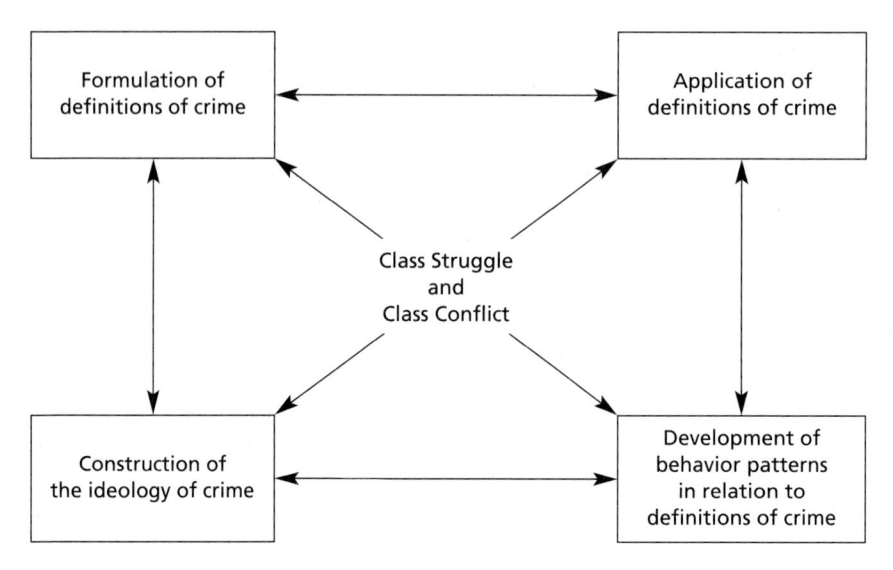

```
                                  Class Struggle
                                      and
                                  Class Conflict
```

Formulation of definitions of crime ⟷ Application of definitions of crime

Construction of the ideology of crime ⟷ Development of behavior patterns in relation to definitions of crime

The Social Reality of Crime

Consequently, the conceptions that are most critical in actually formulating and applying the definitions of crime are those held by the dominant class. These conceptions are certain to be incorporated into the social reality of crime. The more the government acts in reference to crime, the more probable it is that definitions of crime will be created and that behavior patterns will develop in opposition to those definitions. The formulation of definitions of crime, their application, and the development of behavior patterns in relation to the definitions, are thus joined in full circle by the construction of an ideological hegemony toward crime.

VI. Constructing the Social Reality of Crime: The social reality of crime is constructed by the formulation and application of definitions of crime, the development of behavior patterns in relation to these definitions, and the construction of an ideology of crime.

The first five propositions are collected here into a final composition proposition. The theory of the social reality of crime, accordingly, postulates creating a series of phenomena that increase the probability of crime. The result, holistically, is the social reality of crime.

Because the first proposition of the theory is a definition and the sixth is a composite, the body of the theory consists of the four middle propositions. These form a model of crime's social reality. The model, as diagrammed, relates the proposition units into a theoretical system (see figure above). Each unit is related to the others. The theory is thus a system of interacting developmental propositions. The phenomena denoted in the propositions and their relationships culminate in what is regarded as the amount and character of crime at any time—that is, in the social reality of crime.

The theory of the social reality of crime as I have formulated it is inspired by a change that is occurring in our view of the world. This change, pervading all levels of society, pertains to the world that we all construct and from which, at the same time, we pretend to separate ourselves in our human experiences. For the study of crime, a revision in thought has directed attention to the criminal process: All relevant phenomena contribute to creating definitions of crime, development of behaviors by those involved in criminal-defining situations, and constructing an ideology of crime. The result is the social reality of crime that is constantly being constructed in society.

10

Deviance: The Constructionist Stance

JOEL BEST

Best, one of the leading practitioners of the social constructionist approach, offers a historical analysis of the way theories of deviance have evolved. He notes the rise and decline in significance of several approaches to deviance theory, showing how social constructionism arose from its early roots in the sociology of deviance and moved into explaining social problems. The constructionist perspective represents a wedding of the views of labeling and conflict theories. It joins the micro analysis of the former by looking at how individuals encounter societal reactions and become labeled as deviants with the broader, more structural, social power contribution of the latter, isolating certain groups as more likely to have social reactions and definitions formed and applied by and against them. It is this social constructionist stance that will frame the organization and selections comprising the remainder of this book, as we examine the process by which groups vie for power in society and try to legislate their views into morality, and then focus on how people develop deviant identities and manage their stigma as a result of those definitions and enforcements.

What does it mean to say that deviance is "socially constructed?" Some people assume that social construction is the opposite of real, but this is a mistake. Reality, that is everything we understand about the world, is socially constructed. The term calls attention to the processes by which people make sense of the world: we create—or construct—meaning. When we define some behavior as deviant, we are socially constructing deviance. The constructionist approach recognizes that people can only understand the world in terms of words and categories that they create and share with one another.

THE EMERGENCE OF CONSTRUCTIONISM

The constructionist stance had its roots in two developments. The first was the publication of Peter L. Berger and Thomas Luckmann's (1966) *The Social Construction of Reality*. Berger and Luckmann were writing about the sociology of knowledge—how social life shapes everything that people know. Their

Reprinted by permission of the author.

book introduced the term "social construction" to a wide sociological audience, and soon other sociologists were writing about the construction of science, news, and other sorts of knowledge, including what we think about deviance.

Second, labeling theory, which had become the leading approach to studying deviance during the 1960s, came under attack from several different directions by the mid-1970s. Conflict theorists charged that labeling theory ignored how elites shaped definitions of deviance and social control policies. Feminists complained that labeling ignored the victimization of women at the hands of both male offenders and male-dominated social control agencies. Activists for gay rights and disability rights insisted that homosexuals and the disabled should be viewed as political minorities, rather than deviants. At the same time, mainstream sociologists began challenging labeling's claims about the ways social control operated and affected deviants' identities.

THE CONSTRUCTIONIST RESPONSE

In response to these attacks, some sociologists sympathetic to the labeling approach moved away from studying deviance. Led by John I. Kitsuse, a sociologist whose work had helped shape labeling theory, these sociologists of deviance turned to studying the sociology of social problems. With Malcolm Spector, Kitsuse published *Constructing Social Problems* (1977)—a book that would inspire many sociologists to begin studying how and why particular social problems emerged as topics of public concern. They argued that sociologists ought to redefine social problems as claims that various conditions constituted social problems; therefore, the constructionist approach involved studying claims and those who made them—the claimsmakers. In this view, sociologists ought to study how and why particular issues such as date rape or binge drinking on college campuses suddenly became the focus of attention and concern. How were these problems constructed?

There were several advantages to studying social problems. First, constructionists had the field virtually to themselves. Although many sociology departments taught social problems courses, there were no rival well-established, coherent theories of social problems. In contrast, labeling had to struggle against functionalism, conflict theory, and other influential approaches to studying deviance.

The constructionist approach was also flexible. Analysts of social problems construction might concentrate on various actors: some examined the power of political and economic elites in shaping definitions of social problems; others focused on the role of activists in bringing attention to problems; and still others concentrated on how media coverage shaped the public's and policymakers' understandings of problems. This flexibility meant that constructionists might criticize some claims as exaggerated, distorted, or unfounded (the sort of critique found in several studies of claims about the menace of

Satanism), but they might also celebrate the efforts of claimsmakers to draw attention to neglected problems (for example, researchers tended to treat claims about domestic violence sympathetically).

Again, it is important to appreciate that "socially constructed" is not a synonym for erroneous or mistaken. All knowledge is socially constructed; to say that a social problem is socially constructed is not to imply that it does not exist, but rather that it is through social interaction that the problem is assigned particular meanings.

THE RETURN TO DEVIANCE

Although constructionists studied the emergence and evolution of many different social problems, ranging from global warming to homelessness, much of their work remained focused on deviance. They studied the construction of rape, child abduction, illicit drugs, family violence, and other forms of deviance.

Closely related to the rise of constructionism were studies of medicalization (Conrad and Schneider 1980). Medicalization—defining deviance as a form of illness requiring medical treatment—was one popular, contemporary way of constructing deviance. By the end of the twentieth century, medical language—"disease," "symptom," "therapy," and so on—was used, not only by medical authorities, but even by amateurs (for example, in the many Twelve-Step programs of the recovery movement).

A large share of constructionist studies traced the rise of social problems to national attention; for example, the construction of the federal War on Drugs was studied by several constructionist researchers. However, other sociologists began studying how deviance was constructed in smaller settings, through interpersonal interaction. In particular, they examined social problems work (Holstein and Miller 1993). Even after claimsmakers have managed to draw attention to some social problem and shape the creation of social policies to deal with it, those claims must be translated into action. Police officers, social workers, and other social problems workers must apply broad constructions to particular cases. Thus, after wife abuse is defined as a social problem, it is still necessary for the police officer investigating a domestic disturbance call to define—or construct—these particular events as an instance of wife abuse (Loseke 1992). Studies of this sort of social problems work are a continuation of earlier research on the labeling process.

CONSTRUCTIONISM'S DOMAIN

Social constructionism, then, has become an influential stance for thinking about deviance, particularly for understanding how concerns about particular forms of deviance emerge and evolve, and for studying how social control agents construct particular acts as deviance and individuals as deviants. Constructionism emphasizes the role of interpretation, of people assigning meaning, or making sense of the behaviors they classify as deviant. This can occur at a societal level, as when the mass media draw attention to a new form of deviance and legislators pass laws against it, but it can also occur in face-to-face interaction, when one individual expresses disapproval of another's rule-breaking. Deviance, like all reality, is constantly being constructed.

REFERENCES

Berger, Peter L., and Thomas Luckmann. 1966. *The Social Construction of Reality.* New York: Doubleday.

Conrad, Peter, and Joseph W. Schneider. 1980. *Deviance and Medicalization.* St. Louis: Mosby.

Holstein, James A., and Gale Miller. 1993. "Social Constructionist and Social Problems Work." Pp. 151–72 in *Reconsidering Social Constructionism,* edited by James Holstein and Gale Miller. Hawthorne, NY: Aldine de Gruyter.

Loseke, Donileen R. 1992. *The Battered Woman and Shelters.* Albany: State University of New York Press.

Spector, Malcolm, and John I. Kitsuse. 1977. *Constructing Social Problems.* Menlo Park, CA: Cummings.

PART III

Studying Deviance

Accurate and reliable knowledge about deviance is critically important to many groups in society. First, policy makers are concerned with deviant groups such as the homeless and transient, the chronically mentally ill, high school dropouts, criminal offenders, prostitutes, juvenile delinquents, gang members, runaways, and other members of disadvantaged and disenfranchised populations. These people pose social problems that lawmakers and social welfare agencies want to alleviate. Second, sociologists and other researchers have an interest in deviance based on their goal of understanding human nature, human behavior, and human society. Deviants are a critical group to this enterprise because they reside near the margins of social definition; they help define the boundaries of what is acceptable and unacceptable to given groups.

In this part we will discuss information about deviance coming from three primary data sources. Government officials and employees of social service agencies routinely collect information about their clients as they process them. This information includes arrest data that are compiled by the police and published by the FBI (*Uniform Crime Reports*), census data on various shifting populations (such as the homeless), victim data from helping agencies (such as battered women's shelters), medical data from emergency rooms (such as DAWN, the Drug Abuse Warning Network) or from state public health agencies (such as the coroner or medical examiner offices), and prosecution data on cases that are tried in the courts. These precollected *official statistics* are then compiled by the various government organizations responsible for collecting them and made available to the public.

Another source of statistical data about deviance is *survey research.* Rather than relying on information the government collects, sociologists gather their own data through large-scale questionnaire surveys. Prominent ones include the National Youth Survey, a self-report questionnaire about delinquent behavior, the Annual High School Survey conducted by the National Institute on Drug Abuse on the drug use of high school seniors, and some of the Kinsey surveys about sexual behavior.

A third kind of information, richly descriptive and analytical rather than numerical, comes from sociologists who conduct *participant-observation field research* on deviance. Much like anthropologists who go out to live among native peoples, sociological fieldworkers live among members of deviant groups and become intimately familiar with their lives. This type of research yields information more deeply based on the subjects' own perspectives, detailing how they see the world, the allure of deviance for them, the problems they encounter, the ways they resolve these, the significant individuals and groups in their lives, and their role among these others. Unlike the other types of research, participant observation is generally a longitudinal method, entailing years of involvement with subjects. Researchers must gain acceptance by group members, develop meaningful relationships with them, and learn about their deepest thoughts and emotions.

There are many differences among these types of data and among the methods used to gather them. Each has its advantages and disadvantages, and each may do a better job than the others of answering certain research questions. Thus, depending on researchers' particular needs, they may turn to one type of data or use a mixed-method approach. Official statistics have the advantage of being inexpensive to gather and quick to access, since they are already collected and published. They intend to include the entire population of those they address, not some subsample, e.g., all criminals, all victims, or all emergency room admittees. Records about these occurrences can be accessed back as far as the official statistics have been collected, potentially a rather long time. They have certain validity problems, however, and tend to be inaccurate in patterned and systematic ways. Official police statistics, e.g., *Uniform Crime Reports,* fail to include a host of crimes for several reasons: crimes may be unrecognized by victims who do not notice their occurrence or lack the power to define them as deviant; crimes may be unreported by individuals who see no gain by, or fear embarrassment, censure, or retaliation from calling police attention to their victimization; or crimes may be unrecorded by police officers who use their broad discretion to handle problems informally. Other types of official statistics vastly underrepresent criminal activity for similar and other reasons, and may be problematic because the categories used to conceive of them and the way they are assembled and reported changes over time, making comparisons over the years frustrating. In sum, although official statistics yield information about a broad spectrum of people, they may be fairly shallow and unreliable in nature. Besharov and Laumann's selection on child abuse statistics discusses the dramatic rise in the number of reported cases of child abuse and some of the sociological factors that have accounted for this wild swing in the official statistics. They discuss factors that artificially both inflate and deflate our official estimates of child abuse and their consequences for the protection of abused children.

Survey research lets social scientists collect data on the topics of their choice, but it is a more expensive and time-consuming enterprise. Through careful sampling procedures, researchers can gather data about a smaller population and generalize from it to a much larger group with a high degree of accuracy (external reliability). Strict controls over standardization of procedures and detachment of data-gatherers makes this a relatively objective method. Correlations (although not causal relations) between social factors can be established carefully. But survey research, like official statistics, has internal validity problems; it may not yield an accurate portrait of the sample group it is studying, especially for topics as sensitive as deviance. First, it is unlikely that people, especially deviants, will fill out a questionnaire and readily disclose information about the hidden aspects of their lives. Second, in responding to the questionnaire, subjects may not define their behavior the same way or use the same terms as the researchers who are writing the questions (prostitutes' conceptions of a "date" may be different from those of survey researchers', and runaways may mean different things when they refer to their "home" than researchers intend). Researchers are then likely to misinterpret the nature and extent of behavior from the answers they receive. Third, sometimes the correlational connections of survey research, that is, what trends occur together (such as deviance and divorce, or violence in the media and violence in everyday life) are mistakenly assumed to be causal connections. But survey research cannot tell us *why* or *how* people act; it can only tell us *what* people are doing, even if these trends range across a broad spread of the population. A much heralded study of sexual behavior is featured in Laumann et al.'s description of their survey, a major, highly professionally designed and conducted study that gives us a glimpse into the problems and creative adaptations that can arise when a comprehensive effort is made to conduct large-scale survey research into Americans' sexual practices.

Participant observation, in contrast, cannot reach as many people, but yields more deep and accurate information about research subjects, backed by researchers' own observations, to enhance its internal validity. Participant observers spend long amounts of time in the field becoming close to the people they study and learning how their subjects perceive, interpret, and act upon the complex and often contradictory nature of their social worlds. In contrast to the detached and objective relationships between survey researchers and their subjects, participant observers rely on the subjectivity and strength of the close personal relationships they forge with the people they study to get behind false fronts and to find out what is really going on. Depth understanding is especially important when studying a topic such as deviance, where so much behavior is hidden due to its stigma and illicit status. Also critically important is the ability of participant observation to study deviance, as Polsky (1967) urged, it occurs *in situ,* in its natural setting, not via the structural constraints of police reporting, or the interpretation and recollection of questionnaire research. But although often less costly than survey research to conduct, it is very time consuming, as depth relationships take long to develop. Field research also lacks the generalizability characterizing careful survey research, as subjects tend to be gathered through a referral (snowball) technique or because they are members of a common "scene," thus giving them shared patterns of behavior that may be found with them more strongly than with broader practitioners of the deviance. We share with readers our own experiences with participant observation in the selection on field research where we talk about what it is like to carry out such research with a criminal, and potentially dangerous, group.

The empirical selections that fill the remainder of this book are primarily based on participant observation studies of deviance for two main reasons. First, as Becker (1973) remarked, participation observation is the method of the interactionist perspective; it offers direct access to the way definitions and laws are socially constructed, to the way people's actions are influenced by their associates, and to the way people's identities are affected by the deviant labels cast on them. Second, these types of studies offer a deeper view of people's feelings, experiences, motivations, and social psychological states, which give a richer and more vivid portrayal of deviance than charts of numbers.

11

Child Abuse Reporting

DOUGLAS J. BESHAROV WITH LISA A. LAUMANN

Besharov and Laumann discuss official statistics in our first selection on varieties of ways that deviance is studied. They note the spectacular rise in our official knowledge about the extent of child abuse, with rates increasing by 300 percent over a recent thirty year period. Such a dramatic change cannot be solely attributable to changes in deviant behavior, but must also involve a measurement artifact. They root the increase in the mandatory reporting laws, the media campaigns surrounding child abuse, and the changed social definition of what constitutes abuse. Besharov and Laumann discuss two ironically opposing problems associated with child abuse statistics, the presence of both unreported and unsubstantiated cases. On the one hand, they claim, we are still unaware of many cases of child abuse because it tends to be hidden, defined as a private family matter, and regarded as "normal" childrearing practice. At the same time, the way we as a society tumultuously attacked this "discovered" social problem and deputized numerous social groups to document it resulted in cases that could not be substantiated. Some of these were unsubstantiated because they were investigated and found to be lacking in substance, but others were unfounded because the families could not be located, or the child abuse, when investigated, was able to remain hidden. The huge increase in the number of cases requiring investigation has overburdened the investigatory dockets of social service agencies and diminished their ability to resolve all allegations. Some desperate situations are being attended to, but others are slipping through the cracks due to over-reporting problems. These cases signal continued ambiguity over definitions of child abuse. Together these problems cast light on the work of social welfare agents to gather official statistics and some of the problems these data encompass.

For 30 years, advocates, program administrators, and politicians have joined to encourage even more reports of suspected child abuse and neglect. Their efforts have been spectacularly successful, with about three million cases of suspected child abuse having been reported in 1993. Large numbers of endangered children still go unreported, but an equally serious problem has developed: Upon investigation, as many as 65 percent of the reports now being

---

Reprinted by permission of Transaction Publishers. "Child Abuse Reporting" by Douglas J. Besharov with Lisa A. Lavmann, from *Society*, May/June 1996. Copyright © 1996 by Transaction Publishers.

made are determined to be "unsubstantiated," raising serious civil liberties concerns and placing a heavy burden on already overwhelmed investigative staffs.

These two problems—nonreporting and inappropriate reporting—are linked and must be addressed together before further progress can be made in combating child abuse and neglect. To lessen both problems, there must be a shift in priorities—away from simply seeking more reports and toward encouraging better reports.

REPORTING LAWS

Since the early 1960s, all states have passed laws that require designated professionals to report specified types of child maltreatment. Over the years, both the range of designated professionals and the scope of reportable conditions have been steadily expanded.

Initially, mandatory reporting laws applied only to physicians, who were required to report only "serious physical injuries" and "nonaccidental injuries." In the ensuing years, however, increased public and professional attention, sparked in part by the number of abused children revealed by these initial reporting laws, led many states to expand their reporting requirements. Now almost all states have laws that require the reporting of all forms of suspected child maltreatment, including physical abuse, physical neglect, emotional maltreatment, and of course, sexual abuse and exploitation.

Under threat of civil and criminal penalties, these laws require most professionals who serve children to report suspected child abuse and neglect. About twenty states require all citizens to report, but in every state, any citizen is permitted to report.

These reporting laws, associated public awareness campaigns, and professional education programs have been strikingly successful. In 1993, there were about three million reports of children suspected of being abused or neglected. This is a twenty-fold increase since 1963, when about 150,000 cases were reported to the authorities. (As we will see, however, this figure is bloated by reports that later turn out to be unfounded.)

Many people ask whether this vast increase in reporting signals a rise in the incidence of child maltreatment. Recent increases in social problems such as out-of-wedlock births, inner-city poverty, and drug abuse have probably raised the underlying rates of child maltreatment, at least somewhat. Unfortunately, so many maltreated children previously went unreported that earlier reporting statistics do not provide a reliable baseline against which to make comparisons. One thing is clear, however: The great bulk of reports now received by child protective agencies would not be made but for the passage of mandatory reporting laws and the media campaigns that accompanied them.

This increase in reporting was accompanied by a substantial expansion of prevention and treatment programs. Every community, for example, is now served by specialized child protective agencies that receive and investigate reports. Federal and state expenditures for child protective programs and associated foster care services now exceed $6 billion a year. (Federal expenditures for foster care, child welfare, and related services make up less than 50 percent of total state and federal expenditures for these services; in 1992, they amounted to a total of $2,773.7 million. In addition, states may use a portion of the $2.8 billion federal Social Services Block Grant for such services, though detailed data on these expenditures are not available. Beginning in 1994, additional federal appropriations funded family preservation and support services.)

As a result, many thousands of children have been saved from serious injury and even death. The best estimate is that over the past twenty years, child abuse and neglect deaths have fallen from over 3,000 a year—and perhaps as many as 5,000—to about 1,000 a year. In New York State, for example, within five years of the passage of a comprehensive reporting law, which also created specialized investigative staffs, there was a 50 percent reduction in child fatalities, from about two hundred a year to less than one hundred. (This is not meant to minimize the remaining problem. Even at this level, maltreatment is the sixth largest cause of death for children under fourteen.)

UNREPORTED CASES

Most experts agree that reports have increased over the past thirty years because professionals and laypersons have become more likely to report apparently abusive and neglectful situations. But the question remains: How many more cases still go unreported?

Two studies performed for the National Center on Child Abuse and Neglect by Westat, Inc., provide a partial answer. In 1980 and then again in 1986, Westat conducted national studies of the incidence of child abuse and neglect. (A third Westat incidence study is now underway.) Each study used essentially the same methodology: In a stratified sample of counties, a broadly representative sample of professionals who serve children was asked whether, during the study period, the children they had seen in their professional capacities appeared to have been abused or neglected. (Actually, the professionals were not asked the ultimate question of whether the children appeared to be "abused" or "neglected." Instead, they were asked to identify children with certain specified harms or conditions, which were then decoded into a count of various types of child abuse and neglect.)

Because the information these selected professionals provided could be matched against pending cases in the local child protective agency, Westat was able to estimate rates of nonreporting among the surveyed professionals. It could not, of course, estimate the level of unintentional nonreporting, since there is no way to know of the situations in which professionals did not recognize signs of possible maltreatment. There is also no way to know how many children the professionals recognized as being maltreated but chose not to report to the study. Obviously, since the study methodology involved asking professionals about children they had seen in their professional capacities, it also did not allow Westat to estimate the number of children seen by nonprofessionals, let alone their nonreporting rate.

Westat found that professionals failed to report many of the children they saw who had observable signs of child abuse and neglect. Specifically, it found that in 1986, 56 percent of apparently abused or neglected children, or about 500,000 children, were not reported to the authorities. This figure, however, seems more alarming than it is: Basically, the more serious the case, the more likely the report. For example, the surveyed professionals reported over 85 percent of the fatal or serious physical abuse cases they saw, 72 percent of the sexual abuse cases, and 60 percent of the moderate physical abuse cases. In contrast, they only reported 15 percent of the educational neglect cases they saw, 24 percent of the emotional neglect cases, and 25 percent of the moderate physical neglect cases.

Nevertheless, there is no reason for complacency. Translating these raw percentages into actual cases means that in 1986, about 2,000 children with observable physical injuries severe enough to require hospitalization were not reported and that more than 100,000 children with moderate physical injuries went unreported, as did more than 30,000 apparently sexually abused children. And these are the rates of nonreporting among relatively well-trained professionals. One assumes that nonreporting is higher among less-well-trained professionals and higher still among laypersons.

Obtaining and maintaining a high level of reporting requires a continuation of the public education and professional training begun thirty years ago. But, now, such efforts must also address a problem as serious as nonreporting: inappropriate reporting.

At the same time that many seriously abused children go unreported, an equally serious problem further undercuts efforts to prevent child maltreatment: The nation's child protective agencies are being inundated by inappropriate reports. Although rules, procedures, and even terminology vary—some states use the phrase "unfounded," others "unsubstantiated" or not indicated—an "unfounded" report, in essence, is one that is dismissed after an investigation finds insufficient evidence upon which to proceed.

UNSUBSTANTIATED REPORTS

Nationwide, between 60 and 65 percent of all reports are closed after an initial investigation determines that they are "unfounded" or "unsubstantiated." This is in sharp contrast to 1974, when only about 45 percent of all reports were unfounded.

A few advocates, in a misguided effort to shield child protective programs from criticism, have sought to quarrel with estimates that I and others have made that the national unfounded rate is between 60 and 65 percent. They have grasped at various inconsistencies in the data collected by different organizations to claim either that the problem is not so bad or that it has always been this bad.

To help settle this dispute, the American Public Welfare Association (APWA) conducted a special survey of child welfare agencies in 1989. The APWA researchers found that between fiscal year 1986 and fiscal year 1988, the weighted average for the substantiation rates in thirty-one states declined 6.7 percent—from 41.8 percent in fiscal year 1986 to 39 percent in fiscal year 1988.

Most recently, the existence of this high unfounded rate was reconfirmed by the annual Fifty State Survey of the National Committee to Prevent Child Abuse (NCPCA), which found that in 1993 only about 34 percent of the reports received by child protective agencies were substantiated.

The experience of New York City indicates what these statistics mean in practice. Between 1989 and 1993, as the number of reports received by the city's child welfare agency increased by over 30 percent (from 40,217 to 52,472), the percentage of substantiated reports fell by about 47 percent (from 45 percent to 24 percent). In fact, the number of substantiated cases—a number of families were reported more than once—actually fell by about 41 percent, from 14,026 to 8,326. Thus, 12,255 additional families were investigated, while 5,700 fewer families received child protective help.

The determination that a report is unfounded can only be made after an unavoidably traumatic investigation that is inherently a breach of parental and family privacy. To determine whether a particular child is in danger, caseworkers must inquire into the most intimate personal and family matters. Often it is necessary to question friends, relatives, and neighbors, as well as school teachers, day-care personnel, doctors, clergy, and others who know the family.

Laws against child abuse are an implicit recognition that family privacy must give way to the need to protect helpless children. But in seeking to protect children, it is all too easy to ignore the legitimate rights of parents. Each year, about 700,000 families are put through investigations of unfounded reports. This is a massive and unjustified violation of parental rights.

Few unfounded reports are made maliciously. Studies of sexual abuse reports, for example, suggest that, at most, from 4 to 10 percent of these reports are knowingly false. Many involve situations in which the person reporting, in a well-intentioned effort to protect a child, overreacts to a vague and often misleading possibility that the child may be maltreated. Others involve situations of poor child care that, though of legitimate concern, simply do not amount to child abuse or neglect. In fact, a substantial proportion of unfounded cases are referred to other agencies for them to provide needed services for the family.

Moreover, an unfounded report does not necessarily mean that the child was not actually abused or neglected. Evidence of child maltreatment is hard to obtain and might not be uncovered when agencies lack the time and resources to complete a thorough investigation or when inaccurate information is given to the investigator. Other cases are labeled unfounded when no services are available to help the family. Some cases must be closed because the child or family cannot be located.

A certain proportion of unfounded reports, therefore, is an inherent—and legitimate—aspect of reporting *suspected* child maltreatment and is necessary to ensure adequate child protection. Hundreds of thousands of strangers report their suspicions; they cannot all be right. But unfounded rates of the current magnitude go beyond anything reasonably needed. Worse, they endanger children who are really abused.

The current flood of unfounded reports is overwhelming the limited resources of child protective agencies. For fear of missing even one abused child, workers perform extensive investigations of vague and apparently unsupported reports. Even when a home visit based on an anonymous report turns up no evidence of maltreatment, they usually interview neighbors, school teachers, and day-care personnel to make sure that the child is not abused. And even repeated anonymous and unfounded reports do not prevent a further investigation. But all this takes time.

As a result, children in real danger are getting lost in the press of inappropriate cases. Forced to allocate a substantial portion of their limited resources to unfounded reports, child protective agencies are less able to respond promptly and effectively when children are in serious danger. Some reports are left uninvestigated for a week and even two weeks after they are received. Investigations often miss key facts, as workers rush to clear cases, and dangerous home situations receive inadequate supervision, as workers must ignore pending cases as they investigate the new reports that arrive daily on their desks. Decision making also suffers. With so many cases of unsubstantiated or unproven risk to children, caseworkers are desensitized to the obvious warning signals of immediate and serious danger.

These nationwide conditions help explain why from 25 to 50 percent of child abuse deaths involve children previously known to the authorities. In 1993, the NCPCA reported that of the 1,149 child maltreatment deaths, 42 percent had already been reported to the authorities. Tens of thousands of other children suffer serious injuries short of death while under child protective agency supervision.

In a 1992 New York City case, for example, five-month-old Jeffrey Harden died from burns caused by scalding water and three broken ribs while under the supervision of New York City's Child Welfare Administration. Jeffrey Harden's family had been known to the administration for more than a year and a half. Over this period, the case had been handled by four separate caseworkers, each conducting only partial investigations before resigning or being reassigned to new cases. It is unclear whether Jeffrey's death was caused by his mother or her boyfriend, but because of insufficient time and overburdened caseloads, all four workers failed to pay attention to a whole host of obvious warning signals: Jeffrey's mother had broken her parole for an earlier conviction of child sexual abuse, she had a past record of beating Jeffrey's older sister, and she had a history of crack addiction and past involvement with violent boyfriends.

Here is how two of the Hardens' caseworkers explained what happened: Their first caseworker could not find Ms. Harden at the address she had listed in her files. She commented, "It was an easy case. We couldn't find the mother so we closed it." Their second caseworker stated that he was unable to spend a sufficient amount of time investigating the case, let alone make the minimum monthly visits because he was tied down with an overabundance of cases and paperwork. He stated, "It's impossible to visit these people within a month. They're all over New York City." Just before Jeffrey's death every worker who had been on the case had left the department. Ironically, by weakening the system's ability to respond, unfounded reports actually discourage appropriate ones. The sad fact is that many responsible individuals are not reporting endangered children because they feel that the system's response will be so weak that reporting will do no good or may even make things worse. . . .

12

Survey of Sexual Behavior of Americans

EDWARD O. LAUMANN, JOHN H. GAGNON,
ROBERT T. MICHAEL, AND STUART MICHAELS

A much heralded study of sexual behavior is featured in this selection by Laumann, Gagnon, Michael, and Michaels. This major, highly professionally designed and conducted study gives us a glimpse into the problems and creative adaptations that arise when a comprehensive effort is made to conduct large-scale survey research into Americans' sexual practices. This selection outlines the procedures for conceiving and carrying out the study, from initial conceptualization to sampling, administration, interviewer training, questionnaire design, and issues of privacy and confidentiality. Readers can get some feel for both the generic features of survey research, the specific decisions as to how this project was implemented, and the strengths and weaknesses associated with this mode of gathering data about deviance.

Most people with whom we talked when we first broached the idea of a national survey of sexual behavior were skeptical that it could be done. Scientists and laypeople alike had similar reactions: "Nobody will agree to participate in such a study." "Nobody will answer questions like these, and, even if they do, they won't tell the truth." "People don't know enough about sexual practices as they relate to disease transmission or even to pleasure or physical and emotional satisfaction to be able to answer questions accurately." It would be dishonest to say that we did not share these and other concerns. But our experiences over the past seven years, rooted in extensive pilot work, focus-group discussions, and the fielding of the survey itself, resolved these doubts, fully vindicating our growing conviction that a national survey could be conducted according to high standards of scientific rigor and replicability. . . .

The society in which we live treats sex and everything related to sex in a most ambiguous and ambivalent fashion. Sex is at once highly fascinating, attractive, and, for many at certain stages in their lives, preoccupying, but it can also be frightening, disturbing, or guilt inducing. For many, sex is considered

Reprinted by permission of the University of Chicago Press.

to be an extremely private matter, to be discussed only with one's closest friends or intimates, if at all. And, certainly for most if not all of us, there are elements of our sexual lives never acknowledged to others, reserved for our own personal fantasies and self-contemplation. It is thus hardly surprising that the proposal to study sex scientifically, or any other way for that matter, elicits confounding and confusing reactions. Mass advertising, for example, unremittingly inundates the public with explicit and implicit sexual messages, eroticizing products and using sex to sell. At the same time, participants in political discourse are incredibly squeamish when handling sexual themes, as exemplified in the curious combination of horror and fascination displayed in the public discourse about Long Dong Silver and pubic hairs on pop cans during the Senate hearings in September 1991 on the appointment of Clarence Thomas to the Supreme Court. We suspect, in fact, that with respect to discourse on sexuality there is a major discontinuity between the sensibilities of politicians and other self-appointed guardians of the moral order and those of the public at large, who, on the whole, display few hang-ups in discussing sexual issues in appropriately structured circumstances. This book is a testament to that proposition.

The fact remains that, until quite recently, scientific research on sexuality has been taboo and therefore to be avoided or at best marginalized. While there is a visible tradition of (in)famous sex research, what is, in fact, most striking is how little prior research exists on sexuality in the general population. Aside from the research on adolescence, premarital sex, and problems attendant to sex such as fertility, most research attention seems to have been directed toward those believed to be abnormal, deviant, criminal, perverted, rare, or unusual, toward sexual pathology, dysfunction, and sexually transmitted disease—the label used typically reflecting the way in which the behavior or condition in question is to be regarded. "Normal sex" was somehow off limits, perhaps because it was considered too ordinary, trivial, and self-evident to deserve attention. To be fair, then, we cannot blame the public and the politicians entirely for the lack of sustained work on sexuality at large—it also reflects the prejudices and understandings of researchers about what are "interesting" scientific questions. There has simply been a dearth of mainstream scientific thinking and speculation about sexual issues. We have repeatedly encountered this relative lack of systematic thinking about sexuality to guide us in interpreting and understanding the many findings reported in this book.

. . . In order to understand the results of our survey, the National Health and Social Life Survey (NHSLS), one must understand how these results were generated. To construct a questionnaire and field a large-scale survey, many research design decisions must be made. To understand the decisions made, one needs to understand the multiple purposes that underlie this research project. Research design is never just a theoretical exercise. It is a set of practical solutions to a multitude of problems and considerations that are chosen under the constraints of limited resources of money, time, and prior knowledge.

SAMPLE DESIGN

The sample design for the NHSLS is the most straightforward element of our methodology because nothing about probability sampling is specific to or changes in a survey of sexual behavior. . . .

Probability sampling, that is, sampling where every member of a clearly specified population has a known probability of selection—what lay commentators often somewhat inaccurately call random sampling—is the sine qua non of modern survey research (see Kish 1965, the classic text on the subject). There is no other scientifically acceptable way to construct a representative sample and thereby to be able to generalize from the actual sample on which data are collected to the population that that sample is designed to represent. Probability sampling as practiced in survey research is a highly developed practical application of statistical theory to the problem of selecting a sample. Not only does this type of sampling avoid the problems of bias introduced by the researcher or by subject self-selection bias that come from more casual techniques, but it also allows one to quantify the variability in the estimates derived from the sample.

In order to determine how large a sample size for a given study should be, one must first decide how precise the estimates to be derived need to be. To illustrate this reasoning process, let us take one of the simplest and most commonly used statistics in survey research, the proportion. Many of the most important results reported in this book are proportions. For example, what proportion of the population had more than five sex partners in the last year? What proportion engaged in anal intercourse? With condoms? Estimates based on our sample will differ from the true proportion in the population because of sampling error (i.e., the random fluctuations in our estimates that are due to the fact that they are based on samples rather than on complete enumerations or censuses). If one drew repeated samples using the same methodology, each would produce a slightly different estimate. If one looks at the distribution of these *estimates,* it turns out that they will be normally distributed (i.e., will follow the famous bell-shaped curve known as the Gaussian or normal distribution) and centered around the true proportion in the population. The larger the sample size, the tighter the distribution of estimates will be.

This analysis applies to an estimate of a single proportion based on the whole sample. In deciding the sample size needed for a study, one must consider the subpopulations for which one will want to construct estimates. For example, one almost always wants to know not just a single parameter for the whole population but parameters for subpopulations such as men and women, whites, blacks, and Hispanics, and younger people and older people. Furthermore, one is usually interested in the intersections of these various breakdowns of the population, for example, young black women. The size of the interval estimate for a proportion based on a subpopulation depends on the size of that group in the sample (sometimes called the *base* "N," i.e., the number in the sample on which the estimate is based). It is actually this kind of number that one needs to consider in determining the sample size for a study.

When we were designing the national survey of sexual behavior in the United States for the NICHD, we applied just these sorts of considerations to come to the conclusion that we needed a sample size of about 20,000 people. . . .

GAINING COOPERATION: THE RESPONSE RATE

First, let us consider the cooperation or response rate. No survey of any size and complexity is able to get every sampling-designated respondent to complete an interview. Individuals can have many perfectly valid reasons why they cannot participate in the survey: being too ill, too busy, or always absent when an effort to schedule an interview is made or simply being unwilling to grant an interview. While the face-to-face or in-person survey is considerably more expensive than other techniques, such as mail or telephone surveys, it usually gets the highest response rate. Even so, a face-to-face, household-based survey such as the General Social Survey successfully interviews, on the average, only about 75 percent of the target sample (Davis and Smith 1991). The missing 25 percent pose a serious problem for the reliability and validity of a survey: is there some systematic (i.e., nonrandom) process at work that distinguishes respondents from nonrespondents? That is, if the people who refuse to participate or who can never be reached to be interviewed differ systematically in terms of the issues being researched from those who are interviewed, then one will not have a representative sample of the population from which the sample was drawn. If the respondents and nonrespondents do not differ systematically, then the results will not be affected. Unfortunately, one usually has no (or only minimal) information about nonrespondents. It is thus a challenge to devise ways of evaluating the extent of bias in the selection of respondents and nonrespondents. Experience tells us that, in most well-studied fields in which survey research has been applied, such moderately high response rates as 75 percent do not lead to biased results. And it is difficult and expensive to push response rates much higher than that. Experience suggests that a response rate close to 90 percent may well represent a kind of upper limit.

Because of our subject matter and the widespread skepticism that survey methods would be effective, we set a completion rate of 75 percent as the survey organization's goal. In fact, we did much better than this; our final completion rate was close to 80 percent. We have extensively investigated whether there are detectable participation biases in the final sample. These comparisons are documented in appendixes A and B. To summarize these investigations, we have compared our sample and our results with other surveys of various sorts and have been unable to detect systematic biases of any substantive significance that would lead us to qualify our findings at least with respect to bias due to sampling.

One might well ask what the secret was of our remarkably high response rate, by far the highest of any national sexual behavior survey conducted so far. There is no secret. Working closely with the NORC senior survey and field management team, we proceeded in the same way as one would in any other national area probability survey. We did not scrimp on interviewer training or on securing a highly mobilized field staff that was determined to get respondent participation in a professional and respectful manner. It was an expensive operation: the average cost of a completed interview was approximately $450.

We began with an area probability sample, which is a sample of households, that is, of addresses, not names. Rather than approach a household by knocking on the door without advance warning, we followed NORC's standard practice of sending an advance letter, hand addressed by the interviewer, about a week before the interviewer expected to visit the address. In this case, the letter was signed by the principal investigator, Robert Michael, who was identified as the dean of the Irving B. Harris Graduate School of Public Policy Studies of the University of Chicago. The letter briefly explained the purpose of the survey as helping "doctors, teachers, and counselors better understand and prevent the spread of diseases like AIDS and better understand the nature and extent of harmful and of healthy sexual behavior in our country." The intent was to convince the potential respondent that this was a legitimate scientific study addressing personal and potentially sensitive topics for a socially useful purpose. AIDS was the original impetus for the research, and it certainly seemed to provide a timely justification for the study. But any general purpose approach has drawbacks. One problem that the interviewers frequently encountered was potential respondents who did not think that AIDS affected them and therefore that information about their sex lives would be of little use.

Mode of Administration: Face-to-Face, Telephone, or Self-Administered

Perhaps the most fundamental design decision, one that distinguishes this study from many others, concerned how the interview itself was to be conducted. In survey research, this is usually called the *mode* of interviewing or of questionnaire administration. We chose face-to-face interviewing, the most costly mode, as the primary vehicle for data collection in the NHSLS. What follows is the reasoning behind this decision.

A number of recent sex surveys have been conducted over the telephone, . . . The principal advantage of the telephone survey is its much lower cost. Its major disadvantages are the length and complexity of a questionnaire that can be realistically administered over the telephone and problems of sampling and sample control. . . . The NHSLS, cut to its absolute minimum length, averaged about ninety minutes. Extensive field experience suggests an upper limit of about forty-five minutes for phone interviews of a cross-sectional survey of the population at large. Another disadvantage of phone surveys is that it is more difficult to find people at home by phone and, even once contact has been made, to get them to participate. . . . One further consideration in evaluating the phone as a mode of interviewing is its unknown effect on the quality of responses. Are people more likely to answer questions honestly and candidly or to dissemble on the telephone as opposed to face-to-face? Nobody knows for sure.

The other major mode of interviewing is through self-administered forms distributed either face to face or through the mail.[1] When the survey is conducted by mail, the questions must be self-explanatory, and much prodding is typically required to obtain an acceptable response rate. . . . This procedure has been shown to produce somewhat higher rates of reporting socially undesirable behaviors, such as engaging in criminal acts and substance abuse. We adopted the mixed-mode strategy to a limited extent by using four short, self-administered forms, totaling nine pages altogether, as part of our interview. When filled out, these forms were placed in a "privacy envelope" by the respondent so that the interviewer never saw the answers that were given to these questions. . . .

The fundamental disadvantage of self-administered forms is that the questions must be much simpler in form and language than those that an interviewer can ask. Complex skip patterns must be avoided. Even the simplest skip patterns are usually incorrectly filled out by some respondents on self-administered forms. One has much less control over whether (and therefore much less confidence that) respondents have read and understood the questions on a self-administered form. The NHSLS questionnaire (discussed below) was based on the idea that questions about sexual behavior must be framed as much as possible in the specific contexts of particular patterns and occasions. We found that it is impossible to do this using self-administered questions that are easily and fully comprehensible to people of modest educational attainments.

To summarize, we decided to use face-to-face interviewing as our primary mode of administration of the NHSLS for two principal reasons: it was most likely to yield a substantially higher response rate for a more inclusive cross section of the population at large, and it would permit more complex and detailed questions to be asked. While by far the most expensive approach, such a strategy provides a solid benchmark against which other modes of interviewing can and should be judged. The main unresolved question is whether another mode has an edge over face-to-face interviewing when highly sensitive questions likely to be upsetting or threatening to the respondent are being asked. As a partial control and test of this question, we have asked a number of sensitive questions in both formats so that an individual's responses can be systematically compared. . . . Suffice it to say at this point that there is a stunning consistency in the responses secured by the different modes of administration.

Recruiting and Training Interviewers

Gaining respondents' cooperation requires mastery of a broad spectrum of techniques that successful interviewers develop with experience, guidance from the research team, and careful field supervision. This project required extensive training before entering the field. While interviewers are generally trained to be neutral toward topics covered in the interview, this was especially important when discussing sex, a topic that seems particularly likely to elicit emotionally freighted sensitivities both in the respondents and in the interviewers. Interviewers needed to be fully persuaded about the legitimacy and importance of the research. Toward this end, almost a full day of training was devoted to presentations and discussions with the principal investigators in addition to the extensive advance study materials to read and comprehend. Sample answers to frequently asked questions by skeptical respondents and brainstorming about strategies to convert reluctant respondents were part of the training exercises. A set of endorsement letters from prominent local and national notables and refusal conversion letters were also provided to interviewers. A hotline to the research office at the University of Chicago was set up to allow potential respondents to call in with their concerns. Concerns ranged from those about the legitimacy of the survey, most fearing that it was a commercial ploy to sell them something, to fears that the interviewers were interested in robbing them. Ironically, the fact that the interviewer initially did not know the name of the respondent (all he or she knew was the address) often led to behavior by the interviewer that appeared suspicious to the respondent. For example, asking neighbors for the name of the family in the selected household and/or questions about when the potential respondent was likely to be home induced worries that had to be assuaged. Another major concern was confidentiality—respondents wanted to know how they had come to be selected and how their answers were going to be kept anonymous.

THE QUESTIONNAIRE

The questionnaire itself is probably the most important element of the study design. It determines the content and quality of the information gathered for analysis. Unlike issues related to sample design, the construction of a questionnaire is driven less by technical precepts and more by the concepts and ideas motivating the research. It demands even more art than applied sampling design requires.

Before turning to the specific forms that this took in the NHSLS, we should first discuss several general problems that any survey questionnaire must address. The essence of survey research is to ask a large sample of people from a defined population the *same set of questions*. To do this in a relatively short period of time, many interviewers are needed. In our case, about 220 interviewers from all over the country collected the NHSLS data. The field period, beginning on 14 February 1992 and ending in September, was a time in which over 7,800 households were contacted (many of which turned out to be ineligible for the study) and 3,432 interviews were completed. Central to this effort was gathering comparable information on the same attributes from each and every one of these respondents. The attributes measured by the questionnaire become the variables used in the data analysis. They range from demographic characteristics (e.g., gender, age, and race/ethnicity) to sexual experience measures (e.g., numbers of sex partners in given time periods, frequency of particular practices, and timing of various sexual events) to measures of mental states (e.g., attitudes toward premarital sex, the appeal of particular techniques like oral sex, and levels of satisfaction with particular sexual relationships).

Very early in the design of a national sexual behavior survey, in line with our goal of not reducing this research to a simple behavioral risk inventory, we faced the issue of where to draw the boundaries in defining the behavioral domain that would be encompassed by the concept of sex. This was particularly crucial in defining sexual activity that would lead to the enumeration of a set of sex partners. There are a number of activities that commonly serve as markers for sex and the status of sex partner, especially intercourse and orgasm. While we certainly wanted to include these events and their extent in given relationships and events, we also felt that using them to define and ask about sexual activity might exclude transactions or partners that should be included. Since the common meaning and uses of the term *intercourse* involve the idea of the intromission of a penis, intercourse in that sense as a defining act would at the very least exclude a sexual relationship between two women. There are also many events that we would call sexual that may not involve orgasm on the part of either or both partners.

Another major issue is what sort of language is appropriate in asking questions about sex. It seemed obvious that one should avoid highly technical language because it is unlikely to be understood by many people. One tempting alternative is to use colloquial language and even slang since that is the only language that some people ever use in discussing sexual matters. There is even some evidence that one can improve reporting somewhat by allowing respondents to select their own preferred terminology (Blair et al. 1977; Bradburn et al. 1978; Bradburn and Sudman 1983). Slang and other forms of colloquial speech, however, are likely to be problematic in several ways. First, the use of slang can produce a tone in the interview that is counterproductive because it downplays the distinctiveness of the interviewing situation itself. An essential goal in survey interviewing, especially on sensitive topics like sex, is to create a neutral, nonjudgmental, and confiding atmosphere and to maintain a certain professional distance between the interviewer and the respondent. A key advantage that the interviewer has in initiating a topic for discussion is being a stranger or an outsider who is highly unlikely to come in contact with the respondent again. It is not intended that a longer-term bond between the interviewer and the respondent be formed, whether as an advice giver or a counselor or as a potential sex partner.[2]

The second major shortcoming of slang is that it is highly variable across class and education levels, ages, regions, and other social groupings. It changes meanings rapidly and is often imprecise. Our solution was to seek the simplest possible language—standard English—that was neither colloquial nor highly technical. For example, we chose to use the term *oral sex* rather than the slang *blow job* and *eating pussy* or the precise technical but unfamiliar terms *fellatio* and *cunnilingus*. Whenever possible, we provided definitions when terms were first introduced in a questionnaire—that is, we tried to train our respondents to speak about sex in our terms. Many terms that seemed clear to us may not, of course, be universally understood; for example, terms like *vaginal* or *heterosexual* are not understood very well by substantial portions of the population. Coming up with simple and direct speech was quite a challenge because most of the people working on the questionnaire were highly educated, with strong inclinations toward the circumlocutions and indirections of middle-class discourse on sexual themes. Detailed reactions from field interviewers and managers and extensive pilot testing with a broad cross section of recruited subjects helped minimize these language problems.

ON PRIVACY, CONFIDENTIALITY, AND SECURITY

Issues of respondent confidentiality are at the very heart of survey research. The willingness of respondents to report their views and experiences fully and honestly depends on the rationale offered for why the study is important and on the assurance that the information provided will be treated as confidential. We offered respondents a strong rationale for the study, our interviewers made great efforts to conduct the interview in a manner that protected respondents' privacy, and we went to great lengths to honor the assurances that the information would be treated confidentially. The subject matter of the NHSLS makes the issues of confidentiality especially salient and problematic because there are so many easily imagined ways in which information voluntarily disclosed in an interview might be useful to interested parties in civil and criminal cases involving wrongful harm, divorce proceedings, criminal behavior, or similar matters.

NOTES

1. We ruled out the idea of a mail survey because its response rate is likely to be very much lower than any other mode of interviewing (see Bradburn, Sudman, et al. 1979).

2. Interviewers are not there to give information or to correct misinformation. But such information is often requested in the course of an interview. Interviewers are given training in how to avoid answering such questions (other than clarification of the meaning of particular questions). They are not themselves experts on the topics raised and often do not know the correct answers to questions. For this reason, and also in case emotionally freighted issues for the respondent were raised during the interview process, we provided interviewers with a list of toll-free phone

numbers for a variety of professional sex- and health-related referral services (e.g., the National AIDS Hotline, an STD hotline, the National Child Abuse Hotline, a domestic violence hotline, and the phone number of a national rape and sexual assault organization able to provide local referrals).

REFERENCES

Blair, Ellen, Seymour Sudman, Norman M. Bradburn, and Carol Stacking. 1977. "How to Ask Questions About Drinking and Sex: Response Effects in Measuring Consumer Behavior." *Journal of Marketing Research* 14: 316–321.

Bradburn, Norman M., and Seymour Sudman. 1983. *Asking Questions: A Practical Guide to Questionnaire Design.* San Francisco: Jossey-Bass.

Bradburn, Norman M., Seymour Sudman, Ed Blair, and Carol Stacking. 1979. *Improving Interview Method and Design.* San Francisco: Jossey-Bass.

Davis, James Allan, and Tom W. Smith. 1991. *General Social Surveys, 1972–1991: Cumulative Codebook.* Chicago: National Opinion Research Center.

Kish, Leslie. 1965. *Survey Sampling.* New York: Wiley.

13

Researching Dealers and Smugglers

PATRICIA A. ADLER

Adler offers us a glimpse of what it is like to carry out participant observation research with a deviant group in this description of her study of upper-level drug traffickers. This natural history carefully explains the process used in field research, the relationships formed with setting members, and the feelings researchers experience. We see that when researchers place themselves inside a deviant world it can profoundly affect them as well as cause them serious potential dangers, but we also see that only from this vantage point can they fully comprehend the forces at play in deviant worlds. This article offers readers a greater understanding of some of the research roles, research concerns, and problems and issues that may arise in field research.

I strongly believe that investigative field research (Douglas 1976), with emphasis on direct personal observation, interaction, and experience, is the only way to acquire accurate knowledge about deviant behavior. Investigative techniques are especially necessary for studying groups such as drug dealers and smugglers because the highly illegal nature of their occupation makes them secretive, deceitful, mistrustful, and paranoid. To insulate themselves from the straight world, they construct multiple false fronts, offer lies and misinformation, and withdraw into their group. In fact, detailed, scientific information about upper-level drug dealers and smugglers is lacking precisely because of the difficulty sociological researchers have had in penetrating into their midst. As a result, the only way I could possibly get close enough to these individuals to discover what they were doing and to understand their world from their perspectives (Blumer 1969) was to take a membership role in the setting. While my different values and goals precluded my becoming converted to complete membership in the subculture, and my fears prevented my ever becoming "actively" involved in their trafficking activities, I was able to assume a "peripheral" membership role (Adler and Adler 1987). I became a member of the dealers' and smugglers' social world and participated in their daily activities on that basis. In this chapter, I discuss how I gained access to

---

From Patricia A. Adler, *Wheeling and Dealing* (New York: Columbia University Press, 1985). © Columbia University Press. Reprinted by permission of the publisher.

this group, established research relations with members, and how personally involved I became in their activities.

GETTING IN

When I moved to Southwest County [California] in the summer of 1974, I had no idea that I would soon be swept up in a subculture of vast drug trafficking and unending partying, mixed with occasional cloak-and-dagger subterfuge. I had moved to California with my husband, Peter, to attend graduate school in sociology. We rented a condominium townhouse near the beach and started taking classes in the fall. We had always felt that socializing exclusively with academicians left us nowhere to escape from our work, so we tried to meet people in the nearby community. One of the first friends we made was our closest neighbor, a fellow in his late twenties with a tall, hulking frame and gentle expression. Dave, as he introduced himself, was always dressed rather casually, if not sloppily, in T-shirts and jeans. He spent most of his time hanging out or walking on the beach with a variety of friends who visited his house, and taking care of his two young boys, who lived alternately with him and his estranged wife. He also went out of town a lot. We started spending much of our free time over at his house, talking, playing board games late into the night, and smoking marijuana together. We were glad to find someone from whom we could buy marijuana in this new place, since we did not know too many people. He also began treating us to a fairly regular supply of cocaine, which was a thrill because this was a drug we could rarely afford on our student budgets. We noticed right away, however, that there was something unusual about his use and knowledge of drugs: while he always had a plentiful supply and was fairly expert about marijuana and cocaine, when we tried to buy a small bag of marijuana from him he had little idea of the going price. This incongruity piqued our curiosity and raised suspicion. We wondered if he might be dealing in larger quantities. Keeping our suspicions to ourselves, we began observing Dave's activities a little more closely. Most of his friends were in their late twenties and early thirties and, judging by their lifestyles and automobiles, rather wealthy. They came and left his house at all hours, occasionally extending their parties through the night and the next day into the following night. Yet throughout this time we never saw Dave or any of his friends engage in any activity that resembled a legitimate job. In most places this might have evoked community suspicion, but few of the people we encountered in Southwest County seemed to hold traditionally structured jobs. Dave, in fact, had no visible means of financial support. When we asked him what he did for a living, he said something vague about being a real estate speculator, and we let it go at that. We never voiced our suspicions directly since he chose not to broach the subject with us.

We did discuss the subject with our mentor, Jack Douglas, however. He was excited by the prospect that we might be living among a group of big dealers, and urged us to follow our instincts and develop leads into the group. He knew that the local area was rife with drug trafficking, since he had begun a life history case study of two drug dealers with another graduate student several years previously. That earlier study was aborted when the graduate student quit school, but Jack still had many hours of taped interviews he had conducted with them, as well as an interview that he had done with an undergraduate student who had known the two dealers independently, to serve as a cross-check on their accounts. He therefore encouraged us to become friendlier with Dave and his friends. We decided that if anything did develop out of our observations of Dave, it might make a nice paper for a field methods class or independent study.

Our interests and background made us well suited to study drug dealing. First, we had already done research in the field of drugs. As undergraduates at Washington University we had participated in a nationally funded project on urban heroin use (see Cummins et al. 1972). Our role in the study involved using fieldwork techniques to investigate the extent of heroin use and distribution in St. Louis. In talking with heroin users, dealers, and rehabilitation personnel, we acquired a base of knowledge about the drug world and the subculture of drug trafficking. Second, we had a generally open view toward soft drug use, considering moderate consumption of marijuana and cocaine to be generally nondeviant. This outlook was partially etched by our 1960s-formed attitudes, as we had first been introduced to drug use in an environment of communal friendship, sharing, and counterculture ideology. It also partially reflected the widespread acceptance accorded to marijuana and cocaine use in the surrounding local culture. Third, our age (mid-twenties at the start of the study) and general appearance gave us compatibility with most of the people we were observing.

We thus watched Dave and continued to develop our friendship with him. We also watched his friends and got to know a few of his more regular visitors. We continued to build friendly relations by doing, quite naturally, what Becker (1963), Polsky (1969), and Douglas (1972) had advocated for the early stages of field research: we gave them a chance to know us and form judgments about our trustworthiness by jointly pursuing those interests and activities which we had in common.

Then one day something happened which forced a breakthrough in the research. Dave had two guys visiting him from out of town and, after snorting quite a bit of cocaine, they turned their conversation to a trip they had just made from Mexico, where they piloted a load of marijuana back across the border in a small plane. Dave made a few efforts to shift the conversation to another subject, telling them to "button their lips," but they apparently thought that he was joking. They thought that anybody as close to Dave as we seemed to be undoubtedly knew the nature of his business. They made further allusions to his involvement in the operation and discussed the outcome of the sale. We could feel the wave of tension and awkwardness from Dave when this conversation began, as he looked toward us to see if we understood the implications of what was being said, but then he just shrugged it off as done. Later, after the two guys left, he discussed with us what happened. He admitted to us that he was a member of a smuggling crew and a major marijuana dealer on the side. He said that he knew he could trust us, but that it was his practice to say as little as possible to outsiders about his activities. This inadvertent slip, and Dave's subsequent opening up, were highly significant in forging our entry into Southwest County's drug world. From then on he was open in discussing the nature of his dealing and smuggling activities with us.

He was, it turned out, a member of a smuggling crew that was importing a ton of marijuana weekly and 40 kilos of cocaine every few months. During that first winter and spring, we observed Dave at work and also got to know the other members of his crew, including Ben, the smuggler himself. Ben was also very tall and broad shouldered, but his long black hair, now flecked with gray, bespoke his earlier membership in the hippie subculture. A large physical stature, we observed, was common to most of the male participants involved in this drug community. The women also had a unifying physical trait: they were extremely attractive and stylishly dressed. This included Dave's ex-wife, Jean, with whom he reconciled during the spring. We therefore became friendly with Jean and through her met a number of women ("dope chicks") who hung around the dealers and smugglers. As we continued to gain the friendship of Dave and Jean's associates we were progressively admitted into their inner circle and apprised of each person's dealing or smuggling role.

Once we realized the scope of Ben's and his associates' activities, we saw the enormous research potential in studying them. This scene was different from any analysis of drug trafficking that we had read in the sociological literature because of the amounts they were dealing and the fact that they were importing it themselves. We decided that, if it was at all possible, we would capitalize on this situation, to "opportunistically" (Riemer 1977) take advantage of our prior expertise and of the knowledge, entrée, and rapport we had already developed with several key people in this setting. We therefore discussed the idea of doing a study of the general subculture with Dave and several of his closest friends (now becoming our friends). We assured them of the anonymity, confidentiality, and innocuousness of our work. They were happy to reciprocate our friendship by being of help to our professional careers. In fact, they basked in the subsequent attention we gave their lives.

We began by turning first Dave, then others, into key informants and collecting their life histories in detail. We conducted a series of taped, depth interviews with an unstructured, open-ended format. We questioned them about such topics as their backgrounds, their recruitment into the occupation, the stages of their dealing careers, their relations with others, their motivations, their lifestyle, and their general impressions about the community as a whole.

We continued to do taped interviews with key informants for the next six years until 1980, when we moved away from the area. After that, we occasionally did follow-up interviews when we returned for vacation visits. These later interviews focused on recording the continuing unfolding of events and included detailed probing into specific conceptual areas, such as dealing networks, types of dealers, secrecy, trust, paranoia, reputation, the law, occupational mobility, and occupational stratification. The number of taped interviews we did with each key informant varied, ranging between 10 and 30 hours of discussion.

Our relationship with Dave and the others thus took on an added dimension—the research relationship. As Douglas (1976), Henslin (1972), and Wax (1952) have noted, research relationships involve some form of mutual exchange. In our case, we offered everything that friendship could entail. We did routine favors for them in the course of our everyday lives, offered them insights and advice about their lives from the perspective of our more respectable position, wrote letters on their behalf to the authorities when they got in trouble, testified as character witnesses at their non-drug-related trials, and loaned them money when they were down and out. When Dave was arrested and brought to trial for check-kiting, we helped Jean organize his defense and raise the money to pay his fines. We spelled her in taking care of the children so that she could work on his behalf. When he was eventually sent to the state prison we maintained close ties with her and discussed our mutual efforts to buoy Dave up and secure his release. We also visited him in jail. During Dave's incarceration, however, Jean was courted by an old boyfriend and gave up her reconciliation with Dave. This proved to be another significant turning point in our research because, desperate for money, Jean looked up Dave's old dealing connections and went into the business herself. She did not stay with these marijuana dealers and smugglers for long, but soon moved into the cocaine business. Over the next several years her experiences in the world of cocaine dealing brought us into contact with a different group of people. While these people knew Dave and his associates (this was very common in the Southwest County dealing and smuggling community), they did not deal with them directly. We were thus able to gain access to a much wider and more diverse range of subjects than we would have had she not branched out on her own.

Dave's eventual release from prison three months later brought our involvement in the research to an even deeper level. He was broke and had nowhere to go. When he showed up on our doorstep, we took him in. We offered to let him stay with us until he was back on his feet again and could afford a place of his own. He lived with us for seven months, intimately sharing his daily experiences with us. During this time we witnessed, firsthand, his transformation from a scared ex-con who would never break the law again to a hard-working legitimate employee who only dealt to get money for his children's Christmas presents, to a full-time dealer with no pretensions at legitimate work. Both his process of changing attitudes and the community's gradual reacceptance of him proved very revealing.

We socialized with Dave, Jean, and other members of Southwest County's dealing and smuggling community on a near-daily basis, especially during the first four years of the research (before we had a child). We worked in their legitimate businesses, vacationed together, attended their weddings, and cared for their children. Throughout their relationship with us, several participants became co-opted to the researcher's perspective[1] and actively sought out instances of behavior which filled holes in the conceptualizations we were developing. Dave, for one, became so intrigued by our conceptual dilemmas that he undertook a "natural experiment" entirely on his own, offering an unlimited supply of drugs to a lower-level dealer to see if he could work up to higher levels of dealing, and what factors would enhance or impinge upon his upward mobility.

In addition to helping us directly through their own experiences, our key informants aided us in widening our circle of contacts. For instance, they let us know when someone in whom we might be interested was planning on dropping by, vouching for our trustworthiness and reliability as friends who could be included in business conversations. Several times we were even awakened in the night by phone calls informing us that someone had dropped by for a visit, should we want to "casually" drop over too. We rubbed the sleep from our eyes, dressed, and walked or drove over, feeling like sleuths out of a television series. We thus were able to snowball, through the active efforts of our key informants,[2] into an expanded study population. This was supplemented by our own efforts to cast a research net and befriend other dealers, moving from contact to contact slowly and carefully through the domino effect.

THE COVERT ROLE

The highly illegal nature of dealing in illicit drugs and dealers' and smugglers' general level of suspicion made the adoption of an overt research role highly sensitive and problematic. In discussing this issue with our key informants, they all agreed that we should be extremely discreet (for both our sakes and theirs). We carefully approached new individuals before we admitted that we were studying them. With many of these people, then, we took a covert posture in the research setting. As nonparticipants in the business activities which bound members together into the group, it was difficult to become fully accepted as peers. We therefore tried to establish some sort of peripheral, social membership in the general crowd, where we could be accepted as "wise" (Goffman 1963) individuals and granted a courtesy membership. This seemed an attainable goal, since we had begun our involvement by forming such relationships with our key informants. By being introduced to others in this wise rather than overt role, we were able to interact with people who would otherwise have shied away from us. Adopting a courtesy membership caused us to bear a courtesy stigma,[3] however, and we suffered since we, at times, had to disguise the nature of our research from both lay outsiders and academicians.

In our overt posture we showed interest in dealers' and smugglers' activities, encouraged them to talk about themselves (within limits, so as to avoid acting like narcs), and ran home to write field notes. This role offered us the advantage of gaining access to unapproachable people while avoiding researcher effects, but it prevented us from asking some necessary, probing questions and from tape recording conversations.[4] We therefore sought, at all times, to build toward a conversion to the overt role. We did this by working to develop their trust.

DEVELOPING TRUST

Like achieving entrée, the process of developing trust with members of unorganized deviant groups can be slow and difficult. In the absence of a formal structure separating members from outsiders, each individual must form his or her own judgment about whether new persons can be admitted to their confidence. No gatekeeper existed to smooth our path to being trusted, although our key informants acted in this role whenever they could by providing introductions and references. In addition, the unorganized nature of this group meant that we met people at different times and were constantly at different levels in our developing relationships with them. We were thus trusted more by some people than by others, in part because of their greater familiarity with us. But as Douglas (1976) has noted, just because someone knew us or even liked us did not automatically guarantee that they would trust us.

We actively tried to cultivate the trust of our respondents by tying them to us with favors. Small things, like offering the use of our phone, were followed with bigger favors, like offering the use of our car, and finally really meaningful favors, like offering the use of our home. Here we often trod a thin line, trying to ensure our personal safety while putting ourselves in enough of a risk position, along with our research subjects, so that they would trust us. While we were able to build a "web of trust" (Douglas 1976) with some members, we found that trust, in large part, was not a simple status to attain in the drug world. Johnson (1975) has pointed out that trust is not a one-time phenomenon, but an ongoing developmental process. From my experiences in this research I would add that it cannot be simply assumed to be a one-way process either, for it can be diminished, withdrawn, reinstated to varying degrees, and re-questioned at any point. Carey (1972) and Douglas (1972) have remarked on this waxing and waning process, but it was especially pronounced for us because our subjects used large amounts of cocaine over an extended period of time. This tended to make them alternately warm and cold to us. We thus lived through a series of ups and downs with the people we were trying to cultivate as research informants.

THE OVERT ROLE

After this initial covert phase, we began to feel that some new people trusted us. We tried to intuitively feel when the time was right to approach them and go overt. We used two means of approaching people to inform them that we were involved in a study of dealing and smuggling: direct and indirect. In some cases our key informants approached their friends or connections and, after vouching for our absolute trustworthiness, convinced these associates to talk to us. In other instances, we approached people directly, asking for their help with our project. We worked our way through a progression with these secondary contacts, first discussing the dealing scene overtly and later moving to taped life history interviews. Some people reacted well to us, but others responded skittishly, making appointments to do taped interviews only to break them as the day drew near, and going through fluctuating stages of being honest with us or putting up fronts about their dealing activities. This varied, for some, with their degree of active involvement in the business. During the times when they had quit dealing, they would tell us about their present and past activities, but when they became actively involved again, they would hide it from us.

This progression of covert to overt roles generated a number of tactical difficulties. The first was the problem of *coming on too fast* and blowing it. Early in the research we had a dealer's old lady (we thought) all set up for the direct approach. We knew many dealers in common and had discussed many things tangential to dealing with her without actually mentioning the subject. When we asked her to do a taped interview of her bohemian lifestyle, she agreed without hesitation. When the interview began, though, and she found out why we were interested in her, she balked, gave us a lot of incoherent jumble, and ended the session as quickly as possible. Even though she lived only three houses away we never saw her again. We tried to move more slowly after that.

A second problem involved simultaneously *juggling our overt and covert roles* with different people. This created the danger of getting our cover blown with people who did not know about our research (Henslin 1972). It was very confusing to separate the people who knew about our study from those who did not, especially in the minds of our informants. They would make occasional veiled references in front of people, especially when loosened by intoxicants, that made us extremely uncomfortable. We also frequently worried that our snooping would someday be mistaken for police tactics. Fortunately, this never happened.

CROSS-CHECKING

The hidden and conflictual nature of the drug dealing world made me feel the need for extreme certainty about the reliability of my data. I therefore based all my conclusions on independent sources and accounts that we carefully verified. First, we tested information against our own common sense and general knowledge of the scene. We adopted a hard-nosed attitude of suspicion, assuming people were up to more than they would originally admit. We kept our attention especially riveted on "reformed" dealers and smugglers who were living better than they could outwardly afford, and were thereby able to penetrate their public fronts.

Second, we checked out information against a variety of reliable sources. Our own observations of the scene formed a primary reliable source, since we were involved with many of the principals on a daily basis and knew exactly what they were doing. Having Dave live with us was a particular advantage because we could contrast his statements to us with what we could clearly see was happening. Even after he moved out, we knew him so well that we could generally tell when he was lying to us or, more commonly, fooling himself with optimistic dreams. We also observed other dealers' and smugglers' evasions and misperceptions about themselves and their activities. These usually occurred when they broke their own rules by selling to people they did not know, or when they commingled other people's money with their own. We also cross-checked our data against independent, alternative accounts. We were lucky, for this purpose, that Jean got reinvolved in the drug world. By interviewing her, we gained additional insight into Dave's past, his early dealing and smuggling activities, and his ongoing involvement from another person's perspective. Jean (and her connections) also talked to us about Dave's associates, thereby helping us to validate or disprove their statements. We even used this pincer effect to verify information about people we had never directly interviewed. This occurred, for instance, with the tapes that Jack Douglas gave us from his earlier study. After doing our first round of taped interviews with Dave, we discovered that he knew the dealers Jack had interviewed. We were excited by the prospect of finding out what had happened to these people and if their earlier stories checked out. We therefore sent Dave to do some investigative work. Through some mutual friends he got back in touch with them and found out what they had been doing for the past several years.

Finally, wherever possible, we checked out accounts against hard facts: newspaper and magazine reports; arrest records; material possessions; and visible evidence. Throughout the research, we used all these cross-checking measures to evaluate the veracity of new information and to prod our respondents to be more accurate (by abandoning both their lies and their self-deceptions).[5]

After about four years of near-daily participant observation, we began to diminish our involvement in the research. This occurred gradually, as first pregnancy and then a child hindered our ability to follow the scene as intensely and spontaneously as we had before. In addition, after having a child, we were less willing to incur as many risks as we had before; we no longer felt free to make decisions based solely on our own welfare. We thus pulled back from what many have referred to as the "difficult hours and dangerous situations" inevitably present in field research on deviants (see Becker 1963; Carey 1972; Douglas 1972). We did, however, actively maintain close ties with research informants (those with whom we had gone overt), seeing them regularly and periodically doing follow-up interviews.

PROBLEMS AND ISSUES

Reflecting on the research process, I have isolated a number of issues which I believe merit additional discussion. These are rooted in experiences which have the potential for greater generic applicability.

The first is the *effect of drugs on the data-gathering process.* Carey (1972) has elaborated on some of the problems he encountered when trying to interview respondents who used amphetamines, while Wax (1952, 1957) has mentioned the difficulty of trying to record field notes while drinking sake. I found that marijuana and cocaine had nearly opposite effects from each other. The latter helped the interview process, while the former hindered it. Our attempts to interview respondents who were stoned on marijuana were unproductive for a number of reasons. The primary obstacle was the effects of the drug. Often, people became confused, sleepy, or involved in eating to varying degrees. This distracted them from our purpose. At times, people even simulated overreactions to marijuana to hide behind the drug's supposed disorienting influence and thereby avoid divulging information. Cocaine, in contrast, proved to be a research aid. The drug's warming and sociable influence opened people up, diminished their inhibitions, and generally increased their enthusiasm for both the interview experience and us.

A second problem I encountered involved *assuming risks while doing research*. As I noted earlier, dangerous situations are often generic to research on deviant behavior. We were most afraid of the people we studied. As Carey (1972), Henslin (1972), and Whyte (1955) have stated, members of deviant groups can become hostile toward a researcher if they think that they are being treated wrongfully. This could have happened at any time from a simple occurrence, such as a misunderstanding, or from something more serious, such as our covert posture being exposed. Because of the inordinate amount of drugs they consumed, drug dealers and smugglers were particularly volatile, capable of becoming malicious toward each other or us with little warning. They were also likely to behave erratically owing to the great risks they faced from the police and other dealers. These factors made them moody, and they vacillated between trusting us and being suspicious of us.

At various times we also had to protect our research tapes. We encountered several threats to our collection of taped interviews from people who had granted us these interviews. This made us anxious, since we had taken great pains to acquire these tapes and felt strongly about maintaining confidences entrusted to us by our informants. When threatened, we became extremely frightened and shifted the tapes between different hiding places. We even ventured forth one rainy night with our tapes packed in a suitcase to meet a person who was uninvolved in the research at a secret rendezvous so that he could guard the tapes for us.

We were fearful, lastly, of the police. We often worried about local police or drug agents discovering the nature of our study and confiscating or subpoenaing our tapes and field notes. Sociologists have no privileged relationship with their subjects that would enable us legally to withhold evidence from the authorities should they subpoena it.[6] For this reason we studiously avoided any publicity about the research, even holding back on publishing articles in scholarly journals until we were nearly ready to move out of the setting. The closest we came to being publicly exposed as drug researchers came when a former sociology graduate student (turned dealer, we had heard from inside sources) was arrested at the scene of a cocaine deal. His lawyer wanted us to testify about the dangers of doing drug-related research, since he was using his research status as his defense. Fortunately, the crisis was averted when his lawyer succeeded in suppressing evidence and had the case dismissed before the trial was to have begun. Had we been exposed, however, our respondents would have acquired guilt by association through their friendship with us.

Our fear of the police went beyond our concern for protecting our research subjects, however. We risked the danger of arrest ourselves through our own violations of the law. Many sociologists (Becker 1963; Carey 1972; Polsky 1969; Whyte 1955) have remarked that field researchers studying deviance must inevitably break the law in order to acquire valid participant observation data. This occurs in its most innocuous form from having "guilty knowledge": information about crimes that are committed. Being aware of major dealing and smuggling operations made us an accessory to their commission, since we failed to notify the police. We broke the law, secondly, through our "guilty observations," by being present at the scene of a crime and witnessing its occurrence (see also Carey 1972). We knew it was possible to get caught in a bust involving others, yet buying and selling was so pervasive that to leave every time it occurred would have been unnatural and highly suspicious. Sometimes drug transactions even occurred in our home, especially when Dave was living there, but we finally had to put a stop to that because we could not handle the anxiety. Lastly, we broke the law through our "guilty actions," by taking part in illegal behavior ourselves. Although we never dealt drugs (we were too scared to be seriously tempted), we consumed drugs and possessed them in small quantities. Quite frankly, it would have been impossible for a nonuser to have gained access to this group to gather the data presented here. This was the minimum involvement necessary to obtain even the courtesy membership we achieved. Some kind of illegal action was also found to be a necessary or helpful component of the research by Becker (1963), Carey (1972), Johnson (1975), Polsky (1969), and Whyte (1955).

Another methodological issue arose from the *cultural clash between our research subjects and ourselves.* While other sociologists have alluded to these kinds of differences (Humphreys 1970, Whyte 1955), few have discussed how the research relationships affected them. Relationships with research subjects are unique because they involve a bond of intimacy between persons who might not ordinarily associate together, or who might otherwise be no more than casual friends. When fieldworkers undertake a major project, they commit themselves to maintaining a long-term relationship with the people they study. However, as researchers try to get depth involvement, they are apt to come across fundamental differences in character, values, and attitudes between their subjects and themselves. In our case, we were most strongly confronted by differences in present versus future orientations, a desire for risk versus security, and feelings of spontaneity versus self-discipline. These differences often caused us great frustration. We repeatedly saw dealers act irrationally, setting themselves up for failure. We wrestled with our desire to point out their patterns of foolhardy behavior and offer advice, feeling competing pulls between our detached, observer role which advised us not to influence the natural setting, and our involved, participant role which called for us to offer friendly help whenever possible.[7]

Each time these differences struck us anew, we gained deeper insights into our core, existential selves. We suspended our own taken-for-granted feelings and were able to reflect on our culturally formed attitudes, character, and life choices from the perspective of the other. When comparing how we might act in situations faced by our respondents, we realized where our deepest priorities lay. These revelations had the effect of changing our self-conceptions: whereas we, at one time, had thought of ourselves as what Rosenbaum (1981) has called "the hippest of non-addicts" (in this case nondealers), we were suddenly faced with being the straightest members of the crowd. Not only did we not deal, but we had a stable, long-lasting marriage and family life, and needed the security of a reliable monthly paycheck. Self-insights thus emerged as one of the unexpected outcomes of field research with members of a different cultural group.

The final issue I will discuss involved the various *ethical problems* which arose during this research. Many fieldworkers have encountered ethical dilemmas or pangs of guilt during the course of their research experiences (Carey 1972; Douglas 1976; Humphreys 1970; Johnson 1975; Klockars 1977, 1979; Rochford 1985). The researchers' role in the field makes this necessary because they can never fully align themselves with their subjects while maintaining their identity and personal commitment to the scientific community. Ethical dilemmas, then, are directly related to the amount of deception researchers use in gathering the data, and the degree to which they have accepted such acts as necessary and therefore neutralized them.

Throughout the research, we suffered from the burden of intimacies and confidences. Guarding secrets which had been told to us during taped interviews was not always easy or pleasant. Dealers occasionally revealed things about themselves or others that we had to pretend not to know when interacting with their close associates. This sometimes meant that we had to lie or build elaborate stories to cover for some people. Their fronts therefore became our fronts, and we had to weave our own web of deception to guard their performances. This became especially disturbing during the writing of the research report, as I was torn by conflicts between using details to enrich the data and glossing over description to guard confidences.[8]

Using the covert research role generated feelings of guilt, despite the fact that our key informants deemed it necessary, and thereby condoned it. Their own covert experiences were far more deeply entrenched than ours, being a part of their daily existence with non-drug world members. Despite the universal presence of covert behavior throughout the setting, we still felt a sense of betrayal every time we ran home to write research notes on observations we had made under the guise of innocent participants.

We also felt guilty about our efforts to manipulate people. While these were neither massive nor grave manipulations, they involved courting people to procure information about them. Our aggressively friendly postures were based on hidden ulterior motives: we did favors for people with the clear expectation that they could only pay us back with research assistance. Manipulation bothered us in two ways: immediately after it was done, and over the long run. At first, we felt awkward, phony, almost ashamed of ourselves, although we believed our rationalization that the end justified the means. Over the long run, though, our feelings were different. When friendship became intermingled with research goals, we feared that people would later look back on our actions and feel we were exploiting their friendship merely for the sake of our research project.

The last problem we encountered involved our feelings of whoring for data. At times, we felt that we were being exploited by others, that we were putting more into the relationship than they, that they were taking us for granted or using us. We felt that some people used a double standard in their relationship with us: they were allowed to lie to us, borrow money and not repay it, and take advantage of us, but we were at all times expected to behave honorably. This was undoubtedly an outgrowth of our initial research strategy where we did favors for people and expected little in return. But at times this led to our feeling bad. It made us feel like we were selling ourselves, our sincerity, and usually our true friendship, and not getting treated right in return.

CONCLUSIONS

The aggressive research strategy I employed was vital to this study. I could not just walk up to strangers and start hanging out with them as Liebow (1967) did, or be sponsored to a member of this group by a social service or reform organization as Whyte (1955) was, and expect to be accepted, let alone welcomed. Perhaps such a strategy might have worked with a group that had nothing to hide, but I doubt it. Our modern, pluralistic society is so filled with diverse subcultures whose interests compete or conflict with each other that each subculture has a set of knowledge which is reserved exclusively for insiders. In order to serve and prosper, they do not ordinarily show this side to just anyone. To obtain the kind of depth insight and information I needed, I had to become like the members in certain ways. They dealt only with people they knew and trusted, so I had to become known and trusted before I could reveal my true self and my research interests. Confronted with secrecy, danger, hidden alliances, misrepresentations, and unpredictable changes of intent, I had to use a delicate combination of overt and covert roles. Throughout, my deliberate cultivation of the norm of reciprocal exchange enabled me to trade my friendship for their knowledge, rather than waiting for the highly unlikely event that information would be delivered into my lap. I thus actively built a web of research contacts, used them to obtain highly sensitive data, and carefully checked them out to ensure validity.

Throughout this endeavor I profited greatly from the efforts of my husband, Peter, who served as an equal partner in this team field research project. It would have been impossible for me to examine this social world as an unattached female and not fall prey to sex role stereotyping which excluded women from business dealings. As a couple, our different genders allowed us to relate in different ways to both men and women (see Warren and Rasmussen 1977). We also protected each other when we entered the homes of dangerous characters, buoyed each other's initiative and courage, and kept the conversation going when one of us faltered. Conceptually, we helped each other keep a detached and analytical eye on the setting, provided multiperspectival insights, and corroborated, clarified, or (most revealingly) contradicted each other's observations and conclusions.

Finally, I feel strongly that to ensure accuracy, research on deviant groups must be conducted in the settings where it naturally occurs. As Polsky (1969: 115–16) has forcefully asserted:

> This means—there is no getting away from it—the study of career criminals *au natural,* in the field, the study of such criminals as they normally go about their work and play, the study of "uncaught" criminals and the study of others who in the past have been caught but are not caught at the time you study them. . . . Obviously we can no longer afford the convenient fiction that in studying criminals in their natural habitat, we would discover nothing really important that could not be discovered from criminals behind bars.

By studying criminals in their natural habitat I was able to see them in the full variability and complexity of their surrounding subculture, rather than within the artificial environment of a prison. I was thus able to learn about otherwise inaccessible dimensions of their lives, observing and analyzing firsthand the nature of their social organization, social stratification, lifestyle, and motivation.

NOTES

1. Gold (1958) discouraged this methodological strategy, cautioning against overly close friendship or intimacy with informants, lest they lose their ability to act as informants by becoming too much observers. Whyte (1955), in contrast, recommended the use of informants as research aides, not for helping in conceptualizing the data but for their assistance in locating data which supports, contradicts, or fills in the researcher's analysis of the setting.

2. See also Biernacki and Waldorf 1981; Douglas 1976; Henslin 1972; Hoffman 1980; McCall 1980; and West 1980 for discussions of "snowballing" through key informants.

3. See Kirby and Corzine 1981; Birenbaum 1970; and Henslin 1972 for more detailed discussion of the nature, problems, and strategies for dealing with courtesy stigmas.

4. We never considered secret tapings because, aside from the ethical problems involved, it always struck us as too dangerous.

5. See Douglas (1976) for a more detailed account of these procedures.

6. A recent court decision, where a federal judge ruled that a sociologist did not have to turn over his field notes to a grand jury investigating a suspicious fire at a restaurant where he worked, indicates that this situation may be changing (Fried 1984).

7. See Henslin 1972 and Douglas 1972, 1976 for further discussions of this dilemma and various solutions to it.

8. In some cases I resolved this by altering my descriptions of people and their actions as well as their names so that other members of the dealing and smuggling community would not recognize them. In doing this, however, I had to keep a primary concern for maintaining the sociological integrity of my data so that the generic conclusions I drew from them would be accurate. In places, then, where my attempts to conceal people's identities from people who know them have been inadequate, I hope that I caused them no embarrassment. See also Polsky 1969; Rainwater and Pittman 1967; and Humphreys 1970 for discussions of this problem.

REFERENCES

Adler, Patricia A., and Peter Adler. 1987. *Membership Roles in Field Research.* Beverly Hills, CA: Sage.

Becker, Howard. 1963. *Outsiders.* New York: Free Press.

Biernacki, Patrick, and Dan Waldorf. 1981. "Snowball sampling." *Sociological Methods and Research* 10: 141–63.

Birenbaum, Arnold. 1970. "On managing a courtesy stigma." *Journal of Health and Social Behavior* 11: 196–206.

Blumer, Herbert. 1969. *Symbolic Interactionism.* Englewood Cliffs, NJ: Prentice-Hall.

Carey, James T. 1972. "Problems of access and risk in observing drug scenes." In Jack D. Douglas, ed., *Research on Deviance*, pp. 71–92. New York: Random House.

Cummins, Marvin, et al. 1972. *Report of the Student Task Force on Heroin Use in Metropolitan Saint Louis.* Saint Louis: Washington University Social Science Institute.

Douglas, Jack D. 1972. "Observing deviance." In Jack D. Douglas, ed., *Research on Deviance*, pp. 3–34. New York: Random House.

———. 1976. *Investigative Social Research.* Beverly Hills, CA: Sage.

Fried, Joseph P. 1984. "Judge protects waiter's notes on fire inquiry." *New York Times,* April 8: 47.

Goffman, Erving. 1963. *Stigma.* Englewood Cliffs, NJ: Prentice-Hall.

Gold, Raymond. 1958. "Roles in sociological field observations." *Social Forces* 36: 217–23.

Henslin, James M. 1972. "Studying deviance in four settings: research experiences with cabbies, suicides, drug users and abortionees." In Jack D. Douglas, ed., *Research on Deviance*, pp. 35–70. New York: Random House.

Hoffman, Joan E. 1980. "Problems of access in the study of social elites and boards of directors." In William B. Shaffir, Robert A. Stebbins, and Allan Turowetz, eds., *Fieldwork Experience*, pp. 45–56. New York: St. Martin's.

Humphreys, Laud. 1970. *Tearoom Trade.* Chicago: Aldine.

Johnson, John M. 1975. *Doing Field Research.* New York: Free Press.

Kirby, Richard, and Jay Corzine. 1981. "The contagion of stigma." *Qualitative Sociology* 4: 3–20.

Klockars, Carl B. 1977. "Field ethics for the life history." In Robert Weppner, ed., *Street Ethnography*, pp. 201–26. Beverly Hills, CA: Sage.

———. 1979. "Dirty hands and deviant subjects." In Carl B. Klockars and Finnbarr W. O'Connor, eds., *Deviance and Decency,* pp. 261–82. Beverly Hills, CA: Sage.

Liebow, Elliott. 1967. *Tally's Corner.* Boston: Little, Brown.

McCall, Michal. 1980. "Who and where are the artists?" In William B. Shaffir, Robert A. Stebbins, and Allan Turowetz, eds., *Fieldwork Experience,* pp. 145–58. New York: St. Martin's.

Polsky, Ned. 1969. *Hustlers, Beats, and Others.* New York: Doubleday.

Rainwater, Lee R., and David J. Pittman. 1967. "Ethical problems in studying a politically sensitive and deviant community." *Social Problems* 14: 357–66.

Riemer, Jeffrey W. 1977. "Varieties of opportunistic research." *Urban Life* 5: 467–77.

Rochford, E. Burke, Jr. 1985. *Hare Krishna in America.* New Brunswick, NJ: Rutgers University Press.

Rosenbaum, Marsha. 1981. *Women on Heroin.* New Brunswick, NJ: Rutgers University Press.

Warren, Carol A. B., and Paul K. Rasmussen. 1977. "Sex and gender in field research." *Urban Life* 6: 349–69.

Wax, Rosalie. 1952. "Reciprocity as a field technique." *Human Organization* 11: 34–37.

———. 1957. "Twelve years later: An analysis of a field experience." *American Journal of Sociology* 63: 133–42.

West, W. Gordon. 1980. "Access to adolescent deviants and deviance." In William B. Shaffir, Robert A. Stebbins, and Allan Turowetz, eds., *Fieldwork Experience,* pp. 31–44. New York: St. Martin's.

Whyte, William F. 1955. *Street Corner Society.* Chicago: University of Chicago Press.

PART IV

Constructing Deviance

As we noted in the general introduction, the social constructionist perspective suggests that deviance should be regarded as lodged in a process of definition, rather than in some objective feature of an object, person, or act. It therefore guides us to look at the process by which a society constructs definitions of deviance and applies them to specific groups of people associated with these objects or acts. The dynamics of these deviance-defining processes may sometimes eclipse the factual grounding on which they rest in their significance for the rise of collective moral sentiments.

MORAL ENTREPRENEURS

The process of constructing and applying definitions of deviance can be understood as a *moral enterprise*. That is, it involves the constructions of moral meanings and the association of them with specific acts or conditions. The way people "make" deviance is similar to the way they manufacture anything else, but because deviance is an abstract concept rather than a tangible product, this process involves individuals drawing on the power and resources of organizations, institutions, agencies, symbols, ideas, communication, and audiences. Becker (1963) has suggested that we call the people involved in these activities *moral entrepreneurs*.

The deviance-making enterprise has two facets: rule-creating (without which there would be no deviant behavior) and rule-enforcing (applying these rules to specific groups of people). We have two kinds of moral entrepreneurs: rule creators and rule enforcers.

Rule creating can be done by individuals acting either alone or in groups. Prominent individuals who have been influential in campaigning for definitions of deviance include Nancy Reagan and her "Just Say No" anti-drug campaign and former Surgeon General C. Everett Koop's campaign to marginalize cigarette smoking. More commonly, however, individuals band together to use their collective energy and resources to change social definitions and create norms and rules. These groups of moral entrepreneurs represent interest groups that can be galvanized and activated into *pressure groups*. Rule creators ensure that our society is supplied with a constant stock of deviance and deviants by defining the behavior of others as immoral. They do this because they perceive threats in and feel fearful, distrustful, and suspicious of the behavior of these others. In so doing, they seek to transform private troubles into public issues and their private morality into the normative order.

Moral entrepreneurs manufacture public morality through a multi-stage process. Their first goal is to generate broad *awareness* of a problem. They do this through a process of claims-making where they assert "danger messages" about a given issue. Claims-makers use these messages to create a sense that certain conditions are problematic and pose a present or future potential danger to society. Issues about which we have recently seen danger messages raised include second-hand smoke, hate crimes, and the hazards of "enabling" for co-dependency. Because no rules exist to deal with the threatening condition, claims-makers construct the impression that these are necessary. In so doing, they draw on the testimonials of various "experts" in the field, such as scholars, doctors, eyewitnesses, ex-participants, and others with specific knowledge of the situation. These testimonials are disseminated to society via the media as "facts." Several rhetorical techniques are used to package and present these facts in the most compelling way. Statistics may show the rise in incidence of a given behavior or its correlation with another social problem. Dramatic case examples can paint a picture of horror in the public's mind, inspiring fear and loathing. New syndromes can be advanced, packaging different issues together into a behavioral pattern portrayed as dangerous, such as "internet-addiction disorder" (caution: people are abandoning their homes and families to spend all their time and money lost in chat rooms) or "centerfold syndrome" (caution: men who read pornography objectify, commodify, and victimize women, seek unattainable trophy figurines, and are unable to engage in meaningful relationships). Finally, rhetoric requires that each side seek the (usually competing) "moral high ground" in their assertions and attacks on each other, disavowing special interests and pursuing only the purest public good.

Second, rule creators must bring about a *moral conversion,* convincing others of their views. With the problem outlined, they have to convert neutral parties and previous opponents into supporting partisans. Their successful conversion of others further legitimates their own beliefs. To effect a moral conversion, rule creators must compete for space in the "public arena," often a limited resource. Hilgartner and Bosk (1988) have suggested that only so many issues can claim widespread attention, and they do so at the expense of others. As Durkheim noted for deviance, only a limited number of public concerns can be supported in society at any given time. Moral entrepreneurs must draw on elements of drama, novelty, politics, and deep mythic themes of the culture to gain the visibility they need. They must also enlist the support of sponsors (opinion leaders who need not have expert knowledge on any particular subject, but are liked and respected) to provide them with public endorsements. Moral entrepreneurial campaigners often turn to athletes, actors, religious leaders, and media personalities for such endorsements. At times the efforts of moral entrepreneurs are so successful that they create a "moral panic." A threat to society is depicted, and concerned individuals promoting the problem, reacting legislators, and sensationalist news media whip the public into a "feeding frenzy." Moral panics, most recently witnessed in cases of child abuse, pedophilia (often perpetrated on the unsuspecting over the internet), and Satanism, tend to develop a life of their own, often moving in exaggerated propulsion beyond their original impetus.

Once the public viewpoint has been swayed and a majority (or a vocal and powerful enough minority) of people have adopted a social definition, it may remain at the level of a norm or become elevated to the status of law through a legislative effort. In some cases both situations occur. For example, while the anti-smoking campaign has been successful in banning cigarette use in various public places, it most effectively relies on normative informal sanctions.

Once norms or rules have been enacted, rule enforcers ensure that they are applied. In our society this process often tends to be selective. Various individuals or groups have greater or lesser power to resist the enforcement of rules against them due to their socioeconomic, racial, religious, gender, political, or other status. Whole battles may begin anew over individuals' or groups' strength to resist the enforcement of norms and laws, with this arena becoming once again a moral entrepreneurial combat zone. President Clinton's efforts, during the 1998 Lewinsky scandal, to rebuff the investigatory and labeling efforts of Special Prosecutor Kenneth Starr and the Republicans illustrate core issues of social power.

DIFFERENTIAL SOCIAL POWER

Specific behavioral acts are not the only things that can be constructed as deviant; this definition can also be applied to a social status or lifestyle. When entire groups of people become relegated to a deviant status through their social condition (especially if it is ascribed through birth rather than voluntarily achieved), we see the force of inequality and differential social power in operation. This dynamic has been discussed earlier in reference to both conflict theory and social constructionism, as we noted that those who control the resources in society (politics, social status, gender, wealth, religious beliefs, mobilization of the masses) have the ability to dominate, both materially and ideologically, over the subordinate groups. Thus certain kinds of laws and enforcement are a product of political action by moral entrepreneurial interest groups that are connected to society's power base. Dominant groups use their strength and position to subjugate the weak.

One way to do this is to pass and enforce norms and rules that define others' behavior as deviant. Thus, the relative deviance of conditions such as minority ethnic or racial status, feminine gender, lower social class, youthful age, and homosexual orientation (as some of the following readings show), if taken in this light can be seen to reflect the application of differential social power in our society. Individuals in these groups may find themselves discriminated against or blocked from the mainstream of society by virtue of this basic feature of their existence, unrelated to any particular situation or act. This application of the deviant label emphatically illustrates the role of power in the deviance-defining enterprise, as those positioned closer to the center of society, holding the greater social, economic, political, and moral resources, can turn the force of the deviant stigma onto others less fortunately placed. In so doing they use the definition of deviance to reinforce their own favored position. This politicization of deviance and the power associated with its use serve to remind us that deviance is not a category inhabited only by those on the marginal outskirts of society: the exotics, erotics, and neurotics. Instead, any group can be pushed into this category by the exercise of another group's greater power.

MORAL ENTREPRENEURS

14

The Social Construction of Drug Scares

CRAIG REINARMAN

In this overview of America's social policies, Reinarman tackles moral and legal attitudes toward illicit drugs. He briefly offers a history of drug scares, the major players engineering them, and the social contexts that have enhanced their development and growth. He then outlines seven factors common to drug scares. These enable him to dissect the essential processes in the rule creation and enforcement phases of drug scares, despite the contradictory cultural values of temperance and hedonistic consumption. From this selection we can see how drugs have been scapegoated to account for a wide array of social problems and used to keep some groups down by defining their actions as deviant. It is clear that despite our society's views on the negative features associated with all illicit drugs, our moral entrepreneurial and enforcement efforts have been concentrated more stringently against the drugs used by members of the powerless lower class and minority racial groups.

Drug "wars," anti-drug crusades, and other periods of marked public concern about drugs are never merely reactions to the various troubles people can have with drugs. These drug scares are recurring cultural and political phenomena *in their own right* and must, therefore, be understood sociologically on their own terms. It is important to understand why people ingest drugs and why some of them develop problems that have something to do with having ingested them. But the premise of this chapter is that it is equally important to understand patterns of acute societal concern about drug use and drug problems. This seems especially so for U.S. society, which has had *recurring* anti-drug crusades and a *history* of repressive anti-drug laws.

Many well-intentioned drug policy reform efforts in the U.S. have come face to face with staid and stubborn sentiments against consciousness-altering substances. The repeated failures of such reform efforts cannot be explained solely in terms of ill-informed or manipulative leaders. Something deeper is involved, something woven into the very fabric of American culture, something which explains why claims that some drug is the cause of much of what is wrong with the world are *believed* so often by so many. The origins and nature of the *appeal* of anti-drug claims must be confronted if we are ever to understand how "drug problems" are constructed in the U.S. such that more enlightened and effective drug policies have been so difficult to achieve.

In this chapter I take a step in this direction. First, I summarize briefly some of the major periods of anti-drug sentiment in the U.S. Second, I draw from them the basic ingredients of which drug scares and drug laws are made. Third, I offer a beginning interpretation of these scares and laws based on those broad features of American culture that make *self-control* continuously problematic.

DRUG SCARES AND DRUG LAWS

What I have called drug scares (Reinarman and Levine, 1989a) have been a recurring feature of U.S. society for 200 years. They are relatively autonomous from whatever drug-related problems exist or are said to exist.[1] I call them "scares" because, like Red Scares, they are a form of moral panic ideologically constructed so as to construe one or another chemical bogeyman, à la "communists," as the core cause of a wide array of pre-existing public problems.

The first and most significant drug scare was over drink. Temperance movement leaders constructed this scare beginning in the late 18th and early 19th century. It reached its formal end with the passage of Prohibition in 1919.[2] As Gusfield showed in his classic book *Symbolic Crusade* (1963), there was far more to the battle against booze than long-standing drinking problems. Temperance crusaders tended to be native born, middle-class, non-urban Protestants who felt threatened by the working-class, Catholic immigrants who were filling up America's cities during industrialization.[3] The latter were what Gusfield termed "unrepentant deviants" in that they continued their long-standing drinking practices despite middle-class W.A.S.P. norms against them. The battle over booze was the terrain on which was fought a cornucopia of cultural conflicts, particularly over whose morality would be the dominant morality in America.

In the course of this century-long struggle, the often wild claims of Temperance leaders appealed to millions of middle-class people seeking explanations for the pressing social and economic problems of industrializing America. Many corporate supporters of Prohibition threw their financial and ideological weight behind the Anti-Saloon League and other Temperance and Prohibitionist groups because they felt that traditional working-class drinking practices interfered with the new rhythms of the factory, and thus with productivity and profits (Rumbarger, 1989). To the Temperance crusaders' fear of the bar room as a breeding ground of all sorts of tragic immorality, Prohibitionists added the idea of the saloon as an alien, subversive place where unionists organized and where leftists and anarchists found recruits (Levine, 1984).

This convergence of claims and interests rendered alcohol a scapegoat for most of the nation's poverty, crime, moral degeneracy, "broken" families, illegitimacy, unemployment, and personal and business failure—problems whose sources lay in broader economic and political forces. This scare climaxed in the first two decades of this century, a tumultuous period rife with class, racial, cultural, and political conflict brought on by the wrenching changes of industrialization, immigration, and urbanization (Levine, 1984; Levine and Reinarman, 1991).

America's first real drug law was San Francisco's anti-opium den ordinance of 1875. The context of the campaign for this law shared many features with the context of the Temperance movement. Opiates had long been widely and legally available without a prescription in hundreds of medicines (Brecher, 1972; Musto, 1973; Courtwright, 1982; cf. Baumohl, 1992), so neither opiate use nor addiction was really the issue. This campaign focused almost exclusively on what was called the "Mongolian vice" of opium *smoking* by Chinese immigrants (and white "fellow travelers") in dens (Baumohl, 1992). Chinese immigrants came to California as "coolie" labor to build the railroad and dig the gold mines. A small minority of them brought along the practice of smoking opium—a practice originally brought to China by British and American traders in the 19th century. When the railroad was completed and the gold dried up, a decade-long depression ensued. In a tight labor market, Chinese immigrants were a target. The white Workingman's Party fomented racial hatred of the low-wage "coolies" with whom they now had to compete for work. The first law against opium smoking was only one of many laws enacted to harass and control Chinese workers (Morgan, 1978).

By calling attention to this broader political-economic context I do not wish to slight the specifics of the local political-economic context. In addition to the Workingman's Party, downtown businessmen formed merchant associations and urban families formed improvement associations, both of which fought for more than two decades to reduce the impact of San Francisco's vice districts on the order and health of the central business district and on family neighborhoods (Baumohl, 1992).

In this sense, the anti-opium den ordinance was not the clear and direct result of a sudden drug scare alone. The law was passed against a specific form of drug use engaged in by a disreputable group that had come to be seen as threatening in lean economic times. But it passed easily because this new threat was understood against the broader historical backdrop of long-standing local concerns about various vices as threats to public health, public morals, and public order. Moreover, the focus of attention were dens where it was suspected that whites came into intimate contact with "filthy, idolatrous" Chinese (see Baumohl, 1992). Some local law enforcement leaders, for example, complained that Chinese men were using this vice to seduce white women into sexual slavery (Morgan, 1978). Whatever the hazards of opium smoking, its initial criminalization in San Francisco had to do with both a general context of recession, class conflict, and racism, and with specific local interests in the control of vice and the prevention of miscegenation.

A nationwide scare focusing on opiates and cocaine began in the early 20th century. These drugs had been widely used for years, but were first criminalized when the addict population began to shift from predominantly white, middle-class, middle-aged women to young, working-class males, African-Americans in particular. This scare led to the Harrison Narcotics Act of 1914, the first federal anti-drug law (see Duster, 1970).

Many different moral entrepreneurs guided its passage over a six-year campaign: State Department diplomats seeking a drug treaty as a means of expanding trade with China, trade which they felt was crucial for pulling the economy out of recession; the medical and pharmaceutical professions whose interests were threatened by self-medication with unregulated proprietary tonics, many of which contained cocaine or opiates; reformers seeking to control what they saw as the deviance of immigrants and Southern Blacks who were migrating off the farms; and a pliant press which routinely linked drug use with prostitutes, criminals, transient workers (e.g., the Wobblies), and African-Americans (Musto, 1973). In order to gain the support of Southern Congressmen for a new federal law that might infringe on "states' rights," State Department officials and other crusaders repeatedly spread unsubstantiated suspicions, repeated in the press, that, e.g., cocaine induced African-American men to rape white women (Musto, 1973: 6–10, 67). In short, there was more to this drug scare, too, than mere drug problems.

In the Great Depression, Harry Anslinger of the Federal Narcotics Bureau pushed Congress for a federal law against marijuana. He claimed it was a "killer weed" and he spread stories to the press suggesting that it induced violence—especially among Mexican-Americans. Although there was no evidence that marijuana was widely used, much less that it had any untoward effects, his crusade resulted in its criminalization in 1937—and not incidentally a turnaround in his Bureau's fiscal fortunes (Dickson, 1968). In this case, a new drug law was put in place by a militant moral-bureaucratic entrepreneur who played on racial fears and manipulated a press willing to repeat even his most absurd claims in a context of class conflict during the Depression (Becker, 1963). While there was not a marked scare at the time, Anslinger's claims were never contested in Congress because they played upon racial fears and widely held Victorian values against taking drugs solely for pleasure.

In the drug scare of the 1960s, political and moral leaders somehow reconceptualized this same "killer weed" as the "drop out drug" that was leading America's youth to rebellion and ruin (Himmelstein, 1983). Bio-medical scientists also published uncontrolled, retrospective studies of very small numbers of cases suggesting that, in addition to poisoning the minds and morals of youth, LSD produced broken chromosomes and thus genetic damage (Cohen et al., 1967). These studies were soon shown to be seriously misleading if not meaningless (Tjio et al., 1969), but not before the press, politicians, the medical profession, and the National Institute of Mental Health used them to promote a scare (Weil, 1972: 44–46).

I suggest that the reason even supposedly hard-headed scientists were drawn into such propaganda was that dominant groups felt the country was at war—and not merely with Vietnam. In this scare, there was not so much a "dangerous class" or threatening racial group as multi-faceted political and cultural conflict, particularly between generations, which gave rise to the perception that middle-class youth who rejected conventional values were a dangerous threat.[4] This scare resulted in the Comprehensive Drug Abuse Control Act of 1970, which criminalized more forms of drug use and subjected users to harsher penalties.

Most recently we have seen the crack scare, which began in earnest *not* when the prevalence of cocaine use quadrupled in the late 1970s, nor even when thousands of users began to smoke it in the more potent and dangerous form of freebase. Indeed, when this scare was launched, crack was unknown outside of a few neighborhoods in a handful of major cities (Reinarman and Levine, 1989a) and the prevalence of illicit drug use had been dropping for several years (National Institute on Drug Use, 1990). Rather, this most recent scare began in 1986 when freebase cocaine was renamed crack (or "rock") and sold in precooked, inexpensive units on ghetto streetcorners (Reinarman and Levine, 1989b). Once politicians and the media linked this new form of cocaine use to the inner-city, minority poor, a new drug scare was underway and the solution became more prison cells rather than more treatment slots.

The same sorts of wild claims and Draconian policy proposals of Temperance and Prohibition leaders resurfaced in the crack scare. Politicians have so outdone each other in getting "tough on drugs" that each year since crack came on the scene in 1986 they have passed more repressive laws providing billions more for law enforcement, longer sentences, and more drug offenses punishable by death. One result is that the U.S. now has more people in prison than any industrialized nation in the world—about half of them for drug offenses, the majority of whom are racial minorities.

In each of these periods more repressive drug laws were passed on the grounds that they would reduce drug use and drug problems. I have found no evidence that any scare actually accomplished those ends, but they did greatly expand the quantity and quality of social control, particularly over subordinate groups perceived as dangerous or threatening. Reading across these historical episodes one can abstract a recipe for drug scares and repressive drug laws that contains the following *seven ingredients:*

1. **A Kernel of Truth** Humans have ingested fermented beverages at least since human civilization moved from hunting and gathering to primitive agriculture thousands of years ago. The pharmacopoeia has expanded exponentially since then. So, in virtually all cultures and historical epochs, there has been sufficient ingestion of consciousness-altering chemicals to provide some basis for some people to claim that it is a problem.

2. **Media Magnification** In each of the episodes I have summarized and many others, the mass media has engaged in what I call the *routinization of caricature*—rhetorically recrafting worst cases into typical cases and the episodic into the epidemic. The media dramatize drug problems, as they

do other problems, in the course of their routine news-generating and sales-promoting procedures (see Brecher, 1972: 321–34; Reinarman and Duskin, 1992; and Molotch and Lester, 1974).

3. **Politico-Moral Entrepreneurs** I have added the prefix "politico" to Becker's (1963) seminal concept of moral entrepreneur in order to emphasize the fact that the most prominent and powerful moral entrepreneurs in drug scares are often political elites. Otherwise, I employ the term just as he intended: to denote the *enterprise,* the work, of those who create (or enforce) a rule against what they see as a social evil.[5]

In the history of drug problems in the U.S., these entrepreneurs call attention to drug using behavior and define it as a threat about which "something must be done." They also serve as the media's primary source of sound bites on the dangers of this or that drug. In all the scares I have noted, these entrepreneurs had interests of their own (often financial) which had little to do with drugs. Political elites typically find drugs a functional demon in that (like "outside agitators") drugs allow them to deflect attention from other, more systemic sources of public problems for which they would otherwise have to take some responsibility. Unlike almost every other political issue, however, to be "tough on drugs" in American political culture allows a leader to take a firm stand without risking votes or campaign contributions.

4. **Professional Interest Groups** In each drug scare and during the passage of each drug law, various professional interests contended over what Gusfield (1981: 10–15) calls the "ownership" of drug problems—"the ability to create and influence the public definition of a problem" (1981: 10), and thus to define what should be done about it. These groups have included industrialists, churches, the American Medical Association, the American Pharmaceutical Association, various law enforcement agencies, scientists, and most recently the treatment industry and groups of those former addicts converted to disease ideology.[6] These groups claim for themselves, by virtue of their specialized forms of knowledge, the legitimacy and authority to name what is wrong and to prescribe the solution, usually garnering resources as a result.

5. **Historical Context of Conflict** This trinity of the media, moral entrepreneurs, and professional interests typically interact in such a way as to inflate the extant "kernel of truth" about drug use. But this interaction does not by itself give rise to drug scares or drug laws without underlying conflicts which make drugs into functional villains. Although Temperance crusaders persuaded millions to pledge abstinence, they campaigned for years without achieving alcohol control laws. However, in the tumultuous period leading up to Prohibition, there were revolutions in Russia and Mexico, World War I, massive immigration and impoverishment, and socialist, anarchist, and labor movements, to say nothing of increases in routine problems such as crime. I submit that all this conflict made for a level

of cultural anxiety that provided fertile ideological soil for Prohibition. In each of the other scares, similar conflicts—economic, political, cultural, class, racial, or a combination—provided a context in which claims makers could viably construe certain classes of drug users as a threat.

6. **Linking a Form of Drug Use to a "Dangerous Class"** Drug scares are never about drugs *per se*, because drugs are inanimate objects without social consequence until they are ingested by humans. Rather, drug scares are about the use of a drug by particular groups of people who are, typically, *already* perceived by powerful groups as some kind of threat (see Duster, 1970; Himmelstein, 1978). It was not so much alcohol problems *per se* that most animated the drive for Prohibition but the behavior and morality of what dominant groups saw as the "dangerous class" of urban, immigrant, Catholic, working-class drinkers (Gusfield, 1963; Rumbarger, 1989). It was *Chinese* opium smoking dens, not the more widespread use of other opiates, that prompted California's first drug law in the 1870s. It was only when smokable cocaine found its way to the African-American and Latino underclass that it made headlines and prompted calls for a drug war. In each case, politico-moral entrepreneurs were able to construct a "drug problem" by linking a substance to a group of users perceived by the powerful as disreputable, dangerous, or otherwise threatening.

7. **Scapegoating a Drug for a Wide Array of Public Problems** The final ingredient is scapegoating, i.e., blaming a drug or its alleged effects on a group of its users for a variety of preexisting social ills that are typically only indirectly associated with it. Scapegoating may be the most crucial element because it gives great explanatory power and thus broader resonance to claims about the horrors of drugs (particularly in the conflictual historical contexts in which drug scares tend to occur).

Scapegoating was abundant in each of the cases noted previously. To listen to Temperance crusaders, for example, one might have believed that without alcohol use, America would be a land of infinite economic progress with no poverty, crime, mental illness, or even sex outside marriage. To listen to leaders of organized medicine and the government in the 1960s, one might have surmised that without marijuana and LSD there would have been neither conflict between youth and their parents nor opposition to the Vietnam War. And to believe politicians and the media in the past 6 years is to believe that without the scourge of crack the inner cities and the so-called underclass would, if not disappear, at least be far less scarred by poverty, violence, and crime. There is no historical evidence supporting any of this.

In short, drugs are richly functional scapegoats. They provide elites with fig leaves to place over unsightly social ills that are endemic to the social system over which they preside. And they provide the public with a restricted aperture of attribution in which only a chemical bogeyman or the lone deviants who ingest it are seen as the cause of a cornucopia of complex problems.

TOWARD A CULTURALLY SPECIFIC THEORY OF DRUG SCARES

Various forms of drug use have been and are widespread in almost all societies comparable to ours. A few of them have experienced limited drug scares, usually around alcohol decades ago. However, drug scares have been *far* less common in other societies, and never as virulent as they have been in the U.S. (Brecher, 1972; Levine, 1992; MacAndrew and Edgerton, 1969). There has never been a time or place in human history without drunkenness, for example, but in *most* times and places drunkenness has not been nearly as problematic as it has been in the U.S. since the late 18th century. Moreover, in comparable industrial democracies, drug laws are generally less repressive. Why then do claims about the horrors of this or that consciousness-altering chemical have such unusual power in American culture?

Drug scares and other periods of acute public concern about drug use are not just discrete, unrelated episodes. There is a historical pattern in the U.S. that cannot be understood in terms of the moral values and perceptions of individual anti-drug crusaders alone. I have suggested that these crusaders have benefitted in various ways from their crusades. For example, making claims about how a drug is damaging society can help elites increase the social control of groups perceived as threatening (Duster, 1970), establish one class's moral code as dominant (Gusfield, 1963), bolster a bureaucracy's sagging fiscal fortunes (Dickson, 1968), or mobilize voter support (Reinarman and Levine, 1989a, b). However, the recurring character of pharmaco-phobia in U.S. history suggests that there is something about our *culture* which makes citizens more vulnerable to anti-drug crusaders' attempts to demonize drugs. Thus, an answer to the question of America's unusual vulnerability to drug scares must address why the scapegoating of consciousness-altering substances regularly *resonates* with or appeals to substantial portions of the population.

There are three basic parts to my answer. The first is that claims about the evils of drugs are especially viable in American culture in part because they provide a welcome *vocabulary of attribution* (cf. Mills, 1940). Armed with "DRUGS" as a generic scapegoat, citizens gain the cognitive satisfaction of having a folk devil on which to blame a range of bizarre behaviors or other conditions they find troubling but difficult to explain in other terms. This much may be true of a number of other societies, but I hypothesize that this is particularly so in the U.S. because in our political culture individualistic explanations for problems are so much more common than social explanations.

Second, claims about the evils of drugs provide an especially serviceable vocabulary of attribution in the U.S. in part because our society developed from a *temperance culture* (Levine, 1992). American society was forged in the fires of ascetic Protestantism and industrial capitalism, both of which demand *self-control*. U.S. society has long been characterized as the land of the individual "self-made man." In such a land, self-control has had extraordinary importance. For the middle-class Protestants who settled, defined, and still dominate the U.S., self-control was both central to religious world views and a characterological necessity for economic survival and success in the capitalist market (Weber, 1930 [1985]). With Levine (1992), I hypothesize that in a culture in which self-control is inordinately important, drug-induced altered states of consciousness are especially likely to be experienced as "loss of control," and thus to be inordinately feared.[7]

Drunkenness and other forms of drug use have, of course, been present everywhere in the industrialized world. But temperance cultures tend to arise only when industrial capitalism unfolds upon a cultural terrain deeply imbued with the Protestant ethic.[8] This means that only the U.S., England, Canada, and parts of Scandinavia have Temperance cultures, the U.S. being the most extreme case.

It may be objected that the influence of such a Temperance culture was strongest in the 19th and early 20th century and that its grip on the American *zeitgeist* has been loosened by the forces of modernity and now, many say, postmodernity. The third part of my answer, however, is that on the foundation of a Temperance culture, advanced capitalism has built a *postmodern, mass consumption culture* that exacerbates the problem of self-control in new ways.

Early in the 20th century, Henry Ford pioneered the idea that by raising wages he could simultaneously quell worker protests and increase market demand for mass-produced goods. This mass consumption strategy became central to modern American society and one of the reasons for our economic success (Marcuse, 1964; Aronowitz, 1973; Ewen, 1976; Bell, 1978). Our economy is now so fundamentally predicated upon mass consumption that theorists as diverse as Daniel Bell and Herbert Marcuse have observed that we live in a mass consumption culture. Bell (1978), for example, notes that while the Protestant work ethic and deferred gratification may still hold sway in the workplace, Madison Avenue, the media, and malls have inculcated a new indulgence ethic in the leisure sphere in which pleasure-seeking and immediate gratification reign.

Thus, our economy and society have come to depend upon the constant cultivation of new "needs," the production of new desires. Not only the hardware of social life such as food, clothing, and shelter but also the software of the self—excitement, entertainment, even eroticism—have become mass consumption commodities. This means that our society offers an increasing number of incentives for indulgence—more ways to lose self-control—and a decreasing number of countervailing reasons for retaining it.

In short, drug scares continue to occur in American society in part because people must constantly manage the contradiction between a Temperance culture that insists on self-control and a mass consumption culture which renders self-control continuously problematic. In addition to helping explain the recurrence of drug scares, I think this contradiction helps account for why in the last dozen years millions of Americans have joined 12-Step groups, more than 100 of which have nothing whatsoever to do with ingesting a drug (Reinarman, 1995). "Addiction," or the generalized loss of self-control, has become the meta-metaphor for a staggering array of human troubles. And, of course, we also seem to have a staggering array of politicians and other moral entrepreneurs who take advantage of such cultural contradictions to blame new chemical bogeymen for our society's ills.

NOTES

1. In this regard, for example, Robin Room wisely observes "that we are living at a historic moment when the rate of (alcohol) dependence as a cognitive and existential experience is rising, although the rate of alcohol consumption and of heavy drinking is falling." He draws from this a more general hypothesis about "long waves" of drinking and societal reactions to them: "[I]n periods of increased questioning of drinking and heavy drinking, the trends in the two forms of dependence, psychological and physical, will tend to run in opposite directions. Conversely, in periods of a "wettening" of sentiments, with the curve of alcohol consumption beginning to rise, we may expect the rate of physical dependence . . . to rise while the rate of dependence as a cognitive experience falls" (1991: 154).

2. I say "formal end" because Temperance ideology is not merely alive and well in the War on Drugs but is being applied to all manner of human troubles in the burgeoning 12-Step Movement (Reinarman, 1995).

3. From Jim Baumohl I have learned that while the Temperance movement attracted most of its supporters from these groups, it also found supporters among many others (e.g., labor, the Irish, Catholics, former drunkards, women), each of which had its own reading of and folded its own agenda into the movement.

4. This historical sketch of drug scares is obviously not exhaustive. Readers interested in other scares should see, e.g., Brecher's encyclopedic work *Licit and Illicit Drugs* (1972), especially the chapter on glue sniffing, which illustrates how the media actually created a new drug problem by writing hysterical stories about it. There was also a PCP scare in the 1970s in which law enforcement officials claimed that the growing use of this horse tranquilizer was a severe threat because it made users so violent and gave them such super-human strength that stun guns were necessary. This, too, turned out to be unfounded and the "angel dust" scare was short-lived (see Feldman et al., 1979). The best analysis of how new drugs themselves can lead to panic reactions among users is Becker (1967).

5. Becker wisely warns against the "one-sided view" that sees such crusaders as merely imposing their morality on others. Moral entrepreneurs, he notes, do operate "with an absolute ethic," are "fervent and righteous," and will use "any means" necessary to "do away with" what they see as "totally evil." However, they also "typically believe that their mission is a holy one," that if people do what they want it "will be good for them." Thus, as in the case of abolitionists, the crusades of moral entrepreneurs often "have strong humanitarian overtones" (1963: 147–8). This is no less true for those whose moral enterprise promotes drug scares. My analysis, however, concerns the character and consequences of their efforts, not their motives.

6. As Gusfield notes, such ownership sometimes shifts over time, e.g., with alcohol problems, from religion to criminal law to medical science. With other drug problems, the shift in ownership has been away from medical science toward criminal law. The most insightful treatment of the medicalization of alcohol/drug problems is Peele (1989).

7. See Baumohl's (1990) important and erudite analysis of how the human will was valorized in the therapeutic temperance thought of 19th-century inebriate homes.

8. The third central feature of Temperance cultures identified by Levine (1992), which I will not dwell on, is predominance of spirits drinking, i.e., more concentrated alcohol than wine or beer and thus greater likelihood of drunkenness.

REFERENCES

Aronowitz, Stanley. 1973. *False Promises: The Shaping of American Working Class Consciousness.* New York: McGraw-Hill.

Baumohl, Jim. 1990. "Inebriate Institutions in North America, 1840–1920." *British Journal of Addiction* 85: 1187–1204.

Baumohl, Jim. 1992. "The 'Dope Fiend's Paradise' Revisited: Notes from Research in Progress on Drug Law Enforcement in San Francisco, 1875–1915." *Drinking and Drug Practices Surveyor* 24: 3–12.

Becker, Howard S. 1963. *Outsiders: Studies in the Sociology of Deviance.* Glencoe, IL: Free Press.

———. 1967. "History, Culture, and Subjective Experience: An Exploration of the Social Bases of Drug-Induced Experiences." *Journal of Health and Social Behavior* 8: 162–176.

Bell, Daniel. 1978. *The Cultural Contradictions of Capitalism.* New York: Basic Books.

Brecher, Edward M. 1972. *Licit and Illicit Drugs.* Boston: Little Brown.

Cohen, M. M., K. Hirshorn, and W. A. Frosch. 1967. "In Vivo and in Vitro Chromosomal Damage Induced by LSD-25." *New England Journal of Medicine* 227: 1043.

Courtwright, David. 1982. *Dark Paradise: Opiate Addiction in America Before 1940.* Cambridge, MA: Harvard University Press.

Dickson, Donald. 1968. "Bureaucracy and Morality." *Social Problems* 16: 143–156.

Duster, Troy. 1970. *The Legislation of Morality: Law, Drugs, and Moral Judgement.* New York: Free Press.

Ewen, Stuart. 1976. *Captains of Consciousness: Advertising and the Social Roots of Consumer Culture.* New York: McGraw-Hill.

Feldman, Harvey W., Michael H. Agar, and George M. Beschner. 1979. *Angel Dust.* Lexington, MA: Lexington Books.

Gusfield, Joseph R. 1963. *Symbolic Crusade: Status Politics and the American Temperance Movement.* Urbana: University of Illinois Press.

———. 1981. *The Culture of Public Problems: Drinking-Driving and the Symbolic Order.* Chicago: University of Chicago Press.

Himmelstein, Jerome. 1978. "Drug Politics Theory." *Journal of Drug Issues* 8.

———. 1983. *The Strange Career of Marihuana.* Westport, CT: Greenwood Press.

Levine, Harry Gene. 1984. "The Alcohol Problem in America: From Temperance to Alcoholism." *British Journal of Addiction* 84: 109–119.

———. 1992. "Temperance Cultures: Concern About Alcohol Problems in Nordic and English-Speaking Cultures." In G. Edwards et al., eds., *The Nature of Alcohol and Drug Related Problems.* New York: Oxford University Press.

Levine, Harry Gene, and Craig Reinarman. 1991. "From Prohibition to Regulation: Lessons from Alcohol Policy for Drug Policy." *Milbank Quarterly* 69: 461–494.

MacAndrew, Craig, and Robert Edgerton. 1969. *Drunken Comportment.* Chicago: Aldine.

Marcuse, Herbert. 1964. *One-Dimensional Man: Studies in the Ideology of Advanced Industrial Society.* Boston: Beacon Press.

Mills, C. Wright. 1940. "Situated Actions and Vocabularies of Motive." *American Sociological Review* 5: 904–913.

Molotch, Harvey, and Marilyn Lester. 1974. "News as Purposive Behavior: On the Strategic Uses of Routine Events, Accidents, and Scandals." *American Sociological Review* 39: 101–112.

Morgan, Patricia. 1978. "The Legislation of Drug Law: Economic Crisis and Social Control." *Journal of Drug Issues* 8: 53–62.

Musto, David. 1973. *The American Disease: Origins of Narcotic Control.* New Haven, CT: Yale University Press.

National Institute on Drug Abuse. 1990. *National Household Survey on Drug Abuse: Main Findings 1990.* Washington, DC: U.S. Department of Health and Human Services.

Peele, Stanton. 1989. *The Diseasing of America: Addiction Treatment Out of Control.* Lexington, MA: Lexington Books.

Reinarman, Craig. 1995. "The 12-Step Movement and Advanced Capitalist Culture: Notes on the Politics of Self-Control in Postmodernity." In B. Epstein, R. Flacks, and M. Darnovsky, eds., *Contemporary Social Movements and Cultural Politics.* New York: Oxford University Press.

Reinarman, Craig, and Ceres Duskin. 1992. "Dominant Ideology and Drugs in the Media." *International Journal on Drug Policy* 3: 6–15.

Reinarman, Craig, and Harry Gene Levine. 1989a. "Crack in Context: Politics and Media in the Making of a Drug Scare." *Contemporary Drug Problems* 16: 535–577.

———. 1989b. "The Crack Attack: Politics and Media in America's Latest Drug Scare." In Joel Best, ed., *Images of Issues: Typifying Contemporary Social Problems,* pp. 115–137. New York: Aldine de Gruyter.

Room, Robin G. W. 1991. "Cultural Changes in Drinking and Trends in Alcohol Problems Indicators: Recent U.S. Experience." In Walter B. Clark and Michael E. Hilton, eds., *Alcohol in America: Drinking Practices and Problems,* pp. 149–162. Albany: State University of New York Press.

Rumbarger, John J. 1989. *Profits, Power, and Prohibition: Alcohol Reform and the Industrializing of America, 1800–1930.* Albany: State University of New York Press.

Tijo, J. H., W. N. Pahnke, and A. A. Kurland. 1969. "LSD and Chromosomes: A Controlled Experiment." *Journal of the American Medical Association* 210: 849.

Weber, Max. 1985 (1930). *The Protestant Ethic and the Spirit of Capitalism.* London: Unwin.

Weil, Andrew. 1972. *The Natural Mind.* Boston: Houghton Mifflin.

15

Blowing Smoke
Status Politics and the Smoking Ban

JUSTIN L. TUGGLE AND MALCOLM D. HOLMES

Tuggle and Holmes' selection on the struggle over cigarette smoking in America expands Reinarman's consideration to the licit drug realm, examining the struggle and counter-struggle over tobacco between the moral entrepreneurs and the status quo defenders. They note the medical, ethical, and socio-economic arguments raised to sway public opinion and demonize public consumption of cigarettes. They mention the spate of claims put forward pitting anti-smoking groups who have argued that secondhand smoke is toxic and that smokers should not be allowed to inflict their pollution onto others against opponents who have argued that the government should not legislate their morality. These issues illustrate the concern that frequently arises in deviance, where society must balance the right of individual freedoms (the desire to smoke) against the needs of the common good (public health). This reading also shows the relation between claims-making and social power, tracing the status of the social groups on each side of this campaign. Its fundamental message lies in how moral entrepreneurs use their status to attach deviant labels to the behavior of others, and, in so doing, to keep those others in a subordinate status position. They thus show deviance, like Quinney does, as a tool by which higher status groups retain and enforce their interests over subordinate groups.

Over the past half century, perceptions of tobacco and its users have changed dramatically. In the 1940s and 1950s, cigarette smoking was socially accepted and commonly presumed to lack deleterious effects (see, e.g., Ram 1941). Survey data from the early 1950s showed that a minority believed cigarette smoking caused lung cancer (Viscusi 1992). By the late 1970s, however, estimates from survey data revealed that more than 90% of the population thought that this link existed (Roper Organization 1978). This and other harms associated with tobacco consumption have provided the impetus for an antismoking crusade that aims to normatively redefine smoking as deviant behavior (Markle and Troyer 1979).

There seems to be little question that tobacco is a damaging psychoactive substance characterized by highly adverse chronic health effects (Steinfeld 1991). In this regard, the social control movement probably makes considerable

Copyright 1997 from *Deviant Behavior*, Vol. 18, 1. Reproduced by permission of Taylor & Francis, Inc., http://www.routledge-ny.com.

sense in terms of public policy. At the same time, much as ethnicity and religion played a significant role in the prohibition of alcohol (Gusfield 1963), social status may well play a part in this latest crusade.

Historically, attempts to control psychoactive substances have linked their use to categories of relatively powerless people. Marijuana use was associated with Mexican Americans (Bonnie and Whitebread 1970), cocaine with African Americans (Ashley 1975), opiates with Asians (Ben-Yehuda 1990), and alcohol with immigrant Catholics (Gusfield 1963). During the heyday of cigarette smoking, it was thought that

> Tobacco's the one blessing that nature has left for all humans to enjoy. It can be consumed by both the "haves" and "have nots" as a common leveler, one that brings all humans together from all walks of life regardless of class, race, or creed. (Ram, 1941, p. 125)

But in contrast to this earlier view, recent evidence has shown that occupational status (Ferrence 1989; Marcus et al. 1989; Covey et al. 1992), education (Ferrence 1989; Viscusi 1992) and family income (Viscusi 1992) are related negatively to current smoking. Further, the relationships of occupation and education to cigarette smoking have become stronger in later age cohorts (Ferrence 1989). Thus we ask, *is the association of tobacco with lower-status persons a factor in the crusade against smoking in public facilities?* Here we examine that question in a case study of a smoking ban implemented in Shasta County, California.

STATUS POLITICS AND THE CREATION OF DEVIANCE

Deviance is socially constructed. Complex pluralistic societies have multiple, competing symbolic-moral universes that clash and negotiate (Ben-Yehuda 1990). Deviance is relative, and social morality is continually restructured. Moral, power, and stigma contests are ongoing, with competing symbolic-moral universes striving to legitimize particular lifestyles while making others deviant (Schur 1980; Ben-Yehuda 1990).

The ability to define and construct reality is closely connected to the power structure of society (Gusfield 1963). Inevitably, then, the distribution of deviance is associated with the system of stratification. The higher one's social position, the greater one's moral value (Ben-Yehuda 1990). Differences in lifestyles and moral beliefs are corollaries of social stratification (Gusfield 1963; Zurcher and Kirkpatrick 1976; Luker 1984). Accordingly, even though grounded in the system of stratification, status conflicts need not be instrumental; they may also be symbolic. Social stigma may, for instance, attach to behavior thought indicative of a weak will (Goffman 1963). Such moral anomalies occasion status degradation ceremonies, public denunciations expressing indignation not at a behavior per se, but rather against the individual motivational type that produced it (Garfinkel 1956). The denouncers act as public figures, drawing upon communally shared experience and speaking in the name of ultimate values. In this respect, status degradation involves a reciprocal element: Status conflicts and the resultant condemnation of a behavior characteristic of a particular status category symbolically enhances the status of the abstinent through the degradation of the participatory (Garfinkel 1956; Gusfield 1963).

Deviance creation involves political competition in which moral entrepreneurs originate moral crusades aimed at generating reform (Becker 1963; Schur 1980; Ben-Yehuda 1990). The alleged deficiencies of a specific social group are revealed and reviled by those crusading to define their behavior as deviant. As might be expected, successful moral crusades are generally dominated by those in the upper social strata of society (Becker 1963). Research on the antiabortion (Luker 1984) and antipornography (Zurcher and Kirkpatrick 1976) crusades has shown that activists in these movements are of lower socioeconomic status than their opponents, helping explain the limited success of efforts to redefine abortion and pornography as deviance.

Moral entrepreneurs' goals may be either assimilative or coercive reform (Gusfield 1963). In the former instance, sympathy to the deviants' plight engenders integrative efforts aimed at lifting the repentant to the superior moral plane allegedly held by those of higher social status. The latter strategy emerges when deviants are viewed as intractably denying the moral and status superiority of the reformers' symbolic-moral universe. Thus, whereas assimilative reform may employ educative strategies, coercive reform turns to law and force for affirmation.

Regardless of aim, the moral entrepreneur cannot succeed alone. Success in establishing a moral crusade is dependent on acquiring broader public support. To that end, the moral entrepreneur must mobilize power, create a perceived threat potential for the moral issue in question, generate public awareness of the issue, propose a clear and acceptable solution to the problem, and overcome resistance to the crusade (Becker 1963; Ben-Yehuda 1990).

THE STATUS POLITICS OF CIGARETTE SMOKING

The political dynamics underlying the definition of deviant behaviors may be seen clearly in efforts to end smoking in public facilities. Cigarettes were an insignificant product of the tobacco industry until the end of the 19th century, after which they evolved into its staple (U.S. Department of Health and Human Services 1992). Around the turn of the century, 14 states banned cigarette smoking and all but one other regulated sales to and possession by minors (Nuehring and Markle 1974). Yet by its heyday in the 1940s and 1950s, cigarette smoking was almost universally accepted, even considered socially desirable (Nuehring and Markle 1974; Steinfeld 1991). Per capita cigarette consumption in the United States peaked at approximately 4,300 cigarettes per year in the early 1960s, after which it declined to about 2,800 per year by the early 1990s (U.S. Department of Health and Human Services 1992). The beginning of the marked decline in cigarette consumption corresponded to the publication of the report to the surgeon general on the health risks of smoking (U.S. Department of Health, Education and Welfare 1964). Two decades later, the hazards of passive smoking were being publicized (e.g., U.S. Department of Health and Human Services 1986).

Increasingly, the recognition of the apparent relationship of smoking to health risks has socially demarcated the lifestyles of the smoker and nonsmoker, from widespread acceptance of the habit to polarized symbolic-moral universes. Attitudes about smoking are informed partly by medical issues, but perhaps even more critical are normative considerations (Nuehring and Markle 1974); more people have come to see smoking as socially reprehensible and deviant, and smokers as social misfits (Markle and Troyer 1979). Psychological assessments have attributed an array of negative evaluative characteristics to smokers (Markle and Troyer 1979). Their habit is increasingly thought unclean and intrusive.

Abstinence and bodily purity are the cornerstones of the nonsmoker's purported moral superiority (Feinhandler 1986). At the center of their symbolic-moral universe, then, is the idea that people have the right to breathe clean air in public spaces (Goodin 1989). Smokers, on the other hand, stake their claim to legitimacy in a precept of Anglo-Saxon political culture—the right to do whatever one wants unless it harms others (Berger 1986). Those sympathetic to smoking deny that environmental tobacco smoke poses a significant health hazard to the nonsmoker (Aviado 1986). Yet such arguments have held little sway in the face of counterclaims from authoritative governmental agencies and high status moral entrepreneurs.

The development of the antismoking movement has targeted a lifestyle particularly characteristic of the working classes (Berger 1986). Not only has there been an overall decline in cigarette smoking, but, as mentioned above, the negative relationships of occupation and education to cigarette smoking have become more pronounced in later age cohorts (Ferrence 1989). Moreover, moral entrepreneurs crusading against smoking are representatives of a relatively powerful "knowledge class," comprising people employed in areas such as education and the therapeutic and counseling agencies (Berger 1986).

Early remedial efforts focused on publicizing the perils of cigarette smokers, reflecting a strategy of assimilative reform (Neuhring and Markle 1974; Markle and Troyer 1979). Even many smokers expressed opposition to cigarettes and a generally repentant attitude. Early educative efforts were thus successful in decreasing cigarette consumption, despite resistance from the tobacco industry. Then, recognition of the adverse effects of smoking on nonusers helped precipitate a turn to coercive reform measures during the mid 1970s (Markle and Troyer 1979). Rather than a repentant friend in need of help, a new definition of the smoker as enemy emerged. Legal abolition of smoking in public facilities became one focus of social control efforts, and smoking bans in public spaces have been widely adopted in recent years (Markle and Troyer 1979; Goodin 1989).

The success of the antismoking crusade has been grounded in moral entrepreneurs' proficiency at mobilizing power, a mobilization made possible by highly visible governmental campaigns, the widely publicized health risks of smoking, and the proposal of workable and generally acceptable policies to ameliorate the problem. The success of this moral crusade has been further facilitated by the association of deviant characteristics with those in lower social strata, whose stigmatization reinforces existing relations of power and prestige. Despite the formidable resources and staunch opposition of the tobacco industry, the tide of public opinion and policy continues to move toward an anti-smoking stance.

RESEARCH PROBLEM

The study presented below is an exploratory examination of the link between social status and support for a smoking ban in public facilities. Based on theorizing about status politics, as well as evidence about patterns of cigarette use, it was predicted that supporters of the smoking ban would be of higher status than those who opposed it. Further, it was anticipated that supporters of the ban would be more likely to make negative normative claims denouncing the allegedly deviant qualities of smoking, symbolically enhancing their own status while lowering that of their opponents.

The site of this research was Shasta County, California. The population of Shasta County is 147,036, of whom 66,462 reside in its only city, Redding (U.S. Bureau of the Census 1990). This county became the setting for the implementation of a hotly contested ban on smoking in public buildings.

In 1988, California voters passed Proposition 99, increasing cigarette taxes by 25 cents per pack. The purpose of the tax was to fund smoking prevention and treatment programs. Toward that end, Shasta County created the Shasta County Tobacco Education Program. The director of the program formed a coalition with officials of the Shasta County chapters of the American Cancer Society and American Lung Association to propose a smoking ban in all public buildings. The three groups formed an organization to promote that cause, Smoke-Free Air For Everyone (SAFE). Unlike other bans then in effect in California, the proposed ban included restaurants and bars, because its proponents considered these to be places in which people encountered significant amounts of secondhand smoke. They procured sufficient signatures on a petition to place the measure on the county's general ballot in November 1992.

The referendum passed with a 56% majority in an election that saw an 82% turnout. Subsequently, the Shasta County Hospitality and Business Alliance, an antiban coalition, obtained sufficient signatures to force a special election to annul the smoking ban. The special election was held in April 1993. Although the turnout was much lower (48%), again a sizable majority (58.4%) supported the ban. The ordinance went into effect on July 1, 1993.

ANALYTIC STRATEGY

... [D]ata were analyzed in our effort to ascertain the moral and status conflicts underlying the Shasta County smoking ban ... [based on] interviews with five leading moral entrepreneurs and five prominent status quo defenders.[1] These individuals were selected through a snowball sample, with the original respondents identified through interviews with business owners or political advertisements in the local mass media. The selected respondents repeatedly surfaced as the leading figures in their respective coalitions. Semistructured interviews were conducted to determine the reasons underlying their involvement. These data were critical to understanding how the proposed ban was framed by small groups of influential proponents and opponents; it was expected that their concerns would be reflected in the larger public debate about the ban.

FINDINGS

Moral Entrepreneur/Status Quo Defender Interviews

The moral entrepreneurs and status quo defenders interviewed represented clearly different interests. The former group included three high-level administrators in the county's chapters of the American Cancer Society and American Lung Association. A fourth was an administrator for the Shasta County Tobacco Education Project. The last member of this group was a pulmonary physician affiliated with a local hospital. The latter group included four bar and/or restaurant owners and an attorney who had been hired to represent their interests. Thus the status quo defenders were small business owners who might see their economic interests affected adversely by the ban. Importantly, they were representatives of a less prestigious social stratum than the moral entrepreneurs.

The primary concern of the moral entrepreneurs was health. As one stated,

> I supported the initiative to get the smoking ban on the ballot because of all the health implications that secondhand smoke can create. Smoking and secondhand smoke are the most preventable causes of death in this nation.

Another offered that

> On average, secondhand smoke kills 53,000 Americans each year. And think about those that it kills in other countries! It contains 43 cancer-causing chemical agents that have been verified by the Environmental Protection Agency. It is now listed as a Type A carcinogen, which is the same category as asbestos.

Every one of the moral entrepreneurs expressed concern about health issues during the interviews. This was not the only point they raised, however.

Three of the five made negative normative evaluations of smoking, thereby implicitly degrading the status of smokers. They commented that "smoking is no longer an acceptable action," that "smoke stinks," or that "it is just a dirty and annoying habit." Thus, whereas health was their primary concern, such comments revealed the moral entrepreneurs' negative view of smoking irrespective of any medical issues. Smokers were seen as engaging in unclean and objectionable behavior—stigmatized qualities defining their deviant social status.

The stance of the status quo defenders was also grounded in two arguments. All of them expressed concern about individual rights. As one put it,

> I opposed that smoking ban because I personally smoke and feel that it is an infringement of my rights to tell me where I can and cannot smoke. Smoking is a legal activity, and therefore it is unconstitutional to take that right away from me.

Another argued that

> Many people have died for us to have these rights in foreign wars and those also fought on American soil. Hundreds of thousands of people thought that these rights were worth dying for, and now some small group of people believe that they can just vote away these rights.

Such symbolism implies that smoking is virtually a patriotic calling, a venerable habit for which people have been willing to forfeit their lives in time of war. In the status quo defenders' view, smoking is a constitutionally protected right.

At the same time, each of the status quo defenders was concerned about more practical matters, namely business profits. As one stated, "my income was going to be greatly affected." Another argued,

> If these people owned some of the businesses that they are including in this ban, they would not like it either. By taking away the customers that smoke, they are taking away the mainstay of people from a lot of businesses.

The competing viewpoints of the moral entrepreneurs and status quo defenders revealed the moral issues—health versus individual rights—at the heart of political conflict over the smoking ban. Yet it appears that status issues also fueled the conflict. On the one hand, the moral entrepreneurs denigrated smoking, emphasizing the socially unacceptable qualities of the behavior and symbolically degrading smokers' status. On the other hand, status quo defenders were concerned that their livelihood would be affected by the ban. Interestingly, the occupational status of the two groups differed, with the moral entrepreneurs representing the new knowledge class, the status quo defenders a lower stratum of small business owners. Those in the latter group may not have been accorded the prestige and trust granted those in the former (Berger 1986). Moreover, the status quo defenders' concern about business was likely seen as self-aggrandizing.

SUMMARY AND DISCUSSION

This research has examined the moral and status politics underlying the implementation of a smoking ban in Shasta County, California. Moral entrepreneurs crusading for the ban argued that secondhand smoke damages health, implicitly grounding their argument in the principle that people have a right to a smoke-free environment. Status quo defenders countered that smokers have a constitutional right to indulge wherever and whenever they see fit. Public discourse echoed these themes, as seen in the letters to the editor of the local newspaper. Thus debate about the smoking ban focused especially on health versus smokers' rights; yet evidence of social status differences between the competing symbolic-moral universes also surfaced. Competing symbolic-moral universes are defined not only by different ethical viewpoints on a behavior, but also by differences in social power—disparities inevitably linked to the system of stratification (Ben-Yehuda 1990). Those prevailing in moral and stigma contests typically represent the higher socioeconomic echelons of society.

The moral entrepreneurs who engineered the smoking ban campaign were representatives of the prestigious knowledge class, including among their members officials from the local chapters of respected organizations at the forefront of the national antismoking crusade. In contrast, the small business owners who were at the core of the opposing coalition, of status quo defenders, represented the traditional middle class. Clearly, there was an instrumental quality to the restaurant and bar owners' stance, because they saw the ban as potentially damaging to their business interests. But they were unable to shape the public debate, as demonstrated by the letters to the editor.

In many respects, the status conflicts involved in the passage of the Shasta County smoking ban were symbolic. The moral entrepreneurs focused attention on the normatively undesirable qualities of cigarette smoking, and their negative normative evaluations of smoking were reflected in public debate about the ban. Those who wrote in support of the ban more frequently offered negative normative evaluations than antiban writers; their comments degraded smoking and, implicitly, smokers. Since the advent of the antismoking crusade in the United States, smoking has come to be seen as socially reprehensible, smokers as social misfits characterized by negative psychological characteristics (Markle and Troyer 1979).

Ultimately, a lifestyle associated with the less educated, less affluent, lower occupational strata was stigmatized as a public health hazard and targeted for coercive reform. Its deviant status was codified in the ordinance banning smoking in public facilities, including restaurants and bars. The ban symbolized the deviant status of cigarette smokers, the prohibition visibly demonstrating the community's condemnation of their behavior. Further, the smoking ban symbolically amplified the purported virtues of the abstinent lifestyle. A political victory such as the passage of a law is a prestige-enhancing symbolic triumph that is perhaps even more rewarding than its end result (Gusfield 1963). The symbolic nature of the ban serendipitously surfaced in another way during one author's unstructured observations in 42 restaurants and 21 bars in the area: Whereas smoking was not observed in a single restaurant, it occurred without sanction in all but one of the bars. Although not deterring smoking in one of its traditional bastions, the ban called attention to its deviant quality and, instrumentally, effectively halted it in areas more commonly frequented by the abstemious.

Although more systematic research is needed, the findings of this exploratory case study offer a better understanding of the dynamics underlying opposition to smoking and further support to theorizing about the role of status politics in the creation of deviant types. Denunciation of smoking in Shasta County involved not only legitimate allegations about public health, but negative normative evaluations of those engaged in the behavior. In the latter regard, the ban constituted a status degradation ceremony, symbolically differentiating the pure and abstinent from the unclean and intrusive. Not coincidentally, the stigmatized were more likely found among society's lower socioeconomic strata, their denouncers among its higher echelons.

Certainly the class and ethnic antipathies underlying attacks on cocaine and opiate users earlier in the century were more manifest than those revealed in the crusade against cigarette smoking. But neither are there manifest status conflicts in the present crusades against abortion (Luker 1984) and pornography (Zurcher and Kirkpatrick 1976); yet the underlying differences of status between opponents in those movements are reflected in their markedly different symbolic-moral universes, as was the case in the present study.

This is not to suggest that smoking should be an approved behavior. The medical evidence seems compelling: Cigarette smoking is harmful to the individual smoker and to those exposed to secondhand smoke. However, the objective harms of the psychoactive substance in question are irrelevant to the validity of our analysis, just as they were to Gusfield's (1963) analysis of the temperance movement's crusade against alcohol use. Moreover, it is not our intention to imply that the proban supporters consciously intended to degrade those of lower social status. No doubt they were motivated primarily by a sincere belief that smoking constitutes a public health hazard. In the end, however, moral indignation and social control flowed down the social hierarchy. Thus we must ask: Would cigarette smoking be defined as deviant if there were a positive correlation between smoking and socioeconomic status?

NOTE

1. Although the term moral entrepreneur is well established in the literature on deviance, there seems to be little attention to or consistency in a corresponding term for the interest group(s) opposing them. Those that have been employed, such as "forces for the status quo" (Markle and Troyer 1979), tend to be awkward. "Status quo defenders" is used here for lack of a simpler or more common term.

REFERENCES

Ashley, Richard. 1975. *Cocaine: Its History, Uses, and Effects.* New York: St. Martin's Press.

Aviado, Domingo M. 1986. "Health Issues Relating to 'Passive' Smoking." Pp. 137–165 in *Smoking and Society: Toward a More Balanced Assessment,* edited by Robert D. Tollison. Lexington, MA: Lexington Books.

Becker, Howard S. 1963. *Outsiders: Studies in the Sociology of Deviance.* New York: Free Press.

Ben-Yehuda, Nachman. 1990. *The Politics and Morality of Deviance: Moral Panics, Drug Abuse, Deviant Science, and Reversed Stigmatization.* Albany, NY: State University of New York Press.

Berger, Peter L. 1986. "A Sociological View of the Antismoking Phenomenon." Pp. 225–240 in *Smoking and Society: Toward a More Balanced Assessment,* edited by Robert D. Tollison. Lexington, MA: Lexington Books.

Bonnie, Richard J., and Charles H. Whitebread II. 1970. "The Forbidden Fruit and the Tree of Knowledge: An Inquiry into the Legal History of American Marihuana Prohibition." *Virginia Law Review* 56: 971–1203.

Covey, Lirio S., Edith A. Zang, and Ernst L. Wynder. 1992. "Cigarette Smoking and Occupational Status: 1977 to 1990." *American Journal of Public Health* 82: 1230–1234.

Feinhandler, Sherwin J. 1986. *The Social Role of Smoking.* Pp. 167–187 in *Smoking and Society: Toward a More Balanced Assessment,* edited by Robert D. Tollison. Lexington, MA: Lexington Books.

Ferrence, Roberta G. 1989. *Deadly Fashion: The Rise and Fall of Cigarette Smoking in North America.* New York: Garland.

Garfinkel, Harold. 1956. "Conditions of Successful Degradation Ceremonies." *American Journal of Sociology* 61: 402–424.

Goffman, Erving. 1963. *Stigma: Notes on the Management of Spoiled Identity.* Englewood Cliffs, NJ: Prentice-Hall.

Goodin, Robert E. 1989. *No Smoking: The Ethical Issues.* Chicago: University of Chicago Press.

Gusfield, Joseph R. 1963. *Symbolic Crusade: Status Politics and the American Temperance Movement.* Urbana, IL: University of Illinois Press.

Luker, Kristin. 1984. *Abortion and the Politics of Motherhood.* Berkeley, CA: University of California.

Marcus, Alfred C., Donald R. Shopland, Lori A. Crane, and William R. Lynn. 1989. "Prevalence of Cigarette Smoking in United States: Estimates from the 1985 Current Population Survey." *Journal of the National Cancer Institute* 81: 409–414.

Markle, Gerald E., and Ronald J. Troyer. 1979. "Smoke Gets in Your Eyes: Cigarette Smoking as Deviant Behavior." *Social Problems* 26: 611–625.

Neuhring, Elane, and Gerald E. Markle. 1974. "Nicotine and Norms: The Re-Emergence of a Deviant Behavior." *Social Problems* 21: 513–526.

Ram, Sidney P. 1941. *How to Get More Fun Out of Smoking.* Chicago: Cuneo.

Roper Organization. 1978, May. *A Study of Public Attitudes Toward Cigarette Smoking and the Tobacco Industry in 1978, Volume 1.* New York: Roper.

Schur, Edwin M. 1980. *The Politics of Deviance: Stigma Contests and the Uses of Power.* New York: Random House.

Steinfeld, Jesse. 1991. "Combating Smoking in the United States: Progress Through Science and Social Action." *Journal of the National Cancer Institute* 83: 1126–1127.

U.S. Bureau of the Census. 1990. *General Population Characteristics.* Washington, DC: U.S. Government Printing Office.

U.S. Department of Health, Education and Welfare. 1964. *Smoking and Health: Report of the Advisory Committee to the Surgeon General of the Public Health Service.* Washington, DC: U.S. Government Printing Office.

U.S. Department of Health and Human Services. 1986. *The Health Consequences of Involuntary Smoking. A Report of the Surgeon General.* Washington, DC: U.S. Government Printing Office.

U.S. Department of Health and Human Services. 1992. *Smoking and Health in the Americas. A 1992 Report of the Surgeon General, in Collaboration with the Pan American Health Organization.* Washington, DC: U.S. Government Printing Office.

Viscusi, W. Kip. 1992. *Smoking: Making the Risky Decision.* New York: Oxford University Press.

Zurcher, Louis A. Jr., and R. George Kirkpatrick. 1976. *Citizens for Decency: Antipornography Crusades as Status Defense.* Austin, TX: University of Texas Press.

16

Moral Panics
The Case of Satanic Day Care Centers

MARY DEYOUNG

DeYoung takes us through the ups and downs of a moral panic in her analysis of the Satanic day care scare. After describing this concept and lodging it in the deviance literature, she takes us through the sequential escalation, peak, and eventual decline of this scare that gripped the public attention and divided whole communities on themselves, alarming families and ruining the careers of child care professionals. deYoung traces the involvement of various kinds of moral entrepreneurs, legislators, and the media, showing how these intentional actors interfaced with chance elements of timing to move the panic along. In so doing, she notes the role of social workers, mental health professionals, attorneys, and law enforcement officers, moral entrepreneurs also highlighted in the previous two selections: Reinarman's professional interest groups and Tuggle and Holmes' knowledge class. This article shows the fervor that can sweep people up and the way it becomes a monolithic juggernaut, rolling over everything in its path.

The term "moral panic" was coined by Cohen (1972) to describe a collective response, generated by unsettling social strain and incited and spread by interest groups, toward persons who are actively transformed into "folk devils" and then treated as threats to dominant social interests and values. Through the use of highly emotive claims and fear-based appeals, a moral panic tends to orchestrate cultural consent that something must be done, and quickly, to deal with this alleged threat. The increased social control that typically follows from such consent ends up preserving and reasserting the very hegemonic values and interests that purportedly are being undermined by the folk devils. A moral panic, then, serves a distinct stabilizing function at a time of unsettling social strain.

Why and how a moral panic arises, the types of people it demonizes, and the methods by which it ends up defining what Durkheim (1938) refers to as the normative contours and moral boundaries of a given society at any historical moment have been of considerable interest to sociologists. The concept has been used to analyze, among other things, the witch-hunts in Europe and America (Ben-Yehuda 1980; Erikson 1966); temperance movements and drug

---

Copyright 1998 from *Deviant Behavior*, vol. 19, 3. Reproduced by permission of Taylor & Francis, Inc., http://www.routledge-ny.com.

panics (Goode 1990; Gusfield 1963; Wilkins 1994); anti-pornography and censorship crusades (Greek and Thompson 1992; Shuker 1986); the mugging and garrotting scares of the 19th century (Adler 1996; Sindall 1987); and law and order campaigns throughout the western world (King 1987; Williams 1993; Zatz 1987).

The term also is used to describe the collective response to new folk devils who were demonized in the 1980s—day care providers who, it was claimed, were abusing their very young charges in satanic rituals that included such horrific practices as blood-drinking, cannibalism, and human sacrifices. Between 1983 and 1991, in fact, over a hundred day care centers in major urban areas and small towns across the country were investigated for what quickly came to be known as satanic ritual abuse (Nathan and Snedeker 1995). These investigations created deep and often irreparable breaches in the communities where they occurred and resulted in scores of arrests, often long and costly criminal trials, many convictions despite the absence of any corroborating and material evidence, usually draconian prison sentences and, over recent years, many reversals of those convictions upon appeal (deYoung 1994).

The satanic day care scare had all of what Goode and Ben-Yehuda (1994) set out as the defining characteristics of a moral panic: it was widespread, overreactive, volatile, hostile, and largely irrational. But another look at this moral panic, and at the day care cases that are the stuff of it, reveals some interesting refinements that are needed in classical moral panic theory if it is to retain its explanatory and analytical power in the contemporary social world.

It is the purpose of this present article to offer that look. First, the article presents an overview of the satanic day care moral panic. Then, it uses data from a sample of 15 day care cases, some more notorious than others, to advance and illustrate a discussion of the areas of classic moral panic theory in need of refinement and updating. It should be noted that for the purposes of this article, a day care case was included in the sample if it met all of the following criteria: (1) an investigation led to the arrest of one or more day care provider; (2) allegations of satanic ritual abuse; that is, of the sexual abuse of children carried out during, or as part of, the ceremonial worship of Satan were actively investigated and publicly reported, even if they were not introduced into criminal trial; and (3) there are sufficient archival data in the form of court transcripts, legal briefs, investigative reports, interview transcripts, and local and national news articles to assess the case. Information that further identifies the 15 cases in the sample is presented in Table 1.

THE SATANIC DAY CARE MORAL PANIC

A basic analysis of any moral panic must account for its timing, target and trigger, content, spread, and denouement (Goode and Ben-Yehuda 1994). Each of these factors will be examined in turn to provide a necessarily brief overview of the satanic day care moral panic of the 1980s.

Table 1 Day Care Center, Year of Investigation Initiation, and Location

Day Care Center	Year	Location
McMartin	1983	Manhattan Beach, CA
Country Walk	1984	Miami, FL
Small World	1984	Niles, MI
Fells Acres	1984	Malden, MA
Georgian Hills	1984	Memphis, TN
Rogers Park Jewish Community	1984	Chicago, IL
Manhattan Ranch	1984	Manhattan Beach, CA
Craig's Country	1985	Clarksville, MD
Felix's	1985	Carson City, NV
East Valley YMCA	1985	El Paso, TX
Glendale Montessori	1987	Stuart, FL
Old Cutler	1989	Miami, FL
Little Rascals	1989	Edenton, NC
Faith Chapel	1989	San Diego, CA
Fran's	1991	Austin, TX

Timing of the Moral Panic

By the 1980s, a number of social, ideological, professional, and political forces had contributed to a growing cultural anxiety about satanic menaces to children (Richardson, Best, and Bromley 1991; Victor 1993). From concerns about demonic influences in heavy metal music, fantasy role-playing games, tarot cards and ouija boards; to urban legends about mysterious satanists abducting fair-haired, blue-eyed children from shopping malls; to rumors of covert satanic cults filming child pornography; to tales about child sex rings, that decade was rife with "mini-moral panics" about satanic menaces to children, and ripened by them for more.

Target and Trigger of the Moral Panic

Coincident with that concern about the protection of children was another one about their daily care. The economic strains that made participation in the market economy a necessity, and the ideological force of the women's movement that made it an increasingly accessible alternative to unpaid housework, combined in that decade to put more and more women with young children into the labor market. In 1980, in fact, a record 45% of them were working outside of the home and using public and private day care centers for daily child care (Hofferth and Phillips 1987).

Most working parents were doing so with more than a little anxiety, however, and considered day care centers a change for the worse from the stay-at-home child care of their parents' generation (Hutchison 1992). That anxiety was heightened by the impact of other types of economic strains. Deep cuts in federal funding that over half of the public day care centers had received just a few years before closed down many of them, and left the remaining centers with high enrollment fees, too many enrollees and, because of low wages, too few providers and high staff turnover (Hofferth and Phillips 1987). Trapped as they were between necessity and contingency, working parents reluctantly began transforming the almost sacred covenantal duty of caring for their young children into businesslike contractual arrangements with day care providers.

The tension created by this imbrication of covenant and contract made that most innocuous of social institutions, the local day care center, the target of a moral panic. But a trigger was yet needed, some kind of spark that in the words of Adler (1996) "would link ethereal sentiment to focused activity"(p. 262).

That spark was lit in 1983 at the McMartin Preschool. A 2.5 year-old enrollee made a statement, vaguely suggestive of sexual abuse, that eventually was worked into an allegation of satanic ritual abuse by social workers who already had some experience as claims makers in the mini-moral panics about satanic menaces to children (deYoung 1997).

Content of the Moral Panic

Eventually, 369 more current and past enrollees of the McMartin Preschool were identified as victims. Their claims, elicited over repeated and suggestive interviews by social workers now convinced of a satanic influence in the case, came to define the still unfamiliar term of satanic ritual abuse and, in doing so, gave the ensuing moral panic its content. The children described, among other ghastly things, the ritualistic ingestion of urine, feces, blood, semen, and human flesh; the disinterment and mutilation of corpses; the sacrifices of infants; and orgies with their day care providers, costumed as devils and witches, in classrooms, tunnels under the center, and in car washes, airplanes, mansions, cemeteries, hotels, ranches, neighborhood stores, local gyms, churches, and hot air balloons. In the accusatorial atmosphere of this nascent moral panic, they named not only the seven McMartin day care providers as their satanic abusers, but local businesspeople and city officials, world leaders, television and film stars, and even their own family members (Nathan and Snedeker 1995).

Spread of the Moral Panic

The same social strains that accounted for the timing of the onset of the satanic day care moral panic also created an engendering environment for its rapid spread across the country. The role of interest, professional and grass-roots groups in sustaining both the drama and exigency of this moral panic, however, cannot be underestimated.

As in all contemporary moral panics, the news media emerged as a major interest group. The McMartin Preschool case, and all the "little McMartins" as the well over a hundred ensuing cases it triggered were euphemistically called, fit all of the criteria for newsworthiness set out by Soothill and Walby (1991). They had ample complexity to warrant daily coverage from different angles; nearly intolerable horror to evoke and sustain intense emotional responses; enough familiarity in terms of location, key claims-makers, and even prime suspects to spark interest, real enough folk devils in the roles of day care providers to demonize; and sufficient exigency to elicit feelings that something must be done, and to focus action in doing it.

The news media hardly were monolithic as an interest group, however. While the nearly hysterical tone of reportage set in local coverage of the McMartin Preschool case was mimicked in local news media in other cases across the country, as well as in the mass media, it was tempered considerably in the national press and quelled completely in a few investigative reports, in large circulation newspapers (Charlier and Downing 1988; Reinhold 1990). And, as both the McMartin Preschool case and many, although certainly not all, of the other satanic day care cases that followed began to fall apart as criminal charges were dismissed, children recanted, or day care providers were exonerated in courts of law, the tone of even local news coverage changed to one of skepticism, criticism, and even excoriation (Sauer 1993; Shaw 1990).

One eventual target of that excoriation was the very professionals who, in the role of what Becker (1963: 145) so aptly described as "moral entrepreneurs," had triggered and spread the satanic day care moral panic. During its nearly decade-long duration, with many of their activities funded and endorsed by the National Center for Child Abuse and Neglect, these social workers, mental health professionals, attorneys and law enforcement officers acted as the chief claims-makers in not only the local and national news media, but also on network television talk shows and primetime news magazines. The social workers and mental health professionals, in particular, became captains of a burgeoning "sexual abuse industry," as Goodyear-Smith (1993) referred to it, and took to the lecture circuit, addressed child protection conferences, conducted workshops, consulted with professionals involved in other cases, and testified as expert witnesses in the criminal and civil trials of the day care providers.

Their claims about satanic day care centers also were voiced in sworn testimony in high profile government hearings. The social worker who interviewed most of the children in the McMartin Preschool case, for example, added a touch of conspiracy to the satanic day care cases when in late 1984 she testified before Congress that an organized operation of "child predators" was using day care centers "as a ruse for a large, unthinkable network of crimes against children. If such an operation involves child pornography, or selling children, as is frequently alleged, it may have greater financial, legal and community resources at its disposal than those attempting to expose it" (Brozan 1984: A-21).

Rhetoric like that may be enough to ignite a moral panic, but rhetoric backed up by "facts" is more combustible (Best 1990). To make more persuasive claims, both about the satanic day care problem and their expertise in it, professionals developed and widely disseminated a wholly synthetic diabolism out of materials haphazardly borrowed from eclectic sources on satanism, the occult, mysticism, paganism and witchcraft (Mulhern 1991). They constructed "indicator lists" to assist other professionals, both here and abroad, in identifying child victims, and "symptom lists" to guide the course of their therapy. Little of what passed for facts about satanic ritual abuse in day care centers was the result of well designed and controlled empirical studies or even systematic case studies, and little of the exchange of facts was taking place in the formal arena of peer-reviewed professional journals, but a burgeoning critical literature was, thus deeply dividing the professional field into claims-makers and counterclaims-makers, believers, and skeptics.

Parents of the allegedly victimized children were unabashed believers, and they constitute the grassroots group that spread the satanic day care moral panic. Like the professionals, and sometimes in conjunction with them, the parents were very vocal and, given their outrage and grief, very pitiable claims-makers. Many of them became keenly politicized as well. Some of the parents in the Little Rascals case, for example, formed an organization called Citizens Against Child Abuse to raise money to assist prosecutors in bringing the seven day care providers to trial. Believe the Children, a group formed by a coterie of McMartin Preschool parents, established a clearinghouse on satanic ritual abuse replete with a speakers' bureau, support groups for parents, police, and prosecutors involved in other day care cases, and a referral list of sympathetic professionals.

As another example, a legislative group called CLOUT was formed by parents in the aftermath of the Manhattan Ranch case. It successfully lobbied the California legislature for "child-friendly" changes in criminal trial procedures in which young children testified as witnesses, changes that were well into place when the long-delayed McMartin Preschool case finally came to trial. The rapid adoption across the country of these and other courtroom innovations, such as shielding child witnesses by allowing them to testify on videotape, closed-circuit television, behind screens, or with their backs to the defendants have not been without legal controversy since they violate the defendants' First and Sixth Amendment rights to a public trial in which accusatory witnesses can be confronted (Montoya 1995).

Denouement of the Moral Panic

The satanic day care moral panic effectively ended in 1991 but its denouement is no more a matter of coincidence than was its onset nearly a decade before. Several factors contributed to its demise. The overweening cultural anxiety about satanic menaces to children largely had been debunked and many of its most vocal claims-makers had retreated into silence (Victor 1993). Changing economic conditions over the decade only increased the number of women in the labor force and the concomitant increase in the use of day care may have worked to integrate this service even more thoroughly into the culture, thus reducing the conflict associated with its use (Hofferth and Phillips 1987).

Changes in day care over the decade of the moral panic, and largely in reaction to it, also acted to pare down any residual conflict about its use. State licensing agencies tightened day care regulations and by legislative fiat were given more teeth to enforce them. As a result, allegations of any kind were promptly, even aggressively, investigated and the licenses of day care centers in compliance were suspended or revoked. In the immediate wake of the Georgian Hills case, for example, 15 local day care centers were investigated for sexual abuse (Mydans 1994), and in the months after the McMartin Preschool case was made public, nine other centers in the small bedroom community of Manhattan Beach were closed down when investigations were launched (Fisher 1989). Significant day care reforms also took place on a state level. After the Country Walk case, as another example, the Governor of Florida held public hearings on the accessibility and quality of day care, and then asked the state legislature for $30 million to reform the system (Ynclan 1984).

In the accusatorial atmosphere of the moral panic, day care providers also took measures to protect themselves from false allegations (Bordin 1996). They installed video cameras to record their activities, opened up private spaces to public view, and kept physical contact with their young charges to a necessary minimum. They adopted open-door policies and invited parents to drop in without notice to talk with staff, observe their children or even spend time with them. The net effect of these and other changes was not only to make day care centers more accessible to worried parents, but more like families, thus further minifying the anxiety about their use (deYoung 1997).

What also certainly played a role in the denouement of the moral panic was the fact that the satanic day care cases, so reprehensible in the court of public opinion, nonetheless did not fare well in courts of law. . . . [I]n many cases in the sample charges were dropped against day care providers and convictions eventually overturned. Although this is certainly due to the fact that the public requires a lower standard of evidence, for lack of a better term, to develop an impression about a case than a court does to adjudicate it, the considerable clash between public opinion and judicial reaction made it difficult to sustain the drama and the exigency of the satanic day care moral panic.

Finally, the schism within the claims-making professional groups that widened over the years of the moral panic also played a role in its demise. The satanic day care center was the site upon which an almost gothic professional struggle for social, political, and moral meaning had taken place, yet no consensus about that meaning was ever reached. The intra-professional dispute about the satanic day care cases, the inter-professional criticism of how they were handled, coupled with the growing public discontent with the expansion of clinical authority into families, institutions, and courts of law very well may have led to the construction of a new folk devil—the overzealous, short-sighted professional, bent on proving sexual abuse of any kind—and a new moral panic, colloquially known as "the backlash," now being directed against them (Myers 1994).

REFERENCES

Adler, Jeffrey. S. 1996. "The Making of a Moral Panic in 19th Century America: The Boston Garroting Hysteria of 1865." *Deviant Behavior* 17: 259–278.

Becker, Howard. 1963. *Outsiders.* New York: Free Press.

Ben-Yehuda, Nachman. 1980. "The European Witch Craze of the 14th and 17th centuries: A Sociological Perspective." *American Journal of Sociology* 86: 1–31.

Best, Joel. 1990. *Threatened Children: Rhetoric and Concern about Child-Victims.* Chicago, IL: University of Chicago Press.

Bordin, Judith A. 1996. "The Aftermath of Unsubstantiated Child Abuse Allegations in Child Care Centers." *Child and Youth Care Forum* 25: 73–87.

Brozan, Nan. 1984. "Witness Says She Fears 'Child Predator' Network." *New York Times* September 18: A-21.

Charlier, Tom, and Shirley Downing. 1988. "Justice Abused: A 1980's Witchhunt" (Series). *Memphis Commercial Appeal,* January.

Cohen, Stanley. 1972. *Folk Devils and Moral Panics: The Creation of the Mods and the Rockers.* Oxford, England: Basil Blackwell.

deYoung, Mary. 1994. "One Face of the Devil: The Satanic Ritual Abuse Moral Crusade and the Law." *Behavioral Sciences and the Law* 12: 389–407.

———. 1997. "The Devil Goes to Day Care: McMartin and the Making of a Moral Panic." *Journal of American Culture* 20: 19–26.

Durkheim, Emile. 1938. *The Rules of Sociological Method.* Trans. Sarah Solovay and John H. Mueller. Ed. George E. G. Catlin. Chicago, IL: University of Chicago.

Erikson, Kai T. 1966. *Wayward Puritan: A Study in the Sociology of Deviance.* New York: Wiley.

Fisher, Mary A. 1989. "A Case of Dominoes?" *Los Angeles Magazine,* October: 126–135.

Goode, Erich. 1990. "The American Drug Panic of the 1980's: Social Construction or Objective Threat?" *International Journal of the Addictions* 25: 1083–1098.

Goode, Erich, and Nachman Ben-Yehuda. 1994. *Moral Panics.* Cambridge, MA: Blackwell.

Goodyear-Smith, Felicity. 1993. *First Do No Harm: The Sexual Abuse Industry.* Auckland, New Zealand: Benton-Gay.

Greek, Cecil E., and William Thompson. 1992. "Anti-Pornography Campaigns: Saving the Family in America and England." *International Journal of Politics, Culture and Society* 5: 601–616.

Gusfield, Joseph. 1963. *Symbolic Crusade.* Urbana, IL: University of Illinois.

Hofferth, Sandra L., and Deborah A. Phillips. 1987. "Child Care in the United States: 1970–1995." *Journal of Marriage and the Family* 49: 559–571.

Hutchison, Elizabeth D. 1992. "Child Welfare as a Woman's Issue." *Families in Society* 73: 67–77.

King, Peter. 1987. "Newspaper Reporting, Prosecution Practice, and Perceptions of Urban Crime: The Colchester Crime Wave of 1765." *Continuity and Change* 2: 423–454.

Montoya, Jean. 1995. "Lessons from *Akiki* and *Michaels* on Shielding Child Witnesses." *Psychology, Public Policy and Law* 1: 340–369.

Mulhern, Sherril. 1991. "Satanism and Psychotherapy: A Rumor in Search of an Inquisition." Pp. 145–172 in *The Satanism Scare,* edited by J.T. Richardson, J. Best, D. G. Bromley. Harthorne, New York: Aldine deGruyter.

Mydans, Seth. 1994. "Grand Jury Rips Sex-Abuse Prosecution, Says Parents, Therapists Influenced Kids." *Memphis Commercial-Appeal,* June 3: A-1.

Myers, John E. B. 1994. *The Backlash: Child Protection Under Fire.* Newbury Park, CA: Sage.

Nathan, Debbie, and Michael Snedeker. 1995. *Satan's Silence: Ritual Abuse and the Making of a Modern American Witch Hunt.* New York: Basic.

Reinhold, Robert. 1990. "How Lawyers and the Media Turned the McMartin Case into a Tragic Media Circus." *New York Times,* January 25: D-1.

Richardson, James T., Joel Best, and David G. Bromley. 1991. *The Satanism Scare.* Hawthorne, New York: Aldine deGruyter.

Sauer, Mark. 1993. "Decade of Accusations: The McMartin Preschool Case Launched 100 Others—And a Vigorous Debate on How to Question Youngsters." *San Diego Union Tribune,* August 29: D-1.

Shaw, David. 1990. "Reporter's Early Exclusive Triggered a Media Frenzy." *Los Angeles Times,* January 20:A-1.

Shuker, Roy. 1986. "'Video Nasties': Censorship and the Politics of Popular Culture." *New Zealand Sociology* 1: 64–73.

Sindall, Rob. 1987. "The London Garrotting Panics of 1856 and 1862." *Social History* 12: 351–358.

Soothill, Keith, and Sylvia Walby. 1991. *Sex Crime in the News.* London, England: Routledge.

Victor, Jeffrey S. 1993. *Satanic Panic.* Chicago, IL: Open Court.

Wilkins, Leslie. T. 1994. "Don't Alter Your Mind—It's the World that's Out of Joint." *Social Justice* 21: 148–153.

Williams, Brian. 1993. "Bail Bandits: The Contruction of Moral Panics." *Critical Social Policy* 13: 104–112.

Ynclan, Nery. 1984. "Legislature May Convene on Child Abuse." *Miami Herald,* September 16: B-1.

Zatz, Majorie S. 1987. "Chicano Youth Gangs and Crime: The Creation of a Moral Panic." *Contemporary Crises* 11: 129–158.

DIFFERENTIAL SOCIAL POWER

17

The Saints and the Roughnecks

WILLIAM J. CHAMBLISS

Chambliss' description of the Saints and the Roughnecks shows how the power of social class can operate to facilitate the definition of some groups as deviant and others as not. Although members of the former group actually engage in more delinquent acts than the latter, the Saints are perceived as "good boys," merely engaging in typical adolescent hijinks. Because of their higher class background, their behavior is defined as socially normative, enabling the police, teachers, community members, and parents to look the other way. On the other hand, the Roughnecks, who come from the "wrong side of the tracks," are perceived to be troublemakers, rabblerousers, and delinquents. We see conflict and labeling theories in effect here since social class is the determinant of society's reactions. Behavior done by teenagers from upstanding, middle class families is tolerated, while similar behavior engaged in by lower class youth is reinforced as deviant. Once again, labels are applied based on status, not on patterns of behavior. The Roughnecks, thus, live up to society's expectations of them and continue into deviance.

Eight promising young men—children of good, stable, white, upper-middle-class families, active in school affairs, good pre-college students—were some of the most delinquent boys at Hanibal High School. While community residents and parents knew that these boys occasionally sowed a few wild oats, they were totally unaware that sowing wild oats completely occupied the daily routine of these young men. The Saints were constantly occupied with truancy, drinking, wild driving, petty theft, and vandalism. Yet not one was officially arrested for any misdeed during the two years I observed them.

This record was particularly surprising in light of my observations during the same two years of another gang of Hanibal High School students,

---

Reprinted by permission of Transaction Publishers. "The Saints and the Roughnecks" by William J. Chambliss, from *Society*, V. 11, No. 1. 1973. Copyright © 1973 Transaction Publishers.

six lower-class white boys known as the Roughnecks. The Roughnecks were constantly in trouble with police and community even though their rate of delinquency was about equal with that of the Saints. What was the cause of this disparity? the result? The following consideration of the activities, social class, and community perceptions of both gangs may provide some answers.

THE SAINTS FROM MONDAY TO FRIDAY

The Saints' principal daily concern was with getting out of school as early as possible. The boys managed to get out of school with minimum danger that they would be accused of playing hookey through an elaborate procedure for obtaining "legitimate" release from class. The most common procedure was for one boy to obtain the release of another by fabricating a meeting of some committee, program, or recognized club. Charles might raise his hand in his 9:00 chemistry class and ask to be excused—a euphemism for going to the bathroom. Charles would go to Ed's math class and inform the teacher that Ed was needed for a 9:30 rehearsal of the drama club play. The math teacher would recognize Ed and Charles as "good students" involved in numerous school activities and would permit Ed to leave at 9:30. Charles would return to his class, and Ed would go to Tom's English class to obtain his release. Tom would engineer Charles's escape. The strategy would continue until as many of the Saints as possible were freed. After a stealthy trip to the car (which had been parked in a strategic spot), the boys were off for a day of fun.

Over the two years I observed the Saints, this pattern was repeated nearly every day. There were variations on the theme, but in one form or another, the boys used this procedure for getting out of class and then off the school grounds. Rarely did all eight of the Saints manage to leave school at the same time. The average number avoiding school on the days I observed them was five.

Having escaped from the concrete corridors the boys usually went either to a pool hall on the other (lower-class) side of town or to a cafe in the suburbs. Both places were out of the way of people the boys were likely to know (family or school officials), and both provided a source of entertainment. The pool hall entertainment was the generally rough atmosphere, the occasional hustler, the sometimes drunk proprietor, and, of course, the game of pool. The cafe's entertainment was provided by the owner. The boys would "accidentally" knock a glass on the floor or spill cola on the counter—not all the time, but enough to be sporting. They would also bend spoons, put salt in sugar bowls, and generally tease whoever was working in the cafe. The owner had opened the cafe recently and was dependent on the boys' business which was, in fact, substantial since between the horsing around and the teasing they bought food and drinks.

THE SAINTS ON WEEKENDS

On weekends the automobile was even more critical than during the week, for on weekends the Saints went to Big Town—a large city with a population of over a million 25 miles from Hanibal. Every Friday and Saturday night most of the Saints would meet between 8:00 and 8:30 and would go into Big Town. Big Town activities included drinking heavily in taverns or nightclubs, driving drunkenly through the streets, and committing acts of vandalism and playing pranks.

By midnight on Fridays and Saturdays the Saints were usually thoroughly high, and one or two of them were often so drunk they had to be carried to the cars. Then the boys drove around town, calling obscenities to women and girls; occasionally trying (unsuccessfully so far as I could tell) to pick girls up; and driving recklessly through red lights and at high speeds with their lights out. Occasionally they played "chicken." One boy would climb out the back window of the car and across the roof to the driver's side of the car while the car was moving at high speed (between 40 and 50 miles an hour); then the driver would move over and the boy who had just crawled across the car roof would take the driver's seat.

Searching for "fair game" for a prank was the boys' principal activity after they left the tavern. The boys would drive alongside a foot patrolman and ask directions to some street. If the policeman leaned on the car in the course of answering the question, the driver would speed away, causing him to lose his balance. The Saints were careful to play this prank only in an area where they were not going to spend much time and where they could quickly disappear around a corner to avoid having their license plate number taken.

Construction sites and road repair areas were the special province of the Saints' mischief. A soon-to-be-repaired hole in the road inevitably invited the Saints to remove lanterns and wooden barricades and put them in the car, leaving the hole unprotected. The boys would find a safe vantage point and wait for an unsuspecting motorist to drive into the hole. Often, though not always, the boys would go up to the motorist and commiserate with him about the dreadful way the city protected its citizenry.

Leaving the scene of the open hole and the motorist, the boys would then go searching for an appropriate place to erect the stolen barricade. An "appropriate place" was often a spot on a highway near a curve in the road where the barricade would not be seen by an oncoming motorist. The boys would wait to watch an unsuspecting motorist attempt to stop and (usually) crash into the wooden barricade. With saintly bearing the boys might offer help and understanding.

A stolen lantern might well find its way onto the back of a police car or hang from a street lamp. Once a lantern served as a prop for a reenactment of the "midnight ride of Paul Revere" until the "play," which was taking place at 2:00 A.M. in the center of a main street of Big Town, was interrupted by a police car several blocks away. The boys ran, leaving the lanterns on the street, and managed to avoid being apprehended.

Abandoned houses, especially if they were located in out-of-the-way places, were fair game for destruction and spontaneous vandalism. The boys would break windows, remove furniture to the yard and tear it apart, urinate on the walls, and scrawl obscenities inside.

Through all the pranks, drinking, and reckless driving the boys managed miraculously to avoid being stopped by police. Only twice in two years was I aware that they had been stopped by a Big City policeman. Once was for speeding (which they did every time they drove whether they were drunk or sober), and the driver managed to convince the policeman that it was simply an error. The second time they were stopped they had just left a nightclub and were walking through an alley. Aaron stopped to urinate and the boys began making obscene remarks. A foot patrolman came into the alley, lectured the boys, and sent them home. Before the boys got to the car one began talking in a loud voice again. The policeman, who had followed them down the alley, arrested this boy for disturbing the peace and took him to the police station where the other Saints gathered. After paying a $5.00 fine, and with the assurance that there would be no permanent record of the arrest, the boy was released.

The boys had a spirit of frivolity and fun about their escapades. They did not view what they were engaged in as "delinquency," though it surely was by any reasonable definition of that word. They simply viewed themselves as having a little fun and who, they would ask, was really hurt by it? The answer had to be no one, although this fact remains one of the most difficult things to explain about the gang's behavior. Unlikely though it seems, in two years of drinking, driving, carousing, and vandalism no one was seriously injured as a result of the Saints' activities.

THE SAINTS IN SCHOOL

The Saints were highly successful in school. The average grade for the group was "B," with two of the boys having close to a straight "A" average. Almost all of the boys were popular and many of them held offices in the school. One of the boys was vice-president of the student body one year. Six of the boys played on athletic teams.

At the end of their senior year, the student body selected ten seniors for special recognition as the "school wheels"; four of the ten were Saints. Teachers and school officials saw no problem with any of these boys and anticipated that they would all "make something of themselves."

How the boys managed to maintain this impression is surprising in view of their actual behavior while in school. Their technique for covering truancy was so successful that teachers did not even realize that the boys were absent from school much of the time. Occasionally, of course, the system would backfire and then the boy was on his own. A boy who was caught would be most contrite, would plead guilty and ask for mercy. He inevitably got the mercy he sought.

Cheating on examinations was rampant, even to the point of orally communicating answers to exams as well as looking at one another's papers. Since none of the group studied, and since they were primarily dependent on one another for help, it is surprising that grades were so high. Teachers contributed to the deception in their admitted inclination to give these boys (and presumably others like them) the benefit of the doubt. When asked how the boys did in school, and when pressed on specific examinations, teachers might admit that they were disappointed in John's performance, but would quickly add that they "knew that he was capable of doing better," so John was given a higher grade than he had actually earned. How often this happened is impossible to know. During the time that I observed the group, I never saw any of the boys take homework home. Teachers may have been "understanding" very regularly.

One exception to the gang's generally good performance was Jerry, who had a "C" average in his junior year, experienced disaster the next year, and failed to graduate. Jerry had always been a little more nonchalant than the others about the liberties he took in school. Rather than wait for someone to come get him from class, he would offer his own excuse and leave. Although he probably did not miss any more classes than most of the others in the group, he did not take the requisite pains to cover his absences. Jerry was the only Saint whom I ever heard talk back to a teacher. Although teachers often called him a "cut up" or a "smart kid," they never referred to him as a troublemaker or as a kid headed for trouble. It seems likely, then, that Jerry's failure his senior year and his mediocre performance his junior year were consequences of his not playing the game the proper way (possibly because he was disturbed by his parents' divorce). His teachers regarded him as "immature" and not quite ready to get out of high school.

THE POLICE AND THE SAINTS

The local police saw the Saints as good boys who were among the leaders of the youth in the community. Rarely, the boys might be stopped in town for speeding or for running a stop sign. When this happened the boys were always polite, contrite, and pled for mercy. As in school, they received the mercy they asked for. None ever received a ticket or was taken into the precinct by the local police.

The situation in Big City, where the boys engaged in most of their delinquency, was only slightly different. The police there did not know the boys at all, although occasionally the boys were stopped by a patrolman. Once they were caught taking a lantern from a construction site. Another time they were stopped for running a stop sign, and on several occasions they were stopped for speeding. Their behavior was as before: contrite, polite, and penitent. The urban police, like the local police, accepted their demeanor as sincere. More important, the urban police were convinced that these were good boys just out for a lark.

THE ROUGHNECKS

Hanibal townspeople never perceived the Saints' high level of delinquency. The Saints were good boys who just went in for an occasional prank. After all, they were well dressed, well mannered, and had nice cars. The Roughnecks were a different story. Although the two gangs of boys were the same age, and both groups engaged in an equal amount of wild-oat sowing, everyone agreed that the not-so-well-dressed, not-so-well-mannered, not-so-rich boys were heading for trouble. Townspeople would say, "You can see the gang members at the drugstore, night after night, leaning against the storefront (sometimes drunk) or slouching around inside buying cokes, reading magazines, and probably stealing old Mr. Wall blind. When they are outside and girls walk by, even respectable girls, these boys make suggestive remarks. Sometimes their remarks are downright lewd."

From the community's viewpoint, the real indication that these kids were in for trouble was that they were constantly involved with the police. Some of them had been picked up for stealing, mostly small stuff, of course, "but still it's stealing small stuff that leads to big time crimes." "Too bad," people said. "Too bad that these boys couldn't behave like the other kids in town: stay out of trouble, be polite to adults, and look to their future."

The community's impression of the degree to which this group of six boys (ranging in age from 16 to 19) engaged in delinquency was somewhat distorted. In some ways the gang was more delinquent than the community thought; in other ways they were less.

The fighting activities of the group were fairly readily and accurately perceived by almost everyone. At least once a month, the boys would get into some sort of fight, although most fights were scraps between members of the group or involved only one member of the group and some peripheral hanger-on. Only three times in the period of observation did the group fight together: once against a gang from across town, once against two blacks, and once against a group of boys from another school. For the first two fights the group went out "looking for trouble"—and they found it both times. The third fight followed a football game and began spontaneously with an argument on the football field between one of the Roughnecks and a member of the opposition's football team.

Jack had a particular propensity for fighting and was involved in most of the brawls. He was a prime mover of the escalation of arguments into fights.

More serious than fighting, had the community been aware of it, was theft. Although almost everyone was aware that the boys occasionally stole things, they did not realize the extent of the activity. Petty stealing was a frequent event for the Roughnecks. Sometimes they stole as a group and coordinated their efforts; other times they stole in pairs. Rarely did they steal alone.

The thefts ranged from very small things like paperback books, comics, and ballpoint pens to expensive items like watches. The nature of the thefts varied from time to time. The gang would go through a period of systematically shoplifting items from automobiles or school lockers. Types of thievery varied with the whim of the gang. Some forms of thievery were more profitable than others, but all thefts were for profit, not just thrills.

Roughnecks siphoned gasoline from cars as often as they had access to an automobile, which was not very often. Unlike the Saints, who owned their own cars, the Roughnecks would have to borrow their parents' cars, an event which occurred only eight or nine times a year. The boys claimed to have stolen cars for joy rides from time to time.

Ron committed the most serious of the group's offenses. With an unidentified associate the boy attempted to burglarize a gasoline station. Although this station had been robbed twice previously in the same month, Ron denied any involvement in either of the other thefts. When Ron and his accomplice approached the station, the owner was hiding in the bushes beside the station. He fired both barrels of a double-barreled shotgun at the boys. Ron was severely injured; the other boy ran away and was never caught. Though he remained in critical condition for several months, Ron finally recovered and served six months of the following year in reform school. Upon release from reform school, Ron was put back a grade in school, and began running around with a different gang of boys. The Roughnecks considered the new gang less delinquent than themselves, and during the following year Ron had no more trouble with the police.

The Roughnecks, then, engaged mainly in three types of delinquency: theft, drinking, and fighting. Although community members perceived that this gang of kids was delinquent, they mistakenly believed that their illegal activities were primarily drinking, fighting, and being a nuisance to passersby. Drinking was limited among the gang members, although it did occur, and theft was much more prevalent than anyone realized.

Drinking would doubtless have been more prevalent had the boys had ready access to liquor. Since they rarely had automobiles at their disposal, they could not travel very far, and the bars in town would not serve them. Most of the boys had little money, and this, too, inhibited their purchase of alcohol. Their major source of liquor was a local drunk who would buy them a fifth if they would give him enough extra to buy himself a pint of whiskey or a bottle of wine.

The community's perception of drinking as prevalent stemmed from the fact that it was the most obvious delinquency the boys engaged in. When one of the boys had been drinking, even a casual observer seeing him on the corner would suspect that he was high.

There was a high level of mutual distrust and dislike between the Roughnecks and the police. The boys felt very strongly that the police were unfair and corrupt. Some evidence existed that the boys were correct in their perception.

The main source of the boys' dislike for the police undoubtedly stemmed from the fact that the police would sporadically harass the group. From the standpoint of the boys, these acts of occasional enforcement of the law were whimsical and uncalled for. It made no sense to them, for example, that the police would come to the corner occasionally and threaten them with arrest for loitering when the night before the boys had been out siphoning gasoline from cars and the police had been nowhere in sight. To the boys, the police were stupid on the one hand, for not being where they should have been and catching the boys in a serious offense, and unfair on the other hand, for trumping up "loitering" charges against them.

From the viewpoint of the police, the situation was quite different. They knew, with all the confidence necessary to be a policeman, that these boys were engaged in criminal activities. They knew this partly from occasionally catching them, mostly from circumstantial evidence ("the boys were around when those tires were slashed"), and partly because the police shared the view of the community in general that this was a bad bunch of boys. The best the police could hope to do was to be sensitive to the fact that these boys were engaged in illegal acts and arrest them whenever there was some evidence that they had been involved. Whether or not the boys had in fact committed a particular act in a particular way was not especially important. The police had a broader view: their job was to stamp out these kids' crimes; the tactics were not as important as the end result.

Over the period that the group was under observation, each member was arrested at least once. Several of the boys were arrested a number of times and spent at least one night in jail. While most were never taken to court, two of the boys were sentenced to six months' incarceration in boys' schools.

THE ROUGHNECKS IN SCHOOL

The Roughnecks' behavior in school was not particularly disruptive. During school hours they did not all hang around together, but tended instead to spend most of their time with one or two other members of the gang who were their special buddies. Although every member of the gang attempted to avoid school as much as possible, they were not particularly successful and most of them attended school with surprising regularity. They considered school a burden—something to be gotten through with a minimum of conflict. If they were "bugged" by a particular teacher, it could lead to trouble. One of the boys, Al, once threatened to beat up a teacher and, according to the other boys, the teacher hid under a desk to escape him.

Teachers saw the boys the way the general community did, as heading for trouble, as being uninterested in making something of themselves. Some were also seen as being incapable of meeting the academic standards of the school. Most of the teachers expressed concern for this group of boys and were willing to pass them despite poor performance, in the belief that failing them would only aggravate the problem.

The group of boys had a grade point average just slightly above "C." No one in the group failed a grade, and no one had better than a "C" average. They were very consistent in their achievement or, at least, the teachers were consistent in their perception of the boys' achievement.

Two of the boys were good football players. Herb was acknowledged to be the best player in the school and Jack was almost as good. Both boys were criticized for their failure to abide by training rules, for refusing to come to practice as often as they should, and for not playing their best during practice. What they lacked in sportsmanship they made up for in skill, apparently, and played every game no matter how poorly they had performed in practice or how many practice sessions they had missed.

TWO QUESTIONS

Why did the community, the school, and the police react to the Saints as though they were good, upstanding, nondelinquent youths with bright futures but to the Roughnecks as though they were tough, young criminals who were headed for trouble? Why did the Roughnecks and the Saints in fact have quite different careers after high school—careers which, by and large, lived up to the expectations of the community?

The most obvious explanation for the differences in the community's and law enforcement agencies' reactions to the two gangs is that one group of boys was "more delinquent" than the other. Which group *was* more delinquent? The answer to this question will determine in part how we explain the differential responses to these groups by the members of the community and, particularly, by law enforcement and school officials.

In sheer number of illegal acts, the Saints were the more delinquent. They were truant from school for at least part of the day almost every day of the week. In addition, their drinking and vandalism occurred with surprising regularity. The Roughnecks, in contrast, engaged sporadically in delinquent episodes. While these episodes were frequent, they certainly did not occur on a daily or even a weekly basis.

The difference in frequency of offenses was probably caused by the Roughnecks' inability to obtain liquor and to manipulate legitimate excuses from school. Since the Roughnecks had less money than the Saints, and teachers carefully supervised their school activities, the Roughnecks' hearts may have been as black as the Saints', but their misdeeds were not nearly as frequent.

There are really no clear-cut criteria by which to measure qualitative differences in antisocial behavior. The most important dimension of the difference is generally referred to as the "seriousness" of the offenses.

If seriousness encompasses the relative economic costs of delinquent acts, then some assessment can be made. The Roughnecks probably stole an average of about $5.00 worth of goods a week. Some weeks the figure was considerably higher, but these times must be balanced against long periods when almost nothing was stolen.

The Saints were more continuously engaged in delinquency but their acts were not for the most part costly to property. Only their vandalism and occasional theft of gasoline would so qualify. Perhaps once or twice a month they would siphon a tankful of gas. The other costly items were street signs, construction lanterns, and the like. All of these acts combined probably did not quite average $5.00 a week, partly because much of the stolen equipment was abandoned and presumably could be recovered. The difference in cost of stolen property between the two groups was trivial, but the Roughnecks probably had a slightly more expensive set of activities than did the Saints.

Another meaning of seriousness is the potential threat of physical harm to members of the community and to the boys themselves. The Roughnecks were more prone to physical violence; they not only welcomed an opportunity to fight; they went seeking it. In addition, they fought among themselves frequently. Although the fighting never included deadly weapons, it was still a menace, however minor, to the physical safety of those involved.

The Saints never fought. They avoided physical conflict both inside and outside the group. At the same time, though, the Saints frequently endangered their own and other people's lives. They did so almost every time they drove a car, especially if they had been drinking. Sober, their driving was risky; under the influence of alcohol it was horrendous. In addition, the Saints endangered the lives of others with their pranks. Street excavations left unmarked were a very serious hazard.

Evaluating the relative seriousness of the two gangs' activities is difficult. The community reacted as though the behavior of the Roughnecks was a problem, and they reacted as though the behavior of the Saints was not. But the members of the community were ignorant of the array of delinquent acts that characterized the Saints' behavior. Although concerned citizens were unaware of much of the Roughnecks' behavior as well, they were much better informed about the Roughnecks' involvement in delinquency than they were about the Saints'.

Visibility

Differential treatment of the two gangs resulted in part because one gang was infinitely more visible than the other. This differential visibility was a direct function of the economic standing of the families. The Saints had access to automobiles and were able to remove themselves from the sight of the community. In as routine a decision as to where to go to have a milkshake after school, the Saints stayed away from the mainstream of community life. Lacking transportation, the Roughnecks could not make it to the edge of town. The center of town was the only practical place for them to meet since their homes were scattered throughout the town and any noncentral meeting place put an undue hardship on some members. Through necessity the Roughnecks congregated in a crowded area where everyone in the community passed frequently, including teachers and law enforcement officers. They could easily see the Roughnecks hanging around the drugstore.

The Roughnecks, of course, made themselves even more visible by making remarks to passersby and by occasionally getting into fights on the corner. Meanwhile, just as regularly, the Saints were either at the cafe on one edge of town or in the pool hall at the other edge of town. Without any particular realization that they were making themselves inconspicuous, the Saints were able to hide their time-wasting. Not only were they removed from the mainstream of traffic, but they were almost always inside a building.

On their escapades the Saints were also relatively invisible, since they left Hanibal and travelled to Big City. Here, too, they were mobile, roaming the city, rarely going to the same area twice.

Demeanor

To the notion of visibility must be added the difference in the responses of group members to outside intervention with their activities. If one of the Saints was confronted with an accusing policeman, even if he felt he was truly innocent of a wrongdoing, his demeanor was apologetic and penitent. A Roughneck's attitude was almost the polar opposite. When confronted with a threatening adult authority, even one who tried to be pleasant, the Roughneck's hostility and disdain were clearly observable. Sometimes he might attempt to put up a veneer of respect, but it was thin and was not accepted as sincere by the authority.

School was no different from the community at large. The Saints could manipulate the system by feigning compliance with the school norms. The availability of cars at school meant that once free from the immediate sight of the teacher, the boys could disappear rapidly. And this escape was well enough planned that no administrator or teacher was nearby when the boys left. A Roughneck who wished to escape for a few hours was in a bind. If it were possible to get free from class, downtown was still a mile away, and even if he arrived there, he was still very visible. Truancy for the Roughnecks meant almost certain detection, while the Saints enjoyed almost complete immunity from sanctions.

Bias

Community members were not aware of the transgressions of the Saints. Even if the Saints had been less discreet, their favorite delinquencies would have been perceived as less serious than those of the Roughnecks.

In the eyes of the police and school officials, a boy who drinks in an alley and stands intoxicated on the street corner is committing a more serious offense than is a boy who drinks to inebriation in a nightclub or a tavern and drives around afterwards in a car. Similarly, a boy who steals a wallet from a store will be viewed as having committed a more serious offense than a boy who steals a lantern from a construction site.

Perceptual bias also operates with respect to the demeanor of the boys in the two groups when they are confronted by adults. It is not simply that adults dislike the posture affected by boys of the Roughneck ilk; more important is the conviction that the posture adopted by the Roughnecks is an indication of their devotion and commitment to deviance as a way of life. The posture becomes a cue, just as the type of the offense is a cue, to the degree to which the known transgressions are indicators of the youths' potential for other problems.

Visibility, demeanor, and bias are surface variables which explain the day-to-day operations of the police. Why do these surface variables operate as they do? Why did the police choose to disregard the Saints' delinquencies while breathing down the backs of the Roughnecks?

The answer lies in the class structure of American society and the control of legal institutions by those at the top of the class structure. Obviously, no representative of the upper class drew up the operational chart for the police which led them to look in the ghettoes and on streetcorners—which led them to see the demeanor of lower-class youth as troublesome and that of upper middle-class youth as tolerable. Rather, the procedure simply developed from experience—experience with irate and influential upper-middle-class parents insisting that their son's vandalism was simply a prank and his drunkenness only a momentary "sowing of wild oats"—experience with cooperative or indifferent, powerless, lower-class parents who acquiesced to the law's definition of their son's behavior.

ADULT CAREERS OF THE SAINTS AND THE ROUGHNECKS

The community's confidence in the potential of the Saints and the Roughnecks apparently was justified. If anything the community members underestimated the degree to which these youngsters would turn out "good" or "bad."

Seven of the eight members of the Saints went on to college immediately after high school. Five of the boys graduated from college in four years. The sixth one finished college after two years in the army, and the seventh spent four years in the air force before returning to college and receiving a B.A. degree. Of these seven college graduates, three went on for advanced degrees. One finished law school and is now active in state politics, one finished medical school and is practicing near Hanibal, and one boy is now working for a Ph.D. The other four college graduates entered submanagerial, managerial, or executive training positions with larger firms.

The only Saint who did not complete college was Jerry. Jerry had failed to graduate from high school with the other Saints. During his second senior year, after the other Saints had gone on to college, Jerry began to hang around with what several teachers described as a "rough crowd"—the gang that was heir apparent to the Roughnecks. At the end of his second senior year, when he did graduate from high school, Jerry took a job as a used car salesman, got married, and quickly had a child. Although he made several abortive attempts to go to college by attending night school, when I last saw him (ten years after high school) Jerry was unemployed and had been living on unemployment for almost a year. His wife worked as a waitress.

Some of the Roughnecks have lived up to community expectations. A number of them were headed for trouble. A few were not.

Jack and Herb were the athletes among the Roughnecks and their athletic prowess paid off handsomely. Both boys received unsolicited athletic scholarships to college. After Herb received his scholarship (near the end of his senior year), he apparently did an about-face. His demeanor became very similar to that of the Saints. Although he remained a member in good standing of the Roughnecks, he stopped participating in most activities and did not hang on the corner as often.

Jack did not change. If anything, he became more prone to fighting. He even made excuses for accepting the scholarship. He told the other gang members that the school had guaranteed him a "C" average if he would come to play football—an idea that seems far-fetched, even in this day of highly competitive recruiting.

During the summer after graduation from high school, Jack attempted suicide by jumping from a tall building. The jump would certainly have killed most people trying it, but Jack survived. He entered college in the fall and played four years of football. He and Herb graduated in four years, and both are teaching and coaching in high schools. They are married and have stable families. If anything, Jack appears to have a more prestigious position in the community than does Herb, though both are well respected and secure in their positions.

Two of the boys never finished high school. Tommy left at the end of his junior year and went to another state. That summer he was arrested and placed on probation on a manslaughter charge. Three years later he was arrested for murder; he pleaded guilty to second degree murder and is serving a 30-year sentence in the state penitentiary.

Al, the other boy who did not finish high school, also left the state in his senior year. He is serving a life sentence in a state penitentiary for first degree murder.

Wes is a small-time gambler. He finished high school and "bummed around." After several years he made contact with a bookmaker who employed him as a runner. Later he acquired his own area and has been working it ever since. His position among the bookmakers is almost identical to the position he had in the gang; he is always around but no one is really aware of him. He makes no trouble and he does not get into any. Steady, reliable, capable of keeping his mouth closed, he plays the game by the rules, even though the game is an illegal one.

That leaves only Ron. Some of his former friends reported that they had heard he was "driving a truck up north," but no one could provide any concrete information.

REINFORCEMENT

The community responded to the Roughnecks as boys in trouble, and the boys agreed with that perception. Their pattern of deviancy was reinforced, and breaking away from it became increasingly unlikely. Once the boys acquired an image of themselves as deviants, they selected new friends who affirmed that self-image. As that self-conception became more firmly entrenched, they also became willing to try new and more extreme deviances. With their growing alienation came freer expression of disrespect and hostility for representatives of the legitimate society. This disrespect increased the community's negativism, perpetuating the entire process of commitment to deviance. Lack of a commitment to deviance works the same way. In either case, the process will perpetuate itself unless some event (like a scholarship to college or a sudden failure) external to the established relationship intervenes. For two of the Roughnecks (Herb and Jack), receiving college athletic scholarships created new relations and culminated in a break with the established pattern of deviance. In the case of one of the Saints (Jerry), his parents' divorce and his failing to graduate from high school changed some of his other relations. Being held back in school for a year and losing his place among the Saints had sufficient impact on Jerry to alter his self-image and virtually to assure that he would not go on to college as his peers did. Although the experiments of life can rarely be reversed, it seems likely in view of the behavior of the other boys who did not enjoy this special treatment by the school that Jerry, too, would have "become something" had he graduated as anticipated. For Herb and Jack outside intervention worked to their advantage; for Jerry it was his undoing.

Selective perception and labeling—finding, processing, and punishing some kinds of criminality and not others—means that visible, poor, nonmobile, outspoken, undiplomatic "tough" kids will be noticed, whether their actions are seriously delinquent or not. Other kids, who have established a reputation for being bright (even though underachieving), disciplined and involved in respectable activities, who are mobile and monied, will be invisible when they deviate from sanctioned activities. They'll sow their wild oats—perhaps even wider and thicker than their lower-class cohorts—but they won't be noticed. When it's time to leave adolescence most will follow the expected path, settling into the ways of the middle class, remembering fondly the delinquent but unnoticed fling of their youth. The Roughnecks and others like them may turn around, too. It is more likely that their noticeable deviance will have been so reinforced by police and community that their lives will be effectively channeled into careers consistent with their adolescent background.

18

The Police and the Black Male

ELIJAH ANDERSON

In this excerpt from his book, Streetwise, *Anderson gives us a glimpse into the perspective of ghetto inhabitants as they are handled by police: the agents of social control. Race (African-American), gender (male), and age (youth) are the disempowering features of this population. For young African-American men living in America's inner city, life may be made even more difficult by police who assume that, if there is trouble in the neighborhood, it must be caused by these youngsters. Black men, walking alone at night or cruising a neighborhood in a car, may be stopped, harassed, questioned, or beaten, even if they have done nothing wrong. Merely due to their ascribed status as African-American, their activities are scrutinized in different ways than others in society. They may be subject to the "cycle of oppression," as they are color coded (racially profiled), stopped and questioned, arrested without substantive cause, and assigned to a public defender who most often leads them into a plea bargain, resulting in their having a record that shows up and verifies their deviant status the next time they are arbitrarily stopped. Drawing on conflict theory, we see here how deviant status is used by social control agents to subordinate less powerful groups.*

The police, in the Village-Northton [neighborhood] as elsewhere, represent society's formal, legitimate means of social control.[1] Their role includes protecting law-abiding citizens from those who are not law-abiding, by preventing crime and by apprehending likely criminals. Precisely how the police fulfill the public's expectations is strongly related to how they view the neighborhood and the people who live there. On the streets, color-coding often works to confuse race, age, class, gender, incivility, and criminality, and it expresses itself most concretely in the person of the anonymous black male. In doing their job, the police often become willing parties to this general color-coding of the public environment, and related distinctions, particularly those of skin color and gender, come to convey definite meanings. Although such coding may make the work of the police more manageable, it may also fit well with their own presuppositions regarding race and class relations, thus shaping officers' perceptions of crime "in the city." Moreover, the anonymous black male is usually an ambiguous figure who arouses the utmost caution and is generally considered dangerous until he proves he is not. . . .

Elijah Anderson, *Streetwise* (Chicago: The University of Chicago Press, 1990). Reprinted by permission of the publisher and the author.

There are some who charge—... perhaps with good reason—that the police are primarily agents of the middle class who are working to make the area more hospitable to middle-class people at the expense of the lower classes. It is obvious that the police assume whites in the community are at least middle class and are trustworthy on the streets. Hence the police may be seen primarily as protecting "law-abiding" middle-class whites against anonymous "criminal" black males.

To be white is to be seen by the police—at least superficially—as an ally, eligible for consideration and for much more deferential treatment than that accorded blacks in general. This attitude may be grounded in the backgrounds of the police themselves.[2] Many have grown up in Eastern City's "ethnic" neighborhoods. They may serve what they perceive as their own class and neighborhood interests, which often translates as keeping blacks "in their place"—away from neighborhoods that are socially defined as "white." In trying to do their job, the police appear to engage in an informal policy of monitoring young black men as a means of controlling crime, and often they seem to go beyond the bounds of duty. The following field note shows what pressures and racism young black men in the Village may endure at the hands of the police:

> At 8:30 on a Thursday evening in June I saw a police car stopped on a side street near the Village. Beside the car stood a policeman with a young black man. I pulled up behind the police car and waited to see what would happen. When the policeman released the young man, I got out of my car and asked the youth for an interview.
>
> "So what did he say to you when they stopped you? What was the problem?" I asked. "I was just coming around the corner, and he stopped me, asked me what was my name, and all that. And what I had in my bag. And where I was coming from. Where I lived, you know, all the basic stuff, I guess. Then he searched me down and, you know, asked me who were the supposedly tough guys around here? That's about it. I couldn't tell him who they are. How do I know? Other gang members could, but I'm not from a gang, you know. But he tried to put me in a gang bag, though." "How old are you?" I asked. "I'm seventeen, I'll be eighteen next month." "Did he give any reason for stopping you?" "No, he didn't. He just wanted my address, where I lived, where I was coming from, that kind of thing. I don't have no police record or nothin'. I guess he stopped me on principle, 'cause I'm black." "How does that make you feel?" I asked. "Well, it doesn't bother me too much, you know, as long as I know that I hadn't done nothin', but I guess it just happens around here. They just stop young black guys and ask 'em questions, you know. What can you do?"

On the streets late at night, the average young black man is suspicious of others he encounters, and he is particularly wary of the police. If he is dressed in the uniform of the "gangster," such as a black leather jacket, sneakers, and a "gangster cap," if he is carrying a radio or a suspicious bag (which may be confiscated), or if he is moving too fast or too slow, the police may stop him. As part of the routine, they search him and make him sit in the police car while they run a check to see whether there is a "detainer" on him. If there is nothing, he is allowed to go on his way. After this ordeal the youth is often left afraid, sometimes shaking, and uncertain about the area he had previously taken for granted. He is upset in part because he is painfully aware of how close he has come to being in "big trouble." He knows of other youths who have gotten into a "world of trouble" simply by being on the streets at the wrong time or when the police were pursuing a criminal. In these circumstances, particularly at night, it is relatively easy for one black man to be mistaken for another. Over the years, while walking through the neighborhood I have on occasion been stopped and questioned by police chasing a mugger, but after explaining myself I was released.

Many youths, however, have reason to fear such mistaken identity or harassment, since they might be jailed, if only for a short time, and would have to post bail money and pay legal fees to extricate themselves from the mess (Anderson 1986). When law-abiding blacks are ensnared by the criminal justice system, the scenario may proceed as follows. A young man is arbitrarily stopped by the police and questioned. If he cannot effectively negotiate with the officer(s), he may be accused of a crime and arrested. To resolve this situation he needs financial resources, which for him are in short supply. If he does not have money for an attorney, which often happens, he is left to a public defender who may be more interested in going along with the court system than in fighting for a poor black person. Without legal support, he may well wind up "doing time" even if he is innocent of the charges brought against him. The next time he is stopped for questioning he will have a record, which will make detention all the more likely.

Because the young black man is aware of many cases when an "innocent" black person was wrongly accused and detained, he develops an "attitude" toward the police. The street word for police is "the man," signifying a certain machismo, power, and authority. He becomes concerned when he notices "the man" in the community or when the police focus on him because he is outside his own neighborhood. The youth knows, or soon finds out, that he exists in a legally precarious state. Hence he is motivated to avoid the police, and his public life becomes severely circumscribed.

To obtain fair treatment when confronted by the police, the young man may wage a campaign for social regard so intense that at times it borders on obsequiousness. As one streetwise black youth said: "If you show a cop that you nice and not a smartass, they be nice to you. They talk to you like the man you are. You gonna get ignorant like a little kid, they gonna get ignorant with you." Young black males often are particularly deferential toward the police even when they are completely within their rights and have done nothing wrong. Most often this is not out of blind acceptance or respect for the "law," but because they know the police can cause them hardship. When confronted or arrested, they adopt a particular style of behavior to get on the policeman's good side. Some simply "go limp" or politely ask, "What seems to be the trouble, officer?" This pose requires a deference that is in sharp contrast with the youth's more usual image, but many seem to take it in stride or not even to realize it. Because they are concerned primarily with staying out of trouble, and because they perceive the police as arbitrary in their use of power, many defer in an equally arbitrary way. Because of these pressures, however, black youths tend to be especially mindful of the police and, when they are around, to watch their own behavior in public. Many have come to expect harassment and are inured to it; they simply tolerate it as part of living in the Village-Northton.

After a certain age, say twenty-four, a black man may no longer be stopped so often, but he continues to be the object of policy scrutiny. As one twenty-seven-year-old black college graduate speculated:

> I think they see me with my little bag with papers in it. They see me with penny loafers on. I have a tie on, some days. They don't stop me so much now. See, it depends on the circumstances. If something goes down, and they hear that the guy had on a big black coat, I may be the one. But when I was younger, they could just stop me, carte blanche, any old time. Name taken, searched, and this went on endlessly. From the time I was about twelve until I was sixteen or seventeen, endlessly, endlessly. And I come from a lower-middle-class black neighborhood, OK, that borders a white neighborhood. One neighborhood is all black, and one is all white. OK, just because we were so close to that neighborhood, we were stopped endlessly. And it happened even more when we went up into a suburban community. When we would ride up and out to the suburbs, we were stopped every time we did it.
>
> If it happened today, now that I'm older, I would really be upset. In the old days when I was younger, I didn't know any better. You just expected it, you knew it was gonna happen. Cops would come up, "What you doing, where you coming from?" Say things to you. They might even call you nigger.

Such scrutiny and harassment by local police makes black youths see them as a problem to get beyond, to deal with, and their attempts affect their overall behavior. To avoid encounters with "the man," some streetwise young men camouflage themselves, giving up the urban uniform and emblems that identify them as "legitimate" objects of police attention. They may adopt a more conventional presentation of self, wearing chinos, sweat suits, and generally more conservative dress. Some youths have been known to "ditch" a favorite jacket if they see others wearing one like it, because wearing it increases their chances of being mistaken for someone else who may have committed a crime.

But such strategies do not always work over the long run and must be constantly modified. For instance, because so many young ghetto blacks have begun to wear Fila and Adidas sweat suits as status symbols, such dress has become incorporated into the public image generally associated with young black males. These athletic suits, particularly the more expensive and colorful ones, along with high-priced sneakers, have become the leisure dress of successful drug dealers, and other youths will often mimic their wardrobe to "go for bad" in the quest for local esteem. Hence what was once a "square" mark of distinction approximating the conventions of the wider culture has been adopted by a neighborhood group devalued by that same culture. As we saw earlier, the young black male enjoys a certain power over fashion: whatever the collective peer group embraces can become "hip" in a manner the wider society may not desire (see Goffman 1963). These same styles then attract the attention of the agents of social control.

THE IDENTIFICATION CARD

Law-abiding black people, particularly those of the middle class, set out to approximate middle-class whites in styles of self-presentation in public, including dress and bearing. Such middle-class emblems, often viewed as "square," are not usually embraced by young working-class blacks. Instead, their connections with and claims on the institutions of the wider society seem to be symbolized by the identification card. The common identification card associates its holder with a firm, a corporation, a school, a union, or some other institution of substance and influence. Such a card, particularly from a prominent establishment, puts the police and others on notice that the youth is "somebody," thus creating an important distinction between a black man who can claim a connection with the wider society and one who is summarily judged as "deviant." Although blacks who are established in the middle class might take such cards for granted, many lower-class blacks, who continue to find it necessary to campaign for civil rights denied them because of skin color, believe that carrying an identification card brings them better treatment than is meted out to their less fortunate brothers and sisters. For them this link to the wider society, though often tenuous, is psychically and socially important. The young college graduate continues:

> I know [how] I used to feel when I was enrolled in college last year, when I had an ID card. I used to hear stories about the blacks getting stopped over by the dental school, people having trouble sometimes. I would see that all the time. Young black male being stopped by the police. Young black male in handcuffs. But I knew that because I had that ID card that I would not be mistaken for just somebody snatching a pocketbook, or just somebody being where maybe I wasn't expected to be. See, even though I was intimidated by the campus police—I mean, the first time I walked into the security office to get my ID they all gave me the double-take to see if I was somebody they were looking for. See, after I got the card, I was like, well, they can think that now, but I have this [ID card]. Like, see, late at night when I be walking around, and the cops be checking me out, giving me the looks, you know. I mean, I know guys, students, who were

getting stopped all the time, sometimes by the same officer, even though they had the ID. And even they would say, "Hey, I got the ID, so why was I stopped?"

The cardholder may believe he can no longer be treated summarily by the police, that he is no longer likely to be taken as a "no count," to be prejudicially confused with that class of blacks "who are always causing trouble on the trolley." Furthermore, there is a firm belief that if the police stop a person who has a card, they cannot "do away with him without somebody coming to his defense." This concern should not be underestimated. Young black men trade stories about mistreatment at the hands of the police; a common one involves policemen who transport youths into rival gang territories and release them, telling them to get home the best way they can. From the youth's perspective, the card signifies a certain status in circumstances where little recognition was formerly available.

"DOWNTOWN" POLICE AND LOCAL POLICE

In attempting to manage the police—and by implication to manage themselves—some black youths have developed a working connection of the police in certain public areas of the Village-Northton. Those who spend a good amount of their time on these corners, and thus observing the police, have come to distinguish between the "downtown" police and the "regular" local police.

The local police are the ones who spend time in the area; normally they drive around in patrol cars, often one officer to a car. These officers usually make a kind of working peace with the young men on the streets; for example, they know the names of some of them and may even befriend a young boy. Thus they offer an image of the police department different from that displayed by the "downtown" police. The downtown police are distant, impersonal, and often actively looking for "trouble." They are known to swoop down arbitrarily on gatherings of black youths standing on a street corner; they might punch them around, call them names, and administer other kinds of abuse, apparently for sport. A young Northton man gave the following narrative about his experiences with the police.

> And I happen to live in a violent part. There's a real difference between the violence level in the Village and the violence level in Northton. In the nighttime it's more dangerous over there.
>
> It's so bad now, they got downtown cops over there now. They doin' a good job bringin' the highway patrol over there. Regular cops don't like that. You can tell that. They even try to emphasize to us the certain category. Highway patrol come up, he leave, they say somethin' about it.

"We can do our job over here." We call [downtown police] Nazis. They about six feet eight, seven feet. We walkin', they jump out. "You run, and we'll blow your nigger brains out." I hate bein' called a nigger. I want to say somethin' but get myself in trouble.

When a cop do somethin', nothing happen to 'em. They come from downtown. From what I heard some of 'em don't even wear their real badge numbers. So you have to put up with that. Just keep your mouth shut when they stop you, that's all. Forget about questions, get against the wall, just obey 'em. "Put all that out right there"—might get rough with you now. They snatch you by the shirt, throw you against the wall, pat you hard, and grab you by the arms, and say, "Get outta here." They call you nigger this and little black this, and things like that. I take that. Some of the fellas get mad. It's a whole different world.

Yeah, they lookin' for trouble. They gotta look for trouble when you got five, eight police cars together and they laughin' and talkin', start teasin' people. One night we were at a bar, we read in the paper that the downtown cops comin' to straighten things out. Same night, three police cars, downtown cops with their boots on, they pull the sticks out, beatin' around the corner, chase into bars. My friend Todd, one of 'em grabbed him and knocked the shit out of him. He punched 'im, a little short white guy. They start a riot. Cops started that shit. Everybody start seein' how wrong the cops was—they start throwin' bricks and bottles, cussin' 'em out. They lock my boy up; they had to let him go. He was just standin' on the corner, they snatch him like that.

One time one of 'em took a gun and began hittin' people. My boy had a little hickie from that. He didn't know who the cop was, because there was no such thing as a badge number. They have phony badge numbers. You can tell they're tougher, the way they dress, plus they're bigger. They have boots, trooper pants, blond hair, blue eyes, even black [eyes]. And they seven feet tall, and six foot six inches and six foot eight inches. Big! They are the rough cops. You don't get smart with them or they beat the shit out of you *in front of everybody,* they don't care.

We call 'em Nazis. Even the blacks among them. They ride along with 'em. They stand there and watch a white cop beat your brains out. What takes me out is the next day you don't see 'em. Never see 'em again, go down there, come back, and they ride right back downtown, come back, do their little dirty work, go back downtown, and put their real badges on. You see 'em with a forty-five or fifty-five number: "Ain't no such number here, I'm sorry, son." Plus, they got unmarked cars. No sense takin' 'em to court. But when that happened at that bar, another black cop from the sixteenth [local] district, ridin' a real car, came back and said, "Why don't y'all go on over to the sixteenth district and file a complaint? Them musclin' cops was wrong. Beatin' people." So about ten people went over there; sixteenth district knew nothin' about it. They come in unmarked cars, they must have been downtown cops. Some of 'em do it. Some of 'em are off duty, on their way home. District commander told us they do that. They have a patrol over there, but them cops from downtown have control of them cops. Have bigger ranks and bigger guns. They carry .357s and regular cops carry little .38s. Downtown cops are all around. They carry magnums.

Two cars the other night. We sittin' on the steps playing cards. Somebody called the cops. We turn around and see four regular police cars and two highway police cars. We drinkin' beer and playin' cards. Police get out and say you're gamblin'. We say we got nothin' but cards here, we got no money. They said all right, got back in their cars, and drove away. Downtown cops dressed up like troopers. That's intimidation. Damn!

You call a cop, they don't come. My boy got shot, we had to take him to the hospital ourselves. A cop said, "You know who did it?" We said no. He said, "Well, I hope he dies if y'all don't say nothin'." What he say that for? My boy said, "I hope your mother die," he told the cop right to his face. And I was grabbin' another cop, and he made a complaint about that. There were a lot of witnesses. Even the nurse behind the counter said the cop had no business saying nothin' like that. He said it loud, "I hope he dies." Nothin' like that should be comin' from a cop.

Such behavior by formal agents of social control may reduce the crime rate, but it raises questions about social justice and civil rights. Many of the old-time liberal white residents of the Village view the police with some ambivalence. They want their streets and homes defended, but many are convinced that the police manhandle "kids" and mete out an arbitrary form of "justice." These feelings make many of them reluctant to call the police when they are needed, and they may even be less than completely cooperative after a crime has been committed. They know that far too often the police simply "go out and pick up some poor black kid." Yet they do cooperate, if ambivalently, with these agents of social control.

In an effort to gain some balance in the emerging picture of the police in the Village-Northton, I interviewed local officers. The following edited conversation with Officer George Dickens (white) helps place in context the fears and concerns of local residents, including black males:

I'm sympathetic with the people who live in this neighborhood [the Village-Northton], who I feel are victims of drugs. There are a tremendous number of decent, hardworking people who are just trying to live their life in peace and quiet, not cause any problems for their neighbors, not cause any problems for themselves. They just go about their own business and don't bother anyone. The drug situation as it exists in Northton today causes them untold problems. And some of the young kids are involved in one way or another with this drug culture. As a result, they're gonna come into conflict even with the police they respect and have some rapport with.

We just went out last week on Thursday and locked up ten young men on Cherry Street, because over a period of about a week, we had undercover police officers making drug buys from those young men. This was very well documented and detailed. They were videotaped selling the drugs. And as a result, right now, if you walk down Cherry Street, it's pretty much a ghost town; there's nobody out. [Before, Cherry Street was notorious for drug traffic.] Not only were people buying drugs there, but it was a very active street. There's been some shock value as a result of all those arrests at one time.

Now, there's two reactions to that. The [television] reporters went out and interviewed some people who said, "Aw, the police overreacted, they locked up innocent people. It was terrible, it was harassment." One of the neighbors from Cherry Street called me on Thursday, and she was outraged. Because she said, "Officer, it's not fair. We've been working with the district for well over a year trying to solve some of the problems on Cherry Street." But most of the neighbors were thrilled that the police came and locked all those kids up. So you're getting two conflicting reactions here. One from the people that live there that just wanta be left alone, alright? Who are really being harassed by the drug trade and everything that's involved in it. And then you have a reaction from the people that are in one way or another either indirectly connected or directly connected, where they say, "You know, if a young man is selling drugs, to him that's a job." And if he gets arrested, he's out of a job. The family's lost their income. So they're not gonna pretty much want anybody to come in there to make arrests. So you've got contradicting elements of the community there. My philosophy is that we're going to try to make Northton livable. If that means we have to arrest some of the residents of Northton, that's what we have to do.

You talk to Tyrone Pitts, you know the group that they formed was formed because of a reaction to complaints against one of the officers of how the teenagers were being harassed. And it turned out that basically what he [the officer] was doing was harassing drug dealers. When Northton against Drugs actually formed and seemed to jell, they developed a close working relationship with the police here. For that reason, they felt the officer was doing his job.

I've been here eighteen months. I've seen this neighborhood go from . . . let me say, this is the only place I've ever worked where I've seen a rapport between the police department and the general community like the one we have right now. I've never seen it any place else before coming here. And I'm not gonna claim credit because this happened while I happened to be here. I think a lot of different factors were involved. I think the community was ready to work with the police because of the terrible situation in reference to crack. My favorite expression when talking about crack is "crack changed everything." Crack changed the rules of how the police and the community have to interact with each other. Crack changed the rules about how the criminal justice system is gonna work, whether it works well or poorly. Crack is causing the prisons to be overcrowded. Crack is gonna cause the people that do drug rehabilitation to be overworked. It's gonna cause a wide variety of things. And I think the reason the rapport between the police and the community in Northton developed at the time it did is very simply that drugs to a certain extent made many areas in this city unlivable.

In effect the officer is saying that the residents, regardless of former attitudes, are now inclined to be more sympathetic with the police and to work with them. And at the same time, the police are more inclined to work with the residents. Thus, not only are the police and the black residents of Northton working together, but different groups in the Village and Northton are working with each other against drugs. In effect, law-abiding citizens are coming together, regardless of race, ethnicity, and class. He continues:

Both of us [police and the community] are willing to say, "Look, let's try to help each other." The nice thing about what was started here is that it's spreading to the rest of the city. If we don't work together, this problem is gonna devour us. It's gonna eat us alive. It's a state of emergency, more or less.

In the past there was significant negative feeling among young black men about the "downtown" cops coming into the community and harassing them. In large part these feelings continue to run strong, though many young men appear to "know the score" and to be resigned to their situation, accommodating and attempting to live with it. But as the general community feels under attack, some residents are willing to forgo certain legal and civil rights and undergo personal inconvenience in hopes of obtaining a sense of law and order. The officer continues:

Today we don't have too many complaints about police harassment in the community. Historically there were these complaints, and in almost any minority neighborhood in Eastern City where I ever worked there was more or less a feeling of that [harassment]. It wasn't just Northton; it was a feeling that the police were the enemy. I can honestly say that for the first time in my career I don't feel that people look at me like I'm the enemy. And it feels nice; it feels real good not to be the enemy, ha-ha. I think we [the police] realize that a lot of problems here [in the Village-Northton] are related to drugs. I think the neighborhood realizes that too. And it's a matter of "Who are we gonna be angry with? Are we gonna be angry with the police because we feel like they're this army of occupation, or are we gonna argue with these people who are selling drugs to our kids and shooting up our neighborhoods and generally causing havoc in the area? Who deserves the anger more?" And I think, to a large extent, people of the Village-Northton decided it was the drug dealers and not the police.

I would say there are probably isolated incidents where the police would stop a male in an area where there is a lot of drugs, and this guy may be perfectly innocent, not guilty of doing anything at all. And yet he's stopped by the police because he's specifically in that area, on that street corner where we know drugs are going hog wild. So there may be isolated incidents of that. At the same time, I'd say I know for a fact that our complaints against police in this division, the whole division, were down about 45 percent. If there are complaints, if there are instances of abuse by the police, I would expect that our complaints would be going up. But they're not; they're dropping.

Such is the dilemma many Villagers face when they must report a crime or deal in some direct way with the police. Stories about police prejudice against blacks are often traded at Village get-togethers. Cynicism about the effectiveness of the police mixed with community suspicion of their behavior toward blacks keeps middle-class Villagers from embracing the notion that they must rely heavily on the formal means of social control to maintain even the minimum freedom of movement they enjoy on the streets.

Many residents of the Village, especially those who see themselves as the "old guard" or "old-timers," who were around during the good old days when antiwar and antiracist protest was a major concern, sigh and turn their heads when they see the criminal justice system operating in the ways described here. They express hope that "things will work out," that tensions will ease, that crime will decrease and police behavior will improve. Yet as incivility and crime become increasing problems in the neighborhood, whites become less tolerant of anonymous blacks and more inclined to embrace the police as their heroes.

Such criminal and social justice issues, crystallized on the streets, strain relations between the newcomers and many of the old guard, but in the present context of drug-related crime and violence in the Village-Northton, many of the old-timers are adopting a "law and order" approach to crime and public safety, laying blame more directly on those they see as responsible for such crimes, though they retain some ambivalence. Newcomers can share such feelings with an increasing number of old-time "liberal" residents. As one middle-aged white woman who has lived in the Village for fifteen years said:

> When I call the police, they respond. I've got no complaints. They are fine for me. I know they sometimes mistreat black males. But let's face it, most of the crime is committed by them, and so they can simply tolerate more scrutiny. But that's them.

Gentrifiers and the local old-timers who join them, and some traditional residents continue to fear, care more for their own safety and well-being than for the rights of young blacks accused of wrong-doing. Yet reliance on the police, even by an increasing number of former liberals, may be traced to a general feeling of oppression at the hands of street criminals, whom many believe are most often black. As these feelings intensify and as more yuppies and students inhabit the area and press the local government for services, especially police protection, the police may be required to "ride herd" more stringently on the youthful black population. Thus young black males are often singled out as the "bad" element in an otherwise healthy diversity, and the tensions between the lower-class black ghetto and the middle- and upper-class white community increase rather than diminish.

NOTES

1. See Rubinstein (1973); Wilson (1968); Fogelson (1977); Reiss (1971); Bittner (1967); Banton (1964).

2. For an illuminating typology of police work that draws a distinction between "fraternal" and "professional" codes of behavior, see Wilson (1968).

REFERENCES

Anderson, Elijah. 1986. "Of old heads and young boys: Notes on the urban black experience." Unpublished paper commissioned by the National Research Council, Committee on the Status of Black Americans.

Banton, Michael. 1964. *The policeman and the community.* New York: Basic Books.

Bittner, Egon. 1967. The police on Skid Row. *American Sociological Review* 32 (October): 699–715.

Fogelson, Robert. 1977. *Big city police.* Cambridge: Harvard University Press.

Goffman, Erving. 1963. *Behavior in public places.* New York: Free Press.

Reiss, Albert J. 1971. *The police and the public.* New Haven: Yale University Press.

Rubinstein, Jonathan. 1973. *City police.* New York: Farrar, Straus and Giroux.

Wilson, James Q. 1968. "The police and the delinquent in two cities." In *Controlling delinquents,* ed. Stanton Wheeler. New York: John Wiley.

19

Homophobia and Women's Sport

ELAINE M. BLINDE AND DIANE E. TAUB

Blinde and Taub explore the role of attributed sexual orientation in disempowering women who violate gender norms: varsity female collegiate athletes. By challenging the gender order and opposing male domination, these women intrude into a traditional male sanctum and threaten the male domain of physicality and strength. By casting the lesbian label on women athletes, society stigmatizes them as masculine and as sexual perverts. While the homosexual label is routinely used to degrade male athletes who fail to live up to the hyper-masculine ideal, the lesbian label is used to divide and silence female athletes. They may adopt the perspective of their oppressors and demean their teammates as lesbians, thus destroying team solidarity, and/or shun the label, but be forced to acknowledge its demeaning power as they attempt to escape it. The forceful effect of the lesbian label applied to women athletes shows the dominance not only of heterosexuals over homosexuals, but of men over women.

Central to the preservation of a patriarchal and heterosexist society is a well-established gender order with clearly defined norms and sanctions governing the behavior of men and women. This normative gender system is relayed to and installed in members of society through a pervasive socialization network that is evident in both everyday social interaction and social institutions (Schur 1984). Conformity to established gender norms contributes to the reproduction of male dominance and heterosexual privilege (Lenskyj 1991; Stockard and Johnson 1980).

Despite gender role socialization, not all individuals engage in behavior consistent with gender expectations. Recognizing the potential threat of such aberrations, various mechanisms exist that encourage compliance with the normative gender order. Significant in such processes are the stigmatization and devaluation of those whose behavior deviates from the norm (Schur 1984).

Women's violation of traditional gender role norms represents a particularly serious threat to the patriarchal and heterosexist society because this deviant behavior resists women's subordinate status (Schur 1984). When women

---

"Homophobia and Women's Sport: The Disempowerment of Athletes," by Elaine M. Blinde and Diane E. Taub, *Sociological Focus*, Vol. 25, No. 2, May 1992. Reprinted by permission.

engage in behavior that challenges the established gender order, and thus opposes male domination, attempts are often made by those most threatened to devalue these women and ultimately control their actions. One means of discrediting women who violate gender norms and thereby questioning their "womanhood" is to label them lesbian (Griffin 1987).

The accusation of lesbianism is a powerful controlling mechanism given the homophobia that exists within American society. Homophobia, representing a fear of or negative reaction to homosexuality (Pharr 1988), results in stigmatization directed at those assumed to violate sexuality norms. Lesbianism, in particular, is viewed as threatening to the established patriarchal order and heterosexual family structure since lesbians reject their "natural" gender role, as well as resist economic, emotional, and sexual dependence on men (Gartrell 1984; Lenskyj 1991).

As a means for both discouraging homosexuality and maintaining a patriarchal and heterosexist gender order (Pharr 1988), homophobia controls behavior through contempt for purported norm violators (Koedt, Levine, and Rapone 1973). One method of control is the frequent application of the lesbian label to women who move into traditional male-dominated fields such as politics, business, or the military (Lenskyj 1991). This "lesbian baiting" (Pharr 1988:19) suggests that women's advancement into these arenas is inappropriate. Such messages are particularly potent since they are lodged in a society that condemns, devalues, oppresses, and victimizes individuals labeled as homosexuals (Lenskyj 1990).

Another male arena in which women have made significant strides, and thus risk damaging accusation and innuendo, is that of sport (Blinde and Taub 1992; Lenskyj 1990). Sport is a particularly susceptible arena for lesbian labeling due to the historical linkage of masculinity with athleticism (Birrell 1988). When women enter the domain of sport they are viewed as violating the docile female gender role and therefore extending culturally constructed boundaries of femininity (Cobhan 1982; Lenskyj 1986; Watson 1987). The attribution of masculine qualities to women who participate in sport leads to a questioning of their sexuality and subsequently makes athletes targets of homophobic accusations (Lenskyj 1986)....

Therefore, the present study explores the stereotyping of women athletes as lesbians and the accompanying homophobia fostering this label. General themes and processes which inform us of how these individuals handle the lesbian issue are identified. These dynamics are grounded in the contextual experiences of women athletes and relayed through their voices.

Athletic directors at seven large Division I universities were contacted by telephone and asked to participate in a study examining various aspects of the sport experience of female college athletes. These administrators were requested to provide a list of the names and addresses of all varsity women athletes for the purpose of contacting them for telephone interviews.... Interested athletes were encouraged to return an informed consent form indicating their willingness to participate in a tape-recorded telephone interview. Based on this initial contact, a total of 16 athletes agreed to be in the study.

In order to increase the sample size to the desired 20 to 30 respondents, the names of 30 additional athletes were randomly and proportionately selected from the three lists. Eight of these athletes agreed to be interviewed, resulting in a final sample size of 24. Athletes in the sample were currently participating in a variety of women's inter-collegiate varsity sports—basketball (n = 5), track and field (n = 4), volleyball (n = 3), swimming (n = 3), softball (n = 3), tennis (n = 2), diving (n = 2), and gymnastics (n = 2). With an average age of 20.2 years and overwhelmingly Caucasian (92 %), the sample contained 2 freshmen, 9 sophomores, 5 juniors, and 8 seniors. A majority of the athletes (n = 22) were recipients of an athletic scholarship. . . .

Semi-structured telephone interviews were conducted by two trained female interviewers. All interviews were tape-recorded and lasted from 50 to 90 minutes. Questions were open-ended in nature so that athletes would not feel constrained in discussing those issues most relevant to their experiences. Follow-up questions were utilized to probe how societal perceptions of women athletes impact their behavior and experiences.

RESULTS

Examination of the responses of athletes revealed two prevailing themes related to the presence of the lesbian stereotype in women's sport—(a) a silence surrounding the issue of lesbianism in women's sport, and (b) athletes' internalization of societal stereotypes concerning lesbians and women athletes. It is suggested that these two processes disempower women athletes and thus are counterproductive to the self-actualizing capability of sport participation (Theberge 1987).

SILENCE SURROUNDING LESBIANISM IN WOMEN'S SPORT

One of the most pervasive themes throughout the interviews related to the general silence associated with the lesbian stereotype in women's sport. Although a topic of which athletes are cognizant, reluctance to discuss and address lesbianism in women's sport was evident. Based on the responses of athletes, this silence was manifested in several ways: (a) athletes' difficulty in discussing lesbian topic, (b) viewing lesbianism as a personal and irrelevant issue, (c) disguising athletic identity to avoid lesbian label, (d) team difficulty in addressing lesbian issue, and (e) administrative difficulty in addressing lesbian issue.

ATHLETES' DIFFICULTY IN DISCUSSING LESBIAN TOPIC

Initial indication of silencing was illustrated by the difficulty and uneasiness many athletes experienced in discussing the lesbian stereotype. Some respondents were initially reluctant to mention the topic of lesbianism; discussion of the issue was frequently preceded by awkward or long pauses suggesting feelings of uneasiness or discomfort. Athletes were most likely to introduce this topic when questions were asked about societal perceptions of women's sport and female athletes, as well as inquiries about the existence of stereotypes associated with women athletes. Moreover, the lesbian issue was sometimes discussed without specifically using the term lesbian. For example, some athletes evaded the issue by making indirect references to lesbianism (e.g., using the word "it" rather than a more descriptive term). . . .

Respondents' approach to the topic of lesbianism indicates the degree to which women athletes have been socialized into a cycle of silence. Such silence highlights the suppressing effects of homophobia. Moreover, athletes' reluctance to discuss topics openly related to lesbianism may be to avoid what Goffman (1963) has termed "courtesy stigma," a stigma conferred despite the absence of usual qualifying behavior.

VIEWING LESBIANISM AS A PERSONAL AND IRRELEVANT ISSUE

A second indicator of the silence surrounding the lesbian stereotype was reflected in athletes' general comments about lesbianism. Many respondents indicated that sexual orientation was a very personal issue and thus represented a private and extraneous aspect of an individual's life. These athletes felt it was inappropriate for others to be concerned about the sexual orientation of women athletes.

Although such a manifestation of silence might reflect the path of least resistance by relieving athletes of the need to discuss or disclose their sexual orientation (Lenskyj 1991), it does not eliminate the stigma and stress experienced by women athletes. Also, making lesbianism a private issue does not confront or challenge the underlying homophobia that allows the label to carry such significance. The strategy of making sexual orientation a personal issue depoliticizes lesbianism and ignores broader societal issues.

DISGUISING ATHLETIC IDENTITY TO AVOID LESBIAN LABEL

A third form of silence surrounding the lesbian stereotype was the tendency for athletes to hide their athletic identities. Nearly all respondents indicated that despite feeling pride in being an athlete, there were situations where they preferred that others not know their athletic identity. Although not all athletes indicated that this concealment was to prevent being labeled a lesbian, it was obvious that there was a perceived stigma associated with athletics that many women wanted to avoid (e.g., masculine women, women trying to be men, jock image). In most cases, respondents indicated that disguising their athletic identity was either directly or indirectly related to the lesbian stereotype. . . .

Athletes also stated that they (or other athletes they knew) accentuated certain behaviors in order to reduce the possibility of being labeled a lesbian. Being seen with men, having a boyfriend, or even being sexually promiscuous with men were commonly identified strategies to reaffirm an athlete's heterosexuality. As one athlete commented. "If you are a female athlete and do not have a boyfriend, you are labeled [lesbian]."

As reflected in the responses of athletes, the role of sport participant was often intentially de-emphasized in order to reduce the risk of being labeled lesbian. Modification of athletes' behavior, even to the point of denying critical aspects of self, was deemed necessary for protection from the negativism attached to the lesbian label. This disguising of athletic identity exemplifies what Kitzinger (1987:92) termed "role inversion." In such a situation individuals attempt to demonstrate that their group stereotype is inaccurate by accentuating traits that are in opposition to those commonly associated with the group (in the case of women athletes, stressing femininity and heterosexuality).

TEAM DIFFICULTY IN ADDRESSING LESBIAN ISSUE

Not only did the silence surrounding lesbianism impact certain aspects of the lives of individual athletes, but it also affected interpersonal relationships among team members. This silence was often counterproductive to the development of positive group dynamics (e.g., team cohesion, open lines of communication).

As was often true at the individual level, women's sport teams were unable collectively to discuss, confront, or challenge the labeling of women athletes as lesbians. One factor complicating the ability of women athletes to confront the lesbian stereotype was the divisive nature of the label itself (Gentile 1982); the lesbian issue sometimes split teams into factions or served as the basis for clique formation.

Heterosexual and lesbian athletes often had limited interaction with each other outside the sport arena. Moreover, athletes established distance between themselves and those athletes most likely to be labeled lesbian (i.e., those possessing "masculine" physical or personality characteristics). . . .

From the interviews, there was little evidence that lesbian and nonlesbian athletes collectively pooled their efforts to confront or challenge the lesbian stereotype so prevalent in women's sport. The silence surrounding lesbianism creates divisions among women athletes; this dissension has the effect of preventing female bonding and camaraderie (Lenskyj 1986). Rather than recognizing their shared interests, women athletes focus on their differences and thus deny the formation of "alliance" (Pheterson 1986:149). This difficulty in attaining team cohesion is unfortunate since women's sport is an activity where women as a group can strive for common goals (Lenskyj 1990). The lesbian stereotype not only limits female solidarity, but also minimizes women's ability to challenge collectively the patriarchal and heterosexist system in which they reside (Bennett et al. 1987).

ADMINISTRATIVE DIFFICULTY IN ADDRESSING LESBIAN ISSUE

Another manifestation of silence relayed in the responses of athletes was the apparent unwillingness of coaches and athletic directors to confront openly the lesbian stereotype. As was found with individual athletes and teams, those in leadership positions in women's sport refused to address or challenge this stereotype. Reluctance to confront the lesbian issue at the administrative level undoubtedly influenced the manner in which athletes handled the stereotype. . . .

Because the women's intercollegiate sport system is homophobic and predominately male-controlled (i.e., over half of coaches and four-fifths of administrators are men) (Acosta and Carpenter 1992), it is assumed that survival in women's sport requires collusion in a collective strategy of silence about and denial of lesbianism (Griffin 1987). Coaches and administrators fear that openly addressing the lesbian issue may result in women's sport losing the recent gains made in such areas as fan support, budgets, sponsorship, and credibility (Griffin 1987). Therefore, leaders yield to this fear as they strive to achieve acceptability for women's sport. Such accommodation to the patriarchal, heterosexist sport structure not only contributes to isolation as coaches and administrators are afraid to discuss lesbianism, but also limits their identification with feminist and women's issues (Duquin 1981; Hargreaves 1990; Pharr 1988; Zipter 1988).

SUMMARY

Based on athletes' responses, it was evident that the silence surrounding the lesbian issue in women's sport was deeply ingrained at all levels of the women's intercollegiate sport structure. Such widespread silencing reflects the negativism and fear associated with lesbianism that are so prevalent in a homophobic society. This strategy of silence or avoidance, however, is counterproductive to efforts to dispel or minimize the impact of the lesbian stereotype. Not only does silence disallow a direct confrontation with those who label athletes lesbian, but it also perpetuates the power of the label by leaving unchallenged rumors and insinuations. Moreover, the fear, ignorance, and negative images that are frequently associated with women athletes are reinforced by this silence (Zipter 1988).

Numerous aspects of women's experience in sport are ignored due to the silence surrounding the subject of lesbianism. For example, refusing to address this issue has limited understanding of the dimensionality and complexity of women's sport participation. Moreover, since the stigma associated with the lesbian label inhibits athletes from discussing this topic with each other, these women frequently do not realize that they possess shared experiences that would provide the foundation for female bonding. Without an "alliance" among athletes, little progress is made in improving their plight (Pheterson

1986). Finally, as a result of this preoccupation with silence, women athletes often engage in self-denial as they hide their athletic identity.

ATHLETES' INTERNALIZATION OF SOCIETAL STEREOTYPES

A second major theme reflected in the responses of athletes was a general internalization of stereotypic representations of lesbians and women athletes. As argued by Kitzinger (1987) and Pheterson (1986), members of oppressed and socially marginalized groups often find themselves accepting the stereotypes and prejudices held by the dominant society. Representing "internalized oppression" (Pheterson 1986:148), the responses of athletes revealed an identification with the aggressor, self-concealment, and dependence on others for self-definition (Kitzinger 1987; Pheterson 1986). Acceptance of these societal representations by a disadvantaged group (in this case women athletes) grants legitimacy to the position of those who oppress and contributes to the continued subordination of the oppressed (Wolf 1986). Based on our interviews, athletes' internalization of stereotypes and prejudices were reflected by three categories of responses: (1) acceptance of lesbian stereotypes, (2) acceptance of women's sport team stereotypes, and (3) acceptance of negative images of lesbianism.

ACCEPTANCE OF LESBIAN STEREOTYPES

In response to various open-ended questions, it was apparent that athletes were able to identify a variety of factors that they felt led others to label women athletes as lesbians (e.g., physical appearance, dress, personality characteristics, nature of sport activity). Given that the attribution of homosexuality is most likely to be associated with traits and behaviors judged to be more appropriate for members of the opposite sex (Dunbar, Brown and Amoroso 1973; Dunkle and Francis 1990), it was not surprising that athletes' rationale for the lesbian label included such attributes as muscularity, short hair, masculine clothing, etc.

When athletes were asked about the validity of the lesbian label in women's sport, affirmative replies were frequently based on conjecture. For example, to provide support for why they felt there was a basis for labeling women athletes as lesbians, respondents made such comments as "there are masculine girls on some teams." "it is really obvious," or "you can just tell that some athletes are lesbians."

These explanations tend to reflect an acceptance of societal definitions of lesbianism—beliefs that are largely male-centered and supportive of a patriarchal, heterosexist system (e.g., "girls who look like guys"). Indeed, previous research has shown that people associate physical appearance with homosexuality (Levitt and Klassen 1974; McArthur 1982; Unger, Hilderbrand, and Madar 1982). For example, attrativeness is equated with heterosexuality and a larger, muscular body build is identified with lesbianism.

Moreover, the remarks of athletes demonstrate that the very group that is oppressed (in this case women athletes) accepts societal stereotypes about lesbians and has incorporated these images into their managing of the situation. As suggested by Gartrell (1984) and certainly evident in this sample of women athletes, cultural myths about lesbianism perpetuated in a homophobic society are often firmly ingrained in the thinking of affected individuals.

ACCEPTANCE OF WOMEN'S SPORT TEAM STEREOTYPES

Relative to providing a rationale for why the lesbian label was more likely to be associated with athletes in certain sports, respondents again demonstrated an understanding and internalization of societal stereotypes. The sports most commonly identified with the lesbian label were softball, field hockey, and basketball. In attempting to explain why these team sports were singled out, athletes mentioned such factors as the nature of bodily contact or amount of aggression in the sport, as well as the body build, muscularity, or athleticism needed to play the sport.

Respondents often relied on the "masculine" and "feminine" stereotypes to differentiate sports in which participating women were more or less likely to be subjected to the lesbian label. Although participants in team sports were more likely than individual sports (e.g., gymnastics, swimming, tennis, golf) to be associated with the lesbian label, it was interesting to note that volleyball was often exempt from the connotations of lesbianism.

The higher incidence of lesbian labeling found in team sports (as opposed to individual sports) may also be related to the potential that team sports provide for interpersonal interactions. As previously mentioned, emphasizing teamwork and togetherness, team sports allow women rare opportunities to bond collectively in pursuit of a group goal (Lenskyj 1990). Recognition of this power of female bonding is often reflected by male opposition to women-only activities (Lenskyj 1990).

ACCEPTANCE OF NEGATIVE IMAGES OF LESBIANISM

During the course of the interviews, a large majority of athletes made comments about lesbians which reflected an internalization of the negativism associated with lesbianism. Respondents also demonstrated a similar acceptance when they relayed conversations they had had with both teammates and outsiders.

One form of negativism was reflected by statements that specifically "put down" lesbians. Athletes' negative comments about lesbians were included in conversations with outsiders so others would not associate the lesbian label with them. Representing a form of projection (Gross 1978), some athletes attempted to disassociate from traits that they saw in themselves (e.g., strength, muscularity, aggressiveness). . . .

It is ironic that athletes rarely directed their anger or condemnations at the homophobic society that restricts the actions of women athletes, including the nonlesbian athlete. Rather, by focusing on athletes as lesbians, a blame the victim approach diverts attention from the cause of the oppression (Pharr 1988). As is often true of oppressed groups, a blame the victim philosophy results in an acceptance of the belief system of the oppressor (in this case a patriarchal, heterosexist society) (Pharr 1988). Like other marginalized groups, women athletes accept the normative definitions of their deviance (Kitzinger 1987); in effect, such responses represent a form of collusion with the oppressive forces (Pheterson 1986). Interestingly, no mention was made by respondents about attempts to engage the assistance or support of units on campus sympathetic to gay and lesbians issues (e.g., feminist groups, gay and lesbian organization, affirmative action offices).

SUMMARY

From the interview responses, it was evidence that athletes had internalized societal stereotypes related to lesbians and women athletes, as well as the negativism directed toward lesbianism. This acceptance was so ingrained in these athletes that they were generally unaware of the political ramifications of both lesbianism and the accompanying lesbian stereotype as applied to women athletes. Despite their gender norm violation as athletes, these women often had a superficial understanding of gender issues. Such a lack of awareness may be due in part to the absence of a feminist consciousness in athletes (Boutilier and SanGiovanni 1983; Kaplan 1979) and their open disavowal of being a "feminist," "activist," or "preacher of women's liberation." Accepting societal definitions of their deviance, as well as the inability to see their personal experiences as political in nature, attests to this limited consciousness (Boutilier and SanGiovanni 1983). Athletes' responses are indicative of the degree to which they exhibit internal homophobia so common in American society.

Only a few athletes possessed deeper insight into factors that may underlie the labeling of women athletes as lesbians. For example, one respondent felt women athletes were a "threat to men since they can stand on their own feet." Or, in another situation, an athlete viewed lesbian labeling as a means to devalue women athletes or successful women in general. Still another respondent suggested the label stemmed from jealousy and thus was used as a means to "get back" at women athletes. These rare remarks by respondents transcend the blame the victim view held by the majority of athletes. Such commitments indicate a deeper understanding of how homophobia and patriarchal ideology limit or control women's activities and their bodies.

DISEMPOWERMENT

Given the silence surrounding the lesbian issue and the degree to which athletes have internalized societal images of lesbians and women athletes, the presence of the lesbian stereotype has negative ramifications for women athletes. Although sport participation possesses the potential for creativity and physical excellence (Theberge 1987), women modify their behavior so they will not be viewed as "stepping out of line." Women athletes become disempowered (Pharr 1988) through processes that detract from or reduce the self-actualizing potential of the sport experience.

Attaching the label of lesbian to women who engage in sport diminishes the sporting accomplishments of athletes. Women athletes are seen as something less than "real women" because they do not exemplify traditional female qualities (e.g., dependency, weakness, passivity); thus their accomplishments are not viewed as threatening to men (Birrell 1988). Interestingly, the athlete interviewed believed that the specific group most likely to engage in lesbian labeling was male athletes.

Discrediting women with the label of lesbian works further to control the number of females in sport, particularly in a homophobic society where prejudice against lesbians is intense (Birrell 1988; Zipter 1988). Keeping women out of sport, in turn, prevents females from discovering the power and joy of their own physicality (Birrell 1988) and experiencing the potential of their body. Moreover, discouraging women from participating in sport disempowers them by removing an arena where women can bond together (Birrell 1988; Cobhan 1982). . . .

Another form of disempowerment occurs for those athletes who are lesbians. Intense homophobia often forces lesbians to deny their very essence, thus making the lesbian athlete invisible. Concealment, although protecting the lesbian athletes' identity, imposes psychological strain and can undermine positive self-conceptions (Schur 1984). Misrepresenting their sexuality, lesbian athletes are not in a position to confront the homophobia so prevalent in women's sport. Consequently, this ideology not only remains intact, but also is strengthened (Ettore 1980).

REFERENCES

Acosta, R. Vivian and Linda Jean Carpenter. 1992. "Women in Intercollegiate Sport: A Longitudinal Study—Fifteen Year Update 1977-1992." Unpublished manuscript, Brooklyn College, Department of Physical Education, Brooklyn.

Bennett, Roberts S., K. Gail Whitaker, Nina Jo Woolley Smith, and Anne Sablove. 1987. "Changing The Rules of The Game: Reflections Toward A Feminist Analysis of Sport." *Women's Studies International Forum* 10: 369-386.

Birrell, Susan. 1988. "Discourses on The Gender/Sport Relationship: From Women in Sport to Gender Relations." Pp. 459-502 in *Exercise And Sport Science Reviews,* vol. 16, edited by K. B. Pandolf. New York: MacMillan.

Blinda, Elaine M. and Diane E. Taub. 1992. "Women Athletes as Falsely Accused Deviants: Managing the Lesbian Stigma." *The Sociological Quarterly.*

Boutilier, Mary A. and Lucinda SanGiovanni. 1983. *The Sporting Woman.* Champaign, IL: Human Kinetics.

Cobban, Linn Ni. 1982. "Lesbians in Physical Education And Sport. Pp. 179-186 in *Lesbian Studies: Present And Future,* edited by M. Cruikshank. New York: Feminist Press.

Dunbar, John, Marvin Brown and Donald M. Amoroso. 1973. "Some Correlates of Attitudes Toward Homosexuality." *Journal of Social Psychology* 89: 271-279.

Dunkle, John H. and Patricia L. Francis. 1990. "The Role of Facial Masculinity/Feminity in The Attribution of Homosexuality." *Sex Roles* 23: 157-167.

Duquin, Mary E. 1981. "Feminism And Patriarchy in Physical Education." Paper presented at the annual meetings of the North American Society for the Sociology of Sport, Fort Worth, TX.

Ettore, E. M. 1980. *Lesbians, Women and Society.* London: Routledge and Kegan Paul.

Gartrell, Nanette. 1984. "Combating Homophobia in The Psychotherapy of Lesbians." *Women And Therapy* 3: 13-29.

Gentile, S. 1982. "Out of The Kitchen." *City Sports Monthly* 8: 27.

Goffman, Erving. 1963. *Stigma: Notes on The Management of Spoiled Identity.* Englewood Cliffs, N.J.: Prentice-Hall.

Griffin, Patricia S. 1987. "Homophobia, Lesbians, And Women's Sports: An Exploratory Analysis." Paper presented at the annual meetings of the American Psychological Association, New York.

Gross, Martin L. 1978. *The Psychological Society.* New York: Simon and Schuster.

Hargreaves, Jennifer A. 1990. "Gender on The Sports Agenda." *International Review for Sociology of Sport* 25: 287-308.

Kaplan, Janice. 1979. *Women And Sports.* New York: Viking.

Kitzinger, Celia. 1987. *The Social Construction of Lesbianism.* London: Sage.

Koedt, Anne, Ellen Levine and Anita Rapone. 1973. *Radical Feminism.* New York: Quadrangle.

Lenskyj, Helen. 1986. *Out of Bounds: Women, Sport And Sexuality.* Toronto: Women's Press.

———. 1990. "Power And Play: Gender And Sexuality Issues in Sport and Physical Activity." *International Review for Sociology of Sport* 25: 235-245.

———. 1991. "Combating Homophobia in Sport And Physical Education." *Sociology of Sport Journal* 8: 61-69.

Levitt, Eugene E. and Albert D. Klassen, Jr. 1974. "Public Attitudes toward Homosexuality: Part of the 1970 National Survey by The Institute for Sex Research." *Journal of Homosexuality* 1: 29-43.

McArthur, Leslie Z. 1982. "Judging A Book by Its Cover: A Cognitive Analysis of The Relationship between Physical Appearance And Stereotyping." Pp. 149-211 in *Cognitive Social*

Psychology, edited by Albert H. Hastorf and Alice M. Isen. New York: Elsevier/North-Holland.

Pharr, Suzanne. 1988. *Homophobia: A Weapon of Sexism.* Inverness, CA: Clurdon.

Pheterson, Gail. 1986. "Alliances between Women: Overcoming Internalized Oppression And Internalized Domination." *Signs: Journal of Women in Culture And Society* 12: 146–160.

Schur, Edwin M. 1984. *Labeling Women Deviant: Gender, Stigma, And Social Control.* New York: McGraw-Hill.

Stockard, Jean and Miriam M. Johnson, 1980. *Sex Roles: Sex Inequality And Sex Role Development.* Englewood Cliff, NJ: Prentice-Hall.

Theberge, Nancy. 1987. "Sport And Women's Empowerment." *Women's Studies International Forum* 10: 387–393.

Unger, Rhoda K., Marcia Hilderbrand and Theresa Madar. 1982. "Physical Attractiveness And Assumptions about Social Deviance: Some Sex-By-Sex Comparisons." *Personality And Social Psychology Bulletin* 8: 293–301.

Watson, Tracey. 1987. "Women Athletes And Athletic Women: The Dilemmas And Contradictions of Managing Incongruent Identities." *Sociological Inquiry* 57: 431–446.

Wolf, Charlotte. 1986. "Legitimation of Oppression: Response And Reflexivity." *Symbolic Interaction* 9: 217–234.

Zipter, Yvonne. 1988. *Diamonds Are A Dyke's Best Friend: Reflections, Reminiscences, And Reports from The Field on The Lesbian National Pastime.* Ithaca, NY: Firebrand Books.

PART V

Deviant Identity

We have just looked at how some categories of people and behavior become defined as deviant. Yet labeling theory suggests that a deviant classification floating around abstractly in society is not meaningful unless it gets attached to people. Groups in society not only work to create definitions of deviance, but also to create situations where deviance occurs and is labeled. In this part we examine the way deviance is evoked and shaped. Becoming deviant does not only entail having a definition of deviance and an environment in which it can occur; it also requires that people accept the identity and make it their own. In Part V we will examine the process of how the concept of deviance becomes applied to individuals and how it affects their self-conception.

IDENTITY DEVELOPMENT

We mentioned earlier that, although many people engage in deviance, the label is applied to only a small percentage of them. Such labeling is tied to their formerly "secret deviance" (Becker 1963) becoming exposed, or to an abstract status coming to bear on their personal experience. Thus Jews may not feel stigmatized unless they experience anti-Semitism, and embezzlers may not think of themselves as thieves until they are caught. When this happens, they enter the pathway to the deviant identity, a pathway that follows a certain trajectory. Becker (1963) has suggested that we can think of this path as a "deviant career."

Once people are *caught* and *publicly identified* as deviant their lives change in several ways. Others start to think of them differently. For example, suppose there has been a rash of thefts in a college dormitory, and Jessica, a first-year student, is finally caught and identified as the culprit. She may or may not be reported to authorities and charged with theft, but regardless, she will experience an informal labeling process. People will probably change their attitudes toward her, as they find themselves talking about her behind her back. They may look back on her behavior and engage in "retrospective interpretation" (Kitsuse 1962) as they think about her differently in light of their new information. Where did she say she was when the last theft occurred, where did she say she got the money to buy that new sweater? Jessica may develop what Goffman (1963) has called a "spoiled identity," one with a damaged reputation.

Jessica's dormmates and former friends may then engage in what Lemert (1951) has called "the dynamics of exclusion," deriding and ostracizing her from their social group. When she enters the room she may notice that a sudden hush falls over the conversation. People may not feel comfortable leaving her alone in their room. They may exclude her from their meal plans and study groups. She may become progressively shut out from nondeviant activities and circles. At the same time, others may welcome or *include* her in their deviant circles or activities. She may find that she has developed a reputation that, though repelling to some groups, is attractive to others. They may welcome her as "cool" and invite her into their circles. The more people start to *treat* Jessica and interact with her as a deviant, the more she is likely to *internalize the label* and regard herself as a deviant.

Once people are labeled as deviant and accept that label into their self-conceptions, a variety of outcomes may ensue. Hughes (1945) has suggested that we all have a series of statuses or identities through which we relate to people, including those of sibling, child, friend, student, neighbor, and customer. These identities also derive from some of our demographic or occupational features, like race, gender, age, religion, or social class. Hughes asserts that some statuses are very dominant, overpowering others and coloring the way in which people are viewed. Having a deviant identity may become one of these *master statuses,* rising to the top of the hierarchy and infusing people's self-concept and others' reactions. People who are labeled as deviants, like heroin addicts, cult members, or homosexuals, may be viewed by others through this lens no matter what the situation or setting. In addition, every master status has a set of *auxiliary traits* that accompany it, and once people label someone with a deviant master status they will expect to see the relevant auxiliary traits. A heroin addict may be suspected of being a prostitute or thief, and a homosexual may be suspected of being sexually promiscuous or AIDS-infected. This type of identification spreads the image of deviance to cover the person as a whole and not just one part of him or her. Thus people move back and forth between master statuses and auxiliary traits both ways: once they know about someone's master status they may infer the existence of suspected auxiliary traits; at the same time, when they see several characteristics that appear suspicious, they may link them together and infer that they may be the auxiliary traits revealing a hidden deviant master status.

The process of developing a deviant master status and auxiliary traits helps explain the move from *primary deviance* to *secondary deviance*. As Lemert (1967) defined it, primary deviance refers to the initial type of deviance in which people engage, one that has not necessarily yielded them the master status of deviance. People at the primary deviance stage may be doing the deviance, but they are in denial about their deviant identity. Some people remain forever in the primary deviance stage, committing deviant acts, but never getting noticed. Of those in primary deviance, however, a portion of them get caught, and move into secondary deviance. Others discover their secrets and come to regard them differently. Being labeled reinforces the deviant identity, spoils people's reputations, and leads to altered self-concepts. In secondary deviance, individuals are engaged in the deviance, as before, only now others regard them as deviant. They have to accept the deviant identity, possibly even as a master status. Kitsuse (1980) adds onto Lemert's two stages with his concept of *tertiary deviance*. Not all people who have made it to the point of secondary deviance advance to the tertiary stage, but some do. These individuals go past accepting their deviance and embrace it, coming to reject its definition as deviant. They move from an absolutist perspective, that there is something intrinsically or morally wrong with them, to the relativistic perspective, that their definition as deviant is a social and political construct of moral entrepreneurs. Their movement into tertiary deviance involves their rejection of this view and their decision to engage in identity politics to fight it. They may organize, lobby, protest, raise funds, educate, engage in civil disobedience, and conduct other individual or collective activities designed to fight the deviant label and stigma.

Becker's notion of the identity career and Lemert's and Kitsuse's conceptions of the stages of deviance can all be seen to encompass several stages, beginning with the commission of the deviance and leading to individuals' apprehension and public identification, the changing attitudes and expectations of others towards those individuals, shifting social acceptance or rejection by their friends and acquaintances, individuals' internalization of the deviant label and self-identity and interaction through it, movement into groups of differential associates, and the commission of further acts of deviance. Some of these processes are illustrated by the three readings in Part V.

20

The Adoption and Management of a "Fat" Identity

DOUGLAS DEGHER AND GERALD HUGHES

Degher and Hughes's selection on the way people come to think of themselves as fat is a study in identity transformation. They posit a model where individuals align their self-conception with cues that they derive from their external environment. Although the subjects in this study originally do not hold a view of themselves as obese, they receive active status cues (people say things) and passive status cues (their clothes no longer fit) that jar them away from their former self-conceptions. They follow a process of recognizing that they can no longer be considered to have a normal build, and placing themselves with the new category that fits them more appropriately. This leads to their reconceptualizing themselves as fat. The fat status has a new, negative identity that they adopt, devaluing them and locating them within the deviant realm.

The interactionist perspective has come to play an important part in contemporary criminological and deviance theory. Within this approach, deviance is viewed as a subjectively problematic identity rather than an objective condition of behavior. At the core is the emphasis on "process" rather than on viewing deviance as a static entity. To paraphrase vintage Howard Becker, "... social groups create deviance by making the rules whose infraction constitutes deviance. Consequently, deviance is not a quality of the act ... but rather a consequence of the application by others of rules and sanctions to an offender" (Becker, 1963, p. 9). Attention is focused upon the *interaction* between those being labeled deviant and those promoting the deviant label. In the interactionist literature, emphases are in two major areas: (a) the conditions under which the label "deviant" comes to be applied to an individual and the consequences for the individual of having adopted that label (Tannenbaum, 1939, Lemert, 1951; Kitsuse, 1962, p. 247; Goffman, 1963; Baum, 1987, p. 96; Greenberg, 1989, p. 79), and/or (b) the role of social control

---

Copyright 1991 from "The Identity Change Process: A Field Study of Obesity," Douglas Degher and Gerald Hughes, *Deviant Behavior*, V. 12, No. 4. Reproduced by permission of Taylor & Francis, Inc.; http://www.routledge-ny.com

agents[1] in contributing to the application of deviant labels (Becker, 1963; Piliavin & Briar, 1964, p. 206; Cicourel, 1968; Schur, 1971; Conrad, 1975, p. 12).

Much of this literature frequently assumes that once an individual has been labeled, the promoted label and attendant identity is either internalized or rejected. As Lemert proposes, the shift from primary to secondary deviance is a categorical one, and is primarily a response to problems created by the societal reaction (Lemert, 1951, p. 40).

What is most often neglected is an examination of the mechanistic features of this identity shift. Our focus is on this "identity change process," which is what we have chosen to call this identity shift. Of interest is how individuals come to make some personal sense out of proffered labels and their attendant identities.

METHODOLOGY

The primary methodological tool employed to construct our identity change process model comes from "grounded" analysis. . . . The model presented in this paper emerged from comments and codes appearing in interviews with obese members of a weight reduction organization that had weekly meetings. The frequency of attendance allowed us to consider the members typical, and allowed us to suggest that major issues of obesity are trans-situational and temporally durable. If obesity disappeared tomorrow, we would still be able to apply the generic concepts generated from our data to make statements about the process of "identity change." As suggested by Hadden, Degher, and Fernandez, our focus is on process rather than on unit characteristics of social phenomena (Hadden, Degher, & Fernandez, 1989, p. 9). This provides us with insights that have import for major issues in sociological theory.

SITE SELECTION

Because obese individuals suffer both internally (negative self-concepts) and externally (discrimination), they possess what Goffman refers to as a "spoiled identity" (Goffman, 1963). This seems to be the case particularly in contemporary America with what may be described as an almost pathological emphasis on fitness. The boom in health clubs, sales of videotapes on fitness, diet books, and so forth promote a definition of the "healthy" physical presence. As Kelly (1990) sees it, the boom in physical fitness in the mid-1980s is an attempt by many people to create a specific image of an ideal body. Thus, body build becomes a crucial element in self-appraisal. Consequently, fat people are an ideal strategic group within which to study the "identity change process."

Obese people are not only the subject of negative stereotypes, they are also actively discriminated against in college admissions (Canning & Mayer, 1966, p. 1172), pay more for goods and services (Petit, 1974), receive prejudicial medical treatment (Maddox, Back, & Liederman, 1968, p. 287; Maddox & Liederman, 1969, p. 214), are treated less promptly by salespersons (Pauley, 1989, p. 713), have higher rates of unemployment (Laslett & Warren, 1975, p. 69), and receive lower wages (Register, 1990, p. 130). The obese label is one that seems to clearly fit Becker's description of a "master status," that is

> Some statuses in our society, as in others, override all other statuses and have a certain priority . . . the status deviant (depending on the kind of deviance) is this kind of master status . . . one will be identified as a deviant first, before other identifications are made. (Becker, 1963, p. 33).

Obese people are "fat" first, and only secondarily are seen as possessing ancillary characteristics.

The site for the field observations had to meet two requirements: (a) it had to contain a high proportion of obese, or formerly obese individuals; and (b) these individuals had to be identifiable by the observer. The existence of a large number of national weight control organizations (a) whose membership is composed of individuals who have internalized an obese identity, and (b) who emphasize a radical program of identity change, make these organizations an excellent choice as strategic sites for study and analysis. The local franchise chapter of one of these national weight loss organizations satisfied both of our requirements, and was selected as the site for our study.

Attendance at the weekly meetings of this national weight control group is restricted to individuals who are current members of the organization. Since one requirement for membership is that the individual be at least 10 pounds over the maximum weight for his or her sex and height (according to New York Life tables), all of the people attending the meetings are, or were, overweight, and a high proportion of them are, or were, sufficiently overweight to be classified as obese.[2]

During the period of the initial field observations, the weekly membership of the group varied from 30 to 100 members, with an average attendance of around 60 members. Although there was a considerable turnover in membership, the greatest part of this turnover consisted of "rejoins" (individuals who had been members previously, and were joining again).

Although we have no quantitative data from which to generalize, the group membership appeared to represent a cross section of the larger community. The group included both male and female members, although females did constitute about three-fourths of the membership. Although the membership was predominantly white, a range of ethnicities, notably Hispanic and Native American, existed within the group. The majority of the members appeared to fall within the 30 to 50 age range, although there was a member as young as 11, and one over 70.

DATA COLLECTION

Two types of data were gathered for this study: field observations and in-depth interviews. The field observations were performed while attending meetings of a local weight control organization. The insights gained from these observations were used primarily to develop interview guides. There were two major sources of observation during this period: pre-meeting conversations; and exchanges during the meeting itself.[3] The observations were recorded in note form and served to provide an orientation for the subsequent interviews. The goal during this period of observation was to gain insight into the basic processes of obesity and the obese career.

The in-depth interviews were carried out with 29 members from the local group. The interviews were solicited on a voluntary basis, and each individual was assured anonymity. The interviewees were representative of the group membership. Although most were middle-aged, middle-income, white females, various age groups, ethnicities, marital statuses, genders, and social classes were represented.

These interviews lasted in length from ½ to 2½ hours, with the average interview being about 1 hour and 15 minutes in duration. The interviews produced almost 40 hours of taped discussion, which yielded more than 600 pages of typed transcript for coding.

THE IDENTITY CHANGE PROCESS

In conceptualizing the "identity changes" process, the concept of "career" was employed. As Goffman notes, "career" refers, ". . . to any social strand of any person's course through life" (Goffman, 1961, p. 127). In the present paper, our concern is the change process that takes place as individuals come to see definitions of self in light of specific transmitted information.

An important aspect of this career model is what Becker referred to as "career contingencies," or ". . . those factors on which mobility from one position to another depends. Career contingencies include both the objective facts of social structure, and changes in the perspectives, motivations, and desires of the individual" (Becker, 1963, p. 24).

Thus, the "identity change" process must be viewed on two levels: a public (external) and a private (internal) level. As Goffman has stated, "One value of the concept of career is its two-sidedness. One side is linked to internal matters held dearly and closely, such as image of self and felt identity; the other publicly acceptable institutional complex" (Goffman, 1961, p. 127).

On the public level, social status exists as part of the public domain; social status is socially defined and promoted. The social environment not only contains definitions and attendant stereotypes for each status, it also contains information, in the form of *status cues,* about the applicability of that status for the individual.

On the internal level, two distinct cognitive processes must take place for the identity change process to occur: first, the individual must come to recog-

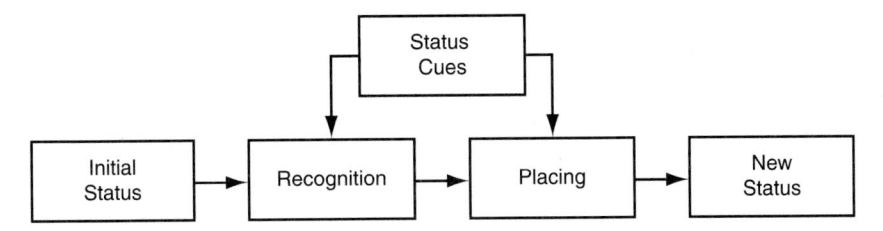

```
                    ┌─────────┐
                    │ Status  │
                    │  Cues   │
                    └────┬────┘
              ┌─────────┴┬──────────┐
              ▼          │          ▼
┌────────┐  ┌──────────┐ │ ┌────────┐  ┌────────┐
│Initial │→ │Recognition│→│ │Placing │→ │  New   │
│ Status │  │          │  │ │        │  │ Status │
└────────┘  └──────────┘  └────────┘  └────────┘
```

Visualization of the Identity Change Process (ICP)

nize that the current status is inappropriate; and second, the individual must locate a new, more appropriate status. Thus, in response to the external status cues, the individual comes to recognize internally that the initial status is inappropriate; and then he or she uses the cues to locate a new, more appropriate status. The identity change occurs in response to, and is mediated through, the status cues that exist in the social environment (see figure).

STATUS CUES: THE EXTERNAL COMPONENT

Status cues make up the public or external component of the identity change process. A status cue is some feature of the social environment that contains information about a particular status or status dimension. Because this paper is about obesity, the cues of interest are about "fatness." Such status cues provide information about whether or not the individual is "fat," and if so, how "fat."

"Recognizing" and "placing" comprise the internal component of the identity change process and occur in response to, and are mediated through, the status cues that exist in the social environment. In order to fully understand the identity change process, it is necessary to explain the interaction between outer and inner processes (Scheff, 1988, p. 396), or in our case, external and internal components of the process.

Status cues are transmitted in two ways: actively and passively. Active cues are communicated through interaction. For example, people are informed by peers, friends, spouses, etc., that they are overweight. The following are some typical comments that occurred repeatedly in the interviews in response to the question, "How did you know that you were fat?"

> I was starting to be called chubby, and being teased in school.

> When my mother would take me shopping, she'd get angry because the clothes that were supposed to be in my age group wouldn't fit me. She would yell at me.

> Well, people would say, "When did you put on all your weight, Bob?" You know, something like that. You know, you kind of get the message, that, you know, I did put on weight.

A second category of cues might accurately be described as passive in form. The information in these cues exists within the environment, but the individual must in some way be sensitized to that information. For example, standing on a scale will provide an individual with information about weight. It is up to the individual to get on the scale, look at the numbers, and then make some sense out of them. Other passive cues might involve seeing one's reflection in a mirror, standing next to others, fitting in chairs, or, as frequently mentioned by respondents, the sizing of clothes. The comments below, all made in response to the question, "How did you know that you were fat?" are representative of passive cue statements.

> I think that it was not being able to wear the clothes that the other kids wore.

> How did I know? Because when we went to get weighed, I weighed more than my, uh, a girl my height should have weighed, according to the chart, according to all the charts that I used to read. That's when I first noticed that I was overweight.

> I would see all these ladies come in and they could wear size 11 and 12, and I thought, Why can't I do that? I should be able to do that.

Both active and passive cues serve as mechanisms for communicating information about a specific status. As can be seen from the data, events occur that force the individual to evaluate his or her conceptions of self.

RECOGNIZING

The term "recognizing" refers to the cognitive process by which an individual becomes aware that a particular status is no longer appropriate. As shown in the figure, the process assumes the individual's acceptance of some initial status. For obese individuals, the initial status is that of "normal body build."[4] This assumption is based on the observation that none of our interviewees assumed that they were "always fat." Even those who were fat as children could identify when they became aware that they were "fat." Through the perception of discrepant status cues, the individual comes to recognize that the initial status is inappropriate. It is possible that the person will perceive the discrepant cues and will either ignore or reject them, in which case the initial status is retained. The factors regulating such a failure to recognize are important, but are not dealt with in this paper. Further research on this point is called for.

Status cues are the external mechanisms through which the recognizing takes place, but it is paying attention to the information contained in these cues that triggers the internal cognitive process of recognizing.

An important point is that the acceptance (or rejection) of a particular status does not occur simply because the individual possesses a set of objective characteristics. For example, two people may have similar body builds, but one may have a self-definition of "fat" whereas the other may not. There appears to be a rather tenuous connection between objective condition and subjective definition. The following comments are supportive of this disjunction.

> I was really, as far as pounds go, very thin, but I had a feeling about myself that I was huge.

> Well, I don't remember ever thinking about it until I was about in eighth grade. But I was looking back at pictures when I was little. I was always chunky, chubby.

This lack of necessary connection between objective condition and subjective definition points up an important and frequently overlooked feature of social statuses: the extent to which they are *self-evident*. Self-evidentiality refers to the degree to which a person who possesses certain objective status characteristics is *aware* that a particular status label applies to them.[5]

Some statuses possess a high degree of evidentiality: gender identification is one of these.[6] On the other hand, being beautiful or intelligent is somewhat non-self-evident. This is not to imply that individuals are either ignorant of these statuses or of the characteristics upon which they are assigned. People may know that other people are intelligent, but they may be unaware that the label is equally applicable to them.

One idea that emerged quite early from the interviews was that being "fat" is a relatively non-self-evident status. Individuals do not recognize that "fat" is a description that applies to them.[7] The objective condition of being overweight is not sufficient, in itself, to promote the adoption of a "fat" identity. This non-self-evidentiality is demonstrated in the following excerpts.

> I think that I just thought that it was a little bit here and there. I didn't think of it, and I didn't think of myself as looking bad. But you know, I must have.

> I have pictures of me right after the baby was born. I had no idea that I was that fat.

The self-evidentiality of a status is important in the discussion of the identity change process. The less self-evident a status, the more difficult the recognizing process becomes. Further, because recognition occurs in response to status cues, the self-evidentiality of a status will influence the type of cues that play the most prominent role in identity change.

A somewhat speculative observation should be made about status cues in the recognizing process. For our subjects, recognizing occurred primarily throught active cues. When passive cues were involved, they typically were highly visible and unambiguous. In general, active cues appear to be more potent in forcing the individual's attention to the information that the current status is inappropriate. The predominance of these active cues is possibly a consequence of the relatively non-self-evident character of the "fat" status. It is probable that the less self-evident a status, the more likely that the recognizing process will occur through active rather than passive cues.

Once the individual comes to recognize the inappropriateness of the initial status, it becomes necessary to locate a new, more appropriate status. This search for a more appropriate status is referred to as the "placing" process.[8]

PLACING

Placing refers to a cognitive process whereby an individual comes to identify an appropriate status from among those available. The number of status categories along a status dimension influences the placing process. A status dimension may contain any number of status categories. If there are only two status dimensional categories, such as in the case of gender, the placement process is more or less automatic. When individuals recognize that they do not belong in one category, the remaining category becomes the obvious alternative. The greater the number of status categories, the more difficult the placing process becomes.

The body build dimension contains an extremely large number of categories. When an individual recognizes that he or she does not possess a "normal" body build, there are innumerable alternatives open. The knowledge that one's status lies toward the "fat" rather than the "thin" end of the continuum still presents a wide range of choices. In everyday conversation, we hear terms that describe these alternatives: chubby, porky, plump, hefty, full-figured, beer belly, etc. All are informal descriptions reflecting the myriad categories along the body build dimension.

I wasn't real fat in my eyes. I don't think. I was just chunky.

Not fat. I didn't exactly classify it as fat. I just thought, I'm, you know, I am a pudgy lady.

I don't think that I have ever called myself fat. I have called myself heavy.

Even when individuals adopt a "fat" identity, they attempt to make distinctions about how fat they are. Because being fat is a devalued status, individuals attempt to escape the full weight of its negative attributes while still acknowledging the nonnormal status. The following responses exemplify this attempt to neutralize the pejorative connotations of having a "fat" status. The practice of differentiating one's status from others becomes vital in managing a fat identity.

Q. How did you know that you weren't *that* fat?

A. Well, comparing myself to others at the time, I didn't really feel that I was that fat. But I knew maybe because they didn't treat me the same way they treated people who were heavier than me. You know, I got teased lightly, but I was still liked by a lot of people, and the people that were heavy weren't.

As is apparent from this excerpt, the individual neutralized self-image by linking "fatness" with the level of teasing done by peers.

NEW STATUS

The final phase of the identity change process involves the acceptance of a new status. For our informants, it was the acceptance of a "fat" status, along with its previously mentioned pejorative characterizations.[9]

> I hate to look in mirrors. I hate that. It makes me feel so self-conscious. If I walk into a store, and I see my reflection in the glass, I just look away.

> We'd go somewhere and I would think, "I never look as good as everybody else." You know everybody always looks better. I'd cry before we'd go bowling because I'd think, "Oh, I just look awful."

As is clear, the final phase of the identity change process involves the internalization of a negative (deviant) definition of self. For many fat people accepting a new status means starting on the merry-go-round of weight reduction programs.[10] Many of these programs or organizations attempt to get members to accept a devalued status fully, and then work to change it. Consequently, individuals are forced to "admit" that they are fat and to "witness" in front of others.[11] The new identity becomes that of a "fat" person, which the weight reduction programs then attempt to transform. A further analysis of the impact of informal organizations on the identity change process will be attempted in another paper.

CONCLUSION

In this paper, we have attempted to fill a void within the interactionist literature by presenting an inductively generated model of the identity change process. The proposed model treats the change process from a career focus, and thus addresses both the external (public) and the internal (cognitive) features of the identity change.

We have suggested that the adoption of a new status takes place through two sequential cognitive processes, "recognizing" and "placing." First, the individual must come to recognize that a current status is no longer appropriate. Second, the individual must locate a new, more appropriate status from among those available. We have further suggested that these internal or cognitive processes are triggered by and mediated through status cues, which exist in the external environment. These cues can be either active or passive. Active cues are transmitted through interaction, whereas passive cues must be sought out by the individual.

We also found a relationship between the evidentiality of the status, that is, how obvious that status is to the individual, and the role of the different types of cues in the identity change process. Finally, we have suggested that the adoption of a new status is a trigger for further career changes.

Although the model presented in this paper was generated inductively from field data on obese individuals, we are confident that it may be fruitfully applied to the study of other deviant careers. It seems particularly appropriate where the identity involved has a low degree of self-evidentiality.

In addition, we feel that the focus upon the different types of status cues and their differing roles in the recognizing and placing processes can lead to a better understanding of how institutionally promoted identity change occur.

NOTES

1. Included here is research on both rule creators and rule enforcers. We have not made an attempt to analytically separate the two types of investigation.

2. Some of the members had successfully lost their excess weight. When these people were present at a meeting, a leader was careful to introduce them to the other members of the class and to tell how much weight they had lost. This was done to uphold their claim to acceptance by the other group members.

3. Access to this information was gained from an "insider" perspective because one of the researchers was well known among the membership, being an "off and on" member of the organization for 3 years. Thus, he was not confronted with the problem of gaining entry into a semi-closed social setting. Similarly, because the observer had "been an ongoing participant of the group," he did not have to desensitize the other members of the group to his presence.

4. It is important to note that this process can operate generically. That is, it is not only applicable to the "identity change" from a "normal" to a "deviant" identity, but can encompass the reverse process as well. In a forthcoming project, we will use the process to analyze how various rehabilitation programs attempt to get individuals back to the initial status.

5. This concept is different in an important way from what Goffman calls "visibility." He uses the term to refer to ". . . how well or how badly the stigma is adapted to provide means of communicating that the individual possesses it" (Goffman, 1963, p. 48). The focus of the concept is on how readily the social environment can identify that the individual possesses a stigmatized trait. The concept of self-evidentiality deals with how readily the individual can internalize possession of the stigmatized trait. The focus is upon the actor's perceptions, not on the audience.

6. We are referring here to the physiological description of being male or female. We realize that sex roles are much less self-evident.

7. Conversely, a number of individuals thought of themselves as "fat" or "obese," and were objectively "normal." In this case, the existence of objective indicators was insufficient to prevent the individual from adopting a "fat" identity.

8. In some instances, recognizing and placing occur simultaneously. This is especially true when the cue involved is an active one, and contains information about both the initial and new statuses. For example, if peers call a child "fatty," this interaction informs the child that the "normal" status is inappropriate. At the same time, it informs the child that being "fat" is the appropriate status. Even here however, the individual must recognize before it is possible to place.

9. This phase corresponds closely to that presented in much of the "subcultural" research. (See Schur, 1971; Becker, 1963; Sykes & Matza, 1957, p. 664.)

10. Weight Watchers, TOPS, Overeaters Anonymous, Diet Center, and OptiFast are typical examples of this type of program.

11. By witnessing, we are referring to the process whereby individuals come to renounce, in front of others, a former self and former behaviors associated with that self. Some religious groups, Synanon, Alcoholics Anonymous, etc., seem to encourage this type of degradation of self.

REFERENCES

Baum, L. (1987, August 3). Extra pounds can weigh down your career. *Business Week,* p. 96.

Becker, H. S. (1963). *Outsiders: Studies in the sociology of deviance.* New York: Free Press.

Canning, H., & Mayers, J. (1966). Obesity: Its possible effects on college acceptance. *New England Journal of Medicine,* 275(24); November, 1172–1174.

Cicourel, A. (1968). *The social organization of juvenile justice.* New York: Wiley.

Conrad, P. (1975). The discovery of hyperkinesis: Notes on the medicalization of deviant behavior. *Social Problems,* 23(1); October, 12–21.

Goffman, E. (1961). *Asylums.* Garden City, NY: Anchor.

———. (1963). *Stigma: Notes on the management of spoiled identity.* Englewood Cliffs, NJ: Prentice-Hall.

Greenberg, D. (1989). The antifat conspiracy. *New Scientist,* 22 (April 22): 79.

Hadden, S. C., Degher, D., & Fernandez, R. (1989). Sports as a strategic ethnographic arena. *Arena Review,* 13(1), 9–19.

Kelly, J. R. (1990). *Leisure* (2nd ed.). Englewood Cliffs, NJ: Prentice-Hall.

Kitsuse, J. (1962). Societal reactions to deviant behavior: Problems of theory and method. *Social Problems* 9 (Winter): 247–256.

Laslett, B., & Warren, C. A. B. (1975). Losing weight: The organizational promotion of behavior change. *Social Problems,* 23(1), 69–80.

Lemert, E. (1951). *Social pathology,* New York: McGraw-Hill.

Maddox, G. L., Back, K. W., & Liederman, V. (1968). Overweight as social deviance and disability. *Journal of Health and Social Behavior,* 9(4): 287–298.

Maddox, G. L., & Liederman, V. (1969). Overweight as a social disability with medical implications. *Journal of Medical Education,* 9(4): 287–298.

Pauley, L. L. (1989). Customer weights as a variable in salespersons' response time. *Journal of Social Psychology,* 129: 713–714.

Petit, D. W. (1974). The ills of the obese. In G. A. Gray & J. E. Bethune (Eds.), *Treatment and management of obesity.* New York: Harper & Row.

Piliavin, I., & Briar, S. (1964). Police encounters with juveniles. *American Journal of Sociology,* (September): 206–214.

Register, C. A. (1990). Wage effects of obesity among young workers. *Social Science Quarterly,* 71 (March): 130–141.

Scheff, T. (1988). Shame and conformity: The deference emotion system. *American Journal of Sociology,* 53 (June): 395–406.

Schur, E. M. (1971). *Labeling deviant behavior: Its sociological implications.* New York: Harper & Row.

Sykes, G., & Matza, D. (1957). Techniques of neutralization: A theory of delinquency. *American Sociological Review,* (December): 664–670.

Tannenbaum, F. (1939). *Crime and the community.* New York: Columbia University Press.

21

Becoming Bisexual

MARTIN S. WEINBERG, COLIN J. WILLIAMS, AND DOUGLAS W. PRYOR

Weinberg, Williams, and Pryor's study of the identity career followed by individuals who become bisexual illustrates a much more complex identity trajectory than Degher and Hughes' portrayal of their fat subjects. The attraction of individuals in this selection to members of the same sex leads them to question and reject their affiliation with the heterosexual identity. But their continuing attraction to members of the opposite sex leads them to question the appropriateness, for them, of the homosexual designation. Aware that they do not fit comfortably into either of these common sexual identity labels, they are confused and troubled. After struggling with their ambivalences about their dual sexual orientations, they discover the bisexual option and begin the process of reconceptualizing their selves. Although finding this label and discovering other established bisexuals aids their process of self-acceptance, problems remain for people trying to establish identities outside of the more common sexual norms. They remain doubly stigmatized due to their rejection by both the heterosexual and homosexual communities, and never feel completely comfortable with this liminal identity.

Becoming bisexual involves the rejection of not one but two recognized categories of sexual identity: heterosexual and homosexual. Most people settle into the status of heterosexual without any struggle over the identity. There is not much concern with explaining how this occurs; that people are heterosexual is simply taken for granted. For those who find heterosexuality unfulfilling, however, developing a sexual identity is more difficult.

How is it then that some people come to identify themselves as "bisexuals"? As a point of departure we take the process through which people come to identify themselves as "homosexual." A number of models have been formulated that chart the development of a homosexual identity through a series of stages. While each model involves a different number of stages, the models all share three elements. The process begins with the person in a state of identity confusion—feeling different from others, struggling with the acknowledgment of same-sex attractions. Then there is a period of thinking about possibly being homosexual—involving associating with self-identified homosexuals, sexual experimentation, forays into the ho-

---

From *Dual Attraction: Understanding Bisexuality* by Martin S. Weinberg, Colin J. Williams, and Doug Pryor. © 1995 by Oxford University Press, Inc. Used by permission of Oxford University Press, Inc.

mosexual subculture. Last is the attempt to integrate one's self-concept and social identity as homosexual—acceptance of the label, disclosure about being homosexual, acculturation to a homosexual way of life, and the development of love relationships. Not every person follows through each stage. Some remain locked in at a certain point. Others move back and forth between stages.

To our knowledge, no previous model of *bisexual* identity formation exists. In this chapter we present such a model based on the following questions: To what extent is there overlap with the process involved in becoming homosexual? How far is the label "bisexual" clearly recognized, understood, and available to people as an identity? Does the absence of a bisexual subculture in most locales affect the information and support needed for sustaining a commitment to the identity? For our subjects, then, what are the problems in finding the "bisexual" label, understanding what the label means, dealing with social disapproval from two directions, and continuing to use the label once it is adopted? From our fieldwork and interviews, we found that four stages captured our respondents' most common experiences when dealing with questions of identity: initial confusion, finding and applying the label, settling into the identity, and continued uncertainty.

THE STAGES

Initial Confusion

Many of the people interviewed said that they had experienced a period of considerable confusion, doubt, and struggle regarding their sexual identity before defining themselves as bisexual. This was ordinarily the first step in the process of becoming bisexual.

They described a number of major sources of early confusion about their sexual identity. For some, it was the experience of having strong sexual feelings for both sexes that was unsettling, disorienting, and sometimes frightening. Often these were sexual feelings that they said they did not know how to easily handle or resolve.

> In the past, I couldn't reconcile different desires I had. I didn't understand them. I didn't know what I was. And I ended up feeling really mixed up, unsure, and kind of frightened. (F)

> I thought I was gay, and yet I was having these intense fantasies and feelings about fucking women. I went through a long period of confusion. (M)

Others were confused because they thought strong sexual feelings for, or sexual behavior with, the same sex meant an end to their long-standing heterosexuality.

> I was afraid of my sexual feelings for men and . . . that if I acted on them, that would negate my sexual feelings for women. I knew absolutely no

one else who had . . . sexual feelings for both men and women, and didn't realize that was an option. (M)

When I first had sexual feelings for females, I had the sense I should give up my feelings for men. I think it would have been easier to give up men. (F)

A third source of confusion in this initial stage stemmed from attempts by respondents trying to categorize their feelings for, and/or behaviors with, both sexes, yet not being able to do so. Unaware of the term "bisexual," some tried to organize their sexuality by using readily available labels of "heterosexual" or "homosexual"—but these did not seem to fit. No sense of sexual identity jelled; an aspect of themselves remained unclassifiable.

When I was young, I didn't know what I was. I knew there were people like Mom and Dad—heterosexual and married—and that there were "queens." I knew I wasn't like either one. (M)

I thought I had to be either gay or straight. That was the big lie. It was confusing. . . . That all began to change in the late 60s. It was a long and slow process. . . . (F)

Finally, others suggested they experienced a great deal of confusion because of their "homophobia"—their difficulty in facing up to the same-sex component of their sexuality. The consequence was often long-term denial. This was more common among the men than the women, but not exclusively so.

At age seventeen, I became close to a woman who was gay. She had sexual feelings for me. I had some . . . for her but I didn't respond. Between the ages of seventeen and twenty-six I met another gay woman. She also had sexual feelings towards me. I had the same for her but I didn't act on . . . or acknowledge them. . . . I was scared. . . . I was also attracted to men at the same time. . . . I denied that I was sexually attracted to women. I was afraid that if they knew the feelings were mutual they would act on them . . . and put pressure on me. (F)

I thought I might be able to get rid of my homosexual tendencies through religious means—prayer, belief, counseling—before I came to accept it as part of me. (M)

The intensity of the confusion and the extent to which it existed in the lives of the people we met at the Bisexual Center, whatever its particular source, was summed up by two men who spoke with us informally. As paraphrased in our field notes:

The identity issue for him was a very confusing one. At one point, he almost had a nervous breakdown, and when he finally entered college, he sought psychiatric help.

Bill said he thinks this sort of thing happens a lot at the Bi Center. People come in "very confused" and experience some really painful stress.

Finding and Applying the Label

Following this initial period of confusion, which often spanned years, was the experience of finding and applying the label. We asked the people we interviewed for specific factors or events in their lives that led them to define themselves as bisexual. There were a number of common experiences.

For many who were unfamiliar with the term bisexual, the discovery that the category in fact existed was a turning point. This happened by simply hearing the word, reading about it somewhere, or learning of a place called the Bisexual Center. The discovery provided a means of making sense of long-standing feelings for both sexes.

> Early on I thought I was just gay, because I was not aware there was another category, bisexual. I always knew I was interested in men and women. But I did not realize there was a name for these feelings and behaviors until I took Psychology 101 and read about it, heard about it there. That was in college.(F)

> The first time I heard the word, which was not until I was twenty-six, I realized that was what fit for me. What it fit was that I had sexual feelings for both men and women. Up until that point, the only way that I could define my sexual feelings was that I was either a latent homosexual or a confused heterosexual. (M)

> Going to a party at someone's house, and finding out there that the party was to benefit the Bisexual Center. I guess at that point I began to define myself as bisexual. I never knew there was such a word. If I had heard the word earlier on, for example as a kid, I might have been bisexual then. My feelings had always been bisexual. I just did not know how to define them. (F)

> Reading *The Bisexual Option* . . . I realized then that bisexuality really existed and that's what I was. (M)

In the case of others the turning point was their first homosexual or heterosexual experience coupled with the recognition that sex was pleasurable with both sexes. These were people who already seemed to have knowledge of the label "bisexual," yet without experiences with both men and women, could not label themselves accordingly.

> The first time I had actual intercourse, an orgasm with a woman, it led me to realize I was bisexual, because I enjoyed it as much as I did with a man, although the former occurred much later on in my sexual experiences. . . . I didn't have an orgasm with a woman until twenty-two, while with males, that had been going on since the age of thirteen. (M)

> Having homosexual fantasies and acting those out. . . . I would not identify as bi if I only had fantasies and they were mild. But since my fantasies were intensely erotic, and I acted them out, these two things led me to believe I was really bisexual. . . . (M)

After my first involved sexual affair with a woman, I also had feelings for a man, and I knew I did not fit the category dyke. I was also dating gay-identified males. So I began looking at gay/lesbian and heterosexual labels as not fitting my situation. (F)

Still others reported not so much a specific experience as a turning point, but emphasized the recognition that their sexual feelings for both sexes were simply too strong to deny. They eventually came to the conclusion that it was unnecessary to choose between them.

I found myself with men but couldn't completely ignore my feelings for women. When involved with a man I always had a close female relationship. When one or the other didn't exist at any given time, I felt I was really lacking something. I seem to like both. (F)

The last factor that was instrumental in leading people to initially adopt the label bisexual was the encouragement and support of others. Encouragement sometimes came from a partner who already defined himself or herself as bisexual.

Encouragement from a man I was in a relationship with. We had been together two or three years at the time—he began to define as bisexual.... [He] encouraged me to do so as well. He engineered a couple of threesomes with another woman. Seeing one other person who had bisexuality as an identity that fit them seemed to be a real encouragement. (F)

Encouragement from a partner seemed to matter more for women. Occasionally the "encouragement" bordered on coercion as the men in their lives wanted to engage in a *ménage à trois* or group sex.

I had a male lover for a year and a half who was familiar with bisexuality and pushed me towards it. My relationship with him brought it up in me. He wanted me to be bisexual because he wanted to be in a threesome. He was also insanely jealous of my attractions to men, and did everything in his power to suppress my opposite-sex attractions. He showed me a lot of pictures of naked women and played on my reactions. He could tell that I was aroused by pictures of women and would talk about my attractions while we were having sex.... He was twenty years older than me. He was very manipulative in a way. My feelings for females were there and [he was] almost forcing me to act on my attractions.... (F)

Encouragement also came from sex-positive organizations, primarily the Bisexual Center, but also places like San Francisco Sex Information (SFSI), the Pacific Center, and the Institute for Advanced Study of Human Sexuality, ...

At the gay pride parade I had seen the brochures for the Bisexual Center. Two years later I went to a Tuesday night meeting. I immediately felt that I belonged and that if I had to define myself that this was what I would use. (M)

Through SFSI and the Bi Center, I found a community of people ... [who] were more comfortable for me than were the exclusive gay or heterosexual communities. . . . [It was] beneficial for myself to be . . . in a sex-positive community. I got more strokes and came to understand myself better. . . . I felt it was necessary to express my feelings for males and females without having to censor them, which is what the gay and straight communities pressured me to do. (F)

Thus our respondents became familiar with and came to the point of adopting the label bisexual in a variety of ways: through reading about it on their own, being in therapy, talking to friends, having experiences with sex partners, learning about the Bi Center, visiting SFSI or the Pacific Center, and coming to accept their sexual feelings.

Settling into the Identity

Usually it took years from the time of first sexual attractions to, or behaviors with, both sexes before people came to think of themselves as bisexual. The next stage then was one of settling into the identity, which was characterized by a more complete transition in self-labeling.

Most reported that this settling-in stage was the consequence of becoming more self-accepting. They became less concerned with the negative attitudes of others about their sexual preferences.

> I realized that the problem of bisexuality isn't mine. It's society's. They are having problems dealing with my bisexuality. So I was then thinking if they had a problem dealing with it, so should I. But I don't. (F)

> I learned to accept the fact that there are a lot of people out there who aren't accepting. They can be intolerant, selfish, shortsighted and so on. Finally, in growing up, I learned to say "So what, I don't care what others think." (M)

> I just decided I was bi. I trusted my own sense of self. I stopped listening to others tell me what I could or couldn't be. (F)

The increase in self-acceptance was often attributed to the continuing support from friends, counselors, and the Bi Center, through reading, and just being in San Francisco.

> Fred Klein's *The Bisexual Option* book and meeting more and more bisexual people . . . helped me feel more normal. . . . There were other human beings who felt like I did on a consistent basis. (M)

> I think going to the Bi Center really helped a lot. I think going to the gay baths and realizing there were a lot of men who sought the same outlet I did really helped. Talking about it with friends has been helpful and being validated by female lovers that approve of my bisexuality. Also the reaction of people who I've told, many of whom weren't even surprised. (M)

> The most important thing was counseling. Having the support of a bisexual counselor. Someone who acted as somewhat of a mentor. [He] validated my frustration . . ., helped me do problem solving, and guide[d]

me to other supportive experiences like SFSI. Just engaging myself in a supportive social community. (M)

The majority of the people we came to know through the interviews seemed settled in their sexual identity. We tapped this through a variety of questions. . . . Ninety percent said that they did not think they were currently in transition from being homosexual to being heterosexual or from being heterosexual to being homosexual. However, when we probed further by asking this group "Is it possible, though, that someday you could define yourself as either lesbian/gay or heterosexual?" about 40 percent answered yes. About two-thirds of these indicated that the change could be in either direction, though almost 70 percent said that such a change was not probable.

We asked those who thought a change was possible what it might take to bring it about. The most common responses referred to becoming involved in a meaningful relationship that was monogamous or very intense. Often the sex of the hypothetical partner was not specified, underscoring that the overall quality of the relationship was what really mattered.

> Love. I think if I feel insanely in love with some person, it could possibly happen. (M)
>
> If I should meet a woman and want to get married, and if she was not open to my relating to men, I might become heterosexual again. (M)
>
> Getting involved in a longer-term relationship like marriage where I wouldn't need a sexual involvement with anyone else. The sex of the . . . partner wouldn't matter. It would have to be someone who I could commit my whole life to exclusively, a lifelong relationship. (F)

A few mentioned the breaking up of a relationship and how this would incline them to look toward the other sex.

> Steve is one of the few men I feel completely comfortable with. If anything happened to him, I don't know if I'd want to try and build up a similar relationship with another man. I'd be more inclined to look towards women for support. (F)

Changes in sexual behavior seemed more likely for the people we interviewed . . . than changes in how they defined themselves. We asked "Is it possible that someday you could behave either exclusively homosexual or exclusively heterosexual?" Over 80 percent answered yes. This is over twice as many as those who saw a possible change in how they defined themselves, again showing that a wide range of behaviors can be subsumed under the same label. Of this particular group, the majority (almost 60 percent) felt that there was nothing inevitable about how they might change, indicating that it could be in either a homosexual or a heterosexual direction. Around a quarter, though, said the change would be to exclusive heterosexual behavior and 15 percent to exclusive homosexual behavior. (Twice as many women noted the homosexual direction, while many more men than women said the heterosexual direction.) Just over 40 percent responded that a change to exclusive heterosexuality or homosexuality was not very probable, about a third somewhat probable, and about a quarter very probable.

Again, we asked what it would take to bring about such a change in behavior. Once more the answers centered on achieving long-term monogamous and involved relationship, often with no reference to a specific sex.

> For me to behave exclusively heterosexual or homosexual would require that I find a lifetime commitment from another person with a damn good argument of why I should not go to bed with somebody else. (F)

> I am a romantic. If I fell in love with a man, and our relationship was developing that way, I might become strictly homosexual. The same possibility exists with a woman. (M)

Thus "settling into the identity" must be seen in relative terms. Some of the people we interviewed did seem to accept the identity completely. When we compared our subjects' experiences with those characteristic of homosexuals, however, we were struck by the absence of closure that characterized our bisexual respondents—even those who appeared most committed to the identity. This led us to posit a final stage in the formation of sexual identity, one that seems unique to bisexuals.

Continued Uncertainty

The belief that bisexuals are confused about their sexual identity is quite common. The conception has been promoted especially by those lesbians and gays who see bisexuality as being in and of itself a pathological state. From their point of view, "confusion" is literally a built-in feature of "being" bisexual. As expressed in one study:

> While appearing to encompass a wider choice of love objects . . . [the bisexual] actually becomes a product of abject confusion; his self-image is that of an overgrown young adolescent whose ability to differentiate one form of sexuality from another has never developed. He lacks above all a sense of identity. . . . [He] cannot answer the question: What am I?

One evening a facilitator at a Bisexual Center rap group put this belief in a slightly different and more contemporary form:

> One of the myths about bisexuality is that you can't be bisexual without somehow being "schizoid." The lesbian and gay communities do not see being bisexual as a crystallized or complete sexual identity. The homosexual community believes there is no such thing as bisexuality. They think that bisexuals are people who are in transition [to becoming homosexual] or that they are people afraid of being stigmatized [as homosexual] by the heterosexual majority.

We addressed the issue directly in the interviews with two questions: "Do you *presently* feel confused about your bisexuality?" and "Have you ever felt confused . . . ?" . . . For the men, a quarter and 84 percent answered "yes," respectively. For the women, it was about a quarter and 56 percent.

When asked to provide details about this uncertainty, the primary response was that *even after having discovered and applied the label "bisexual" to themselves, and having come to the point of apparent self-acceptance, they still experienced continued intermittent periods of doubt and uncertainty regarding their sexual identity.* One reason was the lack of social validation and support that came with being a self-identified bisexual. The social reaction people received made it difficult to sustain the identity over the long haul.

While the heterosexual world was said to be completely intolerant of any degree of homosexuality, the reaction of the homosexual world mattered more. Many bisexuals referred to the persistent pressures they experienced to relabel themselves "gay" or "lesbian" and to engage in sexual activity exclusively with the same sex. It was asserted that no one was *really* bisexual, and that calling oneself "bisexual" was a politically incorrect and unauthentic identity. Given that our respondents were living in San Francisco (which has such a large homosexual population) and that they frequently moved in and out of the homosexual world (to whom they often looked for support) this could be particularly distressing.

> Sometimes the repeated denial the gay community directs at us. Their negation of the concept and the term bisexual has sometimes made me wonder whether I was just imagining the whole thing. (M)

> My involvement with the gay community. There was extreme political pressure. The lesbians said bisexuals didn't exist. To them, I had to make up my mind and identify as lesbian. . . . I was really questioning my identity, that is, about defining myself as bisexual. . . . (F)

For the women, the invalidation carried over to their feminist identity (which most had). They sometimes felt that being with men meant they were selling out the world of women.

> I was involved with a woman for several years. She was straight when I met her but became a lesbian. She tried to "win me back" to lesbianism. She tried to tell me that if I really loved her, I would leave Bill. I did love her, but I could not deny how I felt about him either. So she left me and that hurt. I wondered if I was selling out my woman identity and if it [being bisexual] was worth it. (F)

A few wondered whether they were lying to themselves about their heterosexual side. One woman questioned whether her heterosexual desires were a result of "acculturation" rather than being her own choice. Another woman suggested a similar social dimension to her homosexual component:

> There was one period when I was trying to be gay because of the political thing of being totally woman-identified rather than being with men. The Women's Culture Center in college had a women's studies minor, so I was totally immersed in women's culture. . . . (F)

Lack of support also came from the absence of bisexual role models, no real bisexual community aside from the Bisexual Center, and nothing in the way of public recognition of bisexuality, which bred uncertainty and confusion.

> I went through a period of dissociation, of being very alone and isolated. That was due to my bisexuality. People would ask, well, what was I? I wasn't gay and I wasn't straight. So I didn't fit. (F)

> I don't feel like I belong in a lot of situations because society is so polarized as heterosexual or homosexual. There are not enough bi organizations or public places to go to like bars, restaurants, clubs. . . . (F)

For some, continuing uncertainty about their sexual identity was related to their inability to translate their sexual feelings into sexual behaviors. (Some of the women had *never* engaged in homosexual sex.)

> Should I try to have a sexual relationship with a woman? . . . Should I just back off and keep my distance, just try to maintain a friendship? I question whether I am really bisexual because I don't know if I will ever act on my physical attractions for females. (F)

> I know I have strong sexual feelings towards men, but then I don't know how to get close to or be sexual with a man. I guess that what happens is I start wondering how genuine my feelings are. . . . (M)

For the men, confusion stemmed more from the practical concerns of implementing and managing multiple partners or from questions about how to find an involved homosexual relationship and what that might mean on a social and personal level.

> I felt very confused about how I was going to manage my life in terms of developing relationships with both men and women. I still see it as a difficult lifestyle to create for myself because it involves a lot of hard work and understanding on my part and that of the men and women I'm involved with. (M)

> I've thought about trying to have an actual relationship with a man. Some of my confusion revolves around how to find a satisfactory sexual relationship. I do not particularly like gay bars. I have stopped having anonymous sex. . . . (M)

Many men and women felt doubts about their bisexual identity because of being in an exclusive sexual relationship. After being exclusively involved with an opposite-sex partner for a period to time, some of the respondents questioned the homosexual side of their sexuality. Conversely, after being exclusively involved with a partner of the same sex, other respondents called into question the heterosexual component of their sexuality.

> When I'm with a man or a woman sexually for a period of time, then I begin to wonder how attracted I really am to the other sex. (M)

In the last relationship I had with a woman, my heterosexual feelings were very diminished. Being involved in a lesbian lifestyle put stress on my self-identification as a bisexual. It seems confusing to me because I am monogamous for the most part, monogamy determines my lifestyle to the extremes of being heterosexual or homosexual. (F)

Others made reference to a lack of sexual activity with weaker sexual feelings and affections for one sex. Such learning did not fit with the perception that bisexuals should have balanced desires and behaviors. The consequence was doubt about "really" being bisexual.

On the level of sexual arousal and deep romantic feelings, I feel them much more strongly for women than for men. I've gone so far as questioning myself when this is involved. (M)

I definitely am attracted to and it is much easier to deal with males. Also, guilt for my attraction to females has led me to wonder if I am just really toying with the idea. Is the sexual attraction I have for females something I constructed to pass time or what? (F)

Just as "settling into the identity" is a relative phenomenon, so too is "continued uncertainty," which can involve a lack of closure as part and parcel of what it means to be bisexual.

We do not wish to claim too much for our model of bisexual identity formation. There are limits to its general application. The people we interviewed were unique in that not only did *all* the respondents define themselves as bisexual (a consequence of our selection criteria), but they were also all members of a bisexual social organization in a city that perhaps more than any other in the United States could be said to provide a bisexual subculture of some sort. Bisexuals in places other than San Francisco surely must move through the early phases of the identity process with a great deal more difficulty. Many probably never reach the later stages.

Finally, the phases of the model we present are very broad and somewhat simplified. While the particular problems we detail within different phases may be restricted to the type of bisexuals in this study, the broader phases can form the basis for the development of more sophisticated models of bisexual identity formation.

Still, not all bisexuals will follow these patterns. Indeed, given the relative weakness of the bisexual subculture compared with the social pressures toward conformity exhibited in the gay subculture, there may be more varied ways of acquiring a bisexual identity. Also, the involvement of bisexuals in the heterosexual world means that various changes in heterosexual lifestyles (e.g., a decrease in open marriages or swinging) will be a continuing, and as yet unexplored, influence on bisexual identity. Finally, wider societal changes, notably the existence of AIDS, may make for changes in the overall identity process. Being used to choice and being open to both sexes can give bisexuals a range of adaptations in their sexual life that are not available to others.

22

Anorexia Nervosa and Bulimia

PENELOPE A. MCLORG AND DIANE E. TAUB

McLorg and Taub's study of eating disorders describes and analyzes women's progression along an identity career from their initial stage of hyper-conformity through Lemert's stages of primary and secondary deviance. They illustrate how the intense societal preoccupation with weight leads women to the kind of deviant behavior that maintains (initially) their positive external status while they are deteriorating internally. Along the way, these women move through a progression of more common fixations about dieting, to frustration with dieting and movement toward more radical solutions such as bingeing, purging, compulsive exercising, and lack of eating. These behaviors stand apart from their identities, McLorg and Taub argue, since they can avoid the deviant label and self-conception, remaining in the primary deviance stage until they get caught and labeled as having an eating disorder. Once this occurs they move to secondary deviance, reconceptualized by others as anorectic or bulimic. With this label cast on them, they are forced to interact with others through the vehicle of their deviance, thus reinforcing their eating disorders.

Current appearance norms stipulate thinness for women and muscularity for men; these expectations, like any norms, entail rewards for compliance and negative sanctions for violations. Fear of being overweight—of being visually deviant—has led to a striving for thinness, especially among women. In the extreme, this avoidance of overweight engenders eating disorders, which themselves constitute deviance. Anorexia nervosa, or purposeful starvation, embodies visual as well as behavioral deviation; bulimia, binge-eating followed by vomiting and/or laxative abuse, is primarily behaviorally deviant.

Besides a fear of fatness, anorexics and bulimics exhibit distorted body images. In anorexia nervosa, a 20–25 percent loss of initial body weight occurs, resulting from self-starvation alone or in combination with excessive exercising, occasional binge-eating, vomiting and/or laxative abuse. Bulimia denotes cyclical (daily, weekly, for example) binge-eating followed by vomiting or laxative abuse; weight is normal or close to normal (Humphries et al., 1982). Common physical manifestations of these eating disorders include menstrual cessation or irregularities and electrolyte imbalances; among behavioral traits

---

Copyright 1987 from "Anorexia Nervosa and Bulimia: The Development of Deviant Identities," Penelope A. McLorg and Diane E. Taub, *Deviant Behavior*, Vol. 8. Reproduced by permission of Taylor & Francis, Inc ; http://www.routledge-ny.com.

are depression, obsessions/compulsions, and anxiety (Russell, 1979; Thompson and Schwartz, 1982).

Increasingly prevalent in the past two decades, anorexia nervosa and bulimia have emerged as major health and social problems. Termed an epidemic on college campuses (Brody, as quoted in Schur, 1984: 76), bulimia affects 13% of college students (Halmi et al., 1981). Less prevalent, anorexia nervosa was diagnosed in 0.6% of students utilizing a university health center (Stangler and Printz, 1980). However, the overall mortality rate of anorexia nervosa is 6% (Schwartz and Thompson, 1981) to 20% (Humphries et al., 1982); bulimia appears to be less life-threatening (Russell, 1979).

Particularly affecting certain demographic groups, eating disorders are most prevalent among young, white, affluent (upper-middle to upper class) women in modern, industrialized countries (Crisp, 1977; Willi and Grossman, 1983). Combining all of these risk factors (female sex, youth, high socioeconomic status, and residence in an industrialized country), prevalence of anorexia nervosa in upper class English girls' schools is reported at 1 in 100 (Crisp et al., 1976). The age of onset for anorexia nervosa is bimodal at 14.5 and 18 years (Humphries et al., 1982); the most frequent age of onset for bulimia is 18 (Russell, 1979).

Eating disorders have primarily been studied from psychological and medical perspectives.[1] Theories of etiology have generally fallen into three categories: the ego psychological (involving an impaired child-maternal environment); the family systems (implicating enmeshed, rigid families); and the endocrinological (involving a precipitating hormonal defect). Although relatively ignored in previous studies, the sociocultural components of anorexia nervosa and bulimia (the slimness norm and its agents of reinforcement, such as role models) have been postulated as accounting for the recent, dramatic increases in these disorders (Schwartz et al., 1982; Boskind-White, 1985).[2]

Medical and psychological approaches to anorexia nervosa and bulimia obscure the social facets of the disorders and neglect the individuals' own definitions of their situations. Among the social processes involved in the development of an eating disorder is the sequence of conforming behavior, primary deviance, and secondary deviance. Societal reaction is the critical mediator affecting the movement through the deviant career (Becker, 1963). Within a framework of labeling theory, this study focuses on the emergence of anorexic and bulimic identities, as well as on the consequences of being career deviants.

METHODOLOGY

Sampling and Procedures

Most research on eating disorders has utilized clinical subjects or non-clinical respondents completing questionnaires. Such studies can be criticized for simply counting and describing behaviors and/or neglecting the social construction of the disorders. Moreover, the work of clinicians is often limited by therapeutic orientation. Previous research may also have included individuals who were not in therapy on their own volition and who resisted admitting that they had an eating disorder.

Past studies thus disregard the intersubjective meanings respondents attach to their behavior and emphasize researchers' criteria for definition as anorexic or bulimic. In order to supplement these sampling and procedural designs, the present study utilizes participant observation of a group of self-defined anorexics and bulimics.[3] As the individuals had acknowledged their eating disorders, frank discussion and disclosure were facilitated.

Data are derived from a self-help group, BANISH, Bulimics/Anorexics In Self-Help, which met at a university in an urban center of the mid-South. Founded by one of the researchers (D.E.T.), BANISH was advertised in local newspapers as offering a group experience for individuals who were anorexic or bulimic. Despite the local advertisements, the campus location of the meetings may have selectively encouraged university students to attend. Nonetheless, in view of the modal age of onset and socioeconomic status of individuals with eating disorders, college students have been considered target populations (Crisp et al., 1976; Halmi et al., 1981).

The group's weekly two-hour meetings were observed for two years. During the course of this study, thirty individuals attended at least one of the meetings. Attendance at meetings was varied: ten individuals came nearly every Sunday; five attended approximately twice a month; and the remaining fifteen participated once a month or less frequently, often when their eating problems were "more severe" or "bizarre." The modal number of members at meetings was twelve. The diversity in attendance was to be expected in self-help groups of anorexics and bulimics.

> [Most] people's involvement will not be forever or even a long time. Most people get the support they need and drop out. Some take the time to help others after they themselves have been helped but even they may withdraw after a time. It is a natural and in many cases *necessary* process (emphasis in original) (American Anorexia/Bulimia Association, 1983).

Modeled after Alcoholics Anonymous, BANISH allowed participants to discuss their backgrounds and experiences with others who empathized. For many members, the group constituted their only source of help; these respondents were reluctant to contact health professionals because of shame, embarrassment, or financial difficulties.

In addition to field notes from group meetings, records of other encounters with all members were maintained. Participants visited the office of one of the researchers (D.E.T.), called both researchers by phone, and invited them to their homes or out for a cup of coffee. Such interaction facilitated genuine communication and mutual trust. Even among the fifteen individuals who did not attend the meetings regularly, contact was maintained with ten members on a monthly basis.

Supplementing field notes were informal interviews with fifteen group members, lasting from two to four hours. Because they appeared to represent more extensive experience with eating disorders, these interviewees were chosen to amplify their comments about the labeling process, made during group meetings. Conducted near the end of the two-year observation period, the interviews focused on what the respondents thought antedated and maintained their eating disorders. In addition, participants described others' reactions to their behaviors as well as their own interpretations of these reactions. To protect the confidentiality of individuals quoted in the study, pseudonyms are employed.

Description of Members

The demographic composite of the sample typifies what has been found in other studies (Fox and James, 1976; Crisp, 1977; Herzog, 1982; Schlesier-Stropp, 1984). Group members' ages ranged from nineteen to thirty-six, with the modal age being twenty-one. The respondents were white, and all but one were female. The sole male and three of the females were anorexic; the remaining females were bulimic.[4]

Primarily composed of college students, the group included four non-students, three of whom had college degrees. Nearly all members derived from upper-middle or lower-upper class households. Eighteen students and two non-students were never-married and uninvolved in serious relationships; two non-students were married (one with two children); two students were divorced (one with two children); and six students were involved in serious relationships. The duration of eating disorders ranged from three to fifteen years.

CONFORMING BEHAVIOR

In the backgrounds of most anorexics and bulimics, dieting figures prominently, beginning in the teen years (Crisp, 1977; Johnson et al., 1982; Lacey et al., 1986). As dieters, these individuals are conformist in their adherence to the cultural norms emphasizing thinness (Garner et al., 1980; Schwartz et al., 1982). In our society, slim bodies are regarded as the most worthy and attractive; overweight is viewed as physically and morally unhealthy—"obscene," "lazy," "slothful," and "gluttonous" (DeJong, 1980; Ritenbaugh, 1982; Schwartz et al., 1982).

Among the agents of socialization promoting the slimness norm is advertising. Female models in newspaper, magazine, and television advertisements are uniformly slender. In addition, product names and slogans exploit the thin orientation; examples include "Ultra Slim Lipstick," "Miller Lite," and "Virginia Slims." While retaining pressures toward thinness, an Ayds commercial attempts a compromise for those wanting to savor food: "Ayds . . . so you can taste, chew, and enjoy, while you lose weight." Appealing particularly to women, a nationwide fast-food restaurant chain offers low-calorie selections, so individuals can have a "license to eat." In the latter two examples, the notion of enjoying food is combined with the message to be slim. Food and restaurant advertisements overall convey the pleasures of eating, whereas advertisements for other products, such as fashions and diet aids, reinforce the idea that fatness is undesirable.

Emphasis on being slim affects everyone in our culture, but it influences women especially because of society's traditional emphasis on women's appearance. The slimness norm and its concomitant narrow beauty standards exacerbate the objectification of women (Schur, 1984). Women view themselves as visual entities and recognize that conforming to appearance expectations and "becoming attractive object[s] [are] role obligation[s]" (Laws, as quoted in Schur, 1984: 66). Demonstrating the beauty motivation behind dieting, a recent Nielsen survey indicated that of the 56 percent of all women aged 24 to 54 who dieted during the previous year, 76 percent did so for cosmetic, rather than health, reasons (Schwartz et al., 1982). For most female group members, dieting was viewed as a means of gaining attractiveness and appeal to the opposite sex. The male respondent, as well, indicated that "when I was fat, girls didn't look at me, but when I got thinner, I was suddenly popular."

In addition to responding to the specter of obesity, individuals who develop anorexia nervosa and bulimia are conformist in their strong commitment to other conventional norms and goals. They consistently excel at school and work (Russell, 1979; Bruch, 1981; Humphries et al., 1982), maintaining high aspirations in both areas (Theander, 1970; Lacey et al., 1986). Group members generally completed college-preparatory courses in high school, aware from an early age that they would strive for a college degree. Also, in college as well as high school, respondents joined honor societies and academic clubs.

Moreover, pre-anorexics and -bulimics display notable conventionality as "model children" (Humphries et al., 1982: 199), "the pride and joy" of their parents (Bruch, 1981: 215), accommodating themselves to the wishes of others. Parents of these individuals emphasize conformity and value achievement (Bruch, 1981). Respondents felt that perfect or near-perfect grades were expected of them; however, good grades were not rewarded by parents, because "A's" were common for these children. In addition, their parents suppressed conflicts, to preserve the image of the "all-American family" (Humphries et al., 1982). Group members reported that they seldom, if ever, heard their parents argue or raise their voices.

Also conformist in their affective ties, individuals who develop anorexia nervosa and bulimia are strongly, even excessively, attached to their parents. Respondents' families appeared close-knit, demonstrating palpable emotional ties. Several group members, for example, reported habitually calling home at prescribed times, whether or not they had any news. Such families have been termed "enmeshed" and "overprotective," displaying intense interaction and concern for members' welfare (Minuchin et al., 1978; Selvini-Palazzoli, 1978). These qualities could be viewed as marked conformity to the norm of familial closeness.[5]

Another element of notable conformity in the family milieu of pre-anorexics and -bulimics concerns eating, body weight/shape, and exercising

(Kalucy et al., 1977; Humphries et al., 1982). Respondents reported their fathers' preoccupation with exercising and their mothers' engrossment in food preparation. When group members dieted and lost weight, they received an extraordinary amount of approval. Among the family, body size became a matter of "friendly rivalry." One bulimic informant recalled that she, her mother, and her coed sister all strived to wear a size 5, regardless of their heights and body frames. Subsequent to this study, the researchers learned that both the mother and sister had become bulimic.

As pre-anorexics and -bulimics, group members thus exhibited marked conformity to cultural norms of thinness, achievement, compliance, and parental attachment. Their families reinforced their conformity by adherence to norms of family closeness and weight/body shape consciousness.

PRIMARY DEVIANCE

Even with familial encouragement, respondents, like nearly all dieters (Chernin, 1981), failed to maintain their lowered weights. Many cited their lack of willpower to eat only restricted foods. For the emerging anorexics and bulimics, extremes such as purposeful starvation or binging accompanied by vomiting and/or laxative abuse appeared as "obvious solutions" to the problem of retaining weight loss. Associated with these behaviors was a regained feeling of control in lives that had been disrupted by a major crisis. Group members' extreme weight-loss efforts operated as coping mechanisms for entering college, leaving home, or feeling rejected by the opposite sex.

The primary inducement for both eating adaptations was the drive for slimness: with slimness came more self-respect and a feeling of superiority over "unsuccessful dieters." Brian, for example, experienced a "power trip" upon consistent weight loss through starvation. Binges allowed the purging respondents to cope with stress through eating while maintaining a slim appearance. As former strict dieters, Teresa and Jennifer used binging/purging as an alternative to the constant self-denial of starvation. Acknowledging their parents' desires for them to be slim, most respondents still felt it was a conscious choice on their part to continue extreme weight-loss efforts. Being thin became the "most important thing" in their lives—their "greatest ambition."

In explaining the development of an anorexic or bulimic identity, Lemert's (1951; 1967) concept of primary deviance is salient. Primary deviance refers to a transitory period of norm violations which do not affect an individual's self-concept or performance of social roles. Although respondents were exhibiting anorexic or bulimic behavior, they did not consider themselves to be anorexic or bulimic.

At first, anorexics' significant others complimented their weight loss, expounding on their new "sleekness" and "good looks." Branch and Eurman (1980: 631) also found anorexics' families and friends describing them as "well-groomed," "neat," "fashionable," and "victorious." Not until the respondents approached emaciation did some parents or friends become concerned and withdraw their praise. Significant others also became increasingly aware of the anorexics' compulsive exercising, preoccupation with food preparation (but not consumption), and ritualistic eating patterns (such as cutting food into minute pieces and eating only certain foods at prescribed times).

For bulimics, friends or family members began to question how the respondents could eat such large amounts of food (often in excess of 10,000 calories a day) and stay slim. Significant others also noticed calluses across the bulimics' hands, which were caused by repeated inducement of vomiting. Several bulimics were "caught in the act," bent over commodes. Generally, friends and family required substantial evidence before believing that the respondents' binging or purging was no longer sporadic.

SECONDARY DEVIANCE

Heightened awareness of group members' eating behavior ultimately led others to label the respondents "anorexic" or "bulimic." Respondents differed in their histories of being labeled and accepting the labels. Generally first termed anorexic by friends, family, or medical personnel, the anorexics initially vigorously denied the label. They felt they were not "anorexic enough," not skinny enough; Robin did not regard herself as having the "skeletal" appearance she associated with anorexia nervosa. These group members found it difficult to differentiate between socially approved modes of weight loss—eating less and exercising more—and the extremes of those behaviors. In fact, many of their activities—cheerleading, modeling, gymnastics, aerobics—reinforced their pursuit of thinness. Like other anorexics, Chris felt she was being "ultra-healthy," with "total control" over her body.

For several respondents, admitting they were anorexic followed the realization that their lives were disrupted by their eating disorder. Anorexics' inflexible eating patterns unsettled family meals and holiday gatherings. Their regimented lifestyle of compulsively scheduled activities—exercising, school, and meals—precluded any spontaneous social interactions. Realization of their adverse behaviors preceded the anorexics' acknowledgment of their subnormal body weight and size.

Contrasting with anorexics, the binge/purgers, when confronted, more readily admitted that they were bulimic and that their means of weight loss was "abnormal." Teresa, for example, knew "very well" that her bulimic behavior was "wrong and unhealthy," although "worth the physical risks." While the bulimics initially maintained that their purging was only a temporary weight-loss method, they eventually realized that their disorder represented a "loss of control." Although these respondents regretted the self-indulgence, "shame," and "wasted time," they acknowledged their growing dependence on binging/purging for weight management and stress regulation.

The application of anorexic or bulimic labels precipitated secondary deviance, wherein group members internalized these identities. Secondary deviance refers to norm violations which are a response to society's labeling: "secondary deviation . . . becomes a means of social defense, attack or adaptation to the overt and covert problems created by the societal reaction to primary deviance" (Lemert, 1967: 17). In contrast to primary deviance, secondary deviance is generally prolonged, alters the individual's self-concept, and affects the performance of his/her social roles.

As secondary deviants, respondents felt that their disorders "gave a purpose" to their lives. Nicole resisted attaining a normal weight because it was not "her"—she accepted her anorexic weight as her "true" weight. For Teresa, bulimia became a "companion"; and Julie felt "every aspect of her life," including time management and social activities, was affected by her bulimia. Group members' eating disorders became the salient element of their self-concepts, so that they related to familiar people and new acquaintances as anorexics or bulimics. For example, respondents regularly compared their body shapes and sizes with those of others. They also became sensitized to comments about their appearance, whether or not the remarks were made by someone aware of their eating disorder.

With their behavior increasingly attuned to their eating disorders, group members exhibited role engulfment (Schur, 1971). Through accepting anorexic or bulimic identities, individuals centered activities around their deviant role, downgrading other social roles. Their obligations as students, family members, and friends became subordinate to their eating and exercising rituals. Socializing, for example, was gradually curtailed because it interfered with compulsive exercising, binging, or purging.

Labeled anorexic or bulimic, respondents were ascribed a new status with a different set of role expectations. Regardless of other positions the individuals occupied, their deviant status, or master status (Hughes, 1958; Becker, 1973), was identified before all others. Among group members, Nicole, who was known as the "school's brain," became known as the "school's anorexic." No longer viewed as conforming model individuals, some respondents were termed "starving waifs" or "pigs."

Because of their identities as deviants, anorexics' and bulimics' interactions with others were altered. Group members' eating habits were scrutinized by friends and family and used as a "catch-all" for everything negative that happened to them. Respondents felt self-conscious around individuals who knew of their disorders; for example, Robin imagined people "watching and whispering" behind her. In addition, group members believed others expected them to "act" anorexic or bulimic. Friends of some anorexic group members never offered them food or drink, assuming continued disinterest on the respondents' part. While being hospitalized, Denise felt she had to prove to others she was not still vomiting, by keeping her bathroom door open. Other bulimics, who lived in dormitories, were hesitant to use the restroom for normal purposes lest several friends be huddling at the door, listening for vomiting. In general, individuals interacted with the respondents largely on the basis of their eating disorder; in doing so, they reinforced anorexic and bulimic behaviors.

Bulimic respondents, whose weight-loss behavior was not generally detectable from their appearance, tried earnestly to hide their bulimia by binging and purging in secret. Their main purpose in concealment was to avoid the negative consequences of being known as a bulimic. For these individuals, bulimia connoted a "cop-out": like "weak anorexics," bulimics pursued thinness but yielded to urges to eat. Respondents felt other people regarded bulimia as "gross" and had little sympathy for the sufferer. To avoid these stigmas or "spoiled identities," the bulimics shrouded their behaviors.

Distinguishing types of stigma, Goffman (1963) describes discredited (visible) stigmas and discreditable (invisible) stigmas. Bulimics, whose weight was approximately normal or even slightly elevated, harbored discreditable stigmas. Anorexics, on the other hand, suffered both discreditable and discredited stigmas—the latter due to their emaciated appearance. Certain anorexics were more reconciled than the bulimics to their stigmas: for Brian, the "stigma of anorexia was better than the stigma of being fat." Common to the stigmatized individuals was an inability to interact spontaneously with others. Respondents were constantly on guard against topics of eating and body size.

Both anorexics and bulimics were held responsible by others for their behavior and presumed able to "get out of it if they tried." Many anorexics reported being told to "just eat more," while bulimics were enjoined to simply "stop eating so much." Such appeals were made without regard for the complexities of the problem. Ostracized by certain friends and family members, anorexics and bulimics felt increasingly isolated. For respondents, the self-help group presented a non-threatening forum for discussing their disorders. Here, they found mutual understanding, empathy, and support. Many participants viewed BANISH as a haven from stigmatization by "others."

Group members, as secondary deviants, thus endured negative consequences, such as stigmatization, from being labeled. As they internalized the labels anorexic or bulimic, individuals' self-concepts were significantly influenced. When others interacted with the respondents on the basis of their eating disorders, anorexic or bulimic identities were encouraged. Moreover, group members' efforts to counteract the deviant labels were thwarted by their master statuses.

DISCUSSION

Previous research on eating disorders has dwelt almost exclusively on medical and psychological facets. Although necessary for a comprehensive understanding of anorexia nervosa and bulimia, these approaches neglect the social processes involved. The phenomena of eating disorders transcend concrete disease entities and clinical diagnoses. Multifaceted and complex, anorexia nervosa and bulimia require a holistic research design, in which sociological insights must be included.

A limitation of medical/psychiatric studies, in particular, is researchers' use of a priori criteria in establishing salient variables. Rather than utilizing predetermined standards of inclusion, the present study allows respondents to construct their own reality. Concomitant to this innovative approach to eating disorders is the selection of a sample of self-admitted anorexics and bulimics. Individuals' perceptions of what it means to become anorexic or bulimic are explored. Although based on a small sample, findings can be used to guide researchers in other settings.

With only five to ten percent of reported cases appearing in males (Crisp, 1977; Stangler and Printz, 1980), eating disorders are primarily a women's aberrance. The deviance of anorexia nervosa and bulimia is rooted in the visual objectification of women and attendant slimness norm. Indeed, purposeful starvation and binging/purging reinforce the notion that "a society gets the deviance it deserves" (Schur, 1979: 71). As recently noted (Schur, 1984), the sociology of deviance has generally bypassed systematic studies of women's norm violations. Like male deviants, females endure label applications, internalizations, and fulfillments.

The social processes involved in developing anorexic or bulimic identities comprise the sequence of conforming behavior, primary deviance, and secondary deviance. With a background of exceptional adherence to conventional norms, especially the striving for thinness, respondents subsequently exhibit the primary deviance of starving or binging/purging. Societal reaction to these behaviors leads to secondary deviance, wherein respondents' self-concepts and master statuses become anorexic or bulimic. Within this framework of labeling theory, the persistence of eating disorders, as well as the effects of stigmatization, is elucidated.

Although during the course of this research some respondents alleviated their symptoms through psychiatric help or hospital treatment programs, no one was labeled "cured." An anorexic is considered recovered when weight is normal for two years; a bulimic is termed recovered after being symptom-free for one and one-half years (American Anorexia/Bulimia Association Newsletter, 1985). Thus deviance disavowal (Schur, 1971), or efforts after normalization to counteract the deviant labels, remains a topic for future exploration.

NOTES

1. Although instructive, an integration of the medical, psychological, and sociocultural perspectives on eating disorders is beyond the scope of this paper.

2. Exceptions to the neglect of sociocultural factors are discussions of sex-role socialization in the development of eating disorders. Anorexics' girlish appearance has been interpreted as a rejection of femininity and womanhood (Orbach, 1979; Bruch, 1981; Orbach, 1985). In contrast, bulimics have been characterized as over-conforming to traditional female sex roles (Boskind-Lodahl, 1976).

3. Although a group experience for self-defined bulimics has been reported (Boskind-Lodahl, 1976), the researcher,

from the outset, focused on Gestalt and behaviorist techniques within a feminist orientation.

4. One explanation for fewer anorexics than bulimics in the sample is that, in the general population, anorexics are outnumbered by bulimics at 8 or 10 to 1 (Lawson, as reprinted in American Anorexia/Bulimia Association Newsletter, 1985: 1). The proportion of bulimics to anorexics in the sample is 6.5 to 1. In addition, compared to bulimics, anorexics may be less likely to attend a self-help group as they have a greater tendency to deny the existence of an eating problem (Humphries et al., 1982). However, the four anorexics in the present study were among the members who attended the meetings most often.

5. Interactions in the families of anorexics and bulimics might seem deviant in being inordinately close. However, in the larger societal context, the family members epitomize the norms of family cohesiveness. Perhaps unusual in their occurrence, these families are still within the realm of conformity. Humphries and colleagues (1982: 202) refer to the "highly enmeshed and protective" family as part of the "idealized family myth."

REFERENCES

American Anorexia/Bulimia Association. 1983. Correspondence. April.

American Anorexia/Bulimia Association Newsletter. 1985. 8(3).

Becker, Howard S. 1963. *Outsiders.* New York: Free Press.

Boskind-Lodahl, Marlene. 1976. "Cinderella's stepsisters: A feminist perspective on anorexia nervosa and bulimia." *Signs, Journal of Women in Culture and Society* 2: 342–56.

Boskind-White, Marlene. 1985. "Bulimarexia: A sociocultural perspective." In S. W. Emmett (ed.), *Theory and Treatment of Anorexia Nervosa and Bulimia: Biomedical, Sociocultural and Psychological Perspectives* (pp. 113–26). New York: Brunner/Mazel.

Branch, C. H. Hardin, and Linda J. Eurman. 1980. "Social attitudes toward patients with anorexia nervosa." *American Journal of Psychiatry* 137: 631–32.

Bruch, Hilde. 1981. "Developmental considerations of anorexia nervosa and obesity." *Canadian Journal of Psychiatry* 26: 212–16.

Chernin, Kim. 1981. *The Obsession: Reflections on the Tyranny of Slenderness.* New York: Harper and Row.

Crisp, A. H. 1977. "The prevalence of anorexia nervosa and some of its associations in the general population." *Advances in Psychosomatic Medicine* 9: 38–47.

Crisp, A. H., R. L. Palmer, and R. S. Kalucy. 1976. "How common is anorexia nervosa? A prevalence study." *British Journal of Psychiatry* 128: 549–54.

DeJong, William. 1980. "The stigma of obesity: The consequences of naive assumptions concerning the causes of physical deviance." *Journal of Health and Social Behavior* 21: 75–87.

Fox, K. C., and N. McI. James. 1976. "Anorexia nervosa: A study of 44 strictly defined cases." *New Zealand Medical Journal* 84: 309–12.

Garner, David M., Paul E. Garfinkel, Donald Schwartz, and Michael Thompson. 1980. "Cultural expectations of thinness in women." *Psychological Reports* 47: 483–91.

Goffman, Erving. 1963. *Stigma.* Englewood Cliffs, NJ: Prentice-Hall.

Halmi, Katherine A., James R. Falk, and Estelle Schwartz. 1981. "Binge-eating and vomiting: A survey of a college population." *Psychological Medicine* 11: 697–706.

Herzog, David B. 1982. "Bulimia: The secretive syndrome." *Psychosomatics* 23: 481–83.

Hughes, Everett C. 1958. *Men and Their Work.* New York: Free Press.

Humphries, Laurie L., Sylvia Wrobel, and H. Thomas Wiegert. 1982. "Anorexia nervosa." *American Family Physician* 26: 199–204.

Johnson, Craig L., Marilyn K. Stuckey, Linda D. Lewis, and Donald M. Schwartz. 1982. "Bulimia: A descriptive survey of 316 cases." *International Journal of Eating Disorders* 2(1): 3–16.

Kalucy, R. S., A. H. Crisp, and Britta Harding. 1977. "A study of 56 families with anorexia nervosa." *British Journal of Medical Psychology* 50: 381–95.

Lacey, Hubert J., Sian Coker, and S. A. Birtchnell. 1986. "Bulimia: Factors associated with its etiology and maintenance." *International Journal of Eating Disorders* 5: 475–87.

Lemert, Edwin M. 1951. *Social Pathology.* New York: McGraw-Hill.

———. 1967. *Human Deviance, Social Problems and Social Control.* Englewood Cliffs, NJ: Prentice-Hall.

Minuchin, Salvador, Bernice L. Rosman, and Lester Baker. 1978. *Psychosomatic Families: Anorexia Nervosa in Context.* Cambridge, MA: Harvard University Press.

Orbach, Susie. 1979. *Fat Is a Feminist Issue.* New York: Berkeley.

———. 1985. "Visibility/invisibility: Social considerations in anorexia nervosa—a feminist perspective." In S. W. Emmett (ed.), *Theory and Treatment of Anorexia Nervosa and Bulimia: Biomedical, Sociocultural, and Psychological Perspectives* (pp. 127–38). New York: Brunner/Mazel.

Ritenbaugh, Cheryl. 1982. "Obesity as a culture-bound syndrome." *Culture, Medicine and Psychiatry* 6: 347–61.

Russell, Gerald. 1979. "Bulimia nervosa: An ominous variant of anorexia nervosa." *Psychological Medicine* 9: 429–48.

Schlesier-Stropp, Barbara. 1984. "Bulimia: A review of the literature." *Psychological Bulletin* 95: 247–57.

Schur, Edwin M. 1971. *Labeling Deviant Behavior.* New York: Harper and Row.

———. 1979. *Interpreting Deviance: A Sociological Introduction.* New York: Harper and Row.

———. 1984. *Labeling Women Deviant: Gender, Stigma, and Social Control.* New York: Random House.

Schwartz, Donald M., and Michael G. Thompson. 1981. "Do anorectics get well? Current research and future needs." *American Journal of Psychiatry* 138: 319–23.

Schwartz, Donald M., Michael G. Thompson, and Craig L. Johnson. 1982. "Anorexia nervosa and bulimia: The socio-cultural context." *International Journal of Eating Disorders* 1(3): 20–36.

Selvini-Palazzoli, Mara. 1978. *Self-Starvation: From Individual to Family Therapy in the Treatment of Anorexia Nervosa.* New York: Jason Aronson.

Stangler, Ronnie S., and Adolph M. Printz. 1980. "DSM-III: Psychiatric diagnosis in a university population." *American Journal of Psychiatry* 137: 937–40.

Theander, Sten. 1970. "Anorexia nervosa." *Acta Psychiatrica Scandinavica Supplement* 214: 24–31.

Thompson, Michael G., and Donald M. Schwartz. 1982. "Life adjustment of women with anorexia nervosa and anorexic-like behavior." *International Journal of Eating Disorders* 1(2): 47–60.

Willi, Jurg, and Samuel Grossmann. 1983. "Epidemiology of anorexia nervosa in a defined region of Switzerland." *American Journal of Psychiatry* 140: 564–67.

PART VI

Accounts

Marvin Scott and Stanford Lyman (1968) have suggested that we all engage in instances of deviant behavior, but that we desire to maintain a positive self image in both our own eyes and the eyes of others. To do so, we offer *accounts* intended to explain and *normalize* our deviant behavior. Two kinds of accounts predominate, excuses and justifications.

In offering *excuses,* individuals admit the wrongfulness of their actions but distance themselves from the blame. These excuses may precede or follow the acts of deviance. They are often fairly standard phrases or ideas designed to soften the deviance and relieve individuals of their accountability. These may include appeals to accidents ("my computer malfunctioned and lost my file"), appeals to defeasibility ("I thought my roommate turned my paper in"), appeals to biological drives ("I couldn't stop myself"), and scapegoating ("she borrowed my notes and I couldn't get them back in time to study for the test").

In offering *justifications,* individuals accept responsibility for their actions but seek to have specific instances excused. In so doing, they try to legitimate the acts or its consequences. In drawing on justifications, individuals may invoke sad tales ("I turn tricks because I was sexually abused as a child") or the need for self-fulfillment ("taking hallucinogenic drugs expands my consciousness"). All accounts can be classified into one of these two types.

In an early analysis of accounts, Sykes and Matza (1957) offered a typology of rationalizations they called *techniques of neutralization*. Individuals using these were attempting to resolve the contradictions between what people say and what they do. Sykes and Matza pointed out that even people who engage in deviance want to feel good about themselves, so they offer neutralizations to allow a conception of themselves as "moral." These techniques include the denial of responsibility ("It wasn't my fault"), denial of injury ("no harm done, it was just a prank"), denial of the victim ("nobody was hurt," "he deserved it"), condemnation of the condemners ("society is unfair"), and the appeal to higher loyalties ("some things are more important"). These techniques, like all others, fall into the category of either excuses or justifications. Careful analysis of these accounts will allow you to determine into which group each one falls.

23

Convicted Rapists' Vocabulary of Motive

DIANA SCULLY AND JOSEPH MAROLLA

Scully and Marolla's study of the way rapists rationalize their behavior offers a fascinating glimpse into the accounts offered by criminals. They interview the most hard-core segment of the rapist population, those sentenced to prison time. In analyzing these men's rationalizations, Scully and Marolla divide their accounts into excuses and justifications. Excuses, they find, where men acknowledge the wrongfulness of the act but deny full responsibility, are primarily used by those who admit to their deviant acts. Men who deny having committed rape (over 80% of the population), are more prone to use justifications, where they accept responsibility for their act but provide reasons that legitimate their behavior as not wrong. Scully and Marolla examine the various disavowal techniques, shed light on the repertoire of culturally available neutralizing accounts, and analyze the connection between types of accounts used and the way offenders locate blame.

Psychiatry has dominated the literature on rapists since "irresistible impulse" (Glueck, 1925: 243) and "disease of the mind" (Glueck, 1925: 243) were introduced as the causes of rape. Research has been based on small samples of men, frequently the clinicians' own patient population. Not surprisingly, the medical model has predominated: rape is viewed as an individualistic, idiosyncratic symptom of a disordered personality. That is, rape is assumed to be a psychopathologic problem and individual rapists are assumed to be "sick." However, advocates of this model have been unable to isolate a typical or even predictable pattern of symptoms that are causally linked to rape. Additionally, research has demonstrated that fewer than 5 percent of rapists were psychotic at the time of their rape (Abel et al., 1980).

We view rape as behavior learned socially through interaction with others; convicted rapists have learned the attitudes and actions consistent with sexual aggression against women. Learning also includes the acquisition of culturally

---

From "Convicted Rapists' Vocabulary of Motive: Excuses and Justifications," Diana Scully and Joseph Marolla, *Social Problems*, Vol. 31, No. 5, 1984. © 1984 Society for the Study of Social Problems. Reprinted by permission of University of California Press Journals and Diana Scully.

derived vocabularies of motive, which can be used to diminish responsibility and to negotiate a non-deviant identity.

Sociologists have long noted that people can, and do, commit acts they define as wrong and, having done so, engage various techniques to disavow deviance and present themselves as normal. Through the concept of "vocabulary of motive," Mills (1940: 904) was among the first to shed light on this seemingly perplexing contradiction. Wrong-doers attempt to reinterpret their actions through the use of a linguistic device by which norm-breaking conduct is socially interpreted. That is, anticipating the negative consequences of their behavior, wrong-doers attempt to present the act in terms that are both culturally appropriate and acceptable.

Following Mills, a number of sociologists have focused on the types of techniques employed by actors in problematic situations (Hall and Hewitt, 1970; Hewitt and Hall, 1973; Hewitt and Stokes, 1975; Sykes and Matza, 1957). Scott and Lyman (1968) describe excuses and justifications, linguistic "accounts" that explain and remove culpability for an untoward act after it has been committed. *Excuses* admit the act was bad or inappropriate but deny full responsibility, often through appeals to accident, or biological drive, or through scapegoating. In contrast, *justifications* accept responsibility for the act but deny that it was wrong—that is, they show in this situation the act was appropriate. *Accounts* are socially approved vocabularies that neutralize an act or its consequences and are always a manifestation of an underlying negotiation of identity.

Stokes and Hewitt (1976: 837) use the term "aligning actions" to refer to those tactics and techniques used by actors when some feature of a situation is problematic. Stated simply, the concept refers to an actor's attempt, through various means, to bring his or her conduct into alignment with culture. Culture in this sense is conceptualized as a "set of cognitive constraints—objects—to which people must relate as they form lines of conduct" (1976: 837), and includes physical constraints, expectations and definitions of others, and personal biography. Carrying out aligning actions implies both awareness of those elements of normative culture that are applicable to the deviant act and, in addition, an actual effort to bring the act into line with this awareness. The result is that deviant behavior is legitimized.

This paper presents an analysis of interviews we conducted with a sample of 114 convicted, incarcerated rapists. We use the concept of accounts (Scott and Lyman, 1968) as a tool to organize and analyze the vocabularies of motive which this group of rapists used to explain themselves and their actions. An analysis of their accounts demonstrates how it was possible for 83 percent (n = 114)[1] of these convicted rapists to view themselves as non-rapists.

When rapists' accounts are examined, a typology emerges that consists of admitters and deniers. Admitters (n = 47) acknowledged that they had forced sexual acts on their victims and defined the behavior as rape. In contrast, deniers[2] either eschewed sexual contact or all association with the victim (n = 35),[3] or admitted to sexual acts but did not define their behavior as rape (n = 32). . . . By and large, the deniers used justifications while the admitters used excuses. In some cases, both groups relied on the same themes, stereotypes, and images: some admitters, like most deniers, claimed that women enjoyed being raped. Some deniers excused their behavior by referring to alcohol or drug use, although they did so quite differently than admitters. Through these narrative accounts, we explore convicted rapists' own perceptions of their crimes. . . .

JUSTIFYING RAPE

Deniers attempted to justify their behavior by presenting the victim in a light that made her appear culpable, regardless of their own actions. Five themes run through attempts to justify their rapes: (1) women as seductresses; (2) women mean "yes" when they say "no"; (3) most women eventually relax and enjoy it; (4) nice girls don't get raped; and (5) guilty of a minor wrongdoing.

(1) Women as Seductresses

Men who rape need not search far for cultural language which supports the premise that women provoke or are responsible for rape. In addition to common cultural stereotypes, the fields of psychiatry and criminology (particularly the subfield of victimology) have traditionally provided justifications for rape, often by portraying raped women as the victims of their own seduction (Albin, 1977; Marolla and Scully, 1979). For example, Hollander (1924: 130) argues:

> Considering the amount of illicit intercourse, rape of women is very rare indeed. Flirtation and provocative conduct, i.e. tacit (if not actual) consent is generally the prelude to intercourse.

Since women are supposed to be coy about their sexual availability, refusal to comply with a man's sexual demands lacks meaning and rape appears normal. The fact that violence and, often, a weapon are used to accomplish the rape is not considered. As an example, Abrahamsen (1960: 61) writes:

> The conscious or unconscious biological or psychological attraction between man and woman does not exist only on the part of the offender toward the woman but, also, on her part toward him, which in many instances may, to some extent, be the impetus for his sexual attack. Often a women [sic] unconsciously wishes to be taken by force—consider the theft of the bride in Peer Gynt.

Like Peer Gynt, the deniers we interviewed tried to demonstrate that their victims were willing and, in some cases, enthusiastic participants. In these accounts, the rape became more dependent upon the victim's behavior than upon their own actions.

Thirty-one percent (n = 10) of the deniers presented an extreme view of the victim. Not only willing, she was the aggressor, a seductress who lured them, unsuspecting, into sexual action. Typical was a denier convicted of his first rape and accompanying crimes of burglary, sodomy, and abduction.

According to the pre-sentence reports, he had broken into the victim's house and raped her at knife point. While he admitted to the breaking and entry, which he claimed was for altruistic purposes ("to pay for the prenatal care of a friend's girlfriend"), he also argued that when the victim discovered him, he had tried to leave but she had asked him to stay. Telling him that she cheated on her husband, she had voluntarily removed her clothes and seduced him. She was, according to him, an exemplary sex partner who "enjoyed it very much and asked for oral sex.[4] Can I have it now?" he reported her as saying. He claimed they had spent hours in bed, after which the victim had told him he was good looking and asked to see him again. "Who would believe I'd meet a fellow like this?" he reported her as saying.

In addition to this extreme group, 25 percent (n = 8) of the deniers said the victim was willing and had made some sexual advances. An additional 9 percent (n = 3) said the victim was willing to have sex for money or drugs. In two of these three cases, the victim had been either an acquaintance or picked up, which the rapists said led them to expect sex.

(2) Women Mean "Yes" When They Say "No"

Thirty-four percent (n = 11) of the deniers described their victim as unwilling, at least initially, indicating either that she had resisted or that she had said no. Despite this, and even though (according to pre-sentence reports) a weapon had been present in 64 percent (n = 7) of these 11 cases, the rapists justified their behavior by arguing that either the victim had not resisted enough or that her "no" had really meant "yes." For example, one denier who was serving time for a previous rape was subsequently convicted of attempting to rape a prison hospital nurse. He insisted he had actually completed the second rape, and said of his victim: "She semi-struggled but deep down inside I think she felt it was a fantasy come true." The nurse, according to him, had asked a question about his conviction for rape, which he interpreted as teasing. "It was like she was saying, 'rape me.'" Further, he stated that she had helped him along with oral sex and "from her actions, she was enjoying it." In another case, a 34-year-old man convicted of abducting and raping a 15-year-old teenager at knife point as she walked on the beach, claimed it was a pickup. This rapist said women like to be overpowered before sex, but to dominate after it begins.

> A man's body is like a Coke bottle, shake it up, put your thumb over the opening and feel the tension. When you take a woman out, woo her, then she says "no, I'm a nice girl," you have to use force. All men do this. She said "no" but it was a societal no, she wanted to be coaxed. All women say "no" when they mean "yes" but it's a societal no, so they won't have to feel responsible later.

Claims that the victim didn't resist or, if she did, didn't resist enough, were also used by 24 percent (n = 11) of admitters to explain why, during the incident, they believed the victim was willing and that they were not raping.

These rapists didn't redefine their acts until some time after the crime. For example, an admitter who used a bayonet to threaten his victim, an employee of the store he had been robbing, stated:

> At the time I didn't think it was rape. I just asked her nicely and she didn't resist. I never considered prison. I just felt like I had met a friend. It took about five years of reading and going to school to change my mind about whether it was rape. I became familiar with the subtlety of violence. But at the time, I believed that as long as I didn't hurt anyone it wasn't wrong. At the time, I didn't think I would go to prison, I thought I would beat it.

Another typical case involved a gang rape in which the victim was abducted at knife point as she walked home about midnight. According to two of the rapists, both of whom were interviewed, at the time they had thought the victim had willingly accepted a ride from the third rapist (who was not interviewed). They claimed the victim didn't resist and one reported her as saying she would do anything if they would take her home. In this rapist's view, "She acted like she enjoyed it, but maybe she was just acting. She wasn't crying, she was engaging in it." He reported that she had been friendly to the rapist who abducted her and, claiming not to have a home phone, she gave him her office number—a tactic eventually used to catch the three. In retrospect, this young man had decided, "She was scared and just relaxed and enjoyed it to avoid getting hurt." Note, however, that while he had redefined the act as rape, he continued to believe she enjoyed it.

Men who claimed to have been unaware that they were raping viewed sexual aggression as a man's prerogative at the time of the rape. Thus they regarded their act as little more than a minor wrongdoing even though most possessed or used a weapon. As long as the victim survived without major physical injury, from their perspective, a rape had not taken place. Indeed, even U.S. courts have often taken the position that physical injury is a necessary ingredient for a rape conviction.

(3) Most Women Eventually Relax and Enjoy It

Many of the rapists expected us to accept the image, drawn from cultural stereotype, that once the rape began, the victim relaxed and enjoyed it.[5] Indeed, 69 percent (n = 22) of deniers justified their behavior by claiming not only that the victim was willing, but also that she enjoyed herself, in some cases to an immense degree. Several men suggested that they had fulfilled their victims' dreams. Additionally, while most admitters used adjectives such as "dirty," "humiliated," and "disgusted" to describe how they thought rape made women feel, 20 percent (n = 9) believed that their victim enjoyed herself. For example, one denier had posed as a salesman to gain entry to his victim's house. But he claimed he had had a previous sexual relationship with the victim, that she agreed to have sex for drugs, and that the opportunity to have sex with him produced "a glow, because she was really into oral stuff and fascinated by the idea of sex with a black man. She felt satisfied, fulfilled, wanted me to stay, but I didn't want her." In another case, a denier who had broken into his victim's house but who insisted the victim was his lover and let him in voluntarily, declared "She felt good, kept kissing me and wanted me to stay the night. She felt proud after sex with me." And another denier, who had hid in his victim's closet and later attacked her while she slept, argued that while she was scared at first, "once we got into it, she was ok." He continued to believe he hadn't committed rape because "she enjoyed it and it was like she consented."

(4) Nice Girls Don't Get Raped

The belief that "nice girls don't get raped" affects perception of fault. The victim's reputation, as well as characteristics or behavior which violate normative sex role expectations, are perceived as contributing to the commission of the crime. For example, Nelson and Amir (1975) defined hitchhike rape as a victim-precipitated offense.

In our study, 69 percent (n = 22) of deniers and 22 percent (n = 10) of admitters referred to their victims' sexual reputation, thereby evoking the stereotype that "nice girls don't get raped." They claimed that the victim was known to have been a prostitute, or a "loose" woman, or to have had a lot of affairs, or to have given birth to a child out of wedlock. For example, a denier who claimed he had picked up his victim while she was hitchhiking stated, "To be honest, we [his family] knew she was a damn whore and whether she screwed one or 50 guys didn't matter." According to pre-sentence reports this victim didn't know her attacker and he abducted her at knife point from the street. In another case, a denier who claimed to have known his victim by reputation stated:

> If you wanted drugs or a quick piece of ass, she would do it. In court she said she was a virgin, but I could tell during sex [rape] that she was very experienced.

When other types of discrediting biographical information were added to these sexual slurs, a total of 78 percent (n = 25) of the deniers used the victim's reputation to substantiate their accounts. Most frequently, they referred to the victim's emotional state or drug use. For example, one denier claimed his victim had been known to be loose and, additionally, had turned state's evidence against her husband to put him in prison and save herself from a burglary conviction. Further, he asserted that she had met her current boyfriend, who was himself in and out of prison, in a drug rehabilitation center where they were both clients.

Evoking the stereotype that women provoke rape by the way they dress, a description of the victim as seductively attired appeared in the accounts of 22 percent (n = 7) of deniers and 17 percent (n = 8) of admitters. Typically, these descriptions were used to substantiate their claims about the victim's reputation. Some men went to extremes to paint a tarnished picture of the victim, describing her as dressed in tight black clothes and without a bra; in one case, the victim was portrayed as sexually provocative in dress and carriage.

Not only did she wear short skirts, but she was observed to "spread her legs while getting out of cars." Not all of the men attempted to assassinate their victim's reputation with equal vengeance. Numerous times they made subtle and off-hand remarks like, "She was a waitress and you know how they are."

The intent of these discrediting statements is clear. Deniers argued that the woman was a "legitimate" victim who got what she deserved. For example, one denier stated that all of his victims had been prostitutes; pre-sentence reports indicated they were not. Several times during his interview, he referred to them as "dirty sluts," and argued "anything I did to them was justified." Deniers also claimed their victim had wrongly accused them and was the type of woman who would perjure herself in court.

(5) Only a Minor Wrongdoing

The majority of deniers did not claim to be completely innocent and they also accepted some accountability for their actions. Only 16 percent (n = 5) of deniers argued that they were totally free of blame. Instead, the majority of deniers pleaded guilty to a lesser charge. That is, they obfuscated the rape by pleading guilty to a less serious, more acceptable charge. They accepted being over-sexed, accused of poor judgment or trickery, even some violence, or guilty of adultery or contributing to the delinquency of a minor, charges that are hardly the equivalent of rape.

Typical of this reasoning is a denier who met his victim in a bar when the bartender asked him if he would try to repair her stalled car. After attempting unsuccessfully, he claimed the victim drank with him and later accepted a ride. Out riding, he pulled into a deserted area "to see how my luck would go." When the victim resisted his advances, he beat her and he stated:

> I did something stupid. I pulled a knife on her and I hit her as hard as I would hit a man. But I shouldn't be in prison for what I did. I shouldn't have all this time [sentence] for going to bed with a broad.

This rapist continued to believe that while the knife was wrong, his sexual behavior was justified.

In another case, the denier claimed he picked up his under-age victim at a party and that she voluntarily went with him to a motel. According to pre-sentence reports, the victim had been abducted at knife point from a party. He explained:

> After I paid for a motel, she would have to have sex but I wouldn't use a weapon. I would have explained. I spent money and, if she still said no, I would have forced her. If it had happened that way, it would have been rape to some people but not to my way of thinking. I've done that kind of thing before. I'm guilty of sex and contributing to the delinquency of a minor, but not rape.

In sum, deniers argued that, while their behavior may not have been completely proper, it should not have been considered rape. To accomplish this, they attempted to discredit and blame the victim while presenting their own actions as justified in the context. Not surprisingly, none of the deniers thought of himself as a rapist. A minority of the admitters attempted to lessen the impact of their crime by claiming the victim enjoyed being raped. But despite this similarity, the nature and tone of admitters' and deniers' accounts were essentially different.

EXCUSING RAPE

In stark contrast to deniers, admitters regarded their behavior as morally wrong and beyond justification. They blamed themselves rather than the victim, although some continued to cling to the belief that the victim had contributed to the crime somewhat, for example, by not resisting enough.

Several of the admitters expressed the view that rape was an act of such moral outrage that it was unforgivable. Several admitters broke into tears at intervals during their interviews. A typical sentiment was,

> I equate rape with someone throwing you up against a wall and tearing your liver and guts out of you. . . . Rape is worse than murder . . . and I'm disgusting.

Another young admitter frequently referred to himself as repulsive and confided:

> I'm in here for rape and in my own mind, it's the most disgusting crime, sickening. When people see me and know, I get sick.

Admitters tried to explain their crime in a way that allowed them to retain a semblance of moral integrity. Thus, in contrast to deniers' justifications, admitters used excuses to explain how they were compelled to rape. These excuses appealed to the existence of forces outside of the rapists' control. Through the use of excuses, they attempted to demonstrate that either intent was absent or responsibility was diminished. This allowed them to admit rape while reducing the threat to their identity as a moral person. Excuses also permitted them to view their behavior as idiosyncratic rather than typical and, thus, to believe they were not "really" rapists. Three themes run through these accounts: (1) the use of alcohol and drugs; (2) emotional problems; and (3) nice guy image.

(1) The Use of Alcohol and Drugs

A number of studies have noted a high incidence of alcohol and drug consumption by convicted rapists prior to their crime (Groth, 1979; Queen's Bench Foundation, 1976). However, more recent research has tentatively concluded that the connection between substance use and crime is not as direct as previously thought (Ladouceur, 1983). Another facet of alcohol and drug use mentioned in the literature is its utility in disavowing deviance. McCaghy (1968) found that child molesters used alcohol as a technique for neutralizing

Rapists' Accounts of Own and Victims' Alcohol and/or Drug (A/D) Use and Effect

	Admitters n = 39 %	Deniers n = 25 %
Neither self nor victim used A/D	23	16
Self used A/D	77	72
Of self used, no victim use	51	12
Self affected by A/D	69	40
Of self affected, no victim use or effect	54	24
Self A/D users who were affected	90	56
Victim used A/D	26	72
Of victim used, no self use	0	0
Victim affected by A/D	15	56
Of victim affected, no self use or effect	0	40
Victim A/D users who were affected	60	78
Both self and victim used and affected by A/D	15	16

their deviant identity. Marolla and Scully (1979), in a review of psychiatric literature, demonstrated how alcohol consumption is applied differently as a vocabulary of motive. Rapists can use alcohol both as an excuse for their behavior and to discredit the victim and make her more responsible. We found the former common among admitters and the latter common among deniers.

Alcohol and/or drugs were mentioned in the accounts of 77 percent (n = 30) of admitters and 84 percent (n = 21) of deniers and both groups were equally likely to have acknowledged consuming a substance—admitters, 77 percent (n = 30); deniers, 72 percent (n = 18). However, admitters said they had been affected by the substance; if not the cause of their behavior, it was at least a contributing factor. For example, an admitter who estimated his consumption to have been eight beers and four "hits of acid" reported:

> Straight, I don't have the guts to rape. I could fight a man but not that. To say, "I'm going to do it to a woman," knowing it will scare and hurt her, takes guts or you have to be sick.

Another admitter believed that his alcohol and drug use,

> ... brought out what was already there but in such intensity it was uncontrollable. Feelings of being dominant, powerful, using someone for my own gratification, all rose to the surface.

In contrast, deniers' justifications required that they not be substantially impaired. To say that they had been drunk or high would cast doubt on their ability to control themselves or to remember events as they actually happened. Consistent with this, when we asked if the alcohol and/or drugs had had an effect on their behavior, 69 percent (n = 27) of admitters, but only 40 percent (n = 10) of deniers, said they had been affected.

Even more interesting were references to the victim's alcohol and/or drug use. Since admitters had already relieved themselves of responsibility through claims of being drunk or high, they had nothing to gain from the assertion that the victim had used or been affected by alcohol and/or drugs. On the other hand, it was very much in the interest of deniers to declare that their victim had been intoxicated or high: that fact lessened her credibility and made her more responsible for the act. Reflecting these observations, 72 percent (n = 18) of deniers and 26 percent (n = 10) of admitters maintained that alcohol or drugs had been consumed by the victim. Further, while 56 percent (n = 14) of deniers declared she had been affected by this use, only 15 percent (n = 6) of admitters made a similar claim. Typically deniers argued that the alcohol and drugs had sexually aroused their victim or rendered her out of control. For example, one denier insisted that his victim had become hysterical from drugs, not from being raped, and it was because of the drugs that she had reported him to the police. In addition, 40 percent (n = 10) of deniers argued that while the victim had been drunk or high, they themselves either hadn't ingested or weren't affected by alcohol and/or drugs. None of the admitters made this claim. In fact, in all of the 15 percent (n = 6) of cases where an admitter said the victim was drunk or high, he also admitted to being similarly affected.

These data strongly suggest that whatever role alcohol and drugs play in sexual and other types of violent crime, rapists have learned the advantage to be gained from using alcohol and drugs as an account. Our sample were aware that their victim would be discredited and their own behavior excused or justified by referring to alcohol and/or drugs.

(2) Emotional Problems

Admitters frequently attributed their acts to emotional problems. Forty percent (n = 19) of admitters said they believed an emotional problem had been at the root of their rape behavior, and 33 percent (n = 15) specifically related the problem to an unhappy, unstable childhood or a marital-domestic situation. Still others claimed to have been in a general state of unease. For example, one admitter said that at the time of the rape he had been depressed, feeling he couldn't do anything right, and that something had been missing from his life. But he also added, "being a rapist is not part of my personality." Even admitters who could locate no source for an emotional problem evoked the popular image of rapists as the product of disordered personalities to argue they also must have problems:

> The fact that I'm a rapist makes me different. Rapists aren't all there. They have problems. It was wrong so there must be a reason why I did it.
> I must have a problem.

Our data do indicate that a precipitating event, involving an upsetting problem of everyday living, appeared in the accounts of 80 percent (n = 38) of admitters and 25 percent (n = 8) of deniers. Of those experiencing a precipitating event, including deniers, 76 percent (n = 35) involved a wife or girlfriend. Over and over, these men described themselves as having been in a rage because of an incident involving a woman with whom they believed they were in love.

Frequently, the upsetting event was related to a rigid and unrealistic double standard for sexual conduct and virtue which they applied to "their" woman but which they didn't expect from men, didn't apply to themselves, and, obviously, didn't honor in other women. To discover that the "pedestal" didn't apply to their wife or girlfriend sent them into a fury. One especially articulate and typical admitter described his feeling as follows. After serving a short prison term for auto theft, he married his "childhood sweetheart" and secured a well-paying job. Between his job and the volunteer work he was doing with an ex-offender group, he was spending long hours away from home, a situation that had bothered his wife. In response to her request, he gave up his volunteer work, though it was clearly meaningful to him. Then, one day, he discovered his wife with her former boyfriend "and my life fell apart." During the next several days, he said his anger had made him withdraw into himself and, after three days of drinking in a motel room, he abducted and raped a stranger. He stated:

> My parents have been married for many years and I had high expectations about marriage. I put my wife on a pedestal. When I walked in on her, I felt like my life had been destroyed, it was such a shock. I was bitter and angry about the fact that I hadn't done anything to my wife for cheating. I didn't want to hurt her [victim], only to scare and degrade her.

It is clear that many admitters, and a minority of deniers, were under stress at the time of their rapes. However, their problems were ordinary—the types of upsetting events that everyone experiences at some point in life. The overwhelming majority of the men were not clinically defined as mentally ill in court-ordered psychiatric examinations prior to their trials. Indeed, our sample is consistent with Abel et al. (1980) who found fewer than 5 percent of rapists were psychotic at the time of their offense.

As with alcohol and drug intoxication, a claim of emotional problems works differently depending upon whether the behavior in question is being justified or excused. It would have been counter-productive for deniers to have claimed to have had emotional problems at the time of the rape. Admitters used psychological explanations to portray themselves as having been temporarily "sick" at the time of the rape. Sick people are usually blamed for neither the cause of their illness nor for acts committed while in that state of diminished capacity. Thus, adopting the sick role removed responsibility by excusing the behavior as having been beyond the ability of the individual to control. Since the rapists were not "themselves," the rape was idiosyncratic rather than typical behavior. Admitters asserted a non-deviant identity despite their self-proclaimed disgust with what they had done. Although admitters were willing to assume the sick role, they did not view their problem as a chronic condition, nor did they believe themselves to be insane or permanently impaired. Said one admitter, who believed that he needed psychological counseling: "I have a mental disorder, but I'm not crazy." Instead, admitters viewed their "problem" as mild, transient, and curable. Indeed, part of the appeal of this excuse was that not only did it relieve responsibility, but, as with alcohol and drug addiction, it allowed the rapist to "recover." Thus, at the time of their interviews, only 31 percent (n = 14) of admitters indicated that "being a rapist" was part of their self-concept. Twenty-eight percent (n = 13) of admitters stated they had never thought of themselves as rapists, 8 percent (n = 4) said they were unsure, and 33 percent (n = 16) asserted they had been a rapist at one time but now were recovered. A multiple "ex-rapist," who believed his "problem" was due to "something buried in my subconscious" that was triggered when his girlfriend broke up with him, expressed a typical opinion:

> I was a rapist, but not now. I've grown up, had to live with it. I've hit the bottom of the well and it can't get worse. I feel born again to deal with my problems.

(3) Nice Guy Image

Admitters attempted to further neutralize their crime and negotiate a non-rapist identity by painting an image of themselves as a "nice guy." Admitters projected the image of someone who had made a serious mistake but, in every other respect, was a decent person. Fifty-seven percent (n = 27) expressed regret and sorrow for their victim indicating that they wished there were a way to apologize for or amend their behavior. For example, a participant in a rape-murder, who insisted his partner did the murder, confided, "I wish there was something I could do besides saying 'I'm sorry, I'm sorry.' I live with it 24 hours a day and, sometimes, I wake up crying in the middle of the night because of it."

Schlenker and Darby (1981) explain the significance of apologies beyond the obvious expression of regret. An apology allows a person to admit guilt while at the same time seeking a pardon by signaling that the event should not be considered a fair representation of what the person is really like. An apology separates the bad self from the good self, and promises more acceptable behavior in the future. When apologizing, an individual is attempting to say: "I have repented and should be forgiven," thus making it appear that no further rehabilitation is required.

The "nice guy" statements of the admitters reflected an attempt to communicate a message consistent with Schlenker's and Darby's analysis of apologies. It was an attempt to convey that rape was not a representation of their "true" self. For example,

> It's different from anything else I've ever done. I feel more guilt about this. It's not consistent with me. When I talk about it, it's like being assaulted myself. I don't know why I did it, but once I started, I got into it. Armed robbery was a way of life for me, but not rape. I feel like I wasn't being myself.

Admitters also used "nice guy" statements to register their moral opposition to violence and harming women, even though, in some cases, they had seriously injured their victims. Such was the case of an admitter convicted of gang rape:

> I'm against hurting women. She should have resisted. None of us were the type of person that would use force on a woman. I never positioned myself on a woman unless she showed an interest in me. They would play to me, not me to them. My weakness is to follow. I never would have stopped, let alone pick her up without the others. I never would have let anyone beat her. I never bothered women who didn't want sex; never had a problem with sex or getting it. I loved her—like all women.

Finally, a number of admitters attempted to improve their self-image by demonstrating that, while they had raped, it could have been worse if they had not been a "nice guy." For example, one admitter professed to being especially gentle with his victim after she told him she had just had a baby. Others claimed to have given the victim money to get home or make a phone call, or to have made sure the victim's children were not in the room. A multiple rapist, whose pattern was to break in and attack sleeping victims in their homes, stated:

> I never beat any of my victims and I told them I wouldn't hurt them if they cooperated. I'm a professional thief. But I never robbed the women I raped because I felt so bad about what I had already done to them.

Even a young man, who raped his five victims at gun point and then stabbed them to death, attempted to improve his image by stating:

> Physically they enjoyed the sex [rape]. Once they got involved, it would be difficult to resist. I was always gentle and kind until I started to kill them. And the killing was always sudden, so they wouldn't know it was coming.

SUMMARY AND CONCLUSIONS

Convicted rapists' accounts of their crimes include both excuses and justifications. Those who deny what they did was rape justify their actions; those who admit it was rape attempt to excuse it or themselves. This study does not address why some men admit while others deny, but future research might address this question. This paper does provide insight on how men who are sexually aggressive or violent construct reality, describing the different strategies of admitters and deniers.

Admitters expressed the belief that rape was morally reprehensible. But they explained themselves and their acts by appealing to forces beyond their control, forces which reduced their capacity to act rationally and thus compelled them to rape. Two types of excuses predominated: alcohol/drug intoxication and emotional problems. Admitters used these excuses to negotiate a moral identity for themselves by viewing rape as idiosyncratic rather than typical behavior. This allowed them to reconceptualize themselves as recovered or "ex-rapists," someone who had made a serious mistake which did not represent their "true" self.

In contrast, deniers' accounts indicate that these men raped because their value system provided no compelling reason not to do so. When sex is viewed as a male entitlement, rape is no longer seen as criminal. However, the deniers had been convicted of rape, and like the admitters, they attempted to negotiate an identity. Through justifications, they constructed a "controversial" rape and attempted to demonstrate how their behavior, even if not quite right, was appropriate in the situation. Their denials, drawn from common cultural rape stereotypes, took two forms, both of which ultimately denied the existence of a victim.

The first form of denial was buttressed by the cultural view of men as sexually masterful and women as coy but seductive. Injury was denied by portraying the victim as willing, even enthusiastic, or as politely resistant at first but eventually yielding to "relax and enjoy it." In these accounts, force appeared merely as a seductive technique. Rape was disclaimed: rather than harm the woman, the rapist had fulfilled her dreams. In the second form of denial, the victim was portrayed as the type of woman who "got what she deserved." Through attacks on the victim's sexual reputation and, to a lesser degree, her emotional state, deniers attempted to demonstrate that since the victim wasn't a "nice girl," they were not rapists. Consistent with both forms of denial was the self-interested use of alcohol and drugs as a justification. Thus, in contrast to admitters, who accentuated their own use as an excuse, deniers emphasized the victim's consumption in an effort to both discredit her and make her appear more responsible for the rape. It is important to remember that deniers did not invent these justifications. Rather, they reflect a belief system which has historically victimized women by promulgating the myth that women both enjoy and are responsible for their own rape.

While admitters and deniers present an essentially contrasting view of men who rape, there were some shared characteristics. Justifications particularly, but also excuses, are buttressed by the cultural view of women as sexual commodities, dehumanized and devoid of autonomy and dignity. In this sense, the sexual objectification of women must be understood as an important factor contributing to an environment that trivializes, neutralizes, and, perhaps, facilitates rape.

Finally, we must comment on the consequences of allowing one perspective to dominate thought on a social problem. Rape, like any complex continuum of behavior, has multiple causes and is influenced by a number of social factors. Yet, dominated by psychiatry and the medical model, the underlying assumption that rapists are "sick" has pervaded research. Although methodologically unsound, conclusions have been based almost exclusively on small clinical populations of rapists—that extreme group of rapists who seek counseling in prison and are the most likely to exhibit psychopathology. From this small, atypical group of men, psychiatric findings have been generalized to all men who rape. Our research, however, based on volunteers from the entire prison population, indicates that some rapists, like deniers, viewed and understood their behavior from a popular cultural perspective. This strongly suggests that cultural perspectives, and not an idiosyncratic illness, motivated their behavior. Indeed, we can argue that the psychiatric perspective has contributed to the vocabulary of motive that rapists use to excuse and justify their behavior (Scully and Marolla, 1984).

Efforts to arrive at a general explanation for rape have been retarded by the narrow focus of the medical model and the preoccupation with clinical populations. The continued reduction of such complex behavior to a singular cause hinders, rather than enhances, our understanding of rape.

NOTES

1. These numbers include pretest interviews. When the analysis involves either questions that were not asked in the pretest or that were changed, they are excluded and thus the number changes.

2. There is, of course, the possibility that some of these men really were innocent of rape. However, while the U.S. criminal justice system is not without flaw, we assume that it is highly unlikely that this many men could have been unjustly convicted of rape, especially since rape is a crime with traditionally low conviction rates. Instead, for purposes of this research, we assume that these men were guilty as charged and that their attempt to maintain an image of non-rapist springs from some psychologically or sociologically interpretable mechanism.

3. Because of their outright denial, interviews with this group of rapists did not contain the data being analyzed here and, consequently, they are not included in this paper.

4. It is worth noting that a number of deniers specifically mentioned the victim's alleged interest in oral sex. Since our interview questions about sexual history indicated that the rapists themselves found oral sex marginally acceptable, the frequent mention is probably another attempt to discredit the victim. However, since a tape recorder could not be used for the interviews and the importance of these claims didn't emerge until the data was being coded and analyzed, it is possible that it was mentioned even more frequently but not recorded.

5. Research shows clearly that women do not enjoy rape. Holmstrom and Burgess (1978) asked 93 adult rape victims, "How did it feel sexually?" Not one said they enjoyed it. Further, the trauma of rape is so great that it disrupts sexual functioning (both frequency and satisfaction) for the overwhelming majority of victims, at least during the period immediately following the rape and, in fewer cases, for an extended period of time (Burgess and Holmstrom, 1979; Feldman-Summers et al., 1979). In addition, a number of studies have shown that rape victims experience adverse consequences prompting some to move, change jobs, or drop out of school (Burgess and Holmstrom, 1974; Kilpatrick et al., 1979; Ruch et al., 1980; Shore, 1979).

REFERENCES

Abel, Gene, Judith Becker, and Linda Skinner. 1980. "Aggressive behavior and sex." *Psychiatric Clinics of North America* 3(2): 133–151.

Abrahamsen, David. 1960. *The Psychology of Crime*. New York: John Wiley.

Albin, Rochelle. 1977. "Psychological studies of rape." *Signs* 3(2): 423–435.

Burgess, Ann Wolbert, and Lynda Lytle Holmstrom. 1974. *Rape: Victims of Crisis*. Bowie. Robert J. Brady.

———. 1979. "Rape: Sexual disruption and recovery." *American Journal of Orthopsychiatry* 49(4): 648–657.

Feldman-Summers, Shirley, Patricia E. Gordon, and Jeanette R. Meagher. 1979. "The impact of rape on sexual satisfaction." *Journal of Abnormal Psychology* 88(1): 101–105.

Glueck, Sheldon. 1925. *Mental Disorders and the Criminal Law.* New York: Little, Brown.

Groth, Nicholas A. 1979. *Men Who Rape.* New York: Plenum Press.

Hall, Peter M., and John P. Hewitt. 1970. "The quasi-theory of communication and the management of dissent." *Social Problems* 18(1): 17–27.

Hewitt, John P., and Peter M. Hall. 1973. "Social problems, problematic situations, and quasi-theories." *American Journal of Sociology* 38(3): 367–374.

Hewitt, John P., and Randall Stokes. 1975. "Disclaimers." *American Sociological Review* 40(1): 1–11.

Hollander, Bernard. 1924. *The Psychology of Misconduct, Vice and Crime.* New York: Macmillan.

Holmstrom, Lynda Lytle, and Ann Wolbert Burgess. 1978. "Sexual behavior of assailant and victim during rape." Paper presented at the annual meetings of the American Sociological Association, San Francisco, September 2–8.

Kilpatrick, Dean G., Lois Veronen, and Patricia A. Resnick. 1979. "The aftermath of rape: Recent empirical findings." *American Journal of Orthopsychiatry* 49(4): 658–669.

Ladouceur, Patricia. 1983. "The relative impact of drugs and alcohol on serious felons." Paper presented at the annual meetings of the American Society of Criminology, Denver, November 9–12.

Marolla, Joseph, and Diana Scully. 1979. "Rape and psychiatric vocabularies of motive." In Edith S. Gomberg and Violet Franks (eds.), *Gender and Disordered Behavior: Sex Differences in Psychopathology* (pp. 301–318). New York: Brunner/Mazel.

McCaghy, Charles. 1968. "Drinking and deviance disavowal: The case of child molesters." *Social Problems* 16(1): 43–49.

Mills, C. Wright. 1940. "Situated actions and vocabularies of motive." *American Sociological Review* 5(6): 904–913.

Nelson, Steve, and Amir Menachem. 1975. "The hitchhike victim of rape: A research report." In Israel Drapkin and Emilio Viano (eds.), *Victimology: A New Focus* (pp. 47–65). Lexington, KY: Lexington Books.

Queen's Bench Foundation. 1976. *Rape: Prevention and Resistance.* San Francisco: Queen's Bench Foundation.

Ruch, Libby O., Susan Meyers Chandler, and Richard A. Harter. 1980. "Life change and rape impact." *Journal of Health and Social Behavior* 21(3): 248–260.

Schlenker, Barry R., and Bruce W. Darby. 1981. "The use of apologies in social predicaments." *Social Psychology Quarterly* 44(3): 271–278.

Scott, Marvin, and Stanford Lyman. 1968. "Accounts." *American Sociological Review* 33(1): 46–62.

Scully, Diana, and Joseph Marolla. 1984. "Rape and psychiatric vocabularies of motive: Alternative perspectives." In Ann Wolbert Burgess (ed.), *Handbook on Rape and Sexual Assault.* New York: Garland Publishing.

Shore, Barbara K. 1979. "An Examination of Critical Process and Outcome Factors in Rape." Rockville, MD: National Institute of Mental Health.

Stokes, Randall, and John P. Hewitt. 1976. "Aligning actions." *American Sociological Review* 41(5): 837–849.

Sykes, Gresham M., and David Matza. 1957. "Techniques of neutralization." *American Sociological Review* 22(6): 664–670.

24

The Influence of Situational Ethics on Cheating

DONALD L. McCABE

McCabe uses Sykes and Matza's classic typology of "techniques of neutralization" to classify some of the rationalizations college students use to legitimate their cheating behavior despite the long-standing presence of honor codes at their schools. Based on a survey of over 6,000 students, McCabe shows which neutralization techniques were more commonly used in various familiar situations. Readers will no doubt recognize some of these rationales as they tie their everyday life surroundings to these deviance concepts.

Numerous studies have demonstrated the pervasive nature of cheating among college students (Baird 1980; Haines, Diekhoff, LaBeff, and Clark 1986; Michaels and Miethe 1989; Davis et al. 1992). This research has examined a variety of factors that help explain cheating behavior, but the strength of the relationships between individual factors and cheating has varied considerably from study to study (Tittle and Rowe 1973; Baird 1980; Eisenberger and Shank 1985; Haines et al. 1986; Ward 1986; Michaels and Miethe 1989; Perry, Kane, Bernesser, and Spicker 1990; Ward and Beck 1990).

Although the factors examined in these studies (for example, personal work ethic, gender, self-esteem, rational choice, social learning, deterrence) are clearly important, the work of LaBeff, Clark, Haines, and Diekhoff (1990) suggests that the concept of situational ethics may be particularly helpful in understanding student rationalizations for cheating. Extending the arguments of Norris and Dodder (1979), LaBeff et al. conclude

> that students hold qualified guidelines for behavior which are situationally determined. As such, the concept of situational ethics might well describe . . . college cheating [as having] rules for behavior [that] may not be considered rigid but depend on the circumstances involved (1990, p. 191).

LaBeff et al. believe a utilitarian calculus of "the end justifies the means" underlies this reasoning process and "what is wrong in most situations might be

---

From "The Influence of Situational Ethics on Cheating Among College Students," by Donald L. McCabe, from *Sociological Inquiry,* 62:3, pp. 365–74. Copyright © 1992 by the University of Texas Press. All rights reserved. (Summer 1992). Reprinted by permission of the University of Texas Press and the author.

considered right or acceptable if the end is defined as appropriate" (1990, p. 191). As argued by Edwards (1967), the situation determines what is right or wrong in this decision-making calculus and also dictates the appropriate principles to be used in guiding and judging behavior.

Sykes and Matza (1957) hypothesize that such rationalizations, that is, "justifications for deviance that are seen as valid by the delinquent but not by the legal system or society at large" (p. 666), are common. However, they challenge conventional wisdom that such rationalizations typically follow deviant behavior as a means of protecting "the individual from self-blame and the blame of others after the act" (p. 666). They develop convincing arguments that these rationalizations may logically precede the deviant behavior and "[d]isapproval from internalized norms and conforming others in the social environment is neutralized, turned back, or deflated in advance. Social controls that serve to check or inhibit deviant motivational patterns are rendered inoperative, and the individual is freed to engage in delinquency without serious damage to his self-image" (pp. 666–667).

Using a sample of 380 undergraduate students at a small southwestern university, LaBeff et al. (1990) attempted to classify techniques employed by students in the neutralization of cheating behavior into the five categories of neutralization proposed by Sykes and Matza (1957): (1) denial of responsibility, (2) condemnation of condemners, (3) appeal to higher loyalties, (4) denial of victim, and (5) denial of injury. Although student responses could easily be classified into three of these techniques, denial of responsibility, appeal to higher loyalties, and condemnation of condemners, LaBeff et al. conclude that "[i]t is unlikely that students will either deny injury or deny the victim since there are no real targets in cheating" (1990, p. 196).

The research described here responds to LaBeff et al. in two ways; first, it answers their call to "test the salience of neutralization . . . in more diverse university environments" (p. 197) and second, it challenges their dismissal of denial of injury and denial of victim as neutralization techniques employed by students in their justification of cheating behavior.

METHODOLOGY

The data discussed here were gathered as part of a study of college cheating conducted during the 1990–1991 academic year. A seventy-two-item questionnaire concerning cheating behavior was administered to students at thirty-one highly selective colleges across the country. Surveys were mailed to a minimum of five hundred students at each school and a total of 6,096 completed surveys were returned (38.3 percent response rate). Eighty-eight percent of the respondents were seniors, nine percent were juniors, and the remaining three percent could not be classified. Survey administration emphasized voluntary participation and assurances of anonymity to help combat issues of non-response bias and the need to accept responses without the chance to question or contest them.

The final sample included 61.2 percent females (which reflects the inclusion of five all-female schools in the sample and a slightly higher return rate among female students) and 95.4 percent U.S. citizens. The sample paralleled the ethnic diversity of the participating schools (85.5 percent Anglo, 7.2 percent Asian, 2.6 percent African American, 2.2 percent Hispanic and 2.5 percent other); their religious diversity (including a large percentage of students who claimed no religious preference, 27.1 percent); and their mix of undergraduate majors (36.0 percent humanities, 28.8 percent social sciences, 26.8 percent natural sciences and engineering, 4.5 percent business, and 3.9 percent other).

RESULTS

Of the 6,096 students participating in this research, over two-thirds (67.4 percent) indicated that they had cheated on a test or major assignment at least once while an undergraduate. This cheating took a variety of different forms, but among the most popular (listed in decreasing order of mention) were: (1) a failure to footnote sources in written work, (2) collaboration on assignments when the instructor specifically asked for individual work, (3) copying from other students on tests and examinations, (4) fabrication of bibliographies, (5) helping someone else cheat on a test, and (6) using unfair methods to learn the content of a test ahead of time. Almost one in five students (19.1 percent) could be classified as active cheaters (five or more self-reported incidents of cheating). This is double the rate reported by LaBeff et al. (1990), but they asked students to report only cheating incidents that had taken place in the last six months. Students in this research were asked to report all cheating in which they had engaged while an undergraduate—a period of three years for most respondents at the time of this survey.

Students admitting to any cheating activity were asked to rate the importance of several specific factors that might have influenced their decisions to cheat. These data establish the importance of denial of responsibility and condemnation of condemners as neutralization techniques. For example, 52.4 percent of the respondents who admitted to cheating rated the pressure to get good grades as an important influence in their decision to cheat, with parental pressures and competition to gain admission into professional schools singled out as the primary grade pressures. Forty-six percent of those who had engaged in cheating cited excessive workloads and an inability to keep up with assignments as important factors in their decisions to cheat.

In addition to rating the importance of such preselected factors, 426 respondents (11.0 percent of the admitted cheaters) offered their own justifications for cheating in response to an open-ended question on motivations for cheating. These responses confirm the importance of denial of responsibility and condemnation of condemners as neutralization techniques. They also support LaBeff et al.'s (1990) claim that appeal to higher loyalties is an

Neutralization Strategies: Self-Admitted Cheaters

Strategy	Number	Percent
Denial of responsibility	216	61.0
Mind block	90	25.4
No understanding of material	31	8.8
Other	95	26.8
Condemnation of condemners	99	28.0
Pointless assignment	35	9.9
No respect for professor	28	7.9
Other	36	10.2
Appeal to higher loyalties	24	6.8
Help a friend	10	2.8
Peer pressure	9	2.5
Other	5	1.5
Denial of injury	15	4.2
Cheating is harmless	9	2.5
Does not matter	6	1.7

important neutralization technique. However, these responses also suggest that LaBeff et al.'s dismissal of denial of injury as a justification for student cheating is arguable.

As shown in the table, denial of responsibility was the technique most frequently cited (216 responses, 61.0 percent of the total) in the 354 responses classified into one of Sykes and Matza's five categories of neutralization. The most common responses in this category were mind block, no understanding of the material, a fear of failing, and unclear explanations of assignments. (Although it is possible that some instances of mind block and a fear of failing included in this summary would be more accurately classified as rationalization, the wording of all responses included here suggests that rationalization preceded the cheating incident. Responses that seem to involve post hoc rationalizations were excluded from this summary.) Condemnation of condemners was the second most popular neutralization technique observed (99 responses, 28.0 percent) and included such explanations as pointless assignments, lack of respect for individual professors, unfair tests, parents' expectations, and unfair professors. Twenty-four respondents (6.8 percent) appealed to higher loyalties to explain their behavior. In particular, helping a friend and responding to peer pressures were influences some students could not ignore. Finally fifteen students (4.2 percent) provided responses that clearly fit into the category of denial of injury. These students dismissed their cheating as harmless since it did not hurt anyone or they felt cheating did not matter in some cases (for example, where an assignment counted for a small percentage of the total course grade).

Detailed examination of selected student responses provides additional insight into the neutralization strategies they employ.

Denial of Responsibility

Denial of responsibility invokes the claim that the act was "due to forces outside of the individual and beyond his control such as unloving parents" (Sykes and Matza 1957, p. 667). For example, many students cite an unreasonable workload and the difficulty of keeping up as ample justification for cheating:

> Here at . . . , you must cheat to stay alive. There's so much work and the quality of materials from which to learn, books, professors, is so bad that there's no other choice.
>
> It's the only way to keep up.
>
> I couldn't do the work myself.

The following descriptions of student cheating confirm fear of failure is also an important form of denial of responsibility:

> . . . a take-home exam in a class I was failing.
>
> . . . was near failing.

Some justified their cheating by citing the behavior of peers:

> Everyone has test files in fraternities, etc. If you don't, you're at a great disadvantage.
>
> When most of the class is cheating on a difficult exam and they will ruin the curve, it influences you to cheat so your grade won't be affected.

All of these responses contain the essence of denial of responsibility: the cheater has deflected blame to others or to a specific situational context.

Denial of Injury

As noted in the table, denial of injury was identified as a neutralization technique employed by some respondents. A key element in denial of injury is whether one feels "anyone has clearly been hurt by [the] deviance." In invoking this defense, a cheater would argue "that his behavior does not really cause any great harm despite the fact that it runs counter to the law" (Sykes and Matza 1957, pp. 667–668). For example, a number of students argued that the assignment or test on which they cheated was so trivial that no one was really hurt by their cheating.

> These grades aren't worth much therefore my copying doesn't mean very much. I am ashamed, but I'd probably do it the same way again.
>
> If I extend the time on a take-home it is because I feel everyone does and the teacher kind of expects it. No one gets hurt.

As suggested earlier, these responses suggest the conclusion of LaBeff et al. that "[i]t is unlikely that students will . . . deny injury" (1990, p. 196) must be re-evaluated.

The Denial of the Victim

LaBeff et al. failed to find any evidence of denial of the victim in their student accounts. Although the student motivations for cheating summarized in the table support this conclusion, at least four students (0.1% of the self-admitted cheaters in this study) provided comments elsewhere on the survey instrument which involved denial of the victim. The common element in these responses was a victim deserving of the consequences of the cheating behavior and cheating was viewed as "a form of rightful retaliation or punishment" (Sykes and Matza 1957, p. 668).

This feeling was extreme in one case, as suggested by the following student who felt her cheating was justified by the

> realization that this school is a manifestation of the bureaucratic capitalist system that systematically keeps the lower classes down, and that adhering to their rules was simply perpetuating the institution.

This "we" versus "they" mentality was raised by many students, but typically in comments about the policing of academic honesty rather than as justification for one's own cheating behavior. When used to justify cheating, the target was almost always an individual teacher rather than the institution and could be more accurately classified as a strategy of condemnation of condemners rather than denial of the victim.

The Condemnation of Condemners

Sykes and Matza describe the condemnation of condemners as an attempt to shift "the focus of attention from [one's] own deviant acts to the motives and behavior of those who disapprove of [the] violations. [B]y attacking others, the wrongfulness of [one's] own behavior is more easily repressed or lost to view" (1957, p. 668). The logic of this strategy for student cheaters focused on issues of favoritism and fairness. Students invoking this rationale describe "uncaring, unprofessional instructors with negative attitudes who were negligent in their behavior" (LaBeff et al. 1990, p. 195). For example:

> In one instance, nothing was done by a professor because the student was a hockey player.
>
> The TAs who graded essays were unduly harsh.
>
> It is known by students that certain professors are more lenient to certain types, e.g., blondes or hockey players.
>
> I would guess that 90% of the students here have seen athletes and/or fraternity members cheating on an exam or papers. If you turn in one of these culprits, and I have, the penalty is a five-minute lecture from a coach and/or administrator. All these add up to a "who cares, they'll never do anything to you anyway" attitude here about cheating.

Concerns about the larger society were an important issue for some students:

When community frowns upon dishonesty, then people will change.

If our leaders can commit heinous acts and then lie before Senate committees about their total ignorance and innocence, *then why can't I cheat a little?*

In today's world you do anything to be above the competition.

In general, students found ready targets on which to blame their behavior and condemnation of the condemners was a popular neutralization strategy.

The Appeal to Higher Loyalties

The appeal to higher loyalties involves neutralizing "internal and external controls . . . by sacrificing the demands of the larger society for the demands of the smaller social groups to which the [offender] belongs. [D]eviation from certain norms may occur not because the norms are rejected but because other norms, held to be more pressing or involving a higher loyalty, are accorded precedence" (Sykes and Matza 1957, p. 669). For example, a difficult conflict for some students is balancing the desire to help a friend against the institution's rules on cheating. The student may not challenge the rules, but rather views the need to help a friend, fellow fraternity/sorority member, or roommate to be a greater obligation which justifies the cheating behavior.

Fraternities and sororities were singled out as a network where such behavior occurs with some frequency. For example, a female student at a small university in New England observed:

There's a lot of cheating within the Greek system. Of all the cheating I've seen, it's often been men and women in fraternities and sororities who exchange information or cheat.

The appeal to higher loyalties was particularly evident in student reactions concerning the reporting of cheating violations. Although fourteen of the thirty-one schools participating in this research had explicit honor codes that generally require students to report cheating violations they observe, less than one-third (32.3 percent) indicated that they were likely to do so. When asked if they would report a friend, only 4 percent said they would and most students felt that they should not be expected to do so. Typical student comments included:

Students should not be sitting in judgment of their own peers.

The university is not a police state.

For some this decision was very practical:

A lot of students, 50 percent, wouldn't because they know they will probably cheat at some time themselves.

For others, the decision would depend on the severity of the violation they observed and many would not report what they consider to be minor violations, even those explicitly covered by the school's honor code or policies on academic honesty. Explicit examination or test cheating was one of the few violations where students exhibited any consensus concerning the need to report violations. Yet even in this case many students felt other factors must be considered. For example, a senior at a woman's college in the Northeast commented:

> It would depend on the circumstances. If someone was hurt, *very likely*. If there was no single victim in the case, if the victim was [the] institution . . . , then *very unlikely*.

Additional evidence of the strength of the appeal to higher loyalties as a neutralization technique is found in the fact that almost one in five respondents (17.8 percent) reported that they had helped someone cheat on an examination or major test. The percentage who have helped others cheat on papers and other assignments is likely much higher. Twenty-six percent of those students who helped someone else cheat on a test reported that they had never cheated on a test themselves, adding support to the argument that peer pressure to help friends is quite strong.

CONCLUSIONS

From this research it is clear that college students use a variety of neutralization techniques to rationalize their cheating behavior, deflecting blame to others and/or the situational context, and the framework of Sykes and Matza (1957) seems well-supported when student explanations of cheating behavior are analyzed. Unlike prior research (LaBeff et al. 1990), however, the present findings suggest that students employ all of the techniques described by Sykes and Matza, including denial of injury and denial of victim. Although there was very limited evidence of the use of denial of victim, denial of injury was not uncommon. Many students felt that some forms of cheating were victimless crimes, particularly on assignments that accounted for a small percentage of the total course grade. The present research does affirm LaBeff et al.'s finding that denial of responsibility and condemnation of condemners are the neutralization techniques most frequently utilized by college students. Appeal to higher loyalties is particularly evident in neutralizing institutional expectations that students report cheating violations they observe.

The present results clearly extend the findings of LaBeff et al. into a much wider range of contexts as this research ultimately involved 6,096 students at thirty-one geographically dispersed institutions ranging from small liberal arts colleges in the Northeast to nationally prominent research universities in the South and West. Fourteen of the thirty-one institutions have long-standing honor-code traditions. The code tradition at five of these schools dates to the late 1800s and all fourteen have codes that survived the student unrest of the 1960s. In such a context, the strength of the appeal to higher loyalties and the denial of responsibility as justifications for cheating is a very persuasive argument that neutralization techniques are salient to today's college student. More importantly, it may suggest fruitful areas of future discourse between faculty, administrators, and students on the question of academic honesty.*

REFERENCES

Baird, John S. 1980. "Current Trends in College Cheating." *Psychology in Schools* 17: 512–522.

Davis, Stephen F., Cathy A. Grover, Angela H. Becker, and Loretta N. McGregor. 1992. "Academic Dishonesty: Prevalence, Determinants, Techniques, and Punishments." *Teaching of Psychology*. In press.

Edwards, Paul. 1967. *The Encyclopedia of Philosophy*, no. 3, Paul Edwards (ed.). New York: Macmillan Company and Free Press.

Eisenberger, Robert, and Dolores M. Shank. 1985. "Personal Work Ethic and Effort Training Affect Cheating." *Journal of Personality and Social Psychology* 49: 520–528.

Haines, Valerie J., George Diekhoff, Emily LaBeff, and Robert Clark. 1986. "College Cheating: Immaturity, Lack of Commitment, and the Neutralizing Attitude." *Research in Higher Education* 25: 342–354.

LaBeff, Emily E., Robert E. Clark, Valerie J. Haines, and George M. Diekhoff. 1990. "Situational Ethics and College Student Cheating." *Sociological Inquiry* 60: 190–198.

Michaels, James W., and Terance Miethe. 1989. "Applying Theories of Deviance to Academic Cheating." *Social Science Quarterly* 70: 870–885.

Norris, Terry D., and Richard A. Dodder. 1979. "A Behavioral Continuum Synthesizing Neutralization Theory, Situational Ethics and Juvenile Delinquency." *Adolescence* 55: 545–555.

Perry, Anthony R., Kevin M. Kane, Kevin J. Bernesser, and Paul T. Spicker. 1990. "Type A Behavior, Competitive Achievement-Striving, and Cheating Among College Students." *Psychological Reports* 66: 459–465.

Sykes, Gresham M., and David Matza. 1957. "Techniques of Neutralization: A Theory of Delinquency." *American Sociological Review* 22: 664–670.

Tittle, Charles, and Alan Rowe. 1973. "Moral Appeal, Sanction Threat, and Deviance: An Experimental Test." *Social Problems* 20: 488–498.

Ward, David. 1986. "Self-Esteem and Dishonest Behavior Revisited." *Journal of Social Psychology* 123: 709–713.

Ward, David, and Wendy L. Beck. 1990. "Gender and Dishonesty." *Journal of Social Psychology* 130: 333–339.

---

*The author would like to acknowledge the support of the Rutgers Graduate School of Management Research Resources Committee, Exxon Corporation, and First Fidelity Bancorporation.

PART VII

Stigma Management

The label of deviant marks people with a stigma in the eyes of society. As we have seen, this label may lead to devaluation and exclusion. Consequently, people with deviant features learn how to "manage" their stigma so that they are not shamed or ostracized. This effort requires considerable social skills.

Goffman (1963) has suggested that people with potential deviant stigma fall into two categories: "the discreditable" and "the discredited." The former are those with concealable deviant traits (ex-convicts, secret homosexuals) who may manage themselves so as to avoid the deviant stigma. The latter are either members of the former category who have revealed their deviance or those who cannot hide their deviance (the obese, the physically disabled). These people's lives are characterized by a constant focus on secrecy and information control. Goffman observed that most discreditables engage in "passing" as "normals" in their everyday lives, concealing their deviance and fitting in with regular people. They may do this by avoiding contacts with "stigma symbols," those objects or behaviors that would tip people off to their deviant condition (an anorectic avoiding family meals, mental patients avoiding their medication). Another technique for passing includes using "disidentifiers" such as props, actions, or verbal expressions to distract and fool people into thinking that they do not have the deviant stigma (homosexuals bragging about heterosexual conquests or taking a date to the company picnic). Finally, they may "lead a double life," maintaining two different lifestyles with two distinct groups of people, one that knows about their deviance and one that does not.

In this endeavor, people may employ the aid of others to help conceal their deviance by "covering" for them. In these team performances, friends and family members may assist the deviants by concealing their identities, their whereabouts, their deficiencies, or their pasts. They may even coach the deviants on how to construct stories designed to hide their deviance.

Another form of stigma management, sometimes adopted when concealment fails, involves disclosing the deviance. People may do this for cathartic reasons (alleviating their burden of secrecy), therapeutic reasons (casting it in a positive light), or preventive reasons (so others don't find out in negative ways later, so that efforts at developing relationships are not totally wasted). Although many people find that their disclosures lead to rejection, others are more fortunate.

Disclosures of deviance can follow two courses. In observing the interactions between deviants and normals, Fred Davis (1961) noted that some nondeviant people go through a normalization process in their relationship with the deviant person largely through failing to acknowledge the deviant trait. Although this usually begins with a conspicuous and stilted ignoring of the individual's deviance (*deviance disavowal*), it progresses through stages where more relaxed interaction begins, interaction is directed at features of the person other than his or her deviant stigma, and finally gets to the point where the deviant stigma is overlooked and almost forgotten.

In contrast, deviant people can strive to normalize their relationships with nondeviants through "deviance avowal" (Turner 1972), in which they openly acknowledge their stigma and try to present themselves in a positive light. This avowal often takes the form of humor, "breaking the ice" by joking about their deviant attribute. In this way they show others that they can take the perspective of the normal and see themselves as deviant too, thus forming a bridge to others. This action further asserts that they have nondeviant aspects and that they can see the world as others do.

Thus far we have considered individual modes of adaptation to deviant stigma. Yet these stigma can also be managed through a group or collective effort. Many voluntary associations of stigmatized individuals exist, from the early organizations of prostitutes (COYOTE—Call Off Your Old Tired Ethics), to more recent ones such as the Gay Liberation Front, the Little People of America, the National Stuttering Project, and the Gray Panthers. Most well known are the 12-step programs modeled after the tremendous success of Alcoholics Anonymous (AA), including such groups as Overeaters Anonymous, Narcotics Anonymous, and Gamblers Anonymous.

These groups vary in character. Some are organized as what Stanford Lyman (1970) has called *expressive* groups, whose primary function is to provide support for their members. This support can take the form of organizing social and recreational activities, dispersing legal or medical information, or offering services such as shopping, meals, or transportation. Expressive groups tend to be apolitical, helping their members adapt to their social stigma rather than evade it. They also serve as places where members can come together in the company of other deviants, avoid the censure of nonstigmatized normals, and seek collective solutions to their common problems. It is here that they can make disclosures to others without fear of rejection.

Lyman has also described *instrumental* groups, where members come together not only to accomplish the expressive functions, but also to organize for political activism. This embodies Kitsuse's tertiary deviation, where individuals reject the societal conception and treatment of their stigma and organize to change social definitions. They fight to get others to modify their views of the status or behavior in question so that society, like they, will no longer regard it as deviant. Examples of such groups include ACT UP, an AIDS organization whose members have tried to change social attitudes toward AIDS patients, the National Organization for Women (NOW), and the Disabled in Action.

25

Identity and Stigma of Women with STDs

ADINA NACK

Nack sensitively explores a highly stigmatized terrain in this analysis of how women deal with their sexually transmitted diseases (STDs). She ties stigma management to the career concept by grouping the ways women try to account for their sexually transmitted diseases, both to themselves and others, into a sequential progression. Initially they deny their status and stigma, hoping to keep it secret, while feeling guilty about lying or withholding information from others. When they acknowledge to themselves that they are infected, they react in frustration and anger, deflecting their stigma by blaming others or transferring the stigma to them. For either preventive or therapeutic reasons, Nack's subjects eventually stop denying and deflecting and come to accept the stigma, finding ways to cope with it to friends, family members, and sexual partners. Only when they are ready to accept the stigma consequences of acknowledging their STDs can these women make peace with their selves and accept their condition.

The HIV/AIDS epidemic has garnered the attention of researchers from a variety of academic disciplines. In contrast, the study of other sexually transmitted diseases (STDs) has attracted limited interest outside of epidemiology and public health. In the United States, an estimated three out of four sexually-active adults have HPV infections (human papillomavirus—the virus that can cause genital warts); one out of five have genital herpes infections (Ackerman 1998; CDC Server 1998). . . . This article focuses on how the sexual self-concept is transformed when the experience of living with a chronic STD casts a shadow of disease on the health and desirability of a woman's body, as well as on her perceived possibilities for future sexual experiences. . . .

STIGMA AND THE SEXUAL SELF

For all but one of the 28 women, their STD diagnoses radically altered the way that they saw themselves as sexual beings. Facing both a daunting medical

---

Copyright 2000 From "Damaged Goods: Women Managing the Stigma of STDs," Adina Nack, *Deviant Behavior*, Vol. 21, No. 2. Reproduced by permission of Taylor & Francis, Inc., http://www.routledge-ny.com.

and social reality, the women employed different strategies to manage their new stigma. Each stigma management strategy had ramifications for the transformation of their sexual selves. . . .

Stigma Nonacceptance

Goffman (1963) proposed that individuals at risk for a deviant stigma are either "the discredited" or "the discreditable." The discrediteds' stigma was known to others either because the individuals revealed the deviance or because the deviance was not concealable. In contrast, the discreditable were able to hide their deviant stigma. Goffman found that the majority of discreditables were "passing" as non-deviants by a avoiding "stigma symbols," anything that would link them to their deviance, and by utilizing "disidentifiers," props or actions that would lead others to believe they had a non-deviant status. Goffman (1963) also noted that individuals bearing deviant stigma might eventually resort to "covering," one form of which he defined as telling deceptive stories. To remain discreditable in their everyday lives, nineteen of the women employed the individual stigma management strategies of passing and/or covering. In contrast, nine women revealed their health status to select friends and family members soon after receiving their diagnoses.

Passing The deviant stigma of women with STDs was essentially concealable, though revealed to the necessary inner circle of health care and health insurance providers. For the majority, passing was an effective means of hiding stigma from others, sometimes, even from themselves.

Hillary, a 22-year-old White senior in college, recalled the justifications she had used to distance herself from the reality of her HPV infection and facilitate passing strategies.

> At the time, I was in denial about it. I told myself that that wasn't what it was because my sister had had a similar thing happen, the dysplasia. So, I just kind of told myself that it was hereditary. That was kinda' funny because I asked the nurse that called if it could be hereditary, and she said "No, this is completely sexually transmitted." . . . I really didn't accept it until a few months after my cryosurgery.

Similarly, Gloria, a Chicana graduate student and mother of four, was not concerned about a previous case of gonorrhea she had cured with antibiotics or her chronic HPV "because the warts went away." Out of sight, out of her sex life: "I never told anybody about them because I figured they had gone away, and they weren't coming back. Even after I had another outbreak, I was still very promiscuous. It still hadn't registered that I needed to always have the guy use a condom."

When the women had temporarily convinced themselves that they did not have a contagious infection, it was common to conceal the health risk with partners because the women, themselves, did not perceive the risk as real. Kayla, a lower-middle class, White college senior, felt justified in passing as healthy with partners who used condoms, even though she knew that condoms could break. . . . Tasha, a White graduate student, found out that she might have inadvertently passed as healthy when her partner was diagnosed with clamydia. "I freaked out—I was like, 'Oh my God! I gave you chlamydia. I am so sorry! I am so sorry!' I felt really horrible, and I felt really awful." Sara, a Jewish, upper-middle class 24-year-old, expressed a similar fear of having passed as healthy and exposed a partner to HPV. "Evan called me after we'd been broken up and told me he had genital warts. And, I was with another guy at the time, doing the kinda-sorta-condom-use thing. It was like, 'Oh, my gosh, am I giving this person something?'" Even if the passing is done unintentionally, it still brings guilt to the passer.

The women also tried to disidentify themselves from sexual disease in their attempts to pass as being sexually healthy. Rather than actively using a verbal or symbolic prop or action that would distance them from the stigma, the women took a passive approach. Some gave nonverbal agreement to put downs of other women who were known to have STDs. For example, Hillary recalled such an interaction. "It's funny being around people that don't know that I have an STD and how they make a comment like, 'That girl, she's such a slut. She's a walking STD.' And how that makes me feel when I'm confronted with that, and having them have no idea that they could be talking about me." Others kept silent about their status and tried to maintain the social status of being sexually healthy and morally pure. Kayla admitted to her charade: "I guess I wanted to come across as like really innocent and everything just so people wouldn't think that I was promiscuous, just because inside I felt like they could see it even though they didn't know about the STD." Putting up the facade of sexual purity, these women distanced themselves from any suspicion of sexual disease.

Covering When passing became too difficult, some women resorted to covering to deflect family and friends from the truth. Cleo summed up the rationale by comparing her behavior to what she had learned growing up with an alcoholic father. "They would lie, and it was obvious that it was a lie. But, I learned that's what you do. Like you don't tell people those things that you consider shameful, and then, if confronted, you know, you lie."

Hillary talked to her parents about the HPV surgery, but never as treatment for an STD. She portrayed her moderate cervical dysplasia as a precancerous scare, unrelated to sex. "We never actually talked about it being a STD, and she kind of thought that it was the same thing that my sister had which wasn't sexually transmitted." When Tasha's sister helped her get a prescription for pubic lice, she actually provided the cover story for her embarrassed younger sister. "She totally took control, and made a personal inquiry: 'So, how did you get this? From a toilet seat.' And, I was like, 'a toilet seat,' and she believed me." When I asked Tasha why she confirmed her sister's misconception, she replied "because I didn't want her to know that I had had sex." For Anne, a 28-year-old lower-middle class graduate student, a painful herpes outbreak almost outed her on a walk with a friend. She was so physically uncomfortable that she was actually "waddling." Noticing the strange behavior, her friend asked what was wrong. Anne told her that it was a hemorrhoid: that was only a partial truth because herpes was the primary cause of her pain. As Anne put it, telling her about the hemorrhoid "was embarrassing enough!"

Deception and Guilt The women who chose to deny, pass as normal, and use disidentifiers or cover stories shared more than the shame of having an STD—they had also told lies. With lying came guilt. Anne, who had used the hemorrhoid cover story, eventually felt extremely guilty. Her desire to conceal the truth was in conflict with her commitment to being an honest person. "I generally don't lie to my friends. And I'm generally very truthful with people and I felt like a sham lying to her." Deborah, a 32-year-old, White professional from the Midwest, only disclosed to her first sexual partner after she had been diagnosed with HPV: she passed as healthy with all other partners. Deborah reflected, "I think my choices not to disclose have hurt my sense of integrity." However, her guilt was resolved during her last gynecological exam when the nurse practitioner confirmed that after years of "clean" pap smear results Deborah was not being "medically unethical" by not disclosing to her partners. In other words, her immune system had probably dealt with the HPV in such a way that she might never have another outbreak or transmit the infection to sexual partners.

When Cleo passed as healthy with a sexual partner, she started, "feeling a little guilty about not having told." However, the consequences of passing as healthy were very severe for Cleo:

> No. I never disclosed it to any future partner. Then, one day, I was having sex with Josh, my current husband, before we were married, and we had been together for a few months, maybe, and I'm like looking at his penis, and I said, "Oh, my goodness! You have a wart on your penis! Ahhh!" All of a sudden, it comes back to me.

Cleo's decision to pass left her with both the guilt of deceiving and infecting her husband.

Surprisingly, those women who had *unintentionally* passed as being sexually healthy (i.e., they had no knowledge of their STD-status at the time) expressed a similar level of guilt as those who had been purposefully deceitful. Violet, a middle class, White 36-year-old, had inadvertently passed as healthy with her current partner. Even after she had preventively disclosed to him, she still had to deal with the guilt over possibly infecting him.

> It hurt so bad that morning when he was basically furious at me thinking I was the one he had gotten those red bumps from. It was the hour from hell! I felt really majorly dirty and stigmatized. I felt like "God, I've done the best I can: if this is really caused by the HPV I have, then I feel terrible."

When employing passing and covering techniques, the women strove to keep their stigma from tainting social interactions. They feared reactions that Lemert (1951) has labeled the *dynamics of exclusion:* rejection from their social circles of friends, family, and most importantly sexual partners. For most of the women, guilt surpassed fear and became the trigger to disclose. Those who had been deceitful in passing or covering had to assuage their guilt: their options were either to remain in nonacceptance, disclose, or transfer their guilt to somebody else.

Stigma Deflection

As the women struggled to manage their individual stigma of being sexually diseased, real and imaginary social interactions became the conduit for the contagious label of "damaged goods." Now that the unthinkable had happened to them, the women began to think of their past and present partners as infected, contagious, and potentially dangerous to themselves or other women. The combination of transferring stigma and assigning blame to others allowed the women to deflect the STD stigma away from themselves.

Stigma Transference I propose the concept of *stigma transference* to capture this element of stigma management that has not been addressed by other deviance theorists. Stigma transference is a specialized case of projection which . . . manifests as a clear expression of anger and fear. The women did not connect this strategy to a reduction in their levels of anxiety; in fact, several discussed it in relation to increased anxiety. . . .

Transference of stigma to a partner became more powerful when the woman felt betrayed by her partner. When Hillary spoke of the "whole trust issue" with her ex-partner, she firmly believed he had lied to her about his sexual health status and he would lie to others. Even though she had neither told him about her diagnosis nor had proof of him being infected, she fully transferred her stigma to him.

> He's the type of person who has no remorse for anything. Even if I did tell him, he wouldn't tell the people that he was dating. So it really seemed pretty pointless to me to let him know because he's not responsible enough to deal with it, and it's too bad knowing that he's out there spreading this to God knows how many other people.

Kayla also transferred the stigma of sexual disease to an ex-partner, never confronting him about whether or not he had tested positive for STDs. The auxiliary trait of promiscuity colored her view of him:" I don't know how sexually promiscuous he was, but I'm sure he had had a lot of partners." Robin, a 21-year-old White undergraduate, went so far as to tell her ex-partner that he needed to see a doctor and "do something about it." He doubted her ability to pinpoint contracting genital warts from him and called her a "slut." Robin believed that *he* was the one with the reputation for promiscuity and decided to "trash" him by telling her two friends who hung out with him. Robin hoped to spoil his sexual reputation and scare off his future partners. In the transference of stigma, the women ascribed the same auxiliary traits onto others that others had previously ascribed to them. . . .

In all cases, it was logical to assume that past and current sexual partners may have also been infected. However, the stigma of being sexually diseased had far-reaching consequences into the imaginations of the women. The traumatic impact on their sexual selves led most to infer that future, as yet unknown, partners were also sexually diseased. Kayla summed up this feeling. "After I was diagnosed, I was a lot more cautious and worried about giving it to other people or getting something else because somebody hadn't told me." They had already been damaged by at least one partner. Therefore, they expected that future partners, ones who had not yet come into their lives, held the threat of also being *damaged goods*.

For Hillary, romantic relationships held no appeal anymore. She had heard of others who also had STDs but stayed in nonacceptance and never changed their lifestyle of having casual, unprotected sex:

> I just didn't want to have anything to do with it. A lot of it was not trusting people. When we broke up, I decided that I was not having sex. Initially, it was because I wanted to get an HIV test. Then, I came to kind of a turning point in my life and realized that I didn't want to do the one-night-stand thing anymore. It just wasn't worth it. It wasn't fun. . . .

Blame The women's uses of stigma transference techniques were attempts to alleviate their emotional burdens. First, the finger of shame and guilt pointed inward, toward the women's core sexual selves. Their sexual selves became tainted, dirty, damaged. In turn, they directed the stigma outward to both real and fictional others. Blaming others was a way for all of the women to alleviate some of the internal pressure and turn the anger outward. This emotional component of the *damaged goods* stage externalized the pain of their stigma.

Francine recalled how she and her first husband dealt with the issue of genital warts. "We kind of both ended up blaming it on the whole fraternity situation. I just remember thinking that it was not so much that we weren't clean, but that he hadn't been at some point, but now he was." Francine's husband had likely contracted genital warts from his wild fraternity parties. "We really thought of it as, that woman who did the trains [serial sexual intercourse]. It was still a girl's fault kind-of-thing." By externalizing the blame to the promiscuous women at fraternity parties, Francine exonerated not only herself, but also her husband. . . .

For Violet, it was impossible to neatly deflect the blame away from both herself and her partner. "I remember at the time just thinking, 'Oh man! He gave it to me!' While, he was thinking, 'God, [Violet]! You gave this to me!' So, we kind of just did a truce in our minds. Like, OK, we don't know who gave it—just as likely both ways. So, let's just get treated. We just kind of dropped it." Clearly the impulse to place blame was strong even when there was no easy target.

Often, the easiest targets were men who exhibited the auxiliary traits of promiscuity and deception. Tasha wasn't sure which ex-partner had transmitted the STD. However, she rationalized blaming a particular guy. "He turned out to be kind of a huge liar, lied to me a lot about different stuff. And, so I blamed him. All the other guys were, like, really nice people, really trustworthy." Likewise, when I asked Violet from whom she believed she had contracted chlamydia, she replied. "Dunno,' it could've been from one guy, because that guy had slept with some unsavory women, so therefore he was unsavory." Later, Violet contracted HPV, and the issue of blame contained more anger: "I don't remember that discussion much other than, being mad over who I got it from: 'oh it must have been Jess because he had been with all those women.' I was mad that he probably never got tested. I was OK before him." The actual guilt or innocence of these blame targets was secondary. What mattered to the women was that they could hold someone else responsible.

Stigma Acceptance

Eventually, every woman in the study stopped denying and deflecting the truth of her sexual health status by disclosing to loved ones. The women disclosed for either preventive or therapeutic reasons. That is, they were either motivated to reveal their STD status to prevent harm to themselves or others, or to gain the emotional support of confidants.

Preventive and Therapeutic Disclosures The decision to make a preventive disclosure was linked to whether or not the STD could be cured. Kayla explained, "Chlamydia went away, and I mean it was really bad to have that, but I mean it's not something that you have to tell people later 'cause you know, in case it comes back. Genital warts, you never know." Kayla knew that her parents would find out about the HPV infection because of insurance connections. Prior to her cryosurgery, Kayla decided to tell her mom about her conditions. "I just told her what [the doctor] had diagnosed me with, and she knew my boyfriend and everything, so—it was kind of hard at first. But, she wasn't upset with me. Main thing, she was disappointed, but I think she blamed my boyfriend more than she blamed me." Sara's parents also reacted to her preventive disclosure by blaming her boyfriend: they were disappointed in their daughter, but angry with her boyfriend.

Preventive disclosures to sexual partners, past and present, were a more problematic situation. The women were choosing to put themselves in a position where they could face blame, disgust, and rejection. For those reasons, the women put off preventive disclosures or partners as long as possible. For example, Anne made it clear that she would not have disclosed her herpes to a female sexual partner had they not been, "about to have sex." After "agonizing weeks and weeks and weeks before trying to figure out how to tell," Diana, a 45-year-old African-American professional, finally shared her HPV and herpes status before her current relationship became sexual. Unfortunately, her boyfriend "had a negative reaction": "he certainly didn't want to touch me anywhere near my genitals." In Cleo's case, she told her partner about her HPV diagnosis because she wasn't going to be able to have sexual intercourse for a while after her cryosurgery. Violet described the thought process that led up to her decision to disclose her HPV status to her current partner:

> That was really scary because once you have [HPV], you can't get rid of the virus. And then having to tell my new partner all this stuff. I just wanted to be totally up front with him; we could use condoms. Chances are he's probably totally clean. I'm like, "Oh my god, here I am tainted because I've been with, at this point, 50 guys, without condoms. Who knows what else I could have gotten (long pause, nervous laugh)?" So, that was tough. . . .

Many of the therapeutic disclosures were done to family members. The women wanted the support of those who had known them the longest. Finally willing to risk criticism and shame, they hoped for positive outcomes: acceptance, empathy, sympathy—any form of nonjudgmental support. Tasha disclosed to her mother right after she was diagnosed with chlamydia. "My family died—'Guess what, mom, I got chlamydia.' She's like, 'Chlamydia? How did you find out you got chlamydia?' I'm like. 'Well, my boyfriend got an eye infection.' (laughter) 'How'd he get it in his eye?' (laughter) So, it was the biggest joke in the family for the longest time." In contrast, Rebecca, a White professional in her mid-fifties, shared her thought process behind *not* disclosing to her adult children. "I wanted to tell my younger one . . . I wanted very much for him to know that people could be asymptomatic carriers because I didn't want him to unjustly suspect somebody of cheating on him . . . and I don't believe I ever managed to do it . . . it's hard to bring something like that up." . . .

Consequences of Disclosure With both therapeutic and preventive disclosure, the women experienced some feelings of relief in being honest with loved ones. However, they still carried the intense shame of being sexually diseased women. The resulting emotion was anxiety over how their confidants would react: rejection, disgust, or betrayal. Francine was extremely anxious about disclosing to her husband. "That was really tough on us because I had to go home and tell Damon that I had this outbreak of herpes." When asked what sorts of feelings that brought up, she immediately answered. "Fear. You know I was really fearful—I didn't think that he would think I had recently had sex with somebody else . . . but, I was still really afraid of what it would do to our relationship." . . .

Overall, disclosing intensified the anxiety of having their secret leaked to others in whom they would have never chosen to confide. In addition, each disclosure brought with it the possibility of rejection and ridicule from the people whose opinions they valued most. For Gloria, disclosing was the right thing to do but had painful consequences when her partner's condom slipped off in the middle of sexual intercourse.

> I told him it doesn't feel right. "You'd better check." And, so he checked, and he just jumped off me and screamed, "Oh fuck!" And, I just thought,

oh no, here we go. He just freaked and went to the bathroom and washed his penis with soap. I just felt so dirty. . . .

Disclosures were the interactional component of self-acceptance. The women became fully grounded in their new reality when they realized that the significant people in their lives were now viewing them through the discolored lenses of sexual disease.

CONCLUSION

The women with STDs went through an emotionally difficult process, testing out stigma management strategies, trying to control the impact of STDs on both their self-concepts and on their relationships with others. In keeping with Cooley's "looking glass self" (1902/1964), the women derived their sexual selves from the imagined and real reactions of others. Unable to immunize themselves from the physical wrath of disease, they focused on mediating the potentially harmful impacts of STDs on their sexual self-concepts and on their intimate relationships.

REFERENCES

Ackerman, S. J. 1998. "HPV: Who's Got It and Why They Don't Know." *HPV News* 8(2), Summer: 1, 5–6.

Centers for Disease Control and Prevention. 1998. "Genital Herpes." *National Center for HIV, STD, & TB Prevention.* Online. Netscape Communicator. 4 February.

Cooley, Charles H. [1902] 1964. *Human Nature and the Social Order.* New York: Schocken.

Goffman, Erving, 1963. *Stigma.* Englewood Cliffs, NJ: Prentice Hall.

Lemert, Edwin. 1951. *Social Pathology.* New York: McGraw-Hill.

26

Stigma Management and Collective Action among the Homeless

LEON ANDERSON, DAVID A. SNOW, AND DANIEL CRESS

Another stigmatized population, the homeless, is discussed by Anderson, Snow, and Cress. The way these street residents account for their behavior, the authors note, varies considerably, depending on whether they are conversing among themselves or with other stigmatized individuals (in-group members) or with those they consider more hostile, domiciled individuals (out-group members). To insiders they are likely to explain frankly some of the factors that accompany their homelessness. To outsiders they must present a more respectable front, engaging in a range of strategies that help them maintain their dignity through avoidance as well as confrontation. While they may initially deflect the stigma through passing or covering, many become more hostile and move to defiance, while others organize into groups that seek to mobilize resources to solve their homelessness condition.

In the past 3 decades symbolic interactionists have written extensively on stigmatized individuals and groups and the ways in which they handle the personal and interactional dilemmas they experience. Most stigmatized groups have varying degrees of access to social resources that offer escape from public interactions where individuals are likely to be treated in terms of the stigmatized social categories with which they are identified rather than as individuals with distinct personalities, relationships and biographies. Although relations with nonstigmatized intimates may be far from problem-free, stigmatized individuals can expect at least a degree of personalized recognition, social validation, normalization of interactions (Davis 1961), and support in managing their stigma (Blum 1991). Some groups of stigmatized people, however, have been largely dislocated from such interpersonal moorings. These individuals provide insight into the management of stigma under particularly difficult circumstances, when even occasional reprieve from stigmatization is limited. In this paper we address the issue of stigma management among such

---

Reprinted from Leon Anderson, David Snow, and Daniel Cress, "Negotiating the Public Realm: Stigma Management and Collective Action Among the Homeless," 1994, pp. 121–143, from *Research in Community Sociology*, Vol. 8, No. 1, with permission from Elsevier Science.

populations by examining the case of homeless street people, whose lives are largely conducted on the streets of urban America. We begin by briefly describing the research from which this essay is derived. Next we provide a background sketch of the context and dilemmas of stigmatization faced by the homeless in the United States in the 1980s and 1990s. We then turn to an examination of specific stigma management strategies we have observed among the homeless. We conclude by elaborating the implications of our findings for understanding the homeless as social actors and the ways in which their experiences can inform us about the "community of the streets."

METHODS

This paper is based on data gathered during two research projects. The first was a multimethod case study of homelessness in Austin, Texas in the mid-1980s. The data from that project most pertinent to this essay came from ethnographic observation of homeless individuals in natural settings. During the course of nearly a year and a half we spent over 400 hours with the homeless in a variety of settings, such as in meal and shelter lines, in parks, under bridges, in plasma centers, on city streets, and at day labor pick up sites. Our aim was to acquire an appreciation for the nature of life on the streets from the standpoint of the homeless themselves. We followed them through their daily routines, observing their behavior and recording both what they told us about their lives and what they told each other. From the pool of homeless people we encountered on the streets, we compiled observational and life history information on a nonprobability sample of 168 individuals.[1]

The second research project was a national study of collective action campaigns by homeless people. The relevant component of the project consists of case studies of homeless activists and their allies, targets, and collective actions in eight U.S. cities that were chosen to reflect a range of protest activity. The primary focus was on the internal workings and structure of homeless social movement organizations and their relationships with other community organizations. Our goal was to discern from a variety of perspectives (e.g., the homeless, their resource suppliers, and the targets of their attacks) how the homeless go about collectively protesting their circumstances. This involved spending time on the streets with the homeless, attending and participating in the meetings and actions of their organizations, and interviewing their members, supporters, and antagonists.

BACKGROUND AND CONTEXT

In the spring of 1990 a national chain of toy stores marketed a new doll that capitalized on contemporary public interest in the homeless. "Steve the Tramp," as the doll was named, was a "Dick Tracy action figure whose package describe[d] him as an 'ignorant bum . . . dirty and scarred from a life on the streets. You'll smell him before you'll see him' " ("Toy Stores Pull 'Tramp' Doll" 1990). Although protesters, consisting of homeless people and their advocates, were able to pressure the toy store chain into pulling the doll from its shelves, the very production of the doll exemplified the widespread negative public perceptions of the homeless. To be visibly homeless in the United States is to live with what Everett Hughes (1945) has referred to as a master status, a powerful social category that is associated with a set of traits that its incumbents are believed to possess. In the case of the "homeless person" these characteristics include filthiness, laziness, helplessness, alcoholism, mental illness, and violence.

Homelessness has not always been so stigmatized. In medieval Europe, for instance, it was associated with holiness, and voluntarily taken on by members of certain mendicant religious orders. Negative stereotypes of the homeless in western societies, however, date back at least to the fourteenth century (Beier 1985). The sordid images of the homeless have had numerous consequences. They have provided ballast for laws and social policies that have sought to control and/or punish the homeless. They have also underpinned most programs that have been developed to "help the homeless overcome their problems." An equally pervasive, if less widely discussed, consequence of the stigmatization of the homeless is that it has massively influenced interactions between them and other individuals in public settings.

This latter issue became particularly salient during the 1980s in the United States when a dramatic increase in homelessness resulted in greater intermingling of homeless people with other segments of the population than had been the case when such increases had occurred in the past. The demolition across the country of cheap skid row housing (Hoch and Slayton 1989; Jackson 1987; Rossi 1989), along with the destruction of and decline in federal support for low income housing during the 1980s (Lamar 1988; Schumer 1988; Wright and Lam 1987), displaced many of the poor and precariously housed without providing alternatives. The encroachment of the middle class into areas formerly populated primarily by the poor broke down the spatial ordering of different social classes, which Lofland (1973) has argued serves as a fundamental bedrock of the modern urban order. One result has been greater contact between more affluent citizens and the poor, particularly the poor who have become homeless. . . .

The homeless face a precarious situation since they have so few alternatives to spending their time in public space. While they are often perceived as a public danger, they are themselves far more at risk from the domiciled, from each other, and from the police than are most of those who fear them.

But physical perils are not the only dangers the homeless face. Life on the streets regularly subjects them to subtle but pervasive threats to their senses of self and self-esteem. Their lack of housing forces them to spend a greater amount of time in the public realm than do others, and requires them to conduct private aspects of their daily lives (such as sleeping and toileting) in public settings, as has been poignantly demonstrated in Ellen Baxter and Kim

Hopper's (1981) aptly titled report, *Private Lives/Public Spaces*. Additionally, by virtue of their social dislocation from family members and other supportive social bonds, the homeless may find themselves particularly dependent on interactions in the public realm to fulfill their needs for sociality. This is in striking contrast to most citizens, who fill their more intimate and interactive social needs in other settings and perceive the public realm as a domain for very limited kinds of social involvement (Goffman 1963a; Lofland 1973).

Furthermore, as members of the lowest social status in American society, the homeless receive what Charles Derber (1979) has referred to as "negative attention," or attention that highlights one's defiled or subservient status. The police engage in systematic and consistent surveillance of homeless individuals and groups, in part to control their behavior and in part to assure other citizens that the streets are safe. The homeless are also subject to a variety of status degradation rituals and public humiliations from other citizens, ranging from staring and taunting (especially common from people driving by) to callous engagements at any time of day or night by media reporters and even researchers interested in probing them about their problems and shortcomings. Such engagements assault individuals' already precarious senses of themselves as socially respectable human beings.

The homeless also suffer from what Derber has called "attention deprivation." Derber (1979, p. 42) observes that "members of the subordinate classes are regarded as less worthy of attention in relations with members of dominant classes and so are subjected to subtle yet systematic face-to-face deprivation." Nowhere is this more true than for the homeless, who are routinely avoided. Pedestrians frequently avert their eyes when passing the homeless on the sidewalk, and hasten their pace or cross the street when they sense they may be targeted for interaction. These avoidance rituals may not be as overtly demeaning as the more direct kinds of negative attention the homeless receive, but they too hammer away at the wedge between the homeless and the domiciled by casting the homeless as objects of contamination.

In short, the situation of the homeless in public places is grim. In addition to material deprivations and physical threats, they must also contend with both intended and unintended social psychological assaults. Such stresses take their toll, and several studies have, not surprisingly, found high levels of depression among the homeless (e.g., LaGory, Ritchey and Mullis 1990; Rossi 1989). But the picture presented so far is only half of the story, for the homeless seldom passively succumb to their stigmatized status. Contrary to many public and academic perceptions of them, most of the homeless are resourceful social actors who endeavor to resist and influence the negative perceptions others have of them. In doing so, they creatively adapt the public realm to their own needs and carve out a meaningful place for themselves in the community of the streets.

How do they accomplish these things? What resources do they manage to acquire and make use of in their efforts? What strategies do they develop? And how do they enlist the support of others in their assertions of more positive identities? Finally, how successful are these strategies and techniques in enabling the homeless to either manage or transcend their stigmatization? These are the questions we address in the following pages. In the process we hope to shed light on both the homeless as social actors and the social dimensions of the public realm.

STIGMA MANAGEMENT AMONG THE HOMELESS

Drawing on a distinction made by Goffman (1963b), we can separate the stigma management efforts of the homeless into "in-group" strategies, which prevail in interaction with stigmatized peers, and "out-group" strategies, which are used by the homeless in their encounters with domiciled "normals." We describe a range of strategies in each category.

In-Group Strategies

The in-group strategies the homeless use among themselves are attempts to achieve some temporary psychological reprieve from their demeaning situation. Four of the most common in-group strategies include drinking, cheap entertainment, "hanging out," and identity work.

Drinking Lay observers, social workers, and academics often cite alcohol as a primary reason individuals become homeless. However, our research suggests that while this may be true for some, heavy drinking is often a result of homelessness rather than its cause. This assertion is supported by three observations. First, those who have been on the streets longer tend to drink more than those who have been there a shorter time. Of course, this does not necessarily imply that increased drinking results from extended exposure to the streets since it is also possible that heavy drinkers may find it especially difficult to get off the streets and are therefore overrepresented among the long-term homeless. However, we also observed that among many of those who had recently become homeless, drinking increased over time, suggesting that heavier drinking is an adaptive response to street life. But perhaps the strongest evidence was provided by the homeless themselves, who consistently told us that they drank largely to escape the tribulations of the streets. As one longtime homeless man who struggled with "a drinking problem" explained, "You're in there [the shelters] with lots of people you don't know. They look like shit and smell like it too. And they remind you of where you are and who you are. It ain't pleasant. So you begin to crave a drink." He continued, "You think about it and you get pissed off about it and you get drunk and forget about it. At least for a while, and then you start all over." The use of drinking to provide relief from daily indignities and painful memories of better days was also expressed by a former painter who hit the streets after a long bout with chemical poisoning from the paints he had worked with. "I can't drink anymore, on account of an ulcer," he told us, "but if I could, I'd sure do it. The Sally [Salvation Army] treats you like a dog and the police run you off the labor corner.[2] My family hasn't wanted me around for five years, ever since I lost my job. At least when I could drink, I could forget about it."

Cheap Entertainment In a similar fashion, our observations suggest that cheap entertainment, such as dollar movies and libraries, in addition to providing shelter for the homeless (Love 1958; Spradley 1970; Wallace 1965; Wiseman 1970), provide psychological retreat from stigmatization. A case in point is Tanner, a badly burn-disfigured man, who often headed directly for the public library in the morning after finishing his chores as a part-time worker at the Salvation Army.[3] For the larger part of the day he pored over books on occult subjects, jotting prolific notes that he kept in a notebook devoted to his "studies." Given the negative reaction of many people, including potential employers, to his disfigurement, Tanner's pursuit of his studies in a back corner of the library provided him with an ecological niche in which he could lose himself in intellectual pursuits, safe from stigmatization.

Paperback novels, newspapers, and comic books can function as portable counterparts to the library, serving as distractions while individuals make the daily rounds of service agencies in the community. The utility of books and magazines for managing stigma was illustrated to us one morning when we were sitting in a parking lot, waiting for a plasma center to open. A group of male university students drove by and yelled out pejoratively, "Eh, dragworms! Get a job!" Several of the homeless men in the parking lot yelled back at the students, but one of the men sat undisturbed, reading a Louis L'Amour novel. When the men continued to yell angrily even after the car was gone, the reader turned to us and said, "Man, that's why I always got a book. That way I don't have to pay attention to that shit." On subsequent occasions we witnessed him and other readers ignore similar derisive taunts that were hurled from passing motorists toward the patrons of the Salvation Army dinner line.

Hanging Out Another form of entertainment the homeless engage in is "hanging out," whiling away the day with street peers in activity characterized by a lack of expectations for accomplishment, on the one hand, and engrossment in the immediate present, on the other. As with other forms of entertainment, hanging out provides a retreat from conscious striving and facing one's situation. This can be a particularly alluring alternative to the frustration and embarrassment of looking for work unsuccessfully, as illustrated by Buddy, a 28-year-old homeless man who briefly "shacked up" with a woman receiving AFDC, who he referred to as his "Sugar Mama." When he showed up one afternoon at a popular street corner near the Sally, he recounted:

> My Sugar Mama's always saying, "Look for a job, look for a job." She never let's up. So, I told her today, "Hey, I don't want to hear it anymore. I looked my ass off for jobs this week and nobody wants me. All I want to do is hang out for a while and forget about it." So I came down here to hang with my best friends.

The support of street peers that Buddy alludes to is a key element of in-group stigma management strategies. Largely stonewalled from positive interactions with other members of society, the homeless respond by turning to each other in an easy conviviality that has been noted by numerous researchers (Cohen and Sokolovsky 1989; Rooney 1973; Rubington 1968; Wallace 1965; Wiseman 1970).

Identity Work The street friendships of the homeless serve both material and emotional functions. On the one hand they offer a nexus for the sharing of meager resources, while on the other they provide an essentially nonstigmatizing reference group and a source of interpersonal validation. This second function is clearly evident in the conversations of the homeless as they hang out together in parking lots, on street corners, and in meal lines. One common aspect of their conversations is what we call "identity talk" (Snow and Anderson 1987), in which the homeless construct and negotiate personal identities, consistently casting themselves in positive ways by referring to personal attributes, past accomplishments and future plans.

While the bulk of identity talk is within the bounds of credibility, it occasionally takes on questionable proportions, as when Buddy returned to the streets after spending several days in jail for sleeping in a parked car without the owner's consent. As a group of his acquaintances passed around a joint to celebrate his return to the streets, Buddy, who occasionally mentioned possessing supernatural powers, claimed to be casting a spell on the police:

> I've got the power to make them pay. Shit's coming down, and I can see it, man. I'm making it happen right now: crashing cars and burning buildings! Cops are dying because they fucked with me. And that's the last time they're going to do it.

Interestingly, Buddy's companions listened to his wild assertions without questioning them. In fact, grandiose claims made on the streets are seldom challenged by other homeless individuals. While there may be a certain amount of fear of confrontation involved, it seemed clear to us that the homeless also avoid calling each other into question out of a sense of reciprocity. In keeping with the commonly heard street adage that "what goes around, comes around," to challenge another's story is to risk having someone challenge yours in return. Alternatively, to tacitly support others' identity assertions serves to build goodwill for when one may later make claims of one's own.[4]

The foregoing, in-group stigma management strategies constitute what Goffman (1961) called "secondary adjustments," ways in which individuals who find themselves trapped in demeaning social contexts can stand "apart from the role and the self" implied. They are activities that provide something for individuals to lose themselves in, at least for a while, thereby blunting awareness of their demeaning situations. Many of these activities, such as drinking, "hanging out," and making bigger-than-life identity assertions, may be cited by observers as evidence of the psychological and moral deficiencies of the homeless: of alcoholism, laziness, and mental illness. From an insider's perspective, however, these behaviors are functional and adaptive responses to demoralizing circumstances. Stigmatized by the larger society, the homeless come together to form their own, admittedly fragile, community on the streets. And within that community they engage in activities that offer them both "time out" from more frustrating endeavors and a reference group in which their claims to social respectability are likely to be accepted. In all of this the homeless stand apart from others in the public realm, associating with and relying on each other. For many reasons, however, the homeless may want to interact with nonhomeless individuals. On such occasions they must turn to a different set of stigma management strategies.

Out-Group Strategies

If the homeless were able to limit their interaction to street peers, they could avoid the most direct consequences of their stigmatization. But they are unable to seal themselves off, in part because they reside primarily in public spaces where contact with domiciled others is unavoidable. Furthermore, since they lack the resources they need to survive, they must turn to others for support. Domiciled others are more likely to assign the homeless a stigmatized status than are street peers, and, as a consequence, their interactions with the homeless are likely to be characterized by avoidance and contempt. When the homeless interact with the domiciled, then, they must contend with their stigmatization. We have observed four ways in which they do so. One is to attempt to pass as nonhomeless. Another is to use "covering" strategies which deflect attention from their stigma. A third is to engage in personal acts of defiance, while the fourth strategy is to engage in collective action.

Passing While the popular stereotype of a homeless man is that of a disheveled dirty panhandler who nurses a bottle of wine wrapped in a paper bag, many of the homeless do not fit this image. They may be able to escape being detected as homeless by presenting an appearance that masks their homeless status. In Goffman's (1963b) terms, they practice "information control," endeavoring to pass in public as domiciled individuals. When attempting to pass they must avoid "stigma symbols" that would reveal their true condition. So Ron, a 22-year-old young man who had come to Austin after being laid off from a restaurant job in Denver, and who occasionally visited the University of Texas campus to "watch the girls, and try to get them to talk to me" explained to us, "If you look like a bum, you can forget it, but if you look like a college kid, well, then at least you got a chance." Similarly, we were cautioned by a man who occasionally spent time dozing at the public library that, "You don't want to go there with a bedroll. They see that and they watch you like a hawk."

Just as the homeless find it useful to avoid stigma symbols when attempting to pass, they may also find it advantageous to display "disidentifiers" (Goffman 1963)—symbols that clearly suggest that they are not homeless. It is rare, however, that they can acquire such consumer items. Ron, for instance, in his efforts to present himself as a "college kid" at the university, spent a considerable amount of time attending to the styles and brand names of male students' attire. He lamented, however, that "The only places I can get clothes are down at St. Vincent's by the Sally or out at the church on Sundays, and they never have the good stuff. I thought about going to a laundromat and grabbing a few things out of a dryer, but with my luck I'd get caught."

Despite some potential interactional advantages, then, passing is difficult for the homeless. And even when one can accomplish it, it is more useful for avoiding negative attention than for fostering social opportunities, since the norms of public interaction generally encourage only superficial contact. Finally, as Ron realized, even if passing enabled him to establish interaction, it would ultimately lead to an impasse. "If a girl did talk to me," he mused, "then what?"

Covering While passing entails trying to hide one's homeless status, and thereby avoid stigmatization, covering offers an alternative approach in which one openly admits his or her stigmatized status, but seeks to minimize its impact. As Goffman (1963b, p. 102) has explained, "The individual's object is to reduce tension, that is, to make it easier for himself and others to withdraw . . . attention from the stigma, and to sustain spontaneous involvement in the official content of the interaction." The covering strategies used by the homeless can be separated into two categories: verbal strategies and the use of props.

Verbal covering is similar to the identity talk described above in that the homeless use it to cast themselves in a positive light. But verbal covering strategies differ from identity talk in that they occur in interactions with the non-homeless rather than among the homeless themselves and are frequently tied to requests for material assistance. Panhandling provides examples of a range of verbal covering strategies. While to casual observers panhandlers may appear comfortable in their role and oblivious to the normative violations it entails, most panhandlers find it awkward.[5] To buttress their public selves against possible negative imputations, and to facilitate the openness of others to their entreaties, many panhandlers offer accounts (Scott and Lyman 1968) and disclaimers (Hewitt and Stokes 1975) that mitigate the implications of their soliciting assistance from anonymous others on city streets.

Accounts, as described by Scott and Lyman, include two types. The first is "excuses," in which it is claimed that one's normative violation is beyond one's control and responsibility. An example from the streets is provided by a man with a broken arm who, when panhandling, rapped on his cast with his knuckles and explained, "I fell off a ladder on the job and broke it, now I can't work for six weeks." The second type of account is "justifications." In contrast to excuses, the underlying rationale for justifications is that while the behavior in question is generally negative, in the current situation it is morally acceptable, and perhaps even positive. Illustrative of this strategy is the case of two young men in their late teens who were on the streets with their 12-year-old brother. While the 12-year-old sat by their bags, the older brothers approached passers-by, saying, "Can you spare some change to get our little brother something to eat?" The moral stigma attached to panhandling in this situation was more than outweighed by their concern for their younger sibling.

Many panhandlers feel a particular need to allay the suspicions of potential givers that they will use the money for what may be perceived as improper purposes, such as buying alcohol. They make sure to mention how they will spend the money, as in the case of many of the homeless who worked the main avenue bordering the campus of the University of Texas. They would often ask for "some help to get a hamburger and some fries, since I haven't had anything to eat all day." Whatever their real intent, they believed this strategy alleviated the misgivings of possible donors. Yet other panhandlers took the opposite tack, proclaiming, "I'll be honest with you, man. I need some money for a six-pack"—an account designed to build moral credit by presenting oneself as trustworthy on the one hand, while admitting moral failure on the other. In each case, however, an effort was made to persuasively present oneself in a way that might neutralize other's resistance to one's entreaties.

Some individuals, especially those who are new to panhandling, preface their requests by disclaimers (Hewitt and Stokes 1975) intended to demonstrate their discomfort with having to engage in such counter-normative behavior. Through remarks such as "Man, I hate to ask you this" and "I never had to do this before, but . . ." the homeless person attempts to establish a common moral ground between himself and the potential donor before negotiating an exception. In a similar vein, a 33-year-old self-proclaimed "poet tramp" who went by the name of Rhymin' Mike regularly approached people near the university by asking, "If I could make a poem about you right here, would you be willing to give me a buck?" His offer of a quid pro quo arrangement served, at least in Rhymin' Mike's own mind, to distinguish his street solicitations from those of "panhandlers who don't give anybody anything for what they get"—a distinction that he found the police unwilling to accept, as they arrested him on several occasions for panhandling while he tried to sell his poems.

Verbal covering strategies by no means insure that the panhandler's audience will accept either the financial request or the identity assertion with which it is buttressed. Many people scurry by, attempting to ignore them entirely. Others may appear to accept identity claims, but decline to give money—offering their own excuses, such as that they are broke or out of change. Yet others may give assistance while refusing to go along with the panhandler's identity assertions, as in the case of a businessman in downtown Austin who we observed give a dollar to a panhandler, then gruffly curtail the panhandler's attempted account by telling him, "You got your money, just forget the bullshit."

The willingness of the domiciled to interact with the homeless is undermined, as the preceding example illustrates, by suspicions that they are being conned—that the accounts and disclaimers of the homeless are creative fabrication in their "genre of performance art amid other lively street phenomena" (Schwartz 1991, p. 64). While the con-game aspects of public solicitation by the homeless have been greatly exaggerated,[6] there are occasions when the homeless engage in a degree of misrepresentation—hardly surprising given the extensiveness of the stigma they must endeavor to overcome. Nonetheless, the giver generally grants a "fictional acceptance" (Davis 1961) of the deviant as normal. The precarious situation of the homeless in comparison to other stigmatized groups is revealed by the fact that, while fictional acceptance is common for other deviants as well, for many it may serve at least occasionally as a springboard to fuller acceptance and normalization (Davis 1961). The homeless, on the other hand, are very seldom able to move beyond fictional acceptance since they do not have the opportunity to build on successive sustained encounters with domiciled others—a prerequisite for moving beyond stereotype-based interactions.

Just as the homeless use verbal accounts and disclaimers in attempts to reduce the salience of their stigma, they may also use props as supportive imagery for positive definitions of themselves that deflect attention from their stigmatized status. A 45-year-old Arkansan who worked for years with traveling carnivals, for instance, carried a banjo in a case with the words "Wealth Means Nothing Without God" painted across it. While he seldom performed in public because he feared arrest for pandhandling, the instrument nonetheless served to announce his special identity as a musician and devout Christian.

Others sometimes used props more interactionally, for both social and material purposes. In the case of social purposes, props were chosen for their ability to create social openings with domiciled individuals. The most common prop that we observed was a small pet, such as a puppy, a kitten, and in one case a ferret. While at first glance it would seem incongruous for homeless individuals, whose own lives are precarious, to take on the care of a pet, two advantages were fairly clear. First, pets may provide a socioemotional bond for their owners (e.g., Corson, Corson, and Gwynne 1975; Friedman, Katcher, and Lynch 1980). For the homeless, who face tremendous social rejection and exclusion, the unconditional affection of a pet may be a very special comfort. Second, and less widely acknowledged, pets may serve as a nexus to social interaction with other people. Just as Robins, Sanders, and Cahill (1991) have documented that pets facilitate social interaction among middle-class pet owners in park settings, so we observed that people who would otherwise hesitate to interact with the homeless will sometimes stop to play with and talk about a pet. One example is the situation of Redbeard, a 40-year-old convicted felon, who after his release from prison lived for nearly a year in tramp jungles and abandoned buildings in Austin. Shortly after his arrival on the streets, Redbeard took the money he made from a day labor job and went to a local animal shelter to purchase a pup, which he frequently adorned with colorful bandanas and took to the university area where it attracted the attention of passersby, who would stop to play with the puppy and talk to Redbeard about it. One afternoon as we sat on a side street near the university with several homeless people, one of the young men we were chatting with watched a number of young women stop to pet the puppy. "Look at how they like that dog," he observed. "I'm gonna get me one. That way they'll hang around me too."

In many cases props are used for more material ends, such as to support public solicitation for assistance. Shotgun, a gray-haired artist and tramp in his mid-fifties who claimed to have a bachelor's degree from a Boston fine arts college, frequently spent several hours a day at a busy downtown intersection, holding up a sign proclaiming "Job Wanted—$3 and Food–*Anything!*" The sign was augmented with other props as well, including a ragged backpack and a bedroll. As Shotgun explained to us one winter morning over coffee as he warmed up from his stint on the corner, "You have to look the part for this to work." While the sign and other props seldom actually netted him a job, they did encourage donations without exposing him directly to the rejection often faced by panhandlers.

Occasionally we observed the creation of more elaborate props. In one case a group of long-term homeless who had established a camp beside a railroad trestle near a busy intersection set up a Christmas display on a small bridge. They hung several tattered wreaths and tinsel chains on the bridge and posted two signs, one proclaiming "Happy Birthday Jesus!" and the other announcing "Your Help Appreciated." The bridge was staffed regularly by two or three individuals who waved to commuters and gratefully accepted the frequently brisk business in donations. Part of the effectiveness of this strategy stemmed from the image the props portrayed of the camp's inhabitants: Christian, goodwilled, and friendly people.

As the preceding discussion demonstrates, the homeless (especially homeless men) use covering strategies such as accounts, disclaimers, and props to support their claims to morally acceptable identities and to announce their availability for interaction.[7] Covering strategies hit directly at the central dilemma of the homeless in public places: in order to maintain their dignity they must somehow overcome their position at the bottom of the status hierarchy. But no matter what they do, the lion's share of their encounters with the nonhomeless either directly or indirectly reflect their powerlessness and stigmatization. The homeless often accept these assaults upon their selves with despondent resignation, displaying a remarkable ability to tolerate rejection. But sometimes they reach their limits and respond with defiance.

Defiance Defiant behavior is essentially reactive, entailing actions and verbalizations meant to reject humiliating moral assaults or ridicule. It involves acts that, while appearing to be spontaneous responses to humiliating interactions, actually entail, as Katz has said of rage, "a precisely articulated leap to a righteousness, which logically resolves, just for the crucial moment, the animating dilemma" (Katz 1990, p. 30).

As Goffman (1961) has observed, defiance may be "open" or "contained." Open defiance is an overt and directly confrontational response to a humiliating encounter or situation. An example of this occurred late one afternoon while we were hanging out with our previously mentioned key informant, Ron, and several other young men in the downtown business area. As an attractive woman in a short dress approached us, Ron called to her, "Hey honey, looking good!" In response, the woman shifted her body and raised her head in a manner that signified her disgust with his remark. Stung by this public humiliation, Ron ran up behind her and yelled, "Rich bitch! Rich bitch! What do you think you're doing? Too good to talk to me, huh? Rich damn bitch!" The woman, obviously embarrassed by Ron's harassment, hurried down the street, and Ron returned to the group, proclaiming triumphantly, "That'll teach her."

Open defiance was usually directed toward the perceived source of a specific humiliating act, as in Ron's case or the numerous instances we observed of panhandlers responding defiantly to snubs and demeaning comments from their targets. Occasionally, however, when the source of humiliation was more diffuse, homeless individuals lashed out in more general defiance that was intended to counter the powerless and humiliating roles in which they had been cast. An example occurred one morning at the Texas Employment Commission as we waited with the homeless for day labor jobs. A young man in his mid-twenties, who suffered from a hearing impairment, was passed over for a job because he had not heard the employment counselor call his name. When he realized what had happened, he became angry and started cursing loudly to himself, which bothered several men who were sitting near him and with whom he got into a heated argument. Soon a security guard was called to order him off the premises. What had begun as a mistake by the employment counselor had quickly escalated to multiple confrontations and the evaporation of any chance to get a job, and the man stomped away from the building, screaming angrily, "You can't do this! You can't motherfucking do this to me! All I want's a fucking job!" Looking back at the security guard as he yelled, the man stepped into the street and was nearly hit by a car that honked as it passed within inches of him. Startled, the man jumped back, then, raising clenched fists, he walked into the middle of the busy street cursing at the cars and shouting, "Come on then, motherfuckers, hit me! Hit me!" For the next five minutes, until the police came to arrest him, he stood in the street, asserting his power to demand recognition. While he may have regretted his defiant performance once he was arrested, its enactment enabled him to transcend his powerlessness, at least for the moment.

Openly defiant acts are intended to counter snubs and imputations of negative identities, but, as noted above, they are likely to do this only temporarily at best. Moreover, defiant displays often reinforce others' negative perceptions of the homeless. In this regard, the situation of the homeless resembles that of another social category that suffers powerlessness in public—children. As Cahill (1990, p. 399) has observed regarding occasions on which older children engage in "pointedly disruptive and offensive acts" in public:

> [B]y pointedly violating their tormentors' injunctions, they can demonstrate—to themselves if not to one another—that they have some selfhood and personal autonomy. . . . Of course, they thereby also inadvertently cooperate with adults in expressively reproducing later childhood as a distinctive biographic state populated by irresponsible malcontents.

Similarly, when the homeless engage in acts of defiance, their actions are usually interpreted as indicators of traits associated with the master status of homelessness: mental illness, alcoholism, improper socialization, and criminality, rather than what Goffman (1971) has referred to as "meaningful nonadherence" to demeaning definitions of the self.

The relative dearth of open defiance among the homeless seems to stem from their sense that there is little to be gained by directly challenging the prevailing order and their place in it. The bulk of defiance that occurs is "contained defiance," in which the homeless more subtly and covertly express anger with and repudiation of their stigmatization. We witnessed far more instances of slighted panhandlers mumbling under their breaths than we did of them aggressively challenging their targets, and innumerably more cases of heated complaining to other homeless about bureaucratic unfairness than actual arguing with agency staff. In such instances we have what Goffman (1971, pp. 152–154) graphically referred to as an "afterburn"—an occasion in which an individual has given up the possibility of receiving appropriate remedial action for an interactional offense, but endeavors to convey to others around him that the conduct was unacceptable. Yet the nonconfrontational nature of such responses should not blind us to the fact that they also serve to mitigate humiliation, rattling the bars of one's cell, albeit not so loud that the guards might feel a need to respond.

Collective Action Collective action or protest constitutes the final means of stigma management employed by some of the homeless to overcome their material deprivations and powerlessness and to neutralize their stigma. Historically, stigmatized groups have often turned to collective action to advance their interests and to challenge stereotypic perceptions of them. Examples abound from the civil rights movement's articulation that "Black is beautiful," to the gay movements call for "gay pride." Because of their disadvantaged position, such stigmatized groups are often seen as being incapable of organizing and mounting effective protest. This is especially true of the homeless, who are typically portrayed as an unmobilizable population riddled by physical and emotional incapacities that minimize their ability to organize.[8] As one set of observers explained:

> Anyone who has worked with the homeless knows that they are not likely to organize well. They are usually physically exhausted, confused, and at least on the surface, apathetic. Few have any experience in organizing, and many have little experience in cooperating on a group level (Hope and Young, 1986, pp. 242–3).

Such characterizations are not fully consistent with much that transpired on the streets during the 1980s, however. In the first half of the decade, for example, homeless mobilization occurred independently in a number of cities across the United States. In some cases the homeless were organized by local activists or service providers. In other cases, homeless people organized other homeless. Collective action strategies included both disruptive protests and more institutional means. In some instances, the collective action was seemingly spontaneous and preceded the development of homeless social movement organizations (SMOs).

In other cases, SMOs directed protest activity in a coordinated fashion across a number of cities. By the end of the decade, incidents of homeless collective action were documented by the media in over fifty cities.

Collective action is often discussed in terms of its instrumental goals, which seek control over material resources, and expressive or symbolic goals, which seek to change public perceptions about a group or issue (Jenkins and Perrow 1977). Most collective action contains elements of both instrumental and expressive goals (Turner and Killian 1987), and that undertaken by the homeless is no exception. Although material goals, such as an increase in affordable housing, are a dominant aim, symbolic protest against stigmatization is also a central component of organizational action and a major motivation for individual participation. This is clearly evident in the names and mottoes the homeless often adopt for their organizations. A case in point is the Committee for Dignity and Fairness to the Homeless, a homeless organization in Philadelphia. The name of the organization, which constitutes a call for non-stigmatized treatment, grew out of the experiences of three men who were so outraged by their homeless ordeal, particularly their treatment by some service providers, that they began to organize other homeless people to address the degradation they suffered. The CDFH not only was instrumental in forming a national organization, the National Union of the Homeless, consisting of local homeless SMOs in over a dozen cities, but it spearheaded the Union's adoption of its motto: "Homeless Not Helpless." This motto has been prominently featured on protest placards and buttons and is an unmistakable symbolic assault on depictions of the homeless as unable to care for themselves or act on their own behalf. In a similar vein, a homeless organization in Minneapolis, the Alliance of the Streets, responded to the view of the homeless as passive by printing buttons with the organization's motto, "Look Out World, The Meek Are Getting Ready."

Since social service agencies usually attribute homelessness to personal failings and disabilities, the homeless rarely have input into the programs upon which they depend for day-to-day survival. Therefore homeless organizations have insisted on the participation of the homeless themselves. This constitutes a fundamental challenge to the stigmatization of the homeless, as articulated by a homeless man in Philadelphia.

> The union came about because we were tired of others speaking for us and setting our agenda. Homelessness is a fast-growing industry. Property pimps have sprung up all over, making money off our misery and we didn't have anything to say about it. We would have to be part of the decision making that governed our lives.

One of the leaders of the Oakland Union of the Homeless, discussing how they would recruit other homeless participants, reiterated similar themes.

> We started going out to the shelters and passing out flyers which basically said, "Are you tired of the dead-end lines at the soup kitchen? Are you tired of being treated like dogs in the shelters? Do you think that homeless

people should have a voice in how their destiny is shaped? If you do, come down to an organizational meeting to form the Oakland Union of the Homeless."

None of the participants who were interviewed saw their involvement with these organizations as a viable means off the streets for them personally. They clearly hoped that their actions would improve the conditions of being homeless and perhaps help to obtain low-cost housing for the homeless population as a whole. But the more pertinent reward for many of them was being involved with a group that made them feel a part of something positive related to homelessness. Indeed, the opportunity to collectively recapture a sense of dignity and power seemed to provide sufficient incentives for participation, as typified by a discussion between two men at a Tucson homeless union meeting on how to proceed with a trial of group members who had been arrested for taking over a house owned by the Department of Housing and Urban Development.

> [First man] No deals with HUD. They'll just use any concessions to hold over our heads. We're stronger when we can say "fuck you, man, we're going out into the streets." We can't let them take that away from us.
>
> [Second man] Yeah. No deals, no lawyers. We'll represent ourselves. We got a good mixture of the people and we can all talk from where we come from. It'll blow them away to see a convergence of the people saying, "It's all of us."

The foregoing exchange demonstrates the difference between acts of personal protest or defiance discussed earlier and collective action. Among many of the homeless who engage in it, collective action results in a sense of empowerment, positive group identity, and what Foss and Larkin (1986) have referred to as "disalienation."[9] At the same time, collective action of the homeless has in many cases led to changes in social policy, increased governmental and service agency responsiveness to the needs of the homeless, and participation of the homeless in institutions affecting their lives.

CONCLUSION

No group stands lower in the social order of urban America than the homeless. Lacking the resources to secure and maintain their own shelter, they are forced to conduct much of their daily lives in public space where they find themselves subject to pervasive stigmatization. In this chapter we have examined a wide range of strategies the homeless use to manage this situation. These include in-group strategies, such as drinking, cheap entertainment, hanging out, and identity talk, that function to take their minds off their situations and to recast their identities in a more positive light. They also negotiate their status with the domiciled through such out-group strategies as passing, covering, defiance, and collective action. . . .

NOTES

1. For more detail regarding all of the methods used in this study, see Snow and Anderson (1991, 1993).

2. The labor corner referred to is an unsupervised day labor exchange at a downtown intersection where many of the homeless in Austin gather in the hope of picking up casual employment, such as yardwork and construction site clean-up, from prospective employers who drive by.

3. This and all other names used in the text are pseudonyms for our street informants.

4. Other factors that may reduce the likelihood that individuals will call each other into question include lack of familiarity with others' biographies and experiences (Snow and Anderson 1987) and transitory semi-intimate encounters (Cavan 1966).

5. As a recent study in Tucson concluded, "the most consistent finding regarding how panhandlers felt about panhandling is that it doesn't come easily, and that it is not an enjoyable activity . . . it was repeatedly stated that panhandling is 'embarrassing' and that it takes 'courage and nerve' " (Costello, Heirling, Hunsaker, and McCulloch 1990, p. 31).

6. An extensive popular literature focused on the con games of the homeless has developed in most eras of widespread homelessness from the Middle Ages (Beier 1985) to the Great Depression in the United States (Anderson 1931) to the present ("Begging" 1988; Schwartz 1991). This literature, however, seems often to be sensationalistic and empirically inaccurate. See, for example, Beier's (1985) critique of the "literature of roguery" in Tudor England.

7. This use of props in public can be compared to that of the young women observed by Lofland (1973, p. 147), who used props such as books and magazines to demonstrate that they were tending their own affairs and explicitly *not* available for interaction. Clearly both class and gender play into the different uses to which props are put by homeless men and young women in public places.

8. See Blau (1992) as well as most descriptions of the homeless (e.g., Rossi 1989).

9. See also Wagner and Cohen (1991), who found that the homeless collective action participants they studied in a tent city protest in Maine tended to develop strong identification with each other. They also found that this identification continued long after the protest ended.

REFERENCES

Anderson, N. 1931. *The Milk and Honey Route: A Handbook for Hoboes.* New York: Vanguard.

Baxter, E. and K. Hopper. 1981. *Private Lives/Public Spaces.* New York: Community Service Society.

"Begging: To Give or Not to Give." 1988. *Time* (September 5), pp. 68–74.

Beier, A. L. 1985. *Masterless Men: The Vagrancy Problem in England 1560–1640.* New York: Methuen.

Blau, J. 1992. *The Visible Poor: Homelessness in America.* New York: Oxford University Press.

Blum, N. S. 1991. "The Management of Stigma by Alzheimer Family Caregivers." *Journal of Contemporary Ethnography* 20: 263–284.

Cahill, S. E. 1990. "Childhood and Public Life: Reaffirming Biographical Divisions." *Social Problems* 37: 390–402.

Cavan, S. 1966. *Liquor Licenses.* Chicago: Aldine.

Cohen, C. I. and J. Sokolovsky. 1989. *Old Men of the Bowery; Strategies for Survival among the Homeless.* New York: The Guilford Press.

Corson, S., E. Corson, and P. Gwynne. 1975. "Pet Facilitated Psychotherapy." Pp. 19–36 in *Pet Animals and Society*, edited by R. Anderson. London: Bailliere-Tindall.

Costello, B., J. Heirling, F. Hunsaker, and T. McCulloch. 1990. "Panhandling in Tucson." Unpublished paper, Department of Sociology, University of Arizona.

Davis, F. 1961. "Deviance Disavowal: The Management of Strained Interaction by the Visibly Handicapped." *Social Problems* 9: 120–132.

Derber, C. 1979. *The Pursuit of Attention: Power and Individualism in Everyday Life.* New York: Oxford University Press.

Foss, D. A. and R. Larkin. 1986. *Beyond Revolution: A New Theory of Social Movements.* South Hadley, MA: Bergin and Garvey.

Friedman, E., A. Katcher, and J. Lynch. 1980. "Animal Companions and One-year Survival Rate of Patients after Discharge from a Coronary Care Unit." *Public Health Reports* 95: 307–312.

Goffman, E. 1961. *Asylums.* Garden City, NY: Anchor.

———. 1963a. *Behavior in Public Places.* New York: Free Press.

———. 1963b. *Stigma.* Englewood Cliffs, NJ: Prentice-Hall.

———. 1971. *Relations in Public.* New York: Basic Books.

Hewitt, J. P. and R. Stokes. 1975. "Disclaimers." *American Sociological Review* 40: 1–11.

Hoch, C. and R. A. Slayton. 1989. *New Homeless and Old: Community and the Skid Row Hotel.* Philadelphia, PA: Temple University Press.

Hope, M. and J. Young. 1986. *The Faces of Homelessness.* Lexington, MA: Lexington Books.

Hughes, E. 1945. "Dilemmas and Contradictions of Status." *American Journal of Sociology* 50: 353–359.

Jackson, J. M. 1987. "The Bowery: From Residential Street to Skid Row." In *On Being Homeless: Historical Perspectives,* edited by R. Beard. New York: Museum of the City of New York.

Jenkins, J. C. and C. Perrow. 1977. "Insurgency of the Powerless: Farm Worker Movements, 1946–1972." *American Sociological Review* 42: 249–268.

Katz, J. 1990. *Seductions of Crime.* New York: Basic Books.

La Gory, M., F. Ritchey and J. Mullis. 1990. "Depression among the Homeless." *Journal of Health and Social Behavior* 31: 87–101.

Lamar, J. 1988. "The Homeless: Brick by Brick," *Time* (October 24), pp. 34–38.

Lofland, L.H. 1973. *A World of Strangers: Order and Action in Urban Public Space.* New York: Basic Books.

Love, E. G. 1958. *Subways Are for Sleeping.* New York: New American Library.

Robins, D. M., C.R. Sanders, and S. Cahill. 1991. "Dogs and Their People: Pet-Facilitated Interaction in a Public Setting." *Journal of Contemporary Ethnography* 20: 3–25.

Rooney, J. F. 1973. *"Friendship and Reference Group Orientation Among Skid Row Men."* Unpublished dissertation, University of Pennsylvania.

Rossi, P. 1989. *Down and Out in America: The Origins of Homelessness.* Chicago: The University of Chicago Press.

Rubington, E. 1968. "The Bottle Gang." *Quarterly Journal of Studies on Alcohol* 29: 943–955.

Schumer, C. E. 1988. "Homelessness: A Side Effect of Reaganomics." *The Arizona Daily Star* (March 13).

Schwartz, L. S. 1991. "Beggaring our Better Selves." *Harpers* 283: 62–67.

Scott, M. B. and S. M. Lyman. 1968. "Accounts." *American Sociological Review* 33: 46–62.

Snow, D. A. and L. Anderson. 1987. "Identity Work among the Homeless: The Verbal Construction and Avowal of Personal Identity." *American Journal of Sociology* 92: 1336–1371.

———. 1991. "Researching the Homeless: The Characteristic Features and Virtues of the Case Study." Pp. 148–173 in *A Case for the Case Study*, edited by J. Feagin, A. Orum, and G. Sjoberg. Chapel Hill, NC: University of North Carolina Press.

———. 1993. *Down on Their Luck: A Study of Homeless Street People*. Berkeley, CA: University of California Press.

Spradley, J. 1970. *You Owe Yourself a Drunk: An Ethnography of Urban Nomads*. Boston, MA: Little, Brown.

"Toy Stores Pull 'Tramp' Doll Seen as Degrading Homeless." 1990. *American Journal of Sociology* 92: 1336–1371.

Turner, R. H. and L. M. Killian. 1987. *Collective Behavior*, 3rd ed. Englewood Cliffs, NJ: Prentice-Hall.

Wagner, D. and M. B. Cohen. 1991. "The Power of the People: Homeless Protesters in the Aftermath of Social Movement Participation." *Social Problems* 38: 543–561.

Wallace, S. E. 1965. *Skid Row as a Way of Life*. Totowa, NJ: Bedminster Press.

Wiseman, J. 1970. *Stations of the Lost: The Treatment of Skid Row Alcoholics*. Chicago: University of Chicago Press.

Wright, J. D., and J. Lam. 1987. "Homelessness and the Low Income Housing Supply." *Social Policy* 17: 48–53.

27

Collective Stigma Management and Shame
Avowal, Management, and Contestation

DANIEL D. MARTIN

The way organizations help people collectively manage their stigma is the focus of Martin's study of three stigma management groups associated with obesity: Overeaters Anonymous (OA), Weight Watchers (WW), and the National Association to Advance Fat Acceptance (NAAFA). Martin compares the way these groups frame their purpose and set their goals. While OA and WW help members navigate their way through the world and conform to the prevailing body norms, NAAFA fights these norms and seeks to redefine them. Martin ties the characteristics of these organizations to the way they deal with shame, avowing it, managing it, and contesting it. The selection offers an excellent comparative view of the varied rationales and strategies associated with differently structured collective stigma management groups.

. . . Both cross-cultural research and feminist studies have observed the connection between shame and gender identity, demonstrating that, while women are more susceptible to body shame, for men the emotion is associated with failure to live up to culturally prescribed norms of masculinity (Gilmore 1987; Horowitz 1983).

Feminist scholars investigating the U.S. beauty culture have observed its emergence in the 1870s (Banner 1983; Schwartz 1986) as new appearance norms were created and distributed by industries such as cinema (Featherstone 1982), advertising (Ewen 1976), and scale and corset companies (Schwartz 1986). As a result, the social formation that exists at the end of the twentieth century is a "cult of thinness" (Hesse-Biber 1996) that promotes and normalizes slenderness and its attendant anxieties (Bordo 1993). This cult has far-reaching effects. Barrie Thorne and Zella Luria's (1986) research on children's games involving cross-sex interaction reveal that girls' bodies are more sexually defined and accrue more penalties for an overweight appearance than do boys'. This pattern is consistent with findings of cross-cultural research on mate selection

---

From "Organizational Approaches to Shame: Avowal, Management, and Contestation," by Daniel D. Martin, *The Sociological Quarterly*, Vol. 41, No. 1, 2000. © 2000 by The Regents of the University of California. Reprinted by permission of University of California Press Journals.

that demonstrate that men evaluate the social value of women more heavily in terms of appearance and that women evaluate the social value of men in terms of their occupational status and earning potential (Buss 1989). In the United States, where fat is stigmatizing (Allon 1982; Cahnman 1968), feelings of shame may result in debilitating eating disorders. Research on adolescent girls has demonstrated that the internalization of beauty norms is responsible for producing high degrees of body dissatisfaction and a profound sense of body shame (McFarland and Baker-Baumann 1990) leading to bulimia and anorexia nervosa (Rodin, Striegel-Moore, and Silberstein 1990, p. 362). Of course, the fat content of bodies, along with the degree to which it might be stigmatized and accompanied by body shame, is variable. A slightly overweight body may be experienced as an object of great shame and the focus of disordered eating, or it may be experienced as only a "minor bodily stigma" (Ellis 1998).

However, . . . the research on . . . feminist studies of body shame or sociological studies on shaming as a form of "strategic interaction" (Goffman 1969) have [not] addressed how organizations attempt to assist their members in managing shame. . . .

Drawing upon Erving Goffman's (1986) frame analysis, I assess how shame is managed within three different "appearance organizations" through the discursive and bodily strategies that they supply their members. By appearance organizations, I mean organizations for whom the physical appearance of members is a primary concern. These organizations include Overeaters Anonymous (OA), Weight Watchers, Inc. (WW), and the National Association to Advance Fat Acceptance (NAAFA). The theoretical and substantive questions concerning these organizations are twofold. First, what kinds of organizational frames do these organizations offer their members to make sense of shame? Second, what strategies are used within the organizations in managing or contesting experiences of shame?

THE ORGANIZATIONS

Given the sex ratio of the organizations, all three of the appearance organizations I studied might properly be recognized as "women's organizations." Therefore, the present analysis of shame is an analysis of women's shame. During interviews and participant observation, I found that men rarely talked about shame or embarrassment over the body while women seemed to be continually, acutely aware of it. Based upon information reported by the organizations, participation rates reveal significant sex differences in membership (Table 1). These differences were also revealed in the accounts that women and men constructed about joining and participating in the organizations and about the gender relations that were negotiated within them—a topic beyond the scope of the present article (Martin 1995).

At the beginning of data collection, Weight Watchers International (WW) was the leading national weight loss organization in the United States, with

Table 1 Organizational Membership

	Sex Differences in Participation		
	WW	NAAFA	OA
Women	95%	78%	86%
Men	5%	22%	14%
Total*	100%	100%	100%

*These figures represent national membership rather than participation rates in local chapters.

annual revenues of approximately $1.3 billion (Weber 1990, p. 86). Of its fifteen million members worldwide, approximately 14.3 million are women. In contrast, the exact size and composition of Overeaters Anonymous (OA), a program for compulsive overeaters, remains undetermined, even though the organization estimates that 86 percent of its U.S. members are female (OA 1992, p. 1). Because anonymity in the twelve-step program is strictly enforced, there is no comprehensive roster of OA members. Currently, the number of OA groups is estimated at over seven thousand worldwide (OA 1987, p. 10), but group size varies from five to twenty-five people, making estimates of total membership difficult.

The National Association to Advance Fat Acceptance (NAAFA) describes itself as a civil rights organization in the "size rights movement." It also defines itself as a social support organization and as a self-help group (NAAFA 1990, pp. 1–6). Unlike members of OA and WW, the NAAFA members in the local chapter I studied engaged in a variety of social and political activities. These ranged from informal parties and organized dances, to speaking at community conferences, to holding protests and conducting write-in campaigns. The national organization, founded in 1969, has headquarters in Sacramento, California, and a membership of 2,500–3,000 members. NAAFA consists of a loosely federated set of local chapters and special interest groups, including gay and lesbian groups, fat admirers groups, couples groups, singles, teen and youth groups, groups for super-sized and mid-sized individuals, and feminist groups.

METHODS

Over the course of approximately two years, I conducted participant observation and in-depth interviews at Weight Watchers and NAAFA. In attempting to gain access to OA, I sought out an acquaintance and long-standing group leader in OA who was centrally located in local OA networks. She was helpful in providing me with extensive information about the organization, its program, and its local chapters. Believing that her interest in my project, her sensitivity to research needs, and the fact that we had worked together indicated an openness to facilitating entree, I directly broached the topic. My

Table 2 Interviews

	Women	Men
WW	20	5
NAAFA*	8	4
OA**	41	8
Total	69	17

*Defines the entire population of active chapter members during the period of my observation.
**Includes recorded personal stories of members from the OA audiotape library.

request for an introduction into the closed meetings and access to OA members was met, instead, by recalcitrance motivated by her desire to ensure the anonymity of local members. I was frozen out of the local chapters and was unable to conduct participant observation in groups other than the publicly available "open meetings." A couple of months later, I learned of a local OA audiotape library. I selectively sampled personal stories and presentations that had been taped at national and local conferences and then were produced for OA members as well as members of Alcoholics Anonymous. I utilized the strategy of "theoretical saturation" recommended by Barney G. Glaser and Anselm L. Strauss (1967), selecting audiotapes for analysis until the additional data yielded few additional insights. I was assisted by the proprietor of the service, who was instrumental in locating tapes containing life stories that were as complete as possible.[1] I conducted interviews with all of the active members of NAAFA after I had established myself with the group and had been conducting participant observation and with members of Weight Watchers who responded to ads that I had run in local newspapers. The total number of interviews for all three organizations is in Table 2. Because the organizations themselves constituted the sampling frames, little variation in race, class, or ethnicity was provided within the samples. All of the informants in this study were white and most were middle income.

DATA COLLECTION, EMBARRASSMENT, AND SHAME

Because my own bodily appearance serves as a "dis-identifier" (Goffman 1967), a symbol belying a legitimate, potential, future claim for group membership in WW, OA, and NAAFA, I intentionally gained about twenty-five pounds before participation. I hoped to learn as much as I could about members' own meanings and feelings of the body and participation as well as how the organizational "frames" that were established shaped emotional experience and participation. Having gained my undergraduate degree on a wrestling scholarship attuned me to the issues of extreme dieting, binge-eating, and weigh-ins, which could similarly be found in the lived experience of Weight Watchers, NAAFA, and OA members. While my own weight gain facilitated somatic insight into the lives and meanings of members of OA and Weight Watchers, my burgeoning body size was clearly not comparable to the experience of members of NAAFA. Because I was not subjected to the stigmatization or public inconvenience suffered daily by NAAFA members (some of whom weighed in excess of three hundred pounds) my own "fat identity" never fully crystalized. Indeed, if anything, I was considered a novelty among friends, peers, and colleagues who issued positive sanctions for my "sociological commitment."

At Weight Watchers, I regularly attended weekly meetings, which entailed a logical stream of activities once members entered the meeting place. These included paying weekly dues, being weighed on physician's scales by WW personnel, receiving organizational literature concerning meal plans along with myriad other promotional materials distributed by the organization (such as the monthly newspaper), and finally selecting a seat in the meeting area. As a participating observer, I experienced the anxiety that members later recounted in interviews about "facing the scale," that is, weigh-ins. Because weigh-ins take place in semipublic space, it is possible that queuing members will learn of one's weight, increasing the anxiety that is already present for some members. Having failed weigh-in several times by gaining weight, I was struck by the capacity for the ritual to evoke, simultaneously, feelings of dependency and embarrassment. WW personnel are aware of these feelings, and they commonly query deviating members about the possible causes of weight gain. They also join in the construction of accounts and remedial work directed at either exonerating deviating members or diminishing their presumed culpability.

ORGANIZATIONAL FRAMES

Experiences of shame, humiliation, and embarrassment can be found in the life stories of members of all three appearance organizations. What establishes the meaning of these experiences is the organizational "frame" (Goffman 1986) within which the experiences of members are socially and discursively organized (Table 3). A "frame," according to Goffman (1986, pp. 10–11), is a set of "definitions of situations" that are "built up in accordance with the principles of organization which govern events—at least social ones—and our subjective involvement in them." *Organizational* frames are those definitions constructed and maintained by organizational actors within which experience, interaction, and communication are structured and rendered both personally and organizationally meaningful. For the programs of organizations to be subjectively meaningful, they must facilitate a linkage of individual interpretations to organizational meanings and definitions, that is, "frame alignment" (Snow, Rochford, Worden, and Benford 1986).

Table 3 Shame Work: A Comparison of Organizational Frames

	Overeaters Anonymous	Weight Watchers	NAAFA
Type of organization	Twelve step program	Multinational corporation	Civil rights organization
Organizational frame	Redemption	Rationality	Activism sociality
Organizational goal	Support members in abstinence from overeating	Help clients lose weight/profit	Organize against "size discrimination"
Source of shame	Compulsive over-eating/fat body	Fat body/body image	Societal definitions of beauty, cultural appearance norms
Consequence of shame	Inhibits recovery	Leads clients to exit program	Inhibits activism
Shame strategy	Shame avowal/expiation	Shame management	Contestation of shame
Approach to shame	Self-transformation	Bodily transformation	Societal transformation
Ritual for shame removal	Shame avowal	Dieting	Identity announcement/public confrontation
Meaning of the body	Body as symptomatic display of spiritual deficit; thinner body as a symbol of regaining control over personal, spiritual, and emotional life	Body as a site of contestation over appetite: locus of self-control and self-indulgence	Body as symbol of self-acceptance; political transformation
Body as medium of communication	"Body announcement" used as evidence of program efficiency; **thin** bodies as part of personal testimonies	"Body announcement" via fashion shows; display and glorification of **thin** bodies	"Body announcement" via fashion shows; display and glorification of **fat** bodies
Vocabulary of motive	Overeating as disease	Overeating due to irrational management and lack of education; program as a skilling process for a total lifestyle change	Genetic accounts: large body a product of nature; program for empowerment; stories as "Oppression Tales"

In the case of these appearance organizations, the meanings, as well as the strategies that are dispensed to members for dealing with shame, reflect the structure, agenda, and ideology of the organization, its "frame." The organizational frames of Weight Watchers, OA, and NAAFA represent three unique sets of social and discursive practices developed in accordance with different historical circumstances and audiences. Thus, while the autobiographical content of participants' shame experiences seemed to bear great similarity across the organizations, the "organizational" meaning of these experiences and the strategies employed to manage or alleviate them differed vastly.

The Organizational Frame of OA: Redemption

The Overeaters Anonymous program is a redemptive model of treatment directed at developing a spiritual consciousness though a therapeutic group process. OA meetings, whether open to the public ("open meetings") or open to OA members only ("closed meeting"), are typically held in public facilities, often churches. Meetings begin with a recitation of OA's preamble and the evening's speaker testifying to the effectiveness of the OA program in her or his life. Most meetings end with members joining in the "serenity prayer."[2] OA explicitly acknowledges its reliance upon the Alcoholics Anonymous program of recovery, including the Twelve Steps and the Twelve Traditions developed by Bill W. and other AA founders. Reproduced in most literature published by the organization, the twelve-step program serves as a blueprint for achieving abstinence from compulsive overeating (OA 1990, p. 114).

The Twelve Steps of Overeaters Anonymous

1. We admitted we were powerless over food—that out lives had become unmanageable.
2. Came to believe that a power higher than ourselves could restore us to sanity.
3. Made a decision to turn our will and our lives over to the care of God *as we understood him*.
4. Made a searching and fearless moral inventory of ourselves.
5. Admitted to God, to ourselves and to another human being the exact nature of our wrongs.
6. Were entirely ready to have God remove all these defects of character.
7. Humbly ask Him to remove our shortcomings.
8. Made a list of all persons we had harmed, and became willing to make amends to them all.
9. Made direct amends to such people wherever possible, except when to do so would injure them or others.
10. Continued to take personal inventory and when we were wrong, promptly admitted it.
11. Sought through prayer and meditation to improve our conscious contact with God *as we understood him,* praying only for knowledge of his will for us and the power to carry that out.
12. Having had a spiritual awakening as the result of these steps, we tried to carry this message to compulsive overeaters and to practice these principles in all our affairs.

Essentially the organization prescribes a spiritual program of recovery for what it considers to be an "incurable disease"—compulsive overeating. The concept of disease represents OA's primary "vocabulary of motive" (Mills 1940) that alleviates members of responsibility for their unhealthy overeating: "OA believes that compulsive overeating is an illness—a progressive illness—which cannot be cured but which, like many other illnesses, can be arrested. . . . Once compulsive overeating as an illness has taken hold will power is no longer involved because the suffering overeater has lost the power of choice over food" (OA 1988, p. 2). In defining "compulsive overeating," OA tells the prospective member that "in OA, compulsive overeaters are described as people whose eating habits have caused growing and continuing problems in their lives. It must be emphasized that only the individuals involved can say whether food has become an unmanageable problem" (OA 1988, p. 2).

The redemptive aspect of OA's program is both ideological and structural, as well as observable in relationships that members have with their sponsors. The alignment of personal with collective definitions of compulsive overeating is accomplished, in part, with the assistance of an OA sponsor, a person with substantial "program experience" who serves as the member's confessor, friend, advocate, and spiritual leader as well as a purveyor and mediator of organizational ideology. Sponsors, along with other OA members, are part of the emotion management team that members may use in expressing and policing emotions and securing social support. Redemption from compulsive overeating and an obsessive-compulsive self, according to OA members, is facilitated by sponsors and fellow group members as one "works through denial"—that is, gives up one's idiopathic justifications for compulsive overeating and the avoidance of unpleasant emotions in favor of collective definitions and meanings found in OA's twelve steps and twelve traditions.

The Organizational Frame of NAAFA: Activism

Throughout its history, NAAFA has mobilized lobbying efforts for antidiscrimination bills with state legislatures,[3] held rallies, conducted protests,[4] and brought cases of discrimination to the attention of the mass media. The objective of such action is the creation and enforcement of human rights provisions under which fat people might be identified as a protected group. Yet a bifurcation of NAAFA's primary objectives can be found in the orientation literature for new members: NAAFA refers to itself as both a "human rights organization" (NAAFA 1990, p. ii) and "a self-help group" (NAAFA 1990, p. 2). For NAAFA members both intensive involvement in social activities and activist participation constitute the dominant frame of meaning or, in the words of NAAFA members, "a way of life." Activism, as defined by the organization, includes a broad array of personal and political activities that are marked by a displayed willingness to engage in confrontation. Such activism, as described by the organization (NAAFA 1990, p. 4-3), may include:

> *Personal Activism:* Educating people around you, not letting negative comments about your weight or preference go unchallenged, and your life as a "role model" are all forms of personal activism.

Legislative Activism: is specific activist work that leads to changes in the law. Considering the amount of time, effort, and expertise required in such undertakings, it deserves a category all its own.

Advocacy Activism: includes letter writing and other forms of communication with the "powers that be" regarding your opinions. Did you like how the fat person was portrayed on the TV show? Are you happy with the article in the magazine? Did the salesperson treat you with disrespect because of your fat? Writing letters of praise or protest to the station, the magazine, or the store owner are examples of advocacy activism.

The organizational frame of activism in NAAFA represents a way in which participants situate themselves, emotionally and practically,[5] within the ideological contours of the organization. It is a "relevance structure" through which not only social or political events but also interpersonal lines of action are defined, evaluated, and constructed.

The Organizational Frame of WW: Rationality

The weekly Weight Watchers meetings were held in weight loss centers located in shopping malls, high-rise business complexes, and basements leased from other companies. In contrast to OA, where developing a meal plan and taking inventory of daily consumption is strongly encouraged though largely voluntary, these activities are preestablished, compulsory features of WW's program. The official purpose of the program is to help clients achieve bodily reduction. Hence, emotional and spiritual "recovery" is not pertinent to WW clients as it is for OA members, and bodily reduction for clients of WW is largely a matter of technocratic administration or "technique,"[6] not spiritual practice.

The materials distributed to new clients during the first week of WW membership include a basic program guidebook (containing a multi-item questionnaire assessing eating patterns, levels of desire to succeed in Weight Watchers, and the degree of stress in one's life), a "food diary" that includes daily menus, and a list of items from various food groups. In each subsequent week, members are given additional diaries to keep along with new daily menus and comprehensive guides on various topics including exercise, eating out, socializing, tips for better eating, and dealing with obstacles to weight loss. "Portion control" (the allotment of measured meals items for which WW markets food scales, frozen entrees, and cookbooks) is the fundamental principle underlying its program.

While the topic of compulsive overeating may receive a modicum of attention within meetings, the term cannot be found within WW's literature. Instead, the term "volume eating" is used, and what is emphasized is the type rather than the amount of food consumed. "Portion control" is to be applied within those food groups that are seen as leading to the creation of fat bodies but can be ignored within food groups whose caloric content is negligible. Thus, according to WW, there are no good or bad foods, only those necessitating more or less portion control. . . .

SHAME WORK

In *The Managed Heart,* Arlie Russell Hochschild (1983, p. 7) described "emotion work" as labor that "requires one to induce or suppress feeling in order to sustain the outward countenance that produces the proper state of mind in others." Hochschild's definition pertains to the self-monitoring, management, and modeling of emotion as a strategy used by service personnel to evoke certain feelings. But other activities in which people engage (to elicit, constrain, and manage the emotions of others or encourage their self-management) might also be classified as emotion work, given a less restrictive definition of the term.[7] For example, confronting other people about the discrepancy between their verbal actions and nonverbal cues, such as the tone in their voice, or pointing out apparent discrepancies between seemingly frustrating situations and a lack of emotional response (as family therapists can attest) are very much work and can be emotionally exhausting. Hence, evoking emotional expression within WW's men's meetings might be considered emotion work, while displacing or channeling such expression constitutes work within women's meetings. Within OA, "working through" denial is a form of emotional labor, while for members of NAAFA, emotion work includes attempting to impassion members over issues of weight-based discrimination.

Within all three organizations, a common form of emotion work is "shame work," that is, emotional labor aimed at evoking, removing, or managing shame—though in the contemporary contexts of OA, WW, and NAAFA only the latter two objectives are sought. Within OA and WW, shame work involves activities directed at mitigating the internalized experience of shame that emanated from the personal and social practices of consuming food. Within NAAFA shame work was directed at removing shame that had been acquired through the stigmatization of a "fat identity." Like other forms of emotional labor, shame work includes communicative and expressive action that may take both discursive and bodily forms. In the following sections, I discuss the use of shame avowal (OA), shame management (WW), and shame contestation (NAAFA) and how these are accompanied by use of "body announcements" in managing and systematizing members' shame.

Shame Work in Overeaters Anonymous

For members of OA, patterns of compulsive overeating and resultant body size are seen as a source of shame. Yet this is "only secondary shame" insofar as these patterns have developed as a result of some earlier "primary shame" experience invoked by others. Both female and male members commonly located the development of primary shame experiences in early childhood, citing failures at living up to adult expectations as the reason for their present "shame-based" state. Primary sources of shame, as OA members define them, are profoundly social, most often residing in interaction episodes with significant others who have knowingly or unintentionally evoked it. One strategy used by OA members engaged in shame work might best be termed "shame avowal," an open acknowledgement of past and present shame experiences that serves to expiate the shame experience. Such avowals are commonly woven into the public testimony of OA members:

> Jason (open OA meeting): I also need to say I've never had one perfect day of abstinence, there is no such thing for me. You know there is no perfect abstinence. I've lost weight in this program. I've gained weight in this program. But I have had enough shame over my food and my weight to last me the rest of my life. I don't need to put more on me. You know my goal weight . . . somebody asked me at a convention what my goal weight was. And I said. "My goal weight, if I'm abstinent, it's exactly what I weigh today."

The acknowledgment of shame is cited by OA members as problematic because it is premised on a self-awareness that, if not already present, must be organizationally manufactured. As indicated by one member:

> Rob (OA conference workshop): The second part [of dealing with shame] is to recognize it. To even find out that we have it. And as far as I'm concerned that is the most difficult part of it. To recognize that we have it is, is the thing that's so easy to cover up with denial. I've been in the OA program for about five years and AA about six years. And it took a long time for me to discover that I had any shame at all.

Determining whether the self exists in a deep shame-based state is a subjective evaluation, but the evaluation is also linked to a collective assessment as OA sponsors and other group members attempt to facilitate the awareness of shame-based actions on the part of nascent members. While such an awareness involves the acquisition and application of an organizational vocabulary of motive that is acquired through socialization in OA, it also necessitates the presence of viable experiences and circumstances to which it can be applied. As in other twelve-step programs (Denzin 1987; Rudy 1986), shame experiences are recognized and formulated as members of Overeaters Anonymous pragmatically invoke the organizational vocabulary for understanding and overcoming patterns of compulsive overeating. As it is applied to shame, the motive of denial is used in retrospectively interpreting compulsive overeating as behavior connected to a shame-based state. Working through denial is facilitated by "hitting bottom." Incidences of binging, stealing food, or lying about consumption, which lead to acute intra-personal states of shame, degradation, and humiliation, may all be defined subjectively as "hitting bottom," that is, reaching the lowest possible point in one's life spiritually, emotionally, and socially.[8] According to the redemptive frame of the OA, salvation from a shame-based state of compulsive overeating is only possible after one acknowledges both to self and others the causes and consequences of compulsive overeating and then turns one's life over to a "higher power." As in Alcoholics Anonymous, the concept of a "higher power" is one that is idiosyncratically constructed,[9] yet members are encouraged to let it intervene in their lives and assist them in working through their denial of compulsive overeating. . . .

Shame Work in Weight Watchers

Within Weight Watchers, NAAFA, and OA, body shame is commonly discussed by women. Female members in OA also cite shame experiences in episodes of emotional, physical, or sexual abuse. By contrast, most male members of Weight Watchers and NAAFA appear to be exempt from experiences of body shame: neither did male members in OA seem to share experiences of body shame though they mentioned shame experiences rooted in a general failure to live up to a multiplicity of social expectations. Weight Watchers group leaders claim body shame as exclusive to the experience of female members:

> Robin: Men are not as shameful about their bodies. They don't care if they are thirty pounds overweight, they go ahead and take off their shirts at the beach anyway. Women just aren't able to do that—we're ashamed to wear bathing suits and show off our cellulite or flabby legs or varicose veins at the beach. We are overwhelming told we can't do that, that we must be slim.

As a former, longtime member of Weight Watchers and now group leader, Robin claims that body shame is not only central to the experience of women vis-à-vis men but that being female makes one susceptible to body shame in ways that men are not. Some men in WW, however, did discuss similar experiences:

> Carl: For phys-ed class we had to go swimming. And I always got the largest swim suit. And people would just laugh at me and call me "fatso" and, you know, names. And then if I went to lunch no one would sit with me.

Carl does not specifically identify a personal sense of shame as a residual of this experience but he cites feelings of profound differences induced by episodes of stigmatization. While it seems reasonable to expect the presence of shame in episodes such as the one above, it remained liminal in the experiences of the men I interviewed. In contrast to the OA strategy of shame avowal, the work of WW personnel consisted of shame management that is, attempts to neither deny, contest, or specifically avow the experience but to contain it in hopes of a future, natural dissipation through dieting and weight loss. The management of shame represents a change in WW's treatment ideology over the course of its thirty-year history. Until the 1970s the organization had relied heavily upon the "shame-aversion" strategy (Serber 1970) of "card calling" in modifying the eating behaviors of clients.[10] Card calling involved placing a client's name along with the number of pounds that the client had either gained or lost during the week on an index card. The cards of all clients would then be read publicly at the beginning of each meeting. Similar instances of shaming appear to have been used in other weight loss organizations during this same period. Marcia Millman's *Such a Pretty Face* (1980) reveals that the shaming ritual used to discipline candy-sneaking "kids at fat camp" (children at summer weight loss camps) was to extract public confessions from these children on stage, humiliating them in front of their peers. Earlier research on the organization Take Off Pounds Sensibly (TOPS) reported the use of an equally drastic strategy for motivating clients to lose weight:

> There is a public announcement of each member's success or failure during the previous week. When a client has lost weight, there is applause and much vocal behavior; if she has gained weight, there is obvious diapproval with booing and derision. . . . In some groups the members who has gained the most weight is given the title of "Pig of the Week" and may be made to wear a bib with a picture of a pig on it or may have to sit in a specified area called the "pig pen." (Wagonfeld and Wolowitz 1968)

The loss of Weight Watchers revenues, which resulted when clients terminated their membership after such experiences, proved to be a compelling reason for abandoning shame aversion strategies in favor of strategies directed at its containment. One former client and now group leader of Weight Watchers recounted the outcome: "Those were the old days. Jeez, can you imagine how uncomfortable everyone was? We found that half the group had left before we even got seated." However, the absence of shame talk in WW meetings also reflects corporate concerns for liability. The training provided group leaders does not equip them to conduct psychotherapy or deal intensively with revelation of traumatic events such as sexual, emotional, or physical abuse that might produce shame. In light of episodes where women may make such disclosures, WW developed a strategy that allows organizational personnel to maneuver efficiently through the most troublesome situations. When I asked a group leader what happens if people divulge very personal or very traumatic information that is related to weight loss, she replied:

> The line is drawn that you never touch anything that is not directly affecting the members' weight. If they can't verbalize it you move on. . . . You don't touch it, you let them tell you. And they'll take care of it. It's just amazing. I've never been as scared as when that woman remembered in the meeting having been raped. Because I didn't know what to do with her. And the question that we tell leaders and that we practice in any group—men, women, any kind of group that we're doing—is, "And how does that affect your weight control?" That's your appropriate follow up, no matter what they say, if you think you're into something that you're not qualified for, you let them answer and then you let the group take care of that person. . . . You know, there's always someone that's gonna say, "Well, gee, you really need counseling for rape." So that's the point, once you've thrown it back to them, to back off. . . . you don't have anything to offer them. You're not qualified. If we could just know what we're doing we could do so much more. And we do have an awful lot of psychology background given to us by the company but we can't use it, we're not really qualified.

Here, the emotion work of leaders must strike a precarious balance between the emotional resolution of members' personal problems and corporate exposure, which writes large for group leaders as "uncertified" professionals.

This is accomplished by displacing the responsibility for resolving emotionally charged situations onto members themselves. Thus, group leaders do shame work as well as other kinds of emotion work but always within limits rationally guided by concerns for organizational liability.

The emotion work now done by WW personnel in women's meetings consists of assisting in the management of client's shame experiences. Shame management in WW largely included remedial work done during weight loss meetings as well as teaching clients intrapersonal strategies for minimizing body shame. Remedial work within the social context of group meetings primarily involved supplying clients with viable disclaimers of excuses (Hewitt and Stokes 1975; Lyman and Scott 1970) for program violation. An example of the latter is demonstrated in the following interaction:

> Leader: How did everyone's week go this week?
>
> Client 2: [female, approximately age 24] I almost didn't come in tonight because I was afraid to face the scale.
>
> Leader: Ohhhh . . . what happened?
>
> Client 2: I was really stressed this week and I ate everything in sight. I went off the plan, I'm afraid. I kept saying to myself [laughs] I wonder if they take your ribbon back [a red ribbon is given to clients after they lose their first ten pounds].
>
> Leader: Okay . . . but the positive thing is that you came back this week, see . . . you could have done a lot of other unhealthy things to yourself than just eat. Does anyone have anything to say to [name]?
>
> Client 3: (Woman looks at client 2) Well, tell her why you were so stressed.
>
> Client 2: Well, I went to visit my mother and there was food every where. My future brother-in-law made it to the finals at the state wrestling tournament, and a cousin who I was very close to died this last week.
>
> Client 3: I think you need to give yourself a break.
>
> Client 4: I think you are being way too hard on yourself, you shouldn't worry about the program [WW] with all that you're going through.
>
> Leader: Like I said, I think it's very positive that you *did* decide to come back this week, knowing that you may have not stuck with the program. I think that's just positive in itself. [To other clients] Don't you agree? [Everyone nods their heads yes.]

Here, the group leader enlists the support of other group members in ratifying the excuse. Because bodily monitoring means heightened awareness of body size, it evokes body shame for some members. In such cases, group leaders also recommended nondiscursive strategies for dealing with shame:

> Margaret (group leader): I had one woman who was ashamed to find out how many inches she had lost every week and so instead of using a measuring tape would use different colored yarns to measure her bust,

hips, waist, thighs, calves, and upper arms. Each week, this member would cut the tails off these pieces of yarn according to how many inches she had reduced and therein came to amass a pile of different colored yarn. She could then hold and feel the cut off yarn which represented her weight loss.

In suggesting that clients objectify their weight loss, the group leader offers a technique by which clients might manage shame and gain a sense of control over body size and weight. The tactile strategy offered above is designed to contain or manage rather than avow shame. . . .

Shame Work in NAAFA

In contrast to avowing or managing shame, NAAFA members contest it. The objective of activisim as presented in NAAFA's national newsletter is to "Change the World, Not Your Body" (Wolfe 1992, p. 5). Such contestation can readily be seen in the communicative but nondiscursive practice of making a "body announcement," where the fat body is displayed as shameless. NAAFA members make a public avowal of a "fat identity" not only through these displays but also through forms of public and private confrontation. Letter-writing campaigns and confrontations with public officials, as well as proprietors of businesses and services who engage in size discrimination, comprise part of the protest activities of NAAFA members. No social context is considered exempt from "fat activism." What is formulated in these activities, according to NAAFA members, is a way of life, an organized set of attitudes, and a mode of responding to myriad situations that members face daily.

> Beth: Like one time they were making cracks in the lounge about eating, well it was at lunch time and, oh God, this one woman said, who's always dieting, "If we eat this we'll all have to shop at Women's World [clothing store for large women]." And everybody laughed. And I said, "Oh pardon me." I said, "Don't make fun of the places where I buy my clothes." I just said it in a nice, lighthearted tone. There was just dead silence. God, everybody at the table was embarrassed to death. I thought, "Hey Beth, good for you for saying that."

> Deborah: So I was shopping and all of a sudden I found myself in front of this huge section of Slimfast products. And all of a sudden the idea came to me, [shouts] gee, I don't have to use the entire card [NAAFA business card]. So I just tore the bottom half of the card off, the part that had the NAAFA address and left the part that said, "Do something about your weight, accept it." And I just stuck it on the shelf, sort of like behind one of those little plastic place cards and said. "There."

Contestation emanates from the lived experiences of members who organize their lives around their organizational "fat"—identity. The identity is viewed as relevant in all of the daily activities of NAAFA members, whether confronting fellow employees like Beth or grocery shopping like Deborah.

According to NAAFA members, shame militates against the active initiation of individual and, hence, collective, political practice. NAAFA sees social stigmas as directly linked to the internalized oppression of shame. Shame contestation is thus a requisite component of identity politics where the aggrieved attempts to transform themselves by transforming both societal definitions of beauty and human value and the feeling rules that govern fat bodies. . . .

CONCLUSION

Sociologists studying formal organizations contend that organizations are boundary maintaining social units where society and culture are both experienced by individuals and reproduced in various sets of social practices (Perrow 1986). Society, it is argued, is now absorbed by organizations that mediate culture and cultural themes for the individuals who participate within them. Rudolph Bell's (1985) analysis of "Holy Anorexia" has delineated the historic role of the Catholic church in the creation and social reproduction of themes of shame and obesity, purity and thinness, as they came to be interwoven with the phenomenon anorexia from the thirteenth century on. Feminist scholars (Bordo 1993: Hesse-Biber 1996; Schwartz 1986; Wolf 1991) have argued that the increasing commercialization of the body and bodily needs for the sake of expanding markets has ensured that themes of shame, slothfulness, loss of control, and embarrassment will continue to be associated with fat bodies, particularly fat female bodies. The present study indicates that the cultural meanings of shame associated with the body are not simply residues that exist as historical abstractions.[11] Rather, it is in interaction that takes into consideration the organizational frames that a shame experience is formulated as a social object and given meaning. Organizational frames may be quite variable depending upon the nature, objectives, and ideology within which the organizations attempts to align members' experiences, yet they provide the template in which shame experiences may be systematized in organizational routines.

NOTES

1. Many of the tapes were truncated or contained several short testimonies by members of how OA had helped restore their lives. Yet the tapes yielded little information about the lived experience of the person or how she or he came to OA.

2. The serenity prayer of OA, "God grant me the serenity to accept the things I cannot change; the courage to change the things I can; and the wisdom to know the difference."

3. One of the more recent efforts was bill A 3484 New York State, sponsored by assemblyman Daniel Feldman. The bill was not supported by Governor Mario Cuomo. The event was reported in NAAFA's national newsletter, which commented, "Asked by his interviewer whether questions on the bill had him skating on thin ice with voters who are fat, Cuomo quipped, 'If you are overweight, you shouldn't do that—you're

liable to fall through the ice' " (NAAFA 1993, p. 1).

4. On May 23, 1992. NAAFA chapters in five cities launched demonstrations against Southwest Airlines for ejecting a very large man from a half-full flight and for disallowing another large passenger to board after she had already completed one leg of her flight. (NAAFA, 1992, p. 1)

5. I use the term "practically" in reference to the "investment model" of participation, discussed by Rudy and Greil (1987), observing that all organizational members mobilize personal resources to be used in and by the organizations. This is clear in the case of WW where both time and money represent "sunken costs." Yet it is also the case in OA and NAAFA where members invest time, money, and intellectual and emotional labor to keep the organization operating.

6. The collection of all shame management strategies may be labeled "technique." I have in mind Jacques Ellul's (1964, p. vi) meaning: a "complex of standardized means for attaining a predetermined result. Thus it converts spontaneous and unreflective behavior into behavior that is deliberate and rationalized." Here, the emphasis must be on the idea of a "complex," that is, not only the measurement of food and calories but the restructuring of social relations and development of adaptive strategies for dealing with unexpected situations.

7. Goffman's notion of "cooling the mark out" (1952) would appear to be a rather pervasive form of emotion management found among all types of service personnel. Yet, curiously, in light of a rather extensive citation of Goffman's works, Hochschild (1983) does not mention it. Beyond the management of emotion, the concept of emotion work suggests a range of interactional strategies that might be used in managing the emotions of others. Lying, for example, might be used to "cool out" airline passengers who might otherwise become disruptive.

8. Rudy (1986), in his analysis of Alcoholics Anonymous, observes that "hitting bottom" is ultimately a subjective perception.

9. While some OA members refer to their higher power as "God," others adopt a Durkheimian (Durkheim 1965) conception of this power, defining it as the social force that exists outside of themselves and abides in the group.

10. Shaming appears to have been a treatment modality that was gaining limited popularity within quite diverse fields. Serber (1970), in his research on behavioral modification among transvestites, referred to the treatment as "shame aversion therapy." Braithwaite (1989) revisited the idea in the concept of "reintegrative shaming" within the context of criminal justice.

11. Pareto observed that "residues" (nonscientific belief systems) rarely mobilize people into action even though they may historically express deep collective sentiments (Coser 1977). My point is that cultural themes such as shame are socially reproduced within concrete episodes of interaction.

REFERENCES

Allon, Natalie. 1982. "The Stigma of Overweight in Everyday Life." Pp. 130–174 in *Aspects of Obesity*, edited by Benjamin B. Wolman. New York: Van Nostrand Reinhold.

Banner, Lois W. 1983. *American Beauty*. New York: Knopf.

Bell, Rudolph. 1985. *Holy Anorexia*. Chicago: University of Chicago Press.

Bordo, Susan. 1993. *Unbearable Weight*. Berkeley: University of California Press.

Braithwaite, John. 1989. *Crime Shame and Reintegration*. New York: Cambridge University Press.

Buss, David M. 1989. "Sex Differences in Human Mate

Preferences: Evolutionary Hypotheses Tested in 37 Cultures." *Behavioral and Brain Sciences* 12:1–49.

Cahnman, Werner. 1968. "The Stigma of Obesity," *The Sociological Quarterly* 9:283–299.

Coser, Lewis A. 1977. *Masters of Sociological Thought: Ideas in Historical and Social Content*. 2d ed. New York: Harcourt, Brace, Jovanovich.

Denzin, Norman K. 1987. *The Recovering Alcoholic*. Beverly Hills, CA: Sage.

Durkheim, Emile. 1965. *The Elementary Forms of Religious Life*. NY: The Free Press.

Ellis, Carolyn. 1998. " 'I Hate My Voice': Coming to Terms with Minor Bodily Stigmas." *The Sociological Quarterly* 39:517–538.

Ellul, Jacques. 1964. *The Technological Society*. New York: Vintage.

Ewen, Stuart. 1976. *Captains of Consciousness: Advertising and the Roots of the Consumer Culture*. New York: McGraw-Hill

Featherstone, Mike. 1982. "The Body in Consumer Culture." *Theory Culture & Society* 1:18–33.

Gilmore, David B. 1987. *Honor and Shame and the Unity of the Mediterranean*. Washington, D.C.: American Anthropological Association.

Glaser, Barney G., and Anselm L. Strauss. 1967. *The Discovery of Grounded Theory*. Chicago: Aldine de Gruyter.

Goffman, Erving. 1952. "On Cooling the Mark Out: Some Aspects of Adaptation to Failure." *Psychiatry* 15:451–463.

———. 1961. *Asylums*. Garden City, NY: Anchor.

———. 1967. *Interaction Ritual*. Garden City, NY: Anchor.

———. 1969. *Strategic Interaction*. Philadelphia: University of Pennsylvania Press.

———. 1986. *Frame Analysis*. Boston, MA: Northeastern University Press.

Hesse-Biber, Sharlene. 1996. *Am I Thin Enough Yet? The Cult of Thinness and the Commercialization of Beauty*. New York: Oxford University Press.

Hewitt, John P., and Randall Stokes. 1975. "Disclaimers." *American Sociological Review*, 60:1–11.

Hochschild, Arlie Russell. 1983. *The Managed Heart: Commercialization of Human Feeling*. Berkeley: University of California Press.

Horowitz, Ruth. 1983. *Honor and the American Dream: Culture and Identity in a Chicano Community*. New Brunswick, NJ: Rutgers University Press.

Lyman, Stanford, and Marvin B. Scott. 1970. *A Sociology of the Absurd*. Pacific Palisades, CA: Goodyear.

Martin, Daniel D. 1995. "The Politics of Appearance: Managing Meanings of the Body, Organizationally." Ph.D. dissertation, University of Minnesota.

McFarland, Barbara, and Tyeis L. Baker-Baumann. 1990. *Shame and the Body: Culture and the Compulsive Eater*. Deerfield Beach, FL: Health Communications.

Millman, Marcia. 1980. *Such a Pretty Face: Being Fat in America*. New York: W. W. Norton.

Mills, C. Wright. 1940. "Situated Actions and Vocabularies of Motive." *American Sociological Review* 5:904–913.

National Association to Advance Fat Acceptance (NAAFA). 1990. *NAAFA Workbook: A Complete Study Guide*. Sacramento, CA: NAAFA.

———. 1992. "Southwest Protest: NAAFAns Demonstrate in Five Cities." *NAAFA Newsletter* XXII:1.

———. 1993. "Cuomo: One Law Too Many!" *NAAFA Newsletter* XXIII:6 (May):1.

Overeater Anonymous. 1987. *To the Newcomer: You're Not Alone Anymore*. Torrance. CA: OA.

———. 1988. *Questions & Answers About Compulsive Overeating and the OA Program*. Torrance, CA: OA.

———. 1990. *The Twelve Steps of Overeaters Anonymous*. Torrance, CA: OA.

———. 1992. *Overeaters Anonymoue: Membership Survey Summary*. Torrance, CA: OA.

Perrow, Charles. 1986. *Complex Organizations.* 3d ed. New York: McGraw-Hill.

Rudy, David R. 1986. *Becoming Alcoholic: Alcoholics Anonymous and the Reality of Alcoholism.* Carbondale: Southern Illinois University Press.

Rudy, David R., and Arthur L. Greil. 1987. "Taking the Pledge: The Commitment Process in Alcoholics Anonymous." *Sociological Focus* 20:45–59.

Schwartz, Hillel. 1986. *Never Satisfied: A Cultural History of Diets, Fantasies and Fat.* New York: Free Press.

Serber, Michael. 1970. "Shame Aversion Therapy." *Experimental Psychiatry* 1:213–215.

Snow, David A., E. Burke Rochford, Jr., Steven K. Worden, and Robert D. Benford. 1986. "Frame Alignment Processes, Micromobilization, and Movement Participation." *American Sociological Review* 51:461–481.

Thorne, Barrie, and Zella Luria. 1986. "Sexuality and Gender in Children's Daily Worlds." *Social Problems* 33:176–190.

Wagonfeld Samuel, and Howard M. Wollowitz. 1968. "Obesity and the Self-Help Group: A Look at TOPS." *American Journal of Psychiatry* 125:249–252.

Weber, Joseph. 1990. "The Diet Business Takes It on the Chin." *Business Week,* April 16, pp. 86–88.

Wolf, Naomi. 1991. *The Beauty Myth: How Images of Beauty Are Used against Women.* New York: William Morrow.

Wolfe, Louise. 1992. "Boston 'Free' Party: The Revolution Within; The Revolution Without." *NAAFA Newsletter* 23:5.

PART VIII

The Social Organization of Deviance

In Part VIII we turn to a closer examination of the lives and activities of deviants. Once they get past dealing with outsiders, they must deal with other members of their deviant communities and with the specifics of accomplishing their deviance. There are several ways of looking at how deviants organize their lives. We start by looking at the relationships among groups of deviants, focusing on the character, structure, and consequences of different types of organizations. These encompass the structure or patterns of relationships in which individuals engage when they enter the pursuit of deviance.

As Best and Luckenbill (1980) have noted in their analysis of the social organization of deviants, relationships among deviants can follow many models. These vary along a dimension of sophistication, involving complexity, coordination, and purpose. Deviant associations vary in their numbers of members, the task specialization among members, the stratification within the group, and the amount of authority concentrated in the hands of a leader or leaders. Some groups of deviants are loose and flexible, with members entering or leaving at their own will, uncounted or monitored by anybody. Others maintain more rigid boundaries, with access granted only by the consent of one or more insiders. Membership rituals may vary from none to highly specific acts that must be performed by prospective inductees, thereby granting them not only membership but also a place in the pecking order once they are inside. In some ways, rigidity inside deviant groups is related to its insulation from conventional society; the more its members withdraw into a social and economic world of their own, the more they will develop norms and rules to guide them, replacing those of the outside order.

Groups of deviants also vary in their organizational sophistication, with the more organized groups capable of more complex activities. Such organized groups provide greater resources and services to their members; they pass on the norms, values, and lore of their deviant subculture; they teach novices specific skills and techniques where necessary; and they help one another out when they get in trouble. As a result, individuals who join more tightly-knit deviant scenes tend to be better protected from the efforts of social control agents and more deeply committed to a deviant identity.

The five selections in Part VIII look at *individuals, subcultures, gangs, formal organizations,* and *corporations* encompassing deviants. These readings are organized along a continuum that rises progressively in organizational sophistication from sexual asphyxiates, to punks, to gangs, to internationally organized criminal cartels, to state-organized crime.

INDIVIDUALS

28

Sexual Asphyxia
A Neglected Area of Study

SHEARON A. LOWERY AND CHARLES V. WETLI

Lowery and Wetli offer a rare insight into a highly secretive and hidden form of deviance, sexual asphyxia, otherwise known as the "ultimate orgasm." A solitary, loner practice, sexual asphyxia is pursued by individuals, primarily young men, who desire to raise their level of orgasmic pleasure through self-strangulation. So deviant that practitioners rarely reveal their interest in this act even to friends or lovers, it is accomplished outside of any deviant support group or subculture. Lowery and Wetli outline some of the features of this highly secretive form of individual deviance, the characteristics of individuals who engage in it, and the driving force behind their involvement.

The sociological study of sexual deviance has traditionally been limited to the examination and explication of prostitution and homosexuality, although some scholars have also considered premarital and extramarital sexual behavior to be deviant. Even when such a "broad" definition of sexual deviance has been employed, however, other types of sexual behavior such as sadomasochism and bondage have been typically excluded. Consequently, such practices have remained virtually ignored in the literature of sociology. In addition, judging from an examination of that literature, more unusual sexual practices such as sexual asphyxia are apparently unknown to sociologists. Consider, for example, the following case:

> The sheriff's office received a call from the frantic wife stating she had found her husband bound, gagged, and murdered. The Police and Medical Investigator rushed to the scene to find a 27-year-old male deceased on the bed. The deceased was clad in brassiere, woman's panties,

---

Copyright 1982 from "Sexual Asphyxia: A Neglected Area of Study," Shearon A. Lowery and Charles V. Wetli, *Deviant Behavior*, V. 3, No. 1. Reproduced by permission of Taylor & Francis, Inc., http://www.routledge-ny.com.

red negligee, and panty hose, all of which the wife identified as hers. He was facedown with his legs flexed at the knees. The ankles were bound with four loops of clothesline, loosely knotted. The clothesline was attached to an elastic strap with metal hooks on both ends. The strap was looped about a dog collar encircling the neck. A bath towel was between the dog collar and the skin of the anterior neck. A handkerchief was tied about the shaft and the end of the penis was stained with seminal fluid. A handkerchief was stuffed in the mouth, and a bandana was tied about the face at mouth level, knotted in the back. A large bathroom mirror had been removed from the door and rested against the dresser where it could be visualized by the deceased. (Lewman, 1978: 11)

Although this scene suggests homicide or suicide, the actual manner of death is accidental. This case is in fact a rather typical example of the sexual asphyxia syndrome familiar to medical examiners and to a lesser extent to some police investigators and psychiatrists.

Sexual asphyxia is an autoerotic activity practiced almost exclusively by males. Lethal cases are almost always characterized by an individual male engaged in solitary sexual activity while simultaneously creating a self-induced mechanical or chemical asphyxia. The purpose of the asphyxia is to heighten the sexual pleasure. Cases of lethal sexual asphyxia among females are extremely rare; to our knowledge, only three instances have been reported (Sass, 1974; Danto, 1980; Sullivan and Wray, 1981).

Among males transvestism is frequently concomitant with the practice and there may be erotic literature or mirrors in view of the deceased. Sometimes the deceased's hands are tied behind his back, but close inspection invariably reveals that the victim himself was responsible for this. In addition, it is done in such a way as to permit a quick escape should that be necessary. Finally, the practitioner will have devised some mechanism to impede the flow of oxygen to the brain. This will usually be a ligature (often with a soft material or padding about the neck to prevent abrasion), a plastic bag, or an inhalant (gas, chemical aerosol, etc.). The subjective result is a giddiness and exhilaration; but the objective consequence is asphyxia resulting from mechanical restriction of the airway or chemical replacement of oxygen. If the process continues unabated the individual loses consciousness as a result of the lack of oxygen and the increased retention of carbon dioxide. Death will ensue unless the individual spontaneously breaks away from the asphyxiating device. If that device is a ligature, there is little likelihood of this happening once consciousness has been lost. Instead, the airway becomes further obstructed, as does the venous return from the brain, and finally the arterial blood supply to the brain is disrupted.

The most frequent method used to produce the asphyxia, and the one most discussed, involves ligatures applied to the neck. Ropes, scarves, and the like are tied about the neck in such a manner that they may be manipulated to control the flow of blood and oxygen to the brain. Such choking creates feelings of pleasure and may induce erections and even orgasms in some individuals

(Resnik, 1972). Various descriptions of sexual asphyxia appear in literary sources; a particularly graphic passage can be found in DeSade's *Justine:*

> He got upon the stool, the rope around his neck. . . . He got ready and beckoned her to pull away the stool. Hanging by his neck for a while, his tongue lolling way out, his eyes bulging; but soon, beginning to swoon away, he motioned feebly to Justine to set him loose. On being revived he said, "Oh Therese! One has no idea of such sensations, what a feeling! It surpasses anything I know." (p. 101)

In addition, anthropologists have pointed out that certain ethnic and cultural groups (e.g., Eskimos) are known to choke each other as part of their sexual activity. In such cultures it is common for the children to suspend themselves by the neck during play (Walsh, Stahl, Unger, Lilienstern, and Stephens, 1977).

The limited research that has been done indicates that it is likely that medical authorities underestimate the frequency of death resulting from sexual asphyxia. Some investigators (Sass, 1974; Litman and Swearingen, 1972; Resnik, 1972) have argued that sexual asphyxial practices are relatively common in the United States. Rosenblum and Faber (1979) think that it is reasonable to estimate that at least 250 deaths occur in the United States each year. This estimate is probably very conservative because it is likely that many sexual asphyxial deaths are judged to be either homicide or suicide because of a lack of awareness of the practice in many localities,[1] the sensitivity of the sexual aspects of the practice, or the social status of the victims.[2] For example, it is possible that some deaths among youths that have been attributed to glue sniffing could actually have been the result of the practice of sexual asphyxia. More research is necessary before we can adequately address the issue of the reliability of these national estimates; however, it is important that the issues be raised and the subject investigated.

Although it is difficult to estimate the number of annual deaths from this cause it is even more difficult to estimate the number of individuals who engage in this sexual practice (Resnik, 1972; Rosenblum and Faber, 1979). Because we know that those practitioners for whom sexual asphyxia proved to be fatal took elaborate precautions not to die, we must take this factor into account in any attempt to estimate the number of practitioners. Thus, if we accept the view that those who engage in the behavior do not intend to die,[3] it becomes evident that the actual number of practitioners of sexual asphyxia may be considerable. Because of the precautions taken by these individuals, it is likely that the number of fatalities is small compared to the total number of incidents. In fact, researchers of sexual asphyxial practices posit that "such practices are common" (Litman and Swearingen, 1972: 11) and certainly not the "oddity or rarity" that many people assume (Enos, 1975: 134). But whatever the number of practitioners or the frequency of performance, sexual asphyxia is always potentially lethal and thus deserves careful sociological and psychological examination as well as legal and medical scrutiny.

SOCIAL CHARACTERISTICS OF THE VICTIMS AND THE DEATH SCENE

Almost all of the handful of scientific studies examining sexual asphyxia deal with cases that were discovered because the practitioners were unfortunate enough to die in the process. These postmortem studies do, however, delineate certain patterns. In contrast to most other forms of sexual deviance, sexual asphyxia appears to be an activity confined to middle- and upper-class white males. These males tend to be young; most of the recorded fatalities have occurred among teenage or young adult males who were unmarried at the time of their death. One review of 43 such deaths (Walsh et al., 1977) found the majority of the victims to have been younger than 25; and only 13 were married at the time of death.

Such behavior usually begins in adolescence, when it is a solitary act (Resnik, 1972). However, as the practitioner matures into adulthood the syndrome may become less lethal. This is because he may be able to involve partners in the process to protect him. Such partners are, however, used solely for the purpose of protection of the practitioner and take no part in the act itself. In all other aspects sexual asphyxia remains a solitary act.

The characteristics of the scene of discovery of the victims of sexual asphyxia lead to the conclusion that the practitioner requires solitude and privacy. The body is usually discovered in the victim's residence; many times it is in a bedroom, bathroom, attic, closet, or basement—somewhere that a door might be locked. If the act is performed outdoors, the most frequent setting is a secluded wooded area or an abandoned or little-used structure. There is an obvious need on the part of the participant to avoid any intrusion or interruption during the act (Sass, 1974).

Danto (1980) reported that the majority of the bodies are found either naked or partially clothed and that there is frequently evidence of transvestism (articles of female clothing such as dresses, brassieres, or panty hose). Moreover, there is usually evidence of penile erection and ejaculation. However, evidence of ejaculation must be carefully evaluated because emission of seminal fluid is a frequent consequence of rigor mortis.

It is estimated that in from one-third to one-half of the cases the hands, body, and feet are bound in some manner. And although some of these binding mechanisms are extremely complex, it can usually be demonstrated that the ligatures and ropes could readily be tied and released by the victim himself. In addition, in some cases the scrotal sac may also be bound with string, thread, or rope. Mirrors are frequently present and it is thought that as many as half of the victims may have been viewing themselves in a mirror and/or using erotic materials found near the body. Instances associated with self-inflicted pain or infibulation have also been reported (Sass, 1974).

INTERVIEWS WITH LIVING PRACTITIONERS

Among the studies that report aspects of the practice of sexual asphyxia, only one has involved the interviewing of current practitioners. One additional study (Rosenblum and Faber, 1979) reports a psychiatric case history in which a 15-year-old boy has successfully been treated for such practice. This relative absence of firsthand information about social, psychological, and emotional aspects of the practice has not occurred because researchers failed to seek out practitioners. Indeed, Resnik (1972) reported that he had been unable to locate a single living practitioner of sexual asphyxia in the 10 years he had been interested in the subject. Medical researchers Litman and Swearingen (1972), however, were able to locate and interview nine such individuals. In an effort to reach such persons, they had an article ("Whips, Chains, and Leather") published in the *Los Angeles Free Press,* a weekly underground newspaper with a circulation of about 90,000. As a result of that article and an advertisement, they received about 30 responses. After screening the respondents, face-to-face interviews were conducted with nine men and three women. Only the men, however, were engaged in the type of behavior that they were trying to study.

All of the subjects of the Litman and Swearingen study were white middle-class males who were characterized as "intelligent, verbal, and cooperative." The researchers felt that their volunteer subjects had been motivated to respond to their advertisement by loneliness, a wish to share their interest with others, and a need to find legitimacy for their underground practices. In addition, the researchers felt that for some of them the response was a cry for help.

In general, the sexual orientation of these men was deemed to be homosexual. Even though many of the men had previous heterosexual experiences and several even indicated that they preferred women, Litman and Swearingen classified only two of the men as heterosexual. They also reported a trend toward increasing homosexuality with increasing age. Indeed, the researchers felt that the majority of the subjects hoped that they would find partners by responding to the ad. Finding such a partner is extremely important to them, because such a partner plays a protective role. This desire for partners, moreover, may help to explain the tendency toward homosexuality, in that homosexual partners are easier to find for unusual activity than are heterosexual ones. But despite this search for partners, the essence of the practice remains narcissistic. The focus of attention remains on themselves, even when others are participating; the practitioner of sexual asphyxia is preoccupied with his own fantasies and with his own view and sensations of the world.

Three of the subjects were married, and each had difficulty with his wife. The authors felt that this was the result of "basic personality flaws" rather than the result of their peculiar sexual behaviors. Although all of the subjects were aware of pornographic literature, only one-third indicated that they were strongly influenced by it. Moreover, transvestite elements were "surprisingly infrequent" in this group.

Litman and Swearingen saw the defining characteristics of this group of men to be extreme loneliness and isolation. They felt that all of these men were deeply depressed and death oriented: Six of the nine gave histories of serious depression—often accompanied by suicide attempts. They further posit that men used their perversion to fight off the death trend and to defend themselves against suicide. In spite of the researchers' belief that these men were experiencing both psychological and emotional pain, their study revealed no consistent patterns of family pathology or specific traumata in childhood.

THE ABSENCE OF A DEVIANT SUBCULTURE

One additional important point should be stressed here: No deviant subculture has grown up around the practice of sexual asphyxia. There is no underlying thread that connects one practitioner with another, and participants do not, however subtly, advertise what they do. The solitary nature of the act means that locating one practitioner, even a deceased one, is not likely to lead us to others among his friends and associates. This is unlike other forms of sexual deviance, such as prostitution and homosexuality, where subcultures are well defined and have been extensively examined in the literature. It is even true of other sexually deviant behaviors such as bondage and sadomasochism that have not been extensively studied by sociologists. People who participate in these subcultures routinely advertise in underground publications for like-minded partners; and browsing any "adult" bookstore leads to the discovery that there is a special B&D (bondage and discipline) section for those who enjoy such practices. Indeed, the coroner of San Francisco, where rates of sexual deviance are high, recently began holding clinics in safe sadomasochistic practices because of concern about the number of homicides related to the activity (*Washington Star,* March 13, 1981: 1).

The additional obstacle that this lack of a subculture presents to the researcher should not, however, preclude systematic and rigorous examination of sexual asphyxia. Even when the researcher is unable to implement "ideal" research design and uses volunteer subjects or must rely on "accidental deaths" that bring practitioners to the attention of local authorities, such studies can be useful so long as their sampling limitations are noted and kept in mind. By using information about those individuals unfortunate enough to die in the process, perhaps the living can learn enough from death investigations to illuminate the scope and dimensions of this potentially lethal sexual practice.

THE DADE COUNTY DATA

It was possible to study a number of cases of lethal sexual asphyxia taken from the files of the Medical Examiner Office of Dade County (Miami), Florida. That office is responsible for the investigation of all deaths within the county that are not obviously the result of natural causes. The purpose of the investigation is to determine the cause, manner, and mechanism of death (see Wright and Wetli, 1981). The scene and circumstances of death are initially evaluated by police agencies and, in cases of apparent lethal sexual asphyxia, by a forensic pathologist as well. Medical and social histories are subsequently obtained by both police and forensic investigators. Finally, a complete autopsy is performed. The data here reported were abstracted from the official reports prepared by these authorities.

Previous medical researchers (e.g., Resnik, 1972; Walsh et al., 1977) have outlined a number of criteria that are believed to indicate that a death may have been the result of the practice of sexual asphyxia. Although each of these indicators may not appear in every case, they collectively define the sexual asphyxia syndrome. These criteria include the following:

1. The act is solitary.
2. There is evidence of sexual activity.
3. There is no well-defined evidence of suicidal intent.
4. The deceased is completely or partially unclothed.
5. Transvestism may be present.
6. There may be evidence of previous episodes.
7. Often the extremities and sometimes the genitals are bound.
8. Erotic materials, especially pictures, are often present.

A detailed examination of the following case from Dade County illustrates the unusual nature of the sexual asphyxia syndrome and provides a concrete example of some of the criteria associated with it.

> A hotel maid discovered a 39-year-old caucasian man hanging by his neck. He was naked, and his feet touched the floor. His hands were looped in a rope behind his back. The ligature consisted of a white nylon rope fashioned as a hangman's noose and threaded through two large eyebolts which had been inserted into the wall. A full-length mirror was positioned in front of and to the side of the victim to permit complete self-viewing. A camera was positioned in front of the victim several feet away, and a remote shutter release device was nearby. A large sheet of dull black paper (approximately six feet wide by ten feet long) was on the floor to one side of the victim, and had apparently fallen from an overhang to which it had been attached with tape. On a nearby table were erotic photographs of women which had been cut out of the latest issue of *Playboy* magazine. On an adjacent chair was an ice bucket partially filled with urine. In the bathroom was a kit for developing 35mm slide film. In the bedroom were some hand tools (pliers, screw driver, etc.) as well as a plaster patching compound.
>
> The film in the camera was removed and subsequently developed. It contained a series of self-taken photographs in which the victim was in various poses. The first revealed the victim clad in a shirt and bath towel

imitating female attire. The subsequent pictures revealed a progressive removal of clothing until he was standing naked and holding the noose-end of the rope. The photographs were taken against a dull black background (the paper found on the floor). The penultimate photograph depicted him standing on a low stool and with the ligature mildly constricting the neck. In the final photograph the victim is pictured in nearly the exact position in which he was found at the scene. In none of the photographs was penile erection evident.

A subsequent investigation revealed that this man was a highly competent professional person with a high degree of intelligence. He was married and had two children. At one time he had been in psychotherapy where it was learned that he had marital problems related in part to his spouse's sexual withholding. His wife was aware of the episodes of photo-fantasies. It was learned that she disapproved of them and hence, for more than twelve years, he engaged in this activity outside the home (usually in hotel rooms). The victim's personality was described as narcissistic. He was obsessive-compulsive ("workaholic") and revealed a self-image of being both heroic and sacrificial, qualities which were evident in the self-taken photographs.

It is evident that the self-taken photographs were to subsequently induce sexual stimulation. It must be presumed that, in this sequence of photographs, death intervened before the intended masturbatory activity could take place.

DISCUSSION

The preceding description of some of the sociological variables in our data associated with lethal sexual asphyxia (in the argot of the bondage underground, terminal sex) leads us to the following conclusions. The data support the observation that, unlike most forms of sexual deviance, sexual asphyxia is practiced by young, white, middle- or upper-middle-class males. It is almost exclusively practiced in the home or some safe place near the home. We do not know the number of practitioners, the frequency with which they perform the act, or the actual distribution of their ages.

Perhaps the young are victims of the practice so frequently because they are, like youth everywhere, often oblivious of the dangers involved in what they do. They are straightforward about it—they simply hang themselves. Moreover, because the act is solitary, when something goes wrong and they lose control, they lose their lives as well.

As practitioners become older, they probably become more cautious; they use more elaborate devices and many search for partners. Perhaps then it is merely adaptive behavior that leads to homosexual liaisons. Such unusual practices are indeed more likely to be accepted or at least tolerated by homosexual partners than by heterosexual ones. Our data indicate that when the practice is fatal among older practitioners some complicating factor such as the use of alcohol or drugs is present that may have impaired their judgment enough to have contributed to the accident costing them their lives.

When we examine the social origin of the act and the actor, it is clear that many of the victims have severe difficulties obtaining sufficient sexual gratifications by other means. They lacked sufficient sexual outlets and generally had difficulty interacting with females. Although some researchers (Resnik, 1972; Rosenblum and Faber, 1979) believe that the act generally becomes more elaborate as the practitioner ages, our data indicate that the introduction of fetishes and transvestism may occur early or late. One plausible explanation for the elements of bondage and sadism that may enter into the practice is that the only source of willing partners the practitioner can find is among members of the bondage community—whether they are heterosexual or homosexual. In order to receive the gratifications of sexual asphyxia with safety, they submit to other forms of sexual deviance, bondage, and so forth. They do not seek the painful pleasures of sadomasochism but rather endure them for the sexual pleasures of sexual asphyxia that they seek. There is a great difference, thus, between the practitioner of sadomasochism and the sexual asphyxiate.

In the end, however, many of these men return to solitary practice or never find willing partners; and when an accident occurs, their search for a satisfactory sexual outlet for their sexual needs becomes terminal. Our data confirm that these deaths were not suicides and that there was no suicidal intent. We caution, however, that the data presented in this paper are limited; in all of the cases we report, the individuals died while practicing sexual asphyxia. This group of practitioners therefore may not be representative of the general population of practitioners of sexual asphyxia. Thus, it is evident that additional research is needed to clarify many of the issues raised in this paper, especially the scope and dimensions of the problem—who practices sexual asphyxia, how often, and how frequently does it result in death.

NOTES

1. Werner Simon (1973) has argued that teenage suicides may not have increased as much as it appears. He posits that a number of these tragedies are actually accidental deaths resulting from the practice of sexual asphyxia.

2. For example, a well-known professional whose death clearly resulted from the practice of sexual asphyxia was reported to have died of a heart attack in his hometown newspaper.

3. It should be noted, however, that if a practitioner of sexual asphyxia decides to commit suicide, it is probable that he will choose this method. Litman and Swearingen (1972) report such a case.

REFERENCES

Danto, Bruce L. 1980. "A case of female auto-erotic death." *American Journal of Forensic Medicine and Pathology* 1: 117–121.

DeSade, Marquis. 1964. *Justine*. New York: Castle Books.

Lewman, Larry V. 1978. "Case of the month." *Office of the Medical Investigator* 5: 11–13.

Litman, Robert E., and Charles Swearingen. 1972. "Bondage and suicide." *Archives of General Psychiatry* 27: 80–85.

———. 1973. "Bondage and suicide." *Medical Aspects of Human Sexuality* 7(11): 164–181.

Resnik, H. L. P. 1972. "Eroticized repetitive hangings: A form of self-destructive behavior." *American Journal of Psychotherapy* 26: 4–21.

Rosenblum, Stephen, and Myron M. Faber. 1979. "The adolescent sexual asphyxia syndrome." *Journal of the American Academy of Child Psychiatry* 19: 546–558.

Sass, F. A. 1974. "Sexual asphyxia in the female." *Journal of Forensic Sciences* 20: 181–185.

Simon, Werner. 1973. "Commentary." *Medical Aspects of Human Sexuality* 7(11): 189–193.

Sullivan, William B., Jr., and Steve Wray. 1981. A Case of Sexual Asphyxia of a Female. Paper presented at the 33rd Annual Meeting of the American Academy of Forensic Sciences, Los Angeles, February.

Walsh, F. M., Charles J. Stahl, H. Thomas Unger, Oscar C. Lilienstern, and Robert G. Stephens. 1977. "Autoerotic asphyxial deaths: A medicolegal analysis of forty-three cases." *Legal Medicine Annual* 1977: 157–182.

Wright, Ronald K., and Charles V. Wetli. 1981. "A guide to the forensic autopsy—conceptual aspects." *Pathology Annual* 16: 273–288.

SUBCULTURES

29

Real Punks and Pretenders
The Social Organization of a Counterculture

KATHRYN J. FOX

Fox's study of a Midwestern city's punk scene, or subculture, examines the social relations found in this group of people who associate together but do not need each other to be punk. She presents four types of punks in the scene and examines the relations among them. Starting at the center, she looks at the hardcore punks, gradually working outward to the softcore punks, the preppie punks, and finally to the spectators. For each group she examines their immersion in the punk lifestyle and commitment to punk ideals, activities, style of dress, and mode of survival. Membership in the hardcore inner group requires a more serious dedication than does participation in the transitory outer fringe, yet these style leaders feed on the presence and adulation of the more peripheral groups for sustenance. Each group fills a distinctly different role in the subculture, and identifies itself largely in relation to the others.

Modern Western society has been characterized by a variety of antiestablishment style countercultures following in succession (i.e., the Teddy Boys of 1953–1957, the Mods and Rockers of 1964–1966, the Skinheads of 1967–1970, and the Punk Rockers of the late 1970s; Taylor, 1982). The punk culture is but the latest in this series. Since most studies of youth- and style-oriented groups are British (Frith, 1982), very little has been written from a sociological perspective about punks in the United States. This study is an attempt to fill that void.

Whether or not the punk scene in the United States could be legitimately classified as a social movement is debatable. Most writers on this contemporary phenomenon agree that American punks have a more amorphous, less articulated ideological agenda than punks elsewhere (Brake, 1985; Street, 1986).

From "Real Punks and Pretenders: The Social Organization of a Counterculture," Kathryn J. Fox, *Journal of Contemporary Ethnography*, Vol. 16, No. 3, 1987. Reprinted by permission of Sage Publications, Inc.

While the punk scene in England responded to youth unemployment and working-class problems, the phenomenon in the United States was more closely connected to style than to politics. Street (1986: 175) notes that even for English punks, "politics was part of the style." I would argue that this was even more the case for American punks. The consciousness of the youth in the United States did not parallel the identification with the plight of youth found in Europe. Nonetheless, the "style" code for punks in the United States contained an insistent element of conflict with the dominant value system. The consensual values among the punks, as ambiguous as they were, could best be understood by their contradictory quality with reference to mainstream society. In this respect, punk in America fit the definition of a "counterculture" offered by Yinger (1982: 22–23). According to this definition, the salient feature of a counterculture is its contrariness. Further, as opposed to individual deviant behavior, punks constituted a counterculture in that they shared a specific normative system. Certain behaviors were considered punk, while others were not. Indeed, style was the message and the means of expression. Observation of behaviors that were consistent with punk sensibilities were viewed as indicative of punk "beliefs." These behaviors, along with verbal pronouncements, verified commitment. Within the groups of punks I studied, the degree of commitment to the counterculture lifestyle was the variable that determined placement within the hierarchy of the local scene.

Previous portrayals of youthful, antiestablishment style cultures have discussed their norms and values (Berger, 1967; Davis, 1970; Hebdige, 1981; Yablonsky, 1968), their relationship to conventional society (Cohen, 1972; Douglas, 1970; Flacks, 1971), their focal concerns and ideology (Flacks, 1967; Miller, 1958), and their relationship to social class (Brake, 1980; Hall and Jefferson, 1975; Mungham and Pearson, 1976). With the exception of Davis and Munoz (1968), Kinsey (1982), and Yablonsky (1959), few of the studies of antiestablishment, countercultural groups discuss their implicit stratification. In this essay I will describe and analyze the various categories of membership in the punk scene and show how members of these strata differ with regard to their ideology, appearance, taste, lifestyle, and commitment.

I begin by discussing how I became interested in the topic and the methods I employed to gain access to the group and to gather data. I then offer a description of the setting and the people who frequented this scene. Next I offer a structural portrayal of the social organization of this punk scene, showing how the layers of membership form. I then examine each of the three membership categories (hardcore punks, softcore punks, and preppie punks), as well as the spectator category, focusing on the differences in their attitudes, behavior, and involvement with this antiestablishment style culture. I conclude by outlining the contributions each of these types of members makes to the continuing existence of the punk movement and, more broadly, by describing the relation between the punk counterculture and conventional society.

METHODS

My interest in the punk movement dates back to 1978. At that time, I attended a local punk bar fairly regularly and wore my hair and clothing in "punk" style, albeit not the radical version. I also visited a major northeastern city at about that same time, when the punk scene was in full flower, and spent several nights visiting what are now famous punk hangouts. My early interest and involvement in this scene laid the groundwork for this study, as it permitted me to gain knowledge of the punk vernacular, styles, and motives. The research continued, with active, weekly participation, through the middle of 1986. I have continued to keep a close watch on national trends and developments in punk culture. Further, I continually frequent local punk bars in an effort to deepen my understanding and to observe the decline of this counterculture. However, I conducted the bulk of my interviews in the fall of 1983 over a period of about two months. During that time, I attended "punk night" at a local bar once a week. In addition, I was invited to other punk functions, such as parties, midnight jam sessions, and public property destruction events. I thereby observed approximately 30 members of this movement with some degree of regularity. I used mainly observational techniques, along with some participation. Although I frequented numerous punk gatherings, my participatory role was constrained by the limited time I spent there. I also had to tread a line between covert and overt roles. While some people knew I was researching this setting, I could not reveal this to others because they might have denied me further access to the group. This created a problem, much the same as that experienced by Henslin (1972) and Adler (1985), in that I had to be careful about what I said and to whom I confided my research interests. This "tightrope effect" severely limited my active participation in the scene.

Nevertheless, by following the investigative research techniques advocated by Douglas (1976), I was able to gain the trust of some key members. I tried to establish friendly relations by running errands for them, buying them drinks and food, and driving them to pick up their welfare checks and food stamps. After several weeks, when I began to be recognized, I was able to broach the topic of doing interviews with several people. I formally interviewed nine people at locations outside the bar. These tape-recorded interviews were unstructured and open-ended. Additionally, I conducted 15 informal interviews at the bar. Finally, I had countless conversations with members, nonmembers, interested bystanders, and social scientists who had an interest in the punk counterculture. In all, my somewhat punk appearance, similar age, regularity at the scene, and apparent acceptance of their lifestyle allowed me to move within the scene freely and easily.

SETTING

The research took place in a small cowboy bar, "The Glass Gun," which was transformed into a punk bar one night a week. The bar was situated in a southwestern city with a population of about 500,000. The city itself is located in the "Bible Belt," characterized by conservative religious and political views. The bar was a small, dark, and dilapidated place. There was a stage area where the bands played, surrounded by a wooden rail. Wobbly tables and torn chairs formed a U-shape around the stage. There was a pool table in the corner, which the punks rarely used. Most of the patrons of the club stood at or near the bar.

For the punks in this city, the Glass Gun was the only place to congregate regularly at that time. On these designated nights, local punk bands played to an audience of about 20 people; some were punks, others were not. The typical audience ranged in age from about 16 to 30, although a few were younger or older. Basically, the punk counterculture was a youth phenomenon. It seemed to attract young, single, mobile people. Snow et al. (1980) have suggested that these characteristics make a person more "structurally available" for movement recruitment. Within the punk scene the number of men and women was fairly equal.

The punk style codes were somewhat diverse. Different styles existed for different kinds of punk. Pfohl (1985: 381) has referred to Hebdige's description of punk style as "the outrageous disfigurement of commonsensical images of aesthetics and beauty and the abrasive, destructive codes of punk style. These are aesthetic inversions of the normal, or consensus-producing, rituals of the dominant culture's style." The basic identifiable element was a subculturally accepted punk hairstyle. These ranged from a very short, uneven haircut, sticking straight up in front, to an American Indian mohawk style, to a shaven head. Along with the haircut, a punk fashion prevailed. The two were inextricably associated. The fashion ranged from torn, faded jeans, T-shirts, and army boots to expensive leather outfits.

The punk dress code was also fairly androgynous. There was no real distinction between male and female fashions. Both men and women wore faded jeans, although leather pants and miniskirts were also quite common among the women. The middle-class punk women, who tended to be students, dressed in a more traditionally feminine manner, glorifying and exaggerating the "glamour girl" image reminiscent of the sixties. This included tight skirts, teased hair, and dark, heavy makeup. The other punk women identified with a more masculine, working-class image, deemphasizing their feminine attributes. Both sexes also wore and admired leather jackets. It was also quite common to see both men and women with multiple pierced earrings all the way around the outside of their ears. Men sometimes wore eye makeup as well. (One man wore miniskirts, makeup, and rhinestones. However, this type of behavior occurred infrequently.) Basically, punk style ran counter to what the dominant culture would deem aesthetically pleasing. One major reason punks dressed as they did was to set themselves apart and to make themselves recognizable. The image consisted of dark, drab clothing, short, spiky, "homemade" haircuts, and blank, bored, expressionless faces reminiscent of those of concentration camp prisoners.[1] The punks created a new aesthetic that revealed their lack of hope, cynicism, and rejection of societal norms.

THE SOCIAL ORGANIZATION OF THE PUNK SCENE

Like the youth gangs that Yablonsky (1962) studied, members of this local punk scene constituted a "near-group." The membership was impermanent and shifting, members' expectations were not always clearly defined, consensus within the group was problematic, and the leadership was vague. Yet out of this uncertainty surfaced an apparent consensus about the stratification of the local community and the roles of the three types of members and peripheral hangers-on who participated in this scene. These four typologies can be hierarchically arranged by the presence (or absence) and intensity of their commitment to the punk counterculture and their consequent display of the punk affectations and belief system. They thus formed a series of outwardly expanding concentric circles, with the most committed members occupying the core, inner roles, and the least involved participants falling around the periphery.

Starting from the center, the number of members occupying each stratum progressively increased as the commitment level of the participants diminished. The categories to which I refer come from the terms used by the participants themselves.[2] The *hardcore punks* were the most involved in the scene, and derived the greatest amount of prestige from their association with it. They set the trends and standards for the rest of the members. The *softcore punks* were less dedicated to the antiestablishment lifestyle and to a permanent association with this counterculture, yet their degree of involvement was still high. They were greater in number and, while highly respected by the less committed participants, did not occupy the same social status within the group as the hardcores. Their roles were, in a sense, dictated by the hardcores, whom they admired, and who defined the acceptable norms and values. The *preppie punks* were only minimally committed, constituting the largest portion of the actual membership. They were held in low esteem by the two core groups, following their lead but lacking the inner conviction and degree of participation necessary to be considered socially desirable within the scene. Finally, the *spectators* made up the largest part of the crowd at any public setting where a punk event transpired. They were not truly members of the group, and therefore did not necessarily revere the actions and dedication of the hardcores as did the two intermediary groups. They did not attempt to follow the standards of those committed to this near-group. They were merely outsiders with an interest in the punk scene.

These four groups constituted the range of participants who attended and were involved with, to varying degrees, the punk counterculture. I will now examine in greater detail their styles, beliefs, practices, intentions, and roles in the scene.

PUNKS AND COMMITMENT

At the time of this study, the group of punks was small and disorganized. The number of people at any given punk event had steadily declined since my first encounters with the scene in 1978. Punk was no longer a new phenomenon, and this particularly conservative community did not provide a very conducive atmosphere for a large countercultural group to flourish. Every member of the group expressed dissatisfaction and boredom with the events (or rather, lack of events) within the scene. Even within the limits necessitated by the relatively small size of the group, there was a great deal of variation in terms of punk roles and characteristics. The qualities attributed to the different roles were based upon commitment to the scene. The punks' perceptions of levels of commitment were based principally on their evaluation of physical appearances and lifestyles. The punks categorized members of the scene on these bases and invented terms to describe them. The four types of participants, described above, varied according to their level of commitment to the scene.

Hardcore Punks

Hardcore punks made up the smallest portion of the scene's membership. In the eyes of the other punks, though, they were the essence of the local movement. The hardcores expressed the greatest loyalty to the punk scene as a whole. Although the hardcores embodied punk fashion and lifestyle codes to the highest degree, their commitment to the counterculture went much deeper than that. As one hardcore punk said:

> There's been so much pure bullshit written about punks. Everyone is shown with a safety pin in their ear or blue hair. The public image is too locked into the fashion. That has nothin' to do with punk, really. . . . For me, it is just my way of life.

The feature that distinguished hardcore punks from other punks was their belief in, and concern for, the punk counterculture. In this sense, the hardcore punks had gone beyond commitment; they had undergone the process of conversion (Snow et al., 1986). In other words, not only did they have membership status, but they believed in and espoused the virtues and ideology of the counterculture. Although many hardcores differed on what the counterculture's core values were, they all expressed some concern with punk ideology. These values were ambiguous at best, but included a distinctly antiestablishment, anarchistic sentiment. Street (1986: 175) has described punk as celebrating chaos and "a life lived only for the moment." The associated value system of punk was understood by the incorporation of cynicism and a distrust of authority. In keeping with other subcultures that intentionally distinguish themselves from the dominant culture, the punk aesthetic, lifestyle, and worldview directly confronted those of the larger society and its traditions. While the other types of punks made no reference to group beliefs or values, the hardcores revered the counterculture. For them, being punk had a profound effect on all aspects of their lives. As one hardcore said:

> There are a lot of punks around, even real punks, who don't mean it. At least, not all the way, like I do. Sometimes I feel so good about punk that I cry. And when I see people getting into some band with real punk lyrics, it's like a religious experience.

It was precisely this belief in "punk" as an external reality, like a higher good, that set the hardcores apart. Similar to Sykes and Matza's (1957) "appeal to higher loyalties," hardcore punks based their rejection of conventional society on their commitment to their antiestablishment lifestyles and beliefs. This imbued their self-identity with a sense of seriousness and purpose. Unlike other punks, they did not view their punk identity as a temporary role or a transitory fashion, but as a permanent way of life. As one hardcore member said:

> Punk didn't influence me to be the way I am much. I was always this way inside. When I came into punk, it was what I needed all my life. I could finally be myself.

Without exception, the hardcores reported having always held the values or qualities associated with the punk counterculture. The local scene, in fact, was just a convenient way of expressing these ideas collectively. Perhaps the most essential value professed by the punks was a genuine disdain for the conventional system. Their use of the term *system* here referred to a general concept of the way the material world works: bureaucracies, power structures, and competition for scarce goods. This "system" further referred to the ethic of deferred gratification, conventional hard work for profit, and the concept of private property. While this bears some similarity to Flacks's (1967) discussion of the student movement and Davis's (1970) portrayal of hippies, hardcore punks generally had a disdain for these earlier youth subcultures. There was a general attitude among punks of the need to create and maintain their own distinctive style. Kinsey found this same feature in the antibourgeois "killum and eatum" subculture. According to Kinsey (1982: 316), "K and E offered an attractive setting as its ideology presented an excellent vehicle for expressing hostility toward conventional society." This contempt for authority and the conventional culture was, in fact, such an essential value for the punks that if one expressed prosystem sentiments or support for the present administration, one could not be considered a committed member, no matter how well one looked the part. Overt behavioral and physical attributes, though, were major ways hardcore punks showed disdain for the system. Particular characteristics were essential for consideration as hardcores. Most fundamentally, a verbal commitment to punk values and the punk scene, in general, was required. For example, John, the epitome of a hardcore punk, claimed to hate the system. He talked about the inequality of the system quite often. In John's words:

> Punk set me free. It let me out of the system. I can walk the streets now and do what I want and not live by the demands of the system. When I walk the streets, I am a punk, not a bum.

However, this verbal pronouncement had to be backed by a certain lifestyle that further indicated commitment to the group. This lifestyle consisted of escaping the system in some way. Almost all of the hardcores were unemployed and lived in old, abandoned houses or moved into the homes of friends for periods of time. Some survived from the charity of sycophantic, less committed punks. Others worked in jobs that they considered to be outside the system, such as musicians in rock bands or artists.

Another central feature of the lifestyle was the hardcores' use of dangerous drugs. Many hardcores indulged heavily in sniffing glue. Glue was inexpensive and readily available to the punks. Its use also symbolized the self-destructive, nihilistic attitude of hardcores and their desire to live outside of society's norms. As one member said:

> It is kind of like a competition, a show-off thing. . . . See who has the most guts by seeing who can burn his brain up first. It is like a total lack of care about anything, really.

This closely corresponds to Davis and Munoz's (1968) description of "freaks." Both punks and freaks were "in search of drug kicks as such, especially if [their] craving carries [them] to the point of drug abuse where [their] health, sanity and relations with intimates are jeopardized" (Davis and Munoz, 1968: 306). Again, here we see a rejection of anything the larger society sees as "sensible."

However, the most salient feature of the hardcore lifestyle was the radical physical appearance. In every case, people who were labeled as hardcore had drastically altered some aspect of their bodies. For example, in addition to the hairstyles discussed earlier, they often had tattoos, such as swastikas, on their arms or faces. Brake (1985: 78) has referred to the use of the swastika as a symbol for punks that was actually devoid of any political significance. Rather, the swastika was a "symbol of contempt" employed as a means of offending the traditional culture. The hardcore punks did their best to alienate themselves from the larger society.

According to Kanter (1972), the first requisite in the principle of a gestalt sociology is that a group forms maximum commitment to this higher ideal by sharply differentiating itself from the larger society. The hardcore members did this by going through the initiation rite of passage: semipermanently altering their appearance. As one hardcore said:

> Did you see Russell's mohawk? I'm so glad for him. He finally decided to go for it. Now he is a punk everywhere . . . no way he can hide it now.

This was similar to certain religious cults, such as the Hare Krishnas, where a drastic change in appearance was required for consideration as a total convert (Rochford, 1985). The punk counterculture informally imposed the same prerequisite. By doing something so out of the ordinary to their appearance, the punks voluntarily deprived themselves of some of the larger society's coveted goods. For example, many of the hardcores were desperately poor. They said that they knew all they would have to do to obtain a job would be to grow their hair into a conventional style; yet they refused. This kind of action based on commitment was what Becker (1960) has called "side bets," where committed people act in such ways that affect their other interests separate and apart from their commitment interests. By making specific choices, people who are committed sacrifice the possible benefits of their other roles. An important characteristic of Becker's notion of side bets is that people are fully aware of the potential ramifications of their actions. This point was illustrated by one punk:

> Some of my friends that aren't punk say, "Why don't you get a job? All you'd have to do is grow your hair out or get a wig and you could get a job." I mean, I know I could. Don't they think I knew that when I did it? It was a big step when I finally cut my hair in a mohawk.

For the hardcore punk, being punk was worth the sacrifices; it was perceived as an inherently good quality. In this respect, the hardcores differed from other types of punks. They held the larger punk scene in esteem. As one loyal member said:

> It really pisses me off when people act like ours is the only punk group in the world. They don't even care what bigger and better groups exist. If this whole thing ended tomorrow, they wouldn't care what happened to the whole punk scene.

The hardcores continually expressed their disgust with the local scene. Much like the hippies studied by Davis (1970), "the scene," in itself, was the message. While not enough people joined the group to satisfy the punks, they were, nonetheless, grateful that they had any kind of group environment to which they could attach themselves. The Glass Gun, with its regularly scheduled "punk night," nonhostile attitude toward them, and coterie of interested bystanders, at least gave them a place to express their values collectively. It was essential in maintaining the group's solidarity and social organization.

Softcore Punks

The softcore punks made up a larger portion of the local scene than the hardcores. There were around fifteen softcore punks. There was one fundamental difference between the hardcore and softcore punks. For the hardcores, it was not sufficient just to be antiestablishment or to wear one's hair in a certain way. Rather, one had to embody the punk lifestyle and ideology in all possible ways. As one hardcore punk put it:

> Everybody thinks she is hardcore because she looks so hardcore. I mean, yeah, she has a mohawk, and she won't get a job and she says she's for anarchy, but she doesn't care that much about being punk. She likes all these different kinds of music and stuff. She seems sometimes like she is just in it for fun. She even says she'll be whatever's in when punk goes out!

The softcore punks lived similar lifestyles to the hardcores. However, the element of "seriousness" about the scene, so pervasive among the hardcores, was absent among the softcores. Visually, the two types were basically indistinguishable. They were different only in their level of commitment. The commitment for softcores was to the lifestyle and the image only, not to "punk" as an ideology or an intrinsically valuable good. The softcores made no pretense of concern for either the larger counterculture or the feeling of permanence about their punk roles. As one softcore, Beth, said:

> Everyone thinks I am so serious about it because I have a mohawk. Some people just can't get past it. Sometimes I get tired of it. Other times, I like to play jokes on people; like another friend of mine who has a mohawk, we'll walk down the street and point at someone with regular hair and say, "Wow! Look at him, he's weird, he doesn't have a mohawk." The fact is, if everyone did have one, I'd do something different to my hair.

The softcores identified with the punk image only temporarily. This distinguished their level of commitment from the *conversion* of the hardcores. The softcores' interest in the scene had only to do with what it could offer them at the present time. While participating, they did what was considered a good job of being "punk." However, if a new cultural trend surfaced, it would be just as likely that they would use their energy effectively to create that particular image. As one softcore punk said:

> I've spent time identifying myself as a hippie, then as a women's libber, then an ecologist, and now as a punk. I'm punk now, but I am in the process of changing into something else. I don't know what. I'm getting bored with this scene. But for now, if I'm gonna do it, I'll do it right.

The softcore punks were somewhat committed in that they participated in some of the more drastic elements of punk lifestyle. For example, softcores had their hair cut in severe ways, just like the hardcores. They were, at least temporarily, committed to being punk (or playing punk) in that they "cut" themselves off from some of society's goods as well. However, the softcores did not share the self-destructive bent of the hardcores. The drugs that they consumed, such as marijuana, alcohol, and amphetamines, were not so potentially dangerous. Yet, because of their apparent visual commitment, and because of the lip service they gave to punk values, softcores were viewed as members in good standing. The hardcores liked and respected the softcores; the two groups associated freely. Some hardcores considered softcores to be simply members in transition. Lofland and Stark (1965) have suggested that movements themselves play a role in promoting the ideology in the new members, rather than the members coming to the movement because its ideology coincides with their own established beliefs. This was the case for the softcore punks. They did not claim to have held punk values before becoming punk. As Anne, a softcore, recalled:

> It was scary to me at first. The hype from the magazines and stuff—all this weird shit, y'know. Then I went there and just hung out. The reason it was frightening is that a lot of people had different ideas about life than

me. And I had to change myself to be with them. I had to be more intense, be an outcast. It was exciting because there seemed to be an element of danger in it—like living on the edge.

A process of simply happening onto the scene was typical of softcore members. Many recounted the feelings of purposelessness that preceded the drift into the punk scene. This drift is similar to the drifts that occur in other deviant lifestyles (Matza, 1964). What Matza called the "mood of desperation" often caused people to drift into delinquency or deviant lifestyles. As Joanie said:

I was really doing nothing with my life and I just kinda accidentally came into the punk scene. I gradually got involved in it that way. The music, and the people to an extent, really raised my consciousness about the system.

Softcores' verbal recognition of punk attitudes, such as awareness of the system, helped to validate their punk performances. Hardcores felt that verbal commitment was an essential first step to further commitment. For this reason, the hardcores accepted the softcores and considered them to be genuine and authentic in their punk identity. Such identification with punk values, along with a typical punk lifestyle, made the distinction between hardcores and softcores difficult. Again, the distinction became clear only with regard to the level of commitment, or seriousness, of the two types. One softcore made this qualification more apparent:

They get mad at me and think I'm insincere or whatever 'cause I like to have fun. I take my politics serious, too, but I feel if you are here, you might as well enjoy it. They think being punk is so serious, they are depressed or stoned all the time.

This statement indicates that the hardcores defined the situation for the local scene. The hardcores decided what differentiated real punks (or committed punks) from pretenders. The hardcores considered only themselves and the softcores to be real. The "realness" of a punk was based on the level of commitment. The level was judged on the basis of willingness to sacrifice other identities for the punk identity. To prove this, a member would have to make permanent his or her punk image. What the punk identity offered was status within its own subculture for those who could not or would not achieve it in conventional society (Cohen, 1955). However, commitment to the deviant identity did not stem from a forced label. On the contrary, commitment to the punk identity was a "self-enhancing attachment" (Goffman, in Stebbins, 1971). The punks' self-esteem was enhanced by the approval they received. It would follow, then, that the more consistent one's behavior was with the superficial signs of commitment, the more prestige one would be able to obtain. Doug, a softcore, commented on this aspect of subcultural prestige among the punks:

In their own way, they're elitist. It's kind of like because they're not part of the general run of things, because they've actually chosen to be rejected in a lot of cases, they've kind of set up their own little social order. It seems to me like it's based on, like a contest, who can be more cool than who. With the really hardcore punks, it's who can self-destruct first; in the name of punk, I guess.

The hardcores and the softcores used the same criteria to judge commitment. Both types agreed that the difference between them was their levels of commitment. Both types fully realized that the softcores did not share the same loyalty to, and identification with, the punk counterculture as a whole. Although both types expressed some commitment to the punk identity and lifestyle, they both realized that the hardcores viewed their own identities as permanent and the softcores' as temporary.

Preppie Punks

The preppie punks made up an even larger portion of the crowd at punk events. The preppies frequented the scene, but approached it similarly to a costume party. They were concerned with the novelty and the fashion. The preppies bore some resemblance to Yablonsky's (1968) "plastic hippies" in that they were drawn to the excitement of the scene. Whereas the core members acted nonchalant and natural about being punk, the preppies could not hide their enthusiasm about being part of the scene. This feature contributed to the core members' perceptions of preppies as "not real" punks. As one core member said:

> It really kills me when these preppie girls come up to me and say "Oh, wow, you're so punk; you're so new wave," like I'm really trying or something.

Preppie punks did not lead the lifestyle of the core members. The preppie punks tended to be from middle-class families, whereas the core punks were generally from lower- or working-class backgrounds.[3] Preppie punks often lived with their parents; they tended to be younger, and were often in school or in respectable, system-sanctioned jobs. This quasi-commitment meant that preppies had to be able to turn the punk image on and off at will. For example, a preppie punk hairstyle, although short, was styled in such a versatile way that it could be manipulated to look punk sometimes and conventional at other times. Preppie fashion was much the same way. Mary, a typical preppie punk, put her regular clothes together in a way she thought would look punk. She ripped up her sorority T-shirt. She bought outfits that were advertised as having the "punk look." Her traditional bangs transformed into "punk" bangs, standing straight up using hair spray or setting gel. The distinguishing feature of preppie punks was the manufactured quality of their punk look. This obvious ability to change roles kept the preppies from being considered real or committed. The preppies were not willing to give anything up for a punk identity. As one softcore said:

> They come in with their little punk outfits from Ms. Jordan's [an exclusive clothing store] and it's written all over 'em: money. They think they can have their nice little jobs and their semipunk hairdo and live with mom and dad and be a real punk, too. Well, they can't.

Another said of preppies:

> It's a little hard to take when you have nothing and they try to have everything. Having all that goes against punk. They gotta choose to not have it. Otherwise, they're just playing a game.

The preppies liked to disavow their punk association in situations that would sanction them negatively for such associations. This state of "dual commitments," in which they never had to reject the conventional world in order to be marginally a part of the group, was characteristic of preppie punks (Cohen, 1955). Kanter (1972) has described a process of conversion and commitment that is commonly found in communes. The first step in the process was the renunciation of previous identities. According to this model, the preppie punks would not be considered committed at all. Thus they could not have been categorized as punks in any meaningful sense. As one core member said, "Being 'punk' to them is like playing cowboys and Indians."

Criticizing and joking about the preppies made up a large portion of core members' conversations. Some truly disliked the preppies and others were flattered by their feeble attempts at imitation of core behavior. For example, when a preppie punk approached one core member, he rolled his eyes and said, "Here comes my fan club," with a half-embarrassed smile and a distinct look of pleasure on his face.

Also, the financial function that the preppies served to core members made them more tolerable. Preppies almost always had jobs or survived by their parents' support. Many of the core members subsisted on the continued generosity of their devout fans. The preppies were more than willing to help the other punks. Preppies sometimes offered hardcores financial help in the form of buying them groceries, driving them places, and providing them with cigarettes, alcohol, and other drugs. Because of this, many punks felt that they could not afford to reject outwardly those who were less committed. As Anne said, "One of these days, this kindness is going to dry up."

Yet the joking and poking fun at preppies was a constant activity. It served to separate, for the committed punks, "us" from "them." It reinforced their sense of being the only real punks. Again the distinction made by core members was grounded in the preppies' attempts to play numerous roles. Haircut and clothing were the decisive clues. The real punks could spot a preppie from a distance; they never had to say a word. As one core member said,

> Oh look, she's punked out her hair. Yes, we're impressed. Tomorrow she'll look just like a Barbie doll again.

Perhaps the most definitive statement separating the real punks from the preppies referred to lifestyle:

> All I know is that I live this seven days a week, and they just do it on weekends.

The preppies, though, while definitely removed from core members, still played an important role in the scene.

Spectators

The category of spectators referred to everyone who observed the scene fairly regularly, but were not punks themselves. This type consisted of, literally, "everyone else." They made up the largest portion of the crowd at the Glass Gun on any given night. They were different from the preppie punks in that they did not try to look punk. They made no pretense of commitment to the scene at all. They did not identify themselves as punks; they had no stake in the scene. Spectators consisted of all different types of people and varied in their occupations, clothes, and reasons for being there. The only common denominator this group shared was the desire to stand back and watch, rather than to participate actively in punk activities. One spectator said of his involvement in this scene:

> People on the fringe are usually voyeurs of a sort. They like to be on the receiver's end of what's happening. Maybe punk is really their alter-ego. And maybe that need is satisfied just by watching and pretending. That's how it is with me, anyway.

The spectators liked to observe the fashion, to listen to the music, and to be "in the know" about the scene. They were, in other words, punk appreciators.

For the most part, spectators on the fringe were ignored by the core members. They never received the attention that preppies did because they made no attempts to "play punk." However, if a spectator appeared on the scene looking completely antithetical to punk, core members would simply laugh or say something derogatory about them and drop the subject. For example, one time a hippie-looking character came in and one punk said, "Oh my God, I think we're in a time warp," to which another punk responded, "Maybe we should tell him that Woodstock's over and that it is 1983."

Following such statements, the punks would watch the spectator's reaction to the scene. For the most part, except as a diversion, the punks were uninterested in the spectators. They did not generally associate with them or talk about them much. Presumably this was the case because of the tremendous turnover in spectator membership.

Most spectators either slowly began to identify with the group (most core members started out as spectators) or stopped frequenting the punk events. There were, however, some loyal spectators. They would frequent the club. They knew most of the punks at least slightly. The punks generally liked this sort of spectator because they provided the punks with an audience. Every type of punk thrived on an audience. The punks needed people to shock. The spectator served that function. The attitude that the members had toward the spectators was one of tolerance and indifference. As one core member said of them:

> They're into it for the novelty. It's like going to the circus for them, to be a part of something new and exciting. But that's okay. I like going to the circus, too; I just like being in it better.

Thus though spectators were only peripheral to the scene they provided an alternative set of norms that functioned to delineate the social boundaries of the counterculture. . . .

NOTES

1. I am indebted to David Matza and John Torpey for this analogy.

2. With the exception of the term *softcore,* all of the distinctions between categories came directly from the participants. The members did make a distinction between hardcore and what I am calling softcore punks. However, the softcores were referred to simply as "punks" by the hardcores, in an effort to distinguish the "hardcore" quality they attributed to themselves. I chose to refrain from using the term punk to apply to one specific category so that I can use the term more freely and generally, and to avoid confusion.

3. Very little information is provided in this text about the class, race, and ethnicity of these participants. The community from which these data come is relatively homogeneous. The few references to class are more impressionistic; that is, based upon knowledge of family occupations, school districts, and so on. However, the dearth of this kind of data stems from the fact that I was more interested in the features the members had in common than in the distinctions between them, with the exception of their differing levels of commitment and their styles.

REFERENCES

Adler, P. A. (1985). *Wheeling and Dealing.* New York: Columbia Univ. Press.

Becker, H. S. (1960). "Notes on the concept of commitment." *Amer. J. of Sociology* 66: 32–40.

Berger, B. (1967). "Hippie morality—more old than new." *Transaction* 5: 19–27.

Brake, M. (1980). *The Sociology of Youth Culture and Youth Subcultures.* London: Routledge.

———. (1985). *Comparative Youth Culture: The Sociology of Youth Culture and Subcultures in America, Britain, and Canada.* London: Routledge.

Cohen, A. (1955). *Delinquent Boys.* Glencoe, IL: Free Press.

Cohen, S. (1972). *Folk Devils and Moral Panics.* New York: St. Martin's.

Davis, F. (1970). "Focus on the flower children: Why all of us may be hippies some day." In J. Douglas (ed.), *Observations of Deviance* (pp. 327–340). New York: Random House.

Davis, F., and L. Munoz. (1968). "Heads and freaks: Patterns and meanings of drug use among hippies." *J. of Health and Social Behavior* 9: 156–164.

Douglas, J. (1970). *Youth in Turmoil,* Washington, DC: National Institute of Mental Health.

———. (1976). *Investigative Social Research.* Newbury Park, CA: Sage.

Flacks, R. (1971). "The liberated generation: An exploration of the roots of student protest." *J. of Social Issues* 23: 52–75.

———. (1971). *Youth and Social Change.* Chicago: Markham.

Frith, S. (1982). *Sound Effects.* New York: Pantheon.

Hall, S., and T. Jefferson (eds.). (1975). *Resistance Through Rituals.* London: Hutchinson.

Hebdige, D. (1981). *Subcultures: The Meaning of Style.* New York: Methuen.

Henslin, J. (1972). "Studying deviance in four settings: Research experiences with cabbies, suicides, drug users, and abortionees." In J. Douglas (ed.), *Research on Deviance* (pp. 35–70). New York: Random House.

Kanter, R. M. (1972). *Commitment and Community: Communes and Utopia in Sociological Perspective.* Cambridge, MA: Harvard Univ. Press.

Kinsey, B. A. (1982). "Killum and eatum: Identity consolidation in a middle class poly-drug abuse subculture." *Symbolic Interaction* 5: 311–324.

Lofland, J., and R. Stark. (1965). "Becoming a world saver: A theory of conversion to a deviant perspective." *Amer. Soc. Rev.* 30: 862–875.

Matza, D. (1964). *Delinquency and Drift.* New York: John Wiley.

Miller, W. (1958). "Lower class culture as a generating milieu of gang delinquency." *J. of Social Issues* 14: 5–19.

Mungham, G., and G. Pearson (eds.). (1976). *Working Class Youth Culture.* London: Routledge.

Pfohl, S. (1985). *Images of Deviance and Social Control.* New York: McGraw-Hill.

Rochford, E. B., Jr., (1985). *Hare Krishnas in America.* New Brunswick, NJ: Rutgers Univ. Press.

Snow, D. A., E. B. Rochford, Jr., S. K. Worden, and R. D. Benford. (1986). "Frame alignment and mobilization." *Amer. Soc. Rev.* 51: 464–481.

Snow, D. A., L. Zurcher, Jr., and S. Ekland-Olson. (1980). "Social networks and social movements: A micro-structural approach to differential recruitment." *Amer. Soc. Rev.* 45: 787–801.

Stebbins, R. A. (1971). *Commitment to Deviance.* Westport, CT: Greenwood.

Street, J. (1986). *Rebel Rock: The Politics of Popular Music.* Oxford: Basil Blackwell.

Sykes, G., and D. Matza. (1957). "Techniques of neutralization." *Amer. Soc. Rev.* 22: 664–670.

Taylor, I. (1982). "Moral enterprise, moral panic, and law-and-order campaigns." In M. M. Rosenberg et al. (eds.), *A Sociology of Deviance* (pp. 123–149). New York: St. Martin's.

Yablonsky, L. (1959). "The delinquent gang as a near-group." *Social Problems* 7: 108–117.

———. (1962). *The Violent Gang.* New York: Macmillan.

———. (1968). *The Hippie Trip.* New York: Pegasus.

Yinger, J. M. (1982). *Countercultures: The Promise and Peril of a World Turned Upside Down.* New York: Free Press.

GANGS

30

Gender and Victimization Risk among Young Women in Gangs

JODY MILLER

Miller offers a glimpse into the contemporary urban world of street gangs in this analysis of the role and dangers faced by female gang members. Gang members not only associate together, they need each others' participation in the deviant act to function (no man or woman is a gang unto himself or herself). Once nearly faded to obscurity, gangs made a rebound in American society in the late 1980s, fueled by the drug economy and the increasing economic plight of urban areas. Since that time they have evolved considerably, adding sophisticated nuances and female members. Miller finds that, while women gain status, social life, and some protection from the hazards of street life in joining gangs, they exchange this for a new set of dangers. By entering the gang world, they are exposing themselves to violence, both from rival gang members and their own homeboys. Miller discusses the particularly gendered status dilemmas and risks for these young women, and how these vary depending on their activities, stance, and associations within the group.

An underdeveloped area in the gang literature is the relationship between gang participation and victimization risk. There are notable reasons to consider the issue significant. We now have strong evidence that delinquent lifestyles are associated with increased risk of victimization (Lauritsen, Sampson, and Laub 1991). Gangs are social groups that are organized around delinquency (see Klein 1995), and participation in gangs has been shown to escalate youth's involvement in crime, including violent crime (Esbensen and Huizinga 1993; Esbensen, Huizinga, and Weiher 1993; Faga 1989, 1990; Thornberry et al. 1993). Moreover, research on gang violence

From Jody Miller, Journal of *Research in Crime and Deliquency*, Vol. 35, No. 4, pp. 430, 434, 436, 438–439, 440–447, 448–449, 450–453, coyright © 1998. Reprinted by permission of Sage Publications.

indicates that the primary targets of this violence are other gang members (Block and Block 1993; Decker 1996; Klein and Maxson 1989; Sanders 1993). As such, gang participation can be recognized as a delinquent lifestyle that is likely to involve high risks of victimization (see Huff 1996:97). Although research on female gang involvement has expanded in recent years and includes the examination of issues such as violence and victimization, the oversight regarding the relationship between gang participation and violent victimization extends to this work as well.

The coalescence of attention to the proliferation of gangs and gang violence (Block and Block 1993; Curry, Ball, and Decker 1996; Decker 1996; Klein 1995; Klein and Maxson 1989; Sanders 1993), and a possible disproportionate rise in female participation in violent crimes more generally (Baskin, Sommers, and Fagan 1993; but see Chesney-Lind, Shelden, and Joe 1996), has led to a specific concern with examining female gang members' violent activities. As a result, some recent research on girls in gangs has examined these young women's participation in violence and other crimes as offenders (Bjerregaard and Smith 1993; Brotherton 1996; Fagan 1990; Lauderback, Hansen, and Waldorf 1992; Taylor, 1993). However, an additional question worth investigation is what relationships exist between young women's gang involvement and their experiences and risk of victimization. Based on in-depth interviews with female gang members, this article examines the ways in which gender shapes victimization risk within street gangs. . . .

METHODOLOGY

Data presented in this article come from survey and semistructured in-depth interviews with 20 female members of mixed-gender gangs in Columbus, Ohio. The interviewees ranged in age from 12 to 17; just over three-quarters were African American or multiracial (16 of 20), and the rest (4 of 20) were White. The sample was drawn primarily from several local agencies in Columbus working with at-risk youths, including the county juvenile detention center, a shelter care facility for adolescent girls, a day school within the same institution, and a local community agency.[1] The project was structured as a gang/nongang comparison, and I interviewed a total of 46 girls. Gang membership was determined during the survey interview by self-definition: About one-quarter of the way through the 50+ page interview, young women were asked a series of questions about the friends they spent time with. They then were asked whether these friends were gang involved and whether they themselves were gang members. Of the 46 girls interviewed, 21 reported that they were gang members[2] and an additional 3 reported being gang involved (hanging out primarily with gangs or gang members) but not gang members. The rest reported no gang involvement.

The survey interview was a variation of several instruments currently being used in research in a number of cities across the United States and included a broad range of questions and scales measuring factors that may be related to gang membership.[3] On issues related to violence, it included questions about peer activities and delinquency, individual delinquent involvement, family violence and abuse, and victimization. When young women responded affirmatively to being gang members, I followed with a series of questions about the nature of their gang, including its size, leadership, activities, symbols, and so on. Girls who admitted gang involvement during the survey participated in a follow-up interview to talk in more depth about their gangs and gang activities. The goal of the in-depth interview was to gain a greater understanding of the nature and meanings of gang life from the point of view of its female members. A strength of qualitative interviewing is its ability to shed light on this aspect of the social world, highlighting the meanings individuals attribute to their experiences (Adler and Adler 1987; Glassner and Loughlin 1987; Miller and Glassner 1997). In addition, using multiple methods, including follow-up interviews, provided me with a means of detecting inconsistencies in young women's accounts of their experiences. Fortunately, no serious contradictions arose. However, a limitation of the data is that only young women were interviewed. Thus, I make inferences about gender dynamics, and young men's behavior, based only on young women's perspectives.

GENDER, GANGS, AND VIOLENCE

Gangs as Protection and Risk

An irony of gang involvement is that although many members suggest one thing they get out of the gang is a sense of protection (see also Decker 1996; Joe and Chesney-Lind 1995; Lauderback et al. 1992), gang membership itself means exposure to victimization risk and even a willingness to be victimized. These contradictions are apparent when girls talk about what they get out of the gang, and what being in the gang means in terms of other members' expectations of their behavior. In general, a number of girls suggested that being a gang member is a source of protection around the neighborhood. Erica,[4] a 17-year-old African American, explained, "It's like people look at us and that's exactly what they think, there's a gang, and they respect us for that. They won't bother us. . . . It's like you put that intimidation in somebody." Likewise, Lisa, a 14-year-old White girl, described being in the gang as empowering: "You just feel like, oh my God, you know, they got my back. I don't need to worry about it." Given the violence endemic in many inner-city communities, these beliefs are understandable, and to a certain extent, accurate.

In addition, some young women articulated a specifically gendered sense of protection that they felt as a result of being a member of a group that was predominantly male. Gangs operate within larger social milieus that are characterized by gender inequality and sexual exploitation. Being in a gang with young men means at least the semblance of protection from, and retaliation against, predatory men in the social environment. Heather, a 15-year-old

White girl, noted," You feel more secure when, you know, a guy's around protectin' you, you know, than you would a girl." She explained that as a gang member, because "you get protected by guys . . . not as many people mess with you." Other young women concurred and also described that male gang members could retaliate against specific acts of violence against girls in the gang. Nikkie, a 13-year-old African American girl, had a friend who was raped by a rival gang member, and she said, "It was a Crab [Crip] that raped my girl in Miller Ales, and um, they was ready to kill him." Keisha, an African American 14-year-old, explained, "If I got beat up by a guy, all I gotta do is go tell one of the niggers, you know what I'm sayin'? Or one of the guys, they'd take care of it."

At the same time, members recognized that they may be targets of rival gang members and were expected to "be down" for their gang at those times even when it meant being physically hurt. In addition, initiation rites and internal rules were structured in ways that required individuals to submit to, and be exposed to, violence. For example, young women's descriptions of the qualities they valued in members revealed the extent to which exposure to violence was an expected element of gang involvement. Potential members, they explained, should be tough, able to fight and to engage in criminal activities, and also should be loyal to the group and willing to put themselves at risk for it. Erica explained that they didn't want "punks" in her gang: "When you join something like that, you might as well expect that there's gonna be fights. . . . And, if you're a punk, or if you're scared of stuff like that, then don't join." Likewise, the following dialogue with Cathy, a White 16-year-old, reveals similar themes. I asked her what her gang expected out of members and she responded, "to be true to our gang and to have our backs." When I asked her to elaborate, she explained,

> **Cathy:** Like, uh, if you say you're a Blood, you be a Blood. You wear your rag even when you're by yourself. You know, don't let anybody intimidate you and be like, "Take that rag off." You know, "you better get with our set." Or something like that.
>
> **JM:** Ok. Anything else that being true to the set means?
>
> **Cathy:** Um. Yeah, I mean, just, just, you know, I mean it's, you got a whole bunch of people comin' up in your face and if you're by yourself they ask you what's your claimin', you tell 'em. Don't say "nothin'."
>
> **JM:** Even if it means getting beat up or something?
>
> **Cathy:** Mmhmm.

One measure of these qualities came through the initiation process, which involved the individual submitting to victimization at the hands of the gang's members. Typically this entailed either taking a fixed number of "blows" to the head and/or chest or being "beatin in" by members for a given duration (e.g., 60 seconds). Heather described the initiation as an important event for determining whether someone would make a good member:

> When you get beat in if you don't fight back and if you just like stop and you start cryin' or somethin' or beggin' 'em to stop and stuff like that, then, they ain't gonna, they'll just stop and they'll say that you're not gang material because you gotta be hard, gotta be able to fight, take punches.

In addition to the initiation, and threats from rival gangs, members were expected to adhere to the gang's internal rules (which included such things as not fighting with one another, being "true" to the gang, respecting the leader, not spreading gang business outside the gang, and not dating members of rival gangs). Breaking the rules was grounds for physical punishment, either in the form of a spontaneous assault or a formal "violation," which involved taking a specified number of blows to the head. For example, Keisha reported that she talked back to the leader of her set and "got slapped pretty hard" for doing so. Likewise, Veronica, an African American 15-year-old, described her leader as "crazy, but we gotta listen to 'im. He's just the type that if you don't listen to 'im, he gonna blow your head off. He's just crazy."

It is clear that regardless of members' perceptions of the gang as a form of "protection," being a gang member also involves a willingness to open oneself up to the possibility of victimization. Gang victimization is governed by rules and expectations, however, and thus does not involve the random vulnerability that being out on the streets without a gang might entail in high-crime neighborhoods. Because of its structured nature, this victimization risk may be perceived as more palatable by gang members. For young women in particular, the gendered nature of the streets may make the empowerment available through gang involvement an appealing alternative to the individualized vulnerability they otherwise would face. However, as the next sections highlight, girls' victimization risks continue to be shaped by gender, even within their gangs, because these groups are structured around gender hierarchies as well.

Gender and Status, Crime and Victimization

Status hierarchies within Columbus gangs, like elsewhere, were male dominated (Bowker et al. 1980; Campbell 1990). Again, it is important to highlight that the structure of the gangs these young women belonged to—that is, male-dominated, integrated mixed-gender gangs—likely shaped the particular ways in which gender dynamics played themselves out. Autonomous female gangs, as well as gangs in which girls are in auxiliary subgroups, may be shaped by different gender relations, as well as differences in orientations toward status, and criminal involvement.

All the young women reported having established leaders in their gang, and this leadership was almost exclusively male. While LaShawna, a 17-year-old African American, reported being the leader of her set (which had a membership that is two-thirds girls, many of whom resided in the same residential facility as her), all the other girls in mixed-gender gangs reported that their OG was male. In fact, a number of young women stated explicitly that only male gang members could be leaders. Leadership qualities, and qualities attributed to high-status members of the gang—being tough, able to fight, and willing to "do dirt" (e.g., commit crime, engage in violence) for the gang— were perceived as characteristically masculine. Keisha noted, "The guys, they just harder." She explained, "Guys is more rougher. We have our G's back but, it ain't gonna be like the guys, they just don't give a fuck. They gonna shoot you in a minute." For the most part, status in the gang was related to traits such as the willingness to use serious violence and commit dangerous crimes and, though not exclusively, these traits were viewed primarily as qualities more likely and more intensely located among male gang members.

Because these respected traits were characterized specifically as masculine, young women actually may have had greater flexibility in their gang involvement than young men. Young women had fewer expectations placed on them—by both their male and female peers—in regard to involvement in criminal activities such as fighting, using weapons, and committing other crimes. This tended to decrease girls' exposure to victimization risk comparable to male members, because they were able to avoid activities likely to place them in danger. Girls *could* gain status in the gang by being particularly hard and true to the set. Heather, for example, described the most influential girl in her set as "the hardest girl, the one that don't take no crap, will stand up to anybody." Likewise, Diane, a White 15-year-old, described a highly respected female member in her set as follows:

> People look up to Janeen just 'cause she's so crazy. People just look up to her 'cause she don't care about nothin'. She don't even care about makin' money. Her, her thing is, "Oh, you're a Slob [Blood]? You're a Slob? You talkin' to me? You talkin' shit to me?" Pow, pow! And that's it. That's it.

However, young women also had a second route to status that was less available to young men. This came via their connections—as sisters, girlfriends, cousins—to influential, high-status young men.[5] In Veronica's set, for example, the girl with the most power was the OG's "sister or his cousin, one of 'em." His girlfriend also had status, although Veronica noted that "most of us just look up to our OG." Monica, a 16-year-old African American, and Tamika, a 15-year-old African American, both had older brothers in their gangs, and both reported getting respect, recognition, and protection because of this connection. This route to status and the masculinization of high-status traits functioned to maintain gender inequality within gangs, but they also could put young women at less risk of victimization than young men. This was both because young women were perceived as less threatening and thus were less likely to be targeted by rivals, and because they were not expected to prove themselves in the ways that young men were, thus decreasing their participation in those delinquent activities likely to increase exposure to violence. Thus, gender inequality could have a protective edge for young women.

Young men's perceptions of girls as lesser members typically functioned to keep girls from being targets of serious violence at the hands of rival young men, who instead left routine confrontations with rival female gang members to the girls in their own gang. Diane said that young men in her gang "don't wanna waste their time hittin' on some little girls. They're gonna go get their little cats [females] to go get 'em." Lisa remarked,

> Girls don't face much violence as [guys]. They see a girl, they say, "we'll just smack her and send her on." They see a guy—'cause guys are like a lot more into it than girls are, I've noticed that—and they like, well, "we'll shoot him."

In addition, the girls I interviewed suggested that, in comparison with young men, young women were less likely to resort to serious violence, such as that involving a weapon, when confronting rivals. Thus, when girls' routine confrontations were more likely to be female on female than male on female, girls' risk of serious victimization was lessened further.

Also, because participation in serious and violent crime was defined primarily as a masculine endeavor, young women could use gender as a means of avoiding participation in those aspects of gang life they found risky, threatening, or morally troubling. Of the young women I interviewed, about one-fifth were involved in serious gang violence: A few had been involved in aggravated assaults on rival gang members, and one admitted to having killed a rival gang member, but they were by far the exception. Most girls tended not to be involved in serious gang crime, and some reported that they chose to exclude themselves because they felt ambivalent about this aspect of gang life. Angie, an African American 15-year-old, explained,

> I don't get involved like that, be out there goin' and just beat up people like that or go stealin', things like that. That's not me. The boys, mostly the boys do all that, the girls we just sit back and chill, you know.

Likewise, Diane noted,

> For maybe a drive-by they might wanna have a bunch of dudes. They might not put the females in that. Maybe the females might be weak inside, not strong enough to do something like that, just on the insides. . . . If a female wants to go forward and doin' that, and she wants to risk her whole life for doin' that, then she can. But the majority of the time, that job is given to a man.

Diane was not just alluding to the idea that young men were stronger than young women. She also inferred that young women were able to get out of committing serious crime, more so than young men, because a girl shouldn't have to "risk her whole life" for the gang. In accepting that young men were more central members of the gang, young women could more easily participate in gangs without putting themselves in jeopardy—they could engage in the more routine, everyday activities of the gang, like hanging out, listening to music, and smoking bud (marijuana). These male-dominated mixed-gender gangs thus appeared to provide young women with flexibility in their involvement in gang activities. As a result, it is likely that their risk of victimization at the hands of rivals was less than that of young men in gangs who were engaged in greater amounts of crime.

Girls' Devaluation and Victimization

In addition to girls choosing not to participate in serious gang crimes, they also faced exclusion at the hands of young men or the gang as a whole (see also Bowker et al. 1980). In particular, the two types of crime mentioned most frequently as "off-limits" for girls were drug sales and drive-by shootings. LaShawna explained, "We don't really let our females [sell drugs] unless they really wanna and they know how to do it and not to get caught and everything." Veronica described a drive-by that her gang participated in and said, "They wouldn't let us [females] go. But we wanted to go, but they wouldn't let us." Often, the exclusion was couched in terms of protection. When I asked Veronica why the girls couldn't go, she said, "so we won't go to jail if they was to get caught. Or if one of 'em was to get shot, they wouldn't want it to happen to us." Likewise, Sonita, a 13-year-old African American, noted, "If they gonna do somethin' bad and they think one of the females gonna get hurt they don't let 'em do it with them. . . . Like if they involved with shooting or whatever, [girls] can't go."

Although girls' exclusion from some gang crime may be framed as protective (and may reduce their victimization risk vis-à-vis rival gangs), it also served to perpetuate the devaluation of female members as less significant to the gang—not as tough, true, or "down" for the gang as male members. When LaShawna said her gang blocked girls' involvement in serious crime, I pointed out that she was actively involved herself. She explained, "Yeah, I do a lot of stuff 'cause I'm tough. I likes, I likes messin' with boys. I fight boys. Girls ain't nothin' to me." Similarly, Tamika said, "girls, they little peons."

Some young women found the perception of them as weak a frustrating one. Brandi, an African American 13-year-old, explained, "Sometimes I dislike that the boys, sometimes, always gotta take charge and they think, sometimes, that the girls don't know how to take charge 'cause we're like girls, we're females, and like that." And Chantell, an African American 14-year-old, noted that rival gang members "think that you're more of a punk." Beliefs that girls were weaker than boys meant that young women had a harder time proving that they were serious about their commitment to the gang. Diane explained,

> A female has to show that she's tough. A guy can just, you can just look at him. But a female, she's gotta show. She's gotta go out and do some dirt. She's gotta go whip some girl's ass, shoot somebody, rob somebody or something. To show that she is tough.

In terms of gender-specific victimization risk, the devaluation of young women suggests several things. It could lead to the mistreatment and victimization of girls by members of their own gang when they didn't have specific male protection (i.e., a brother, boyfriend) in the gang or when they weren't able to stand up for themselves to male members. This was exacerbated by activities that led young women to be viewed as sexually available. In addition, because young women typically were not seen as a threat by young men, when they did pose one, they could be punished even more harshly than young men, not only for having challenged a rival gang or gang member but also for having overstepped "appropriate" gender boundaries.

Monica had status and respect in her gang, both because she had proven herself through fights and criminal activities, and because her older brothers were members of her set. She contrasted her own treatment with that of other young women in the gang:

> They just be puttin' the other girls off. Like Andrea, man. Oh my God, they dog Andrea so bad. They like, "Bitch, go to the store." She like, "All right, I be right back." She will go to the store and go and get them whatever they want and come back with it. If she don't get it right, they be like, "Why you do that bitch?" I mean, and one dude even smacked her. And, I mean, and, I don't, I told my brother once. I was like, "Man, it ain't even like that. If you ever see someone tryin' to disrespect me like that or hit me, if you do not hit them or at least say somethin' to them. . . ." So my brothers, they kinda watch out for me.

However, Monica put the responsibility for Andrea's treatment squarely on the young woman: "I put that on her. They ain't gotta do her like that, but she don't gotta let them do her like that either." Andrea was seen as "weak" because she did not stand up to the male members in the gang; thus, her mistreatment was framed as partially deserved because she did not exhibit the valued trails of toughness and willingness to fight that would allow her to defend herself.

An additional but related problem was when the devaluation of young women within gangs was sexual in nature. Girls, but not boys, could be initiated into the gang by being "sexed in"—having sexual relations with multiple male members of the gang. Other members viewed the young women initiated in this way as sexually available and promiscuous, thus increasing their subsequent mistreatment. In addition, the stigma could extend to female members in general, creating a sexual devaluation that all girls had to contend with.

The dynamics of "sexing in" as a form of gang initiation placed young women in a position that increased their risk of ongoing mistreatment at the hands of their gang peers. According to Keisha, "If you get sexed in, you have no respect. That means you gotta go ho'in' for 'em; when they say you give 'em the pussy, you gotta give it to 'em. If you don't, you gonna get your ass beat. I ain't down for that." One girl in her set was sexed in and Keisha said the girl "just do everything they tell her to do, like a dummy." Nikkie reported that two girls who were sexed into her set eventually quit hanging around with the gang because they were harassed so much. In fact, Veronica said the young men in her set purposely tricked girls into believing they were being sexed into the gang and targeted girls they did not like:

> If some girls wanted to get in, if they don't like the girl they have sex with 'em. They run trains on 'em or either have the girl suck their thang. And then they used to, the girls used to think they was in. So, then the girls used to just come try to hang around us and all this little bull, just 'cause, 'cause they thinkin' they in.

Young women who were sexed into the gang were viewed as sexually promiscuous, weak, and not "true" members. They were subject to revictimization and mistreatment, and were viewed as deserving of abuse by other members, both male and female. Veronica continued, "They [girls who are sexed in] gotta do whatever, whatever the boys tell 'em to do when they want 'em to do it, right then and there, in front of whoever. And, I think, that's just sick. That's nasty, that's dumb." Keisha concurred, "She brought that on herself, by bein' the fact, bein' sexed in." There was evidence, however, that girls could overcome the stigma of having been sexed in through their subsequent behavior, by challenging members that disrespect them and being willing to fight. Tamika described a girl in her set who was sexed in, and stigmatized as a result, but successfully fought to rebuild her reputation:

> Some people, at first, they call her "little ho" and all that. But then, now she startin' to get bold. . . . Like, they be like, "Ooh, look at the little ho. She fucked me and my boy." She be like, "Man, forget y'all. Man, what? What?" She be ready to squat [fight] with 'em. I be like, "Ah, look at her!" Uh huh. . . . At first we looked at her like," Ooh, man, she a ho, man." But now we look at her like she just our kickin' it partner. You know, however she got in that's her business.

The fact that there was such an option as "sexing in" served to keep girls disempowered, because they always faced the question of how they got in and of whether they were "true" members. In addition, it contributed to a milieu in which young women's sexuality was seen as exploitable. This may help explain why young women were so harshly judgmental of those girls who were sexed in. Young women who were privy to male gang members' conversations reported that male members routinely disrespect girls in the gang by disparaging them sexually. Monica explained,

> I mean the guys, they have their little comments about 'em [girls in the gang] because, I hear more because my brothers are all up there with the guys and everything and I hear more just sittin' around, just listenin'. And they'll have their little jokes about "Well, ha I had her," and then and everybody else will jump in and say, "Well, I had her, too." And then they'll laugh about it.

In general, because gender constructions defined young women as weaker than young men, young women were often seen as lesser members of the gang. In addition to the mistreatment these perceptions entailed, young women also faced particularly harsh sanctions for crossing gender boundaries—causing harm to rival male members when they had been viewed as nonthreatening. One young woman[6] participated in the assault of a rival female gang member, who had set up a member of the girl's gang. She explained, "The female was supposingly goin' out with one of ours, went back and told a bunch of [rivals] what was goin' on and got the [rivals] to jump my boy. And he ended up in the hospital." The story she told was unique but nonetheless significant for what it indicates about the gendered nature of gang violence and victimization. Several young men in her set saw the girl walking down the street, kidnapped her, then brought her to a member's house. The young woman I interviewed, along with several other girls in her set, viciously beat the girl, then to their surprise the young men took over the beating, ripped off the girl's clothes, brutally gang-raped her, then dumped her in a park. The interviewee noted, "I don't know what happened to her. Maybe she died. Maybe, maybe someone came and helped her. I mean, I don't know." The experience scared the young woman who told me about it. She explained,

> I don't never want anythin' like that to happen to me. And I pray to God that it doesn't. 'Cause God said that whatever you sow you're gonna reap. And like, you know, beatin' a girl up and then sittin' there watchin' somethin' like that happen, well, Jesus that could come back on me. I mean, I felt, I really did feel sorry for her even though my boy was in the hospital and was really hurt. I mean, we coulda just shot her. You know, and it coulda been just over. We coulda just taken her life. But they went farther than that.

This young woman described the gang rape she witnessed as "the most brutal thing I've ever seen in my life." While the gang rape itself was an unusual event, it remained a specifically gendered act that could take place precisely because young women were not perceived as equals. Had the victim been an "equal," the attack would have remained a physical one. As the interviewee herself noted, "we coulda just shot her." Instead, the young men who gang-raped the girl were not just enacting revenge on a rival but on a *young woman* who had dared to treat a young man in this way. The issue is not the question of which is worse—to be shot and killed, or gang-raped and left for dead. Rather, this particular act sheds light on how gender may function to structure victimization risk within gangs.

DISCUSSION

Gender dynamics in mixed-gender gangs are complex and thus may have multiple and contradictory effects on young women's risk of victimization and repeat victimization. My findings suggest that participation in the delinquent lifestyles associated with gangs clearly places young women at risk for victimization. The act of joining a gang involves the initiate's submission to victimization at the hands of her gang peers. In addition, the rules governing gang members' activities place them in situations in which they are vulnerable to assaults that are specifically gang related. Many acts of violence that girls described would not have occurred had they not been in gangs.

It seems, though, that young women in gangs believed they have traded unknown risks for known ones—that victimization at the hands of friends, or at least under specified conditions, was an alternative preferable to the potential of random, unknown victimization by strangers. Moreover, the gang offered both a semblance of protection from others on the streets, especially young men, and a means of achieving retaliation when victimization did occur. . . .

Girls' gender, as an individual attribute, can function to lessen their exposure to victimization risk by defining them as inappropriate targets of rival male gang members' assaults. The young women I interviewed repeatedly commented that young men were typically not as violent in their routine confrontations with rival young women as with rival young men. On the other hand, when young women are targets of serious assault, they may face brutality that is particularly harsh and sexual in nature because they are female—thus, particular types of assault, such as rape, are deemed more appropriate when young women are the victims.

Gender can also function as a state-dependent factor, because constructions of gender and the enactment of gender identities are fluid. On the one hand, young women can call upon gender as a means of avoiding exposure to activities they find risky, threatening, or morally troubling. Doing so does not expose them to the sanctions likely faced by male gang members who attempt to avoid participation in violence. Although these choices may insulate young women from the risk of assault at the hands of rival gang members, perceptions of female gang members—and of women in general—as weak may contribute to more routinized victimization at the hands of the male members of their gangs. Moreover, sexual exploitation in the form of "sexing in" as an initiation ritual may define young women as sexually available, contributing to a likelihood of repeat victimization unless the young woman can stand up for herself and fight to gain other members' respect.

Finally, given constructions of gender that define young women as nonthreatening, when young women do pose a threat to male gang members, the sanctions they face may be particularly harsh because they not only have caused harm to rival gang members but also have crossed appropriate gender boundaries in doing so. In sum, my findings suggest that gender may function to insulate young women from some types of physical assault and lessen their exposure to risks from rival gang members, but also to make them vulnerable to particular types of violence, including routine victimization by their male peers, sexual exploitation, and sexual assault.

NOTES

1. I contacted numerous additional agency personnel in an effort to draw the sample from a larger population base, but many efforts remained unsuccessful despite repeated attempts and promises of assistance. These included persons at the probation department, a shelter and outreach agency for runaways, police personnel, a private residential facility for juveniles, and three additional community agencies. None of the agencies I contacted openly denied me permission to interview young women; they simply chose not to follow up. I do not believe that much bias resulted from the nonparticipation of these agencies. Each has a client base of "at-risk" youths, and the young women I interviewed report overlap with some of these same agencies. For example, a number had been or were on probation, and several reported staying at the shelter for runaways.

2. One young woman was a member of an all-female gang. Because the focus of this article is gender dynamics in mixed-gender gangs, her interview is not included in the analysis.

3. These include the Gang Membership Resistance Surveys in Long Beach and San Diego, the Denver Youth Survey, and the Rochester Youth Development Study.

4. All names are fictitious.

5. This is not to suggest that male members cannot gain status via their connections to high-status men, but that to maintain status, they will have to successfully exhibit masculine traits such as toughness. Young women appear to be held to more flexible standards.

6. Because this excerpt provides a detailed description of a specific serious crime, and because demographic information on respondents is available, I have chosen to conceal both the pseudonym and gang affiliation of the young woman who told me the story.

REFERENCES

Adler, Patricia A. and Peter Adler. 1987. *Membership Roles in Field Research.* Newbury Park, CA: Sage.

Baskin, Deborah, Ira Sommers, and Jeffrey Fagan. 1993. "The Political Economy of Violent Female Street Crime." *Fordham Urban Law Journal* 20:401–17.

Bjerregaard, Beth and Carolyn Smith. 1993. "Gender Differences in Gang Participation, Delinquency, and Substance Use." *Journal of Quantitative Criminology* 4:329–55.

Block, Carolyn Rebecca and Richard Block. 1993. "Street Gang Crime in Chicago." Research in Brief. Washington, DC: National Institute of Justice.

Bowker, Lee H., Helen Shimota Gross, and Malcolm W. Klein. 1980. "Female Participation in Delinquent Gang Activities." *Adolescence* 15(59): 509–19.

Brotherton, David C. 1996. " 'Smartness,' 'Toughness,' and 'Autonomy': Drug Use in the Context of Gang Female Deliquency." *Journal of Drug Issues* 26 (1): 261–77.

Campbell, Anne. 1984. *The Girls in the Gang.* New York: Basil Blackwell.

———. 1990. "Female Participation in Gangs." Pp. 163–82 in *Gangs in America,* edited by G. Ronald Huff. Beverly Hills, CA: Sage.

Chesney-Lind, Meda, Randall G. Shelden, and Karen A. Joe. 1996. "Girls, Delinquency, and Gang Membership." Pp. 185–204 in *Gangs in America*, 2d ed., edited by C. Ronald Huff. Thousand Oaks, CA: Sage.

Curry, G. David, Richard A. Ball, and Scott H. Decker. 1996. Estimating the National Scope of Gang Crime from Law Enforcement Data Research in Brief. Washington, DC: National Institute of Justice.

Decker, Scott H. 1996. "Collective and Normative Features of Gang Violence." *Justice Quarterly* 13 (2): 243–64.

Decker, Scott H. and Barrik Van Winkle. 1996. *Life in the Gang*. Cambridge, UK: Cambridge University Press.

Esbensen, Finn-Aage and David Huizinga. 1993. "Gangs, Drugs, and Delinquency in a Survey of Urban Youth." *Criminology* 31 (4) 565–89.

Esbensen, Finn-Aage, David Huizinga, and Anne W. Weiher. 1993. "Gang and Non-Gang Youth: Differences in Explanatory Factors." *Journal of Contemporary Criminal Justice* 9 (2): 94–116.

Fagan, Jeffrey. 1989. "The Social Organization of Drug Use and Drug Dealing among Urban Gangs." *Criminology* 27(4): 633–67.

———. 1990. "Social Processes of Delinquency and Drug Use among Urban Gangs." Pp. 183–219 in *Gangs in America*, edited by C. Ronald Huff. Newbury Park, CA: Sage.

Glassner, Barry and Julia Loughlin. 1987. *Drugs in Adolescent Worlds: Burnouts to Straights*. New York: St. Martin's.

Huff, C. Ronald. 1996. "The Criminal Behavior of Gang Members and Nongang At-Risk Youth." Pp. 75–102 in *Gangs in America*, 2d ed., edited by C. Ronald Huff. Thousand Oaks, CA: Sage.

Joe, Karen A. and Meda Chesney-Lind. 1995. "Just Every Mother's Angel: An Analysis of Gender and Ethnic Variations in Youth Gang Membership." *Gender & Society* 9(4): 408–30.

Klein, Malcolm W. 1995. *The American Street Gang: Its Nature, Prevalence and Control*. New York: Oxford University Press.

Klein, Malcolm W. and Cheryl L. Maxson. 1989. "Street Gang Violence." Pp. 198–231 in *Violent Crime, Violent Criminals*, edited by Neil Weiner and Marvin Wolfgang. Newbury Park, CA: Sage.

Lauderback, David, Joy Hansen, and Dan Waldorf. 1992. " 'Sisters Are Doin' It for Themselves': A Black Female Gang in San Francisco." *The Gang Journal* 1 (1): 57–70.

Lauritsen, Janet L., Robert J. Sampson, and John H. Laub. 1991. "The Link between Offending and Victimization among Adolescents." *Criminology* 29 (2): 265–92.

Miller, Jody and Barry Glassner. 1997. "The 'Inside' and the 'Outside': Finding Realities in Interviews." Pp. 99–112 in *Qualitative Research*, edited by David Silverman. London: Sage.

Sanders, William. 1993. *Drive-Bys and Gang Bangs: Gangs and Grounded Culture*. Chicago: Aldine.

Taylor, Carl. 1993. *Girls, Gangs, Women and Drugs*. East Lansing: Michigan State University Press.

Thornberry, Terence P., Marvin D. Krohn, Alan J. Lizotie, and Deborah Chard-Wierschem. 1993. "The Role of Juvenile Gangs in Facilitating Delinquent Behavior." *Journal of Research in Crime and Delinquency* 30 (1): 75–85.

FORMAL ORGANIZATIONS

31

International Organized Crime

ROY GODSON AND WILLIAM J. OLSON

If scene participants have loose relationships characterized by socializing together, and gang participants have tighter relationships based on living, partying, and fighting together, individuals who belong to deviant formal organizations share the bond of membership in large-scale, highly structured deviant associations that extend over time and space. Deviant formal organizations entail more highly sophisticated relationships than the less structure or committed crews or rings, where we first see an emerging division of labor, so that individual members learn specific skills and pursue these specialized tasks within the group. Members of deviant formal organizations work together in enterprises that may be specialized or diverse, tightly connected from top to bottom or portioned out in authority to separate subspheres, ethnically and/or familially homogeneous, and capable of earning vast amounts of money. Godson and Olson offer a rare depiction of the modern breed of international cartels that contains a description and analysis of the nature, scope, type of organization, history, and international consequences for crime, law enforcement, and the erosion of legitimate control over this most sophisticated form of deviant group. In so doing they contrast this level of deviant relationships to the less organizationally sophisticated gangs that operate domestically in our major urban areas.

On August 18, 1989, in the middle of an adoring crowd, a gunman murdered Luis Carlos Galan, the leading candidate for the presidency of Colombia. The act shocked the nation. Political murder was not a new phenomenon in Colombia. What was new were the perpetrators. The assassin was not an individual acting alone, nor was he a member of a guerrilla group, of which Colombia has several violent examples. Instead, the assassin was a hireling, a *sicario*, acting on the orders of the Medellín Cartel, one of the world's major international criminal organizations. The murder was only the beginning. The Medellín Cartel launched a full-scale terrorist assault on

---

Reprinted by permission of Transaction Publishers. "International Organized Crime" by Roy Godson and William J. Olson, from *Society*, Jan./Feb. 1995. Copyright © 1995 by Transaction Publishers.

the country. Public facilities and newspaper offices were bombed. Members of leading families were kidnapped. Hundreds of policemen were murdered. A national airline flight was blown up in mid-flight. The cartel waged war against the Colombian state. Its aim was to force the government to come to terms with the cartels, in effect, to share power with the drug traffickers.

The Colombian government, which had been trying to control not only Medellín groups but also the less prominent Cali group, now faced a far more violent and direct threat to its institutions. It called for help. The United States, which was already trying to support Colombian efforts, came forward with an emergency aid package and promises of more aid to follow. The United States provided upwards of $400 million in police, military, and advisory assistance over five years. These funds were intended to eliminate the major drug trafficking organizations in Colombia, and many millions more were spent to help attack the overseas operations responsible for producing and transshipping almost all the cocaine in the world.

The level of support and cooperation was unprecedented. And the threat? A drug-trafficking organization with the wealth and power to challenge the internal stability of one country while it defied the power and authority of the world's remaining superpower. Hubris? Perhaps, but the nature of the confrontation and the fact that it is not over says something fundamental about the modern world, about the nature of state power, international relations, and the stability of governments. And the situation in Colombia is far from the whole picture.

In 1991 and 1992 the disruptive influence of organized criminal activity rocked the foundations of political stability in Italy. Since World War II organized crime has carried out a wide range of domestic activities, using violence, extortion, bribery, and murder to advance its interests. More recently the Sicilian Mafia specifically has been locked in a murderous struggle with the government, assassinating judges, policemen, and those seen as interfering in its operations, and, more ominously, using its economic power to try to corrupt the political process itself. The scandals surrounding official corruption linked to the Sicilian, Neapolitan, and Calabrian Mafias have touched the very heart of the Italian government, undermining its credibility and effectiveness.

In Burma, large parts of the country are under the control of separatist movements who finance their activities by selling opium on the international market. The government itself engages in or countenances the production and trafficking of opium to help finance its operations. All over Asia, organized criminal groups, Pakistani, Thai, Chinese, or Japanese, operate vast international organizations trafficking in drugs or engaging in a wide variety of other criminal activities, in many cases with the complicity of local government and military officials.

In the former Soviet Union and in the struggling states of Eastern Europe, criminal organizations, long held in check, are beginning to grow. They have developed international links to improve their own organizational abilities and marketing contacts. Of more concern, these criminal groups are penetrating local governments (which are often struggling for cohesion and lacking resources) by using bribery and violence to win protection for their expanding operations. Governmental resources, strained to cope with a wide range of social and economic problems, are completely inadequate to respond; in fact, governments are unable to assess the extent of the problem accurately. Political paralysis and economic hardship have combined to give various criminal organizations considerable freedom to operate, even when local governments are not cooperating with criminal elements. Governments, however, are not the only institutions vulnerable to criminal penetration.

In a New York courtroom in 1992, Clark Clifford, an adviser to presidents and one of the most respected men in the United States, was called as a defendant in a case involving a vast illegal international financial enterprise, the Bank of Credit and Commerce International (BCCI). So far, BCCI is the biggest such case, but it illustrates only too graphically the extent to which banking and financial systems are vulnerable to penetration, manipulation, and fraud by criminal groups. The mechanisms whereby incredible sums of illegal proceeds—perhaps $300 billion in drug money alone—are laundered and massive frauds are perpetrated through the world's financial markets are still only dimly understood, but the realities of the process underscore the permeability of the system. This permeability of governments and private business and the growth of major international organized crime raise concerns for the future.

Whether in the developed or in the developing world, criminal organizations' scope of action and range of capabilities are undergoing a profound change. Decline in political order, deteriorating economic circumstances, a growing underground economy that habituates people to working outside the legal framework, easy access to arms, the massive flow of emigrants and refugees, and the normal difficulties involved in engendering meaningful state-to-state cooperation are working to the advantage of criminal organizations. The rise of better-organized, internationally based criminal groups with vast financial resources is creating a new threat to the stability and security of the international system. As Senator John Kerry noted, "this is new. This is something that none of us has ever experienced before. It is not ideological. It has nothing to do with right or left, but it is money-oriented, greed-based criminal enterprise that has decided to take on the lawful institutions and civilized society." The growth of these organizations presents a major challenge to the quality of life in the United States and to U.S. interests.

ORGANIZED CRIME AND BUSINESS

Although major criminal organizations pose a new and compelling challenge to national and international interests, the extent of the threat should not be exaggerated. It is clear that the wealth and power of individual organizations has grown and there are increasing signs of international links between various criminal organizations. This does not mean, however, that there is an integrated, centrally directed criminal conspiracy. The first business of criminal organizations is usually business, its promotion and protection. In this sense criminal organizations are similar to legitimate enterprises. Like the activities of their legal twins, the activities of separate "corporations" can be cooperative or competitive by turns. The long-term threat from these organizations is subtle and more insidious than images of criminal masterminds seeking to dominate the globe in some vast, shared, and centrally coordinated enterprise. It is important and difficult to measure and to understand the true nature and composition of the threat.

Criminal organizations, of course, are not new. Oliver Twist's Fagin is only one of the many memorials to the possibilities of an organized criminal underground. Nor is there anything particularly new about the ethnic composition of such organizations; the Sicilian Mafia and the Chinese Triads, in particular, are of venerable lineage, and many have had international dimensions. Yet there is something fundamentally different in the threat that such organizations now pose to organized society that must be understood so that the United States and its friends and allies around the world can undertake a reasonable, timely, and effective response. First, however, it is useful to define the nature and contours of the emerging problem.

It is essential to come to terms with what is meant by criminal organization. What distinguishes it from regular criminal activity, from a legitimate business enterprise, or from economic activity in underground or informal economies?

In many parts of the world, there are forms of entrepreneurial activity that are classified as illegitimate if they do not meet the state's test for permits, licenses, and so on. These informal economic activities now account for a significant proportion of all meaningful economic activity in many parts of the Third World. In Peru, for example, upwards of 50 percent of the economically active population are engaged in the informal economy, generating as much as 40 percent of the gross domestic product (GDP), all of it illegal under existing law. (This does not include illegal coca cultivation and cocaine processing.) There is considerable organization in this effort. This does not mean, however, that this activity is indicative of organized crime, although there are some important parallels. It is important to keep this distinction in mind. It was not that long ago, for instance, that all nonstate economic activity was illegal in the former Soviet Union. As the economy of Russia struggles to make the transition from statist control to free market and many of the people formerly engaged in illegal activities begin to emerge into the new economy, it will be important to understand and to be able to draw the distinction between true criminals and those who were criminals by definition. What is involved? There are a variety of definitions and definitional approaches, but several elements are essential.

First, the activities involved must be criminal. They must violate laws for which there is a punishment prescribed by a legal authority capable of enforcing the laws. This raises a problem, as the trouble with informal economies indicates. Clearly, many societies, including our own, make the informal economy illegal and provide punishment. By definition, then, these activities are criminal, and when they become a major component of the GDP outside the formal economy and the tax base these activities can adversely affect the growth of the formal economy. However, the informal economy can also make a major contribution to the quality of life, as in the Soviet case, where the average citizen would have been far worse off far sooner if it had not been for private entrepreneurs providing a wide range of goods and services. The dividing line is blurred. Nevertheless, most societies, past and present, have defined a number of behaviors—such as murder, drug trafficking, prostitution, extortion, kidnapping, and theft—as not only outside the law but fundamentally wrong. By contrast, the informal economy essentially encompasses "normal" economic activity that is legitimate in the formal sector. But the primary concern here is not just with crime, per se, but with major crime, and there are additional features to help sharpen the picture.

A second central characteristic of the crime, which does present a significant threat to established societies, is that the criminals engaged in particular activities must be organized. While trite, it is important to understand that the individuals involved are not acting alone, and the activities in which they are engaged are not random. Further, the economic behavior in mature criminal organizations, unlike that of groups in an illegal, informal economy in which groups may loosely cooperate, is intentional. The activities are usually directed by identifiable leaders. Mature criminal organizations operate with varying organizational structures for a common purpose, which is outside the law.

Organization is key, yet there is no standardized organizational chart for a developed criminal organization. There are varying types of organizational patterns. Like a legitimate business enterprise, a criminal organization may employ various features best designed to carry out its purposes. Thus it can be vertically organized and fairly tightly controlled, as are the Colombian cocaine cartels; or it may be regionally organized, often around functions, as is the American Mafia; or it may be even more loosely organized, as are the Jamaican Yardies in England. It may have a quasi-religious character, as do the Chinese Triads, or a semi-political/military organization, as does the Shan United Army in Burma. The organizational nature, then, while a key component, is not the sole defining feature.

The purposes to which the organization is dedicated ultimately define its legality or criminality. The activities of criminal organizations are equivalent to many of the efforts of legitimate business: export-import, trade in various articles, wholesale and retail sales, services. The members of criminal organizations, however, seek to operate in areas outside legal guidelines and generally trade in items also defined as illegal—such as drugs. However, as in the case of the Chinese Triads, they may use their organization for both legal and illegal activities.

Interestingly, there are indications that many of the present major criminal organizations began not as deliberately criminal groups but as protective-benevolent or secret societies for the welfare of persecuted ethnic or political groups. The Chinese Triads and the Sicilian Mafia began not as bodies to carry out illegal economic activities but as organizations to provide a type of community for their members. They provided aid and comfort to members in tough times. They shielded them against turmoil, anarchy, and arbitrary government. They served as resistance movements. Robin Hood and Jesse James of legend and the movies were the leaders of such groups. In the government's eyes they were thugs. In legend they were heroes. This shadowy area between heroic resistance to injustice and criminal activity will continue to pose difficulties in sorting out criminals from patriots.

Over time, however, these groups, acting outside the law and in opposition to existing authority, typically began to focus increasingly on illegal (and legal) acts for profit rather than for a cause. As the profit motive took precedence, resistance took a back seat. Not all criminal organizations, however, began as failed resistance efforts. Many evolved from groups of thieves or gangs that always sought profit outside the law. Some of these groups claimed political motives. But a gang, while organized, is generally a very small and localized affair with very limited means. In general, major criminal organizations of the type that are of greatest concern here have grown into substantial enterprises, often with transnational connections, involve hundreds if not thousands of "employees," and are no longer confined to localized areas. The third point to note about criminal organizations, then, is that their underlying purpose is to make a profit from illegal acts and that they employ a large number of people whose activities are coordinated over time. The organization usually is neither ephemeral nor temporary.

A fourth characteristic of major criminal organizations is their willingness to use violence to promote and protect their interests. Violence and crime often go together. But criminal organizations use violence deliberately. They control its use and direct it in specific cases to achieve defined goals. Violence is a "business" tool and not a random or individual act. There is sometimes a lack of discipline or acts of individual cruelty inside organized crime, but generally violence is used for business purposes. It is directed outward to intimidate or eliminate rivals and threats, and it is directed inward to enforce discipline within the organization. The level of violence may vary—the Medellín Cartel, for example, is far more prone to resort to violence than is the Cali Cartel—but it remains a consciously controlled instrument.

A corollary to the use of violence is the purposeful use of bribery. It should be clear that criminal organizations are in direct conflict with police and other governmental agencies. In some cases, the criminal organization may have more firepower than the state—as with the Shan United Army—but generally they cannot sustain direct violence against official bodies. Instead, they use bribery in order to corrupt the legal system and evade prosecution. The availability of large amounts of ready cash allow the criminal organizations considerable flexibility in using the power of money to suborn government officials on a large scale, sometimes including government ministers. In many parts of the world bribery and corruption are endemic. In these cases, organized crime is just another business group that uses bribery to go about pursuing its business.

FAMILY TIES

Finally, although it is not true in every case, most major criminal organizations have a family or ethnic base. They operate out of a small, tight-knit community. Family ties, especially in circumstances where a member's family can be held accountable for the member's disloyalty, present a practical solution to the problem of ensuring fidelity and obedience. Codes of allegiance, rituals, ethnic bonds, and quasi-religious ceremonies, whether linked to familial ties or not, also help to engage the compliance and loyalty of individuals within organizations. In addition, they create distinguishing codes of recognition that reduced the chances of hostile penetration. The fact that involvement in such an organization places an individual outside the law and subject to arrest also helps to reinforce the ties that bind, and when all else fails, murder is the final sanction to ensure loyal silence.

This is not hard to understand. Members of criminal organizations need to trust each other. The illegal and dangerous nature of the work plus the fact that law enforcement will seek to penetrate and disrupt the organization make some form of security discrimination vital to survival. It is easier to enforce operations security within a group known to one another and bound by ties of kinship or race. It is also easier to control membership and to spot outsiders. One of the advantages that ethnic-based criminal organizations have when they operate outside of their place of origin is that language and culture offer added barriers to fend off or identify outsiders. It is essential for such organizations to protect themselves from a hostile environment, to guard access to information, to ensure the success of operations, to minimize losses, and to guarantee loyalty.

Characteristically, then, organized crime is defined by a more or less formal structure that endures over time, is directed toward a common purpose by a recognizable leadership operating outside the law, is quite often based on family or ethnic identity, and is prepared to use violence or other means to promote and protect common interests and objectives. As noted at the outset, however, there is nothing particularly new about such organizations; they have existed for centuries. A number of factors are now at work that argue that the nature and role of these organizations are raising to a new level the threat that they pose to social order and the stability of nations.

There are three major new characteristics of organized crime at the end of the twentieth century. First is the broader, global canvas of the traditional criminal activity. Second is the growth of transnational links between criminal organizations and between criminal organizations and other groups. Third is the growing ability and power of international criminal organizations (ICOs) to threaten the stability of states, to undermine democratic institutions, to hinder economic development, to undermine alliance relationships, and to challenge even a superpower.

There are a number of factors that aid the ICOs' growing ability to challenge individual states. These factors are not so much characteristics of ICOs as of the environment in which they exist. They benefit, for example, from weak governments without the will or the resources to cope with rich and powerful ICOs. They enjoy fantastic sums of money generated by illegal activities, especially drug production, which gives them maximum flexibility to employ bribery or violence to achieve their ends. They are able to take advantage of the movement of large numbers of people internationally, which gives various organizations a recruitment base around the world. They are also able to capitalize on a decline in economic and political order, especially thriving on the growth of parallel or informal economies that subvert loyalty from the nation-state and government and habituate people to operating outside the legal framework. The ready availability of sophisticated arms and other technologies gives ICOs better means to protect and promote their interests. And systemic limits on the ability of individual states and international organizations to coordinate effective transnational anticriminal programs provide ICOs with maneuvering room to adjust to enforcement threats.

These features, taken separately or together, mean that today's international criminal syndicates are powerful enough to challenge, sometimes to destabilize, and so far, rarely, to control small, weak states. As a recent U.S. Senate report noted, these "new international criminals" represent a threat to international security that no single state can control alone.

These three major features, global operations, transnational links, and the ability to challenge national authorities, will be considered in turn.

THE NEW INTERNATIONAL CRIMINAL

The major ICOs have all the same characteristics and features of more traditional, nationally based criminal organizations. What distinguishes their ability to conduct global operations on the order of a major multinational corporation is their transnational scale and their ability to challenge national and international authority. Disposing of large quantities of ready cash, diversified into a wide range of activities, and employing a workforce spread around the globe, the ICOs represent a different order of magnitude in criminal operations.

The ICOs differ from traditional criminal organizations in the *global scope of their operations.* Traditional organized crime groups have their roots within individual countries, and although they may have overseas connections they do not operate on an international scale. Their organizations and operations are confined to nations or regions, and cities within nations. The America Mafia, also commonly known as La Cosa Nostra (LCN), is a well-known example of this older type. LCN emerged in the 1930s from conflict among gangs of Sicilian immigrants in U.S. cities. Although its members drew on the traditions of their Sicilian origins, LCN was never a subsidiary or arm of the Sicilian Mafia; it is a distinctly American organization. While LCN has had international connections, these have been largely related to being buyers of alcohol and heroin from foreign groups, not as being part of those groups. LCN's organization and operations are essentially domestic and regional.

Many other organized crime groups operating in the United States are primarily of this type. Black and Hispanic groups, for example, are generally localized gangs. They often dominate criminal activity in their respective ethnic neighborhoods. Typically, they control much street-level drug dealing and some distribution activity above the street level. They often interface with international traffickers who control wholesaling and regional distribution. Motorcycle gangs, such as the Hell's Angels, have also become well organized and engage in a wide variety of criminal activity, sometimes developing international suppliers.

The new international criminal groups differ sharply from these domestic groups. They are organized for and engage in large-scale criminal activity across international boundaries. Perhaps the greatest such organizations are Colombian. The Colombian cartels are vertically integrated global businesses. They have tens of thousands of specialized employees and associated individuals or businesses worldwide. Similarly, the Chinese Triads, although more loosely structured, have also developed extensive overseas operations, often in the wake of Chinese emigration. Such global networks provide mobility, an effective communications infrastructure, and international connections for criminal enterprise and sometimes for noncriminal groups who want to use their services. These structures also enhance the criminal groups' ability to create whole new markets for goods and services. This can be done either by creating new products, like "crack," which revolutionized the U.S. cocaine market in the mid-1980s, or by opening up new market areas, as the cocaine cartels have attempted to do in Europe, establishing links to the Mafia or other European criminal organizations.

International networks also provide flexibility to adapt quickly and creatively to enforcement efforts. For example, after U.S. enforcement efforts shut off the flow of heroin originating from Turkey, new sources developed in Southeast and Southwest Asia and Mexico. U.S. law enforcement agencies noted in 1991 the "inherent flexibility" of traffickers in shifting routes and modes of transport in response to enforcement efforts. When enforcement was stepped up in Florida and the Caribbean, cocaine was routed increasingly through Central America and Mexico. There is another aspect to this flexibility: by operating in the international arena, crossing national boundaries at will, the ICOs are often able to thwart localized law enforcement efforts. Diversification of operations and locale and diffusion of risk greatly enhance the ICOs' ability to recover from losses. They are also better able to adjust to changing situations in one country and exploit gaps in international law enforcement cooperation.

One of the emerging characteristics of the major ICOs is the extent of their *transnational links:* their growing interconnectivity with other transnational, nonstate actors. Although traditional criminal organizations have links to similar groups outside their own countries, they generally do not conduct extensive overseas operations and have only limited contact with other criminal groups. The major ICOs, however, have an international focus and operate vast transnational business empires. Their activities go beyond establishing subsidiaries. In addition to operating across international boundaries, the new ICOs have also begun to establish links to other nonstate groups, such as insurgents and terrorists, and similar criminal organizations. These linkages are diverse and in some cases tenuous and tense, but the trend that is emerging is toward closer cooperative relationships. The nature of the relationship can be quite complex.

Traditional, nationally based criminal organizations pose a variety of threats to public order and legitimate business. The U.S. experience with LCN is typical: corruption of officials, penetration of unions, money laundering, prostitution, street crime, gambling, violent internecine power struggles, drug trafficking; in short, the whole range of criminal activity that can be organized for profit. As rich and powerful as these organizations have been, however, they are generally in no position to challenge political order directly. Indeed, in the case of the LCN, recent law enforcement successes have decimated its ranks and may have dealt it a crippling blow. The major ICOs, however, pose a more dangerous threat, one that is again defined by its scope and sheer audacity: a *challenge to authority.*

The Medellín Cartel felt powerful enough to challenge the sovereignty and integrity of the Colombian state, and along with it the United States. The last five years of struggle have damaged the Medellín Cartel, at considerable loss of life and capital, but the Colombian cocaine entrepreneurs remain in business, not only evading annihilation but continuing to prosper and diversify their operations. Alone or in conjunction with various guerrilla groups in Colombia and Peru, they are able to control large areas of the Andes in defiance of government authority. They are also able to transship immense quantities of drugs through the Caribbean, Central America, and Mexico to the United States. In many cases, even in totalitarian Cuba, the Medellín Cartel has been able to corrupt officials throughout the hierarchy, including ranking military officers and ministers of state. Their ability to penetrate governments has even led to major disruptions in the relations between nations, as the Enrique Camarena case in Mexico showed only too clearly.

There is also mounting evidence to indicate that the major ICOs can have a profound effect on local economies. In the Andes, for example, the diversion of labor into illegal activities, the destruction of land and its use for nonproductive crops, and the generation of inflationary pressures all work to undermine the viability of local economies, already none too strong in Bolivia and Peru. Added to this is the penetration of financial markets and the international banking system. Corruption of financial institutions is now a major and growing concern, as the ICOs use international financial networks to launder money and in some cases to provide cover to further illegal activities. Even major developed countries are not immune. The French government, for example, has become concerned with the penetration of the Italian Mafia in southern France. This is also one of the major reasons the British government established the National Criminal Intelligence Service.

The major ICOs enjoy a number of advantages denied to smaller groups. The scale of their operations, their growing international connectivity, and the power and influence that they have garnered at the expense of governments make them a formidable challenge. They enjoy the advantages of economy of scale in their operations. Their enormous financial resources and diversification operations give them maneuvering room in responding to law enforcement encroachments. Their ability to operate across many legal jurisdictions reinforces this inherent flexibility. Moreover, their contacts with other nonstate actors, such as insurgent groups and terrorists, means that they can call increasingly upon allies to help distract the government or even challenge its authority. The cumulative effect of these advantages puts them in a position of great strength.

THE FUTURE

International criminal groups in the United States and worldwide are likely to expand. A number of factors are likely to aid their growth.

Economics of Production

For small farmers in many countries, choosing to grow drug-related crops makes the most sense economically. Markets for other commodities are less profitable and less stable. In many cases, even where the necessary marketing infrastructure and expertise exist, government controls make entry into those markets difficult or impossible for peasants. At the same time, drug entrepreneurs are expanding into markets where drugs have not been a major problem in the past. Without dramatic and unlikely changes, raw materials for drug production will continue to be readily available.

Furthermore, the United States is one of the world's most lucrative markets for illegal as well as legal enterprise. It will continue to attract trade in illegal products. Europe, too, is a large market, and the creation of the Common Market has significantly reduced the extent of border controls that might have helped to restrain drug trafficking and other transnational criminal activities.

International Ungovernability

The growth of international crime parallels a global trend toward ungovernability, that is, the declining ability of governments to govern, to manage a modern state, and to provide adequate or effective services. In some cases criminal organizations have been able to capitalize on the fact that large areas, such as the Andes and the Amazon regions of Latin America, were never under much central government control. They have moved into these remote regions and have begun to provide the major form of authority in much of the region. In other cases, they have begun to contest local control of areas with the government, as have groups in Burma. There are dozens of places, such as Peru and Burma, where state authority has lapsed in whole or in part. It is not limited to small states, either. There are indications that areas of some of the

Central Asian Republics have been given over to drug cultivation. Similarly, areas in Mexico, Pakistan, and southern China appear to be largely beyond government control. This situation provides favorable conditions for criminal groups and bases of operations and safe havens in areas key to drug trafficking and alien smuggling. Experts in political geography predict continuing global fragmentation. Criminal organizations thrive where governments are weak.

Immigration Streams

Ethnic criminal organizations are likely to follow immigration patterns. They do not always do so, but there has often been a strong correlation between the two. In the 1990s economic pressures and widespread ethnic turmoil are likely to generate refugees and immigrants from regions where international criminal groups are based. Between 1980 and 1990 the Asian population in the United States alone grew by 108 percent, from 3.5 million to 7.3 million. The Chinese population grew by 104 percent, from 806,000 to 1.6 million. While the vast majority of immigrants are law-abiding, criminal organizations tend to exploit immigrant communities in a variety of ways. They provide cover and concealment. Immigrant pools also provide a pool of recruits. In addition, the immigrants are usually fearful of law enforcement. Their recent experience in their country of origin makes them reluctant to cooperate with the police in their new countries. The police, moreover, historically do not provide the same degree of service to immigrants. The immigrants do not have important political connections, and the police find it difficult to cooperate with them because of their strange cultures and languages. Hence many experts anticipate increased international organized criminal activity accompanying the immigration of Russians, East Europeans, Asians, Middle Easterners, Kurds, and others.

Border Porosity

The United States' long open borders with Mexico and Canada provide ready access for criminals and illegal goods, and tens of thousands of miles of U.S. coastline are virtually uncontrollable. The opening of free-trade areas, such as the North American Free Trade Agreement and the EC, will lower many existing safeguards and customs inspections as well.

Trends in Technology

Continued advances in technology and international transportation will facilitate growth in international organized crime. The ease of modern communications makes contact among international criminal organizations easy, fast, and more secure. For example, new digital technologies make it more difficult for law enforcement bodies to intercept their communications. The movement of trillions of dollars in wire transfers each day makes it possible for many actors to evade state monitoring.

Relative Disorganization of Law Enforcement

Preventing, disrupting, and successfully prosecuting organized crime in most parts of the world is difficult enough in the best of times. Many traditional organized criminal organizations have survived the onslaught of law enforcement organizations for decades.

Now, however, the U.S. and other states are faced with international criminal groups. As was described earlier, they are bigger and more powerful than most of their predecessors. They operate globally, making it impossible for law enforcement in any one jurisdiction to neutralize major parts of their activities. While some degree of cooperation exists among law enforcement agencies, and new initiatives are getting under way, many observers believe that it is inadequate to the task. In 1992, for example, a U.S. Senate report noted that there is little evidence to suggest that either U.S. or foreign law enforcement entities are currently equipped to meet the challenge of this new breed of international criminal.

CORPORATIONS

32

The Crash of ValuJet Flight 592
A Case Study in State-Corporate Crime

RICK A. MATTHEWS AND DAVID KAUZLARICH

Matthews and Kauzlarich peer into a flight well known in American history, ValuJet 592, which crashed ten minutes after take-off into the Florida Everglades, killing everyone aboard. While it is easy in our culture to blame the airline company and its subcontracted maintenance company, SabreTech, Matthews and Kauzlarich offer a more penetrating analysis of the joint responsibility these corporations share with the Federal government. Offering a model for state-corporate crime, these authors tell a compelling story, but link it to governmental encouragement of the airlines and the back seat that regulation commonly takes to promoting business when government regulatory agencies are charged with overseeing industries they have a vested interest in promoting. This form of white-collar crime involves people working for the good of their organization, not their individual good, and is framed and encouraged by explicit cultural values and practices. Government's ties to big business make conditions ripe for such things to continue happening in the future.

The study of occupational and corporate crime has become widely accepted within criminology, but the study of state crime has remained on the periphery of the discipline (Tunnell 1993a). Recently, however, a number of scholars (e.g., Barak 1991; 1993; Friedrichs 1996a, 1996b, 1998; Ross 1995; Tunnell 1993b) have attempted to articulate the nature, form, extent, and varieties of state crime. While the labels differ, most working in this area agree that governmental, political, or state crimes are illegal or socially injurious acts committed for the benefit of a state or its agencies, *not* for the personal gain of some individual agent of the state. This way of viewing crimes committed by political actors is consistent with the classic distinction made by Clinard and Quinney (1973) between occupational and corporate

"The Crash of ValuJet 592: A Case Study in State-Corporate Crime," by Rick A. Matthews and David Kauzlarich, *Sociological Focus,* Vol. 33, No. 2, August 2000. Reprinted by permission.

crime, and points to the importance of viewing governmental/state/political crime as a form of organizational crime. The threat to use and use of nuclear weapons (Kauzlarich and Kramer 1998; Kauzlarich 1995), the state's permission of institutionalized racism and sexism (Bohm 1993; Caulfield and Wonders 1993), state suppression of civil, political, and human rights (Hamm 1991; Hazlehurst 1991), and genocide (Friedrichs 19996a; Green 1990) are examples of state crime. Such crime can occur internationally or domestically (Kauzlarich 1995), can be committed by any number of state or state-related agencies (Friedrichs 1998), and may or may not be a violation of codified law (Barak 1991).[1]

State crime is arguably one of the most important *and* complicated types of crime to study. State crime is *important* because it reminds us that the creator and enforcer of law can also be a criminal agent. Such crime may inflict a far greater amount of social injury than that caused by traditional street crime. The study of state crime is *complicated* because actors are generally powerful and privileged, and this power and privilege is formally supported through the authority of the state (Barak 1993). Many forms of state crime might be unrecognizable from what many might consider normative state policy. Legal responses to state crime are even less authoritative and problematic than the reaction to corporate crime. Finally, reconstructing the events of a state crime can easily turn into a methodological nightmare because the state has the power to conceal and classify documents that implicate wrongdoing. Given these complexities, and accompanying political and disciplinary problems (Barak 1993; Tunnell 1993a), it should be expected that the development of a solid criminology of the state will be a difficult task.

While the study of state crime is still in its infancy, a promising and important development has recently been made by Kramer and Michalowski (1990) through the introduction of the concept of *state-corporate crime*. Traditionally, the crimes of the state and the crimes of corporations have been viewed as unique and distinct manifestations of organizational behavior. Thus, a separate body of research and theorizing developed for each of these phenomena. Kramer and Michalowski (1991) point out the linkages between state and corporate goals, be they proximal or distal, and argue that some forms of organizational deviance result from the interaction between governmental agencies and private businesses.

A revised definition of state-corporate crime expanded the concept to include harmful actions that are not directly manifested through active state involvement: "State-corporate crimes are illegal or socially injurious actions that result from a mutually reenforcing interaction between (1) policies and/or practices in pursuit of goals of one or more institutions of political governance and (2) policies and/or practices in pursuit of the goals of one or more institutions of economic production and distribution" (Kramer and Michalowski 1991, p. 5; also see Aulete and Michalowski 1993, p. 175).

The concept of state-corporate crime has been used to examine the space shuttle *Challenger* explosion (Kramer 1992), the environmental devastation caused by U.S. nuclear weapons production (Kauzlarich and Kramer 1993,

1998), and the deadly fire at the Imperial Food Products chicken processing plant in Hamlet, North Carolina (Aulette and Michalowski 1993).[2] Other examples of state-corporate crime include the I.G. Farben Company's collusion with Nazi atrocities (Borkin 1978), the Wedtech case involving defense contractor fraud (Friedrichs 1996a), and as we will argue, the violent and deadly crash of ValuJet flight 592 in May of 1996. While at least two white-collar crime textbooks (Green 1990; Friedrichs 1996a) discuss the concept of state-corporate crime, there is a paucity of research and theorizing on the phenomenon; only three published case studies of state-corporate crime exist in the criminological literature. The present examination of the ValuJet crash, then, can increase understanding of state-corporate crime in two ways: (1) through exposing the varied nature and fort of the relationships between the polity and corporations that may lead to injurious outcomes that violate laws, and (2) exploring the usefulness of the core theoretical concepts in the organizational crime literature by applying them to an instance of state-corporate crime.

State-corporate crime, according to Kramer and Michalowski (1991), is a distinct form of organizational deviance because it involves both vertical and horizontal relationships between business and government, which many have generally viewed as separate, discrete entities. For instance, the fire in Hamlet, North Carolina, killed 25 workers and injured another 56, but we suspect that many would place the responsibility for the fire on the Imperial Food Products Company alone, rather than examining how the state facilitated such a crime (Aulette and Michalowski 1993). The same might be said of the *Challenger* explosion, where many might be tempted to view the disaster as a simple "accident," rather than consider how state and corporate goals interacted with one another to produce the death of six astronauts and schoolteacher Christa McAuliffe. In our case study of the crash of ValuJet flight 592, we also would expect that many would assign culpability for the deaths to ValuJet and SabreTech personnel without recognizing the instrumental role the Federal Aviation Administration (FAA) played in the disaster. . . .

In the case study to follow it will become clear that the crash of ValuJet flight 592 resulted from the "mutually reinforcing interaction" between private corporations (ValuJet and SabreTech), and a governmental agency (the Federal Aviation Administration). As such, the crash represents an example of state-facilitated state-corporate crime in which the pursuit of profit by corporations along with the failure of a state agency to effectively monitor them resulted in the violent deaths of 110 people. We first examine the particular events that led up to the fatal crash of ValuJet flight 592, including the specific actions of ValuJet and SabreTech employees. While it is important to understand these specific events, the reasons why they occurred cannot be understood without placing them within the larger socio-historical contexts of governmental regulation of the airline industry and the broader socio-political context of laissez-faire economics enshrined in the Airline Deregulation Act of 1978 (ADA of 1978). Thus, in order to more fully understand exactly what happened and why, we follow Aulette and Michalowski (1993) and examine the larger "nested contexts" within which state-facilitated crime occurs.

EVENTS LEADING TO THE CRASH

The ValuJet corporation, founded by Robert Priddy, a former baggage handler, had overcome many obstacles and quickly developed its own niche in the airline industry. ValuJet grew from 2 to 50 aircraft (including the acquisition of 48 aircraft in 31 months), and within four years had a profit of $6.8 million dollars (Levinson, Underwood, and Turque 1996; Hosenball and Underwood 1996). Based in Atlanta, Georgia, ValuJet was approaching its fourth year of existence when flight 592 crashed. The early years of ValuJet were characterized by rapid growth and the development of a reputation for providing exceptionally low-priced airfares (as low as $39.00) and staying, in the words of Priddy "lean and mean" (Hosenball and Underwood 1996). The lean and mean aspect of ValuJet meant, among other things, a non-unionized labor force, paying pilots about half of the industry average, having pilots pay for their own training, and outsourcing maintenance (Hosenball and Underwood 1996). Like many late-20th-century corporations, ValuJet viewed outsourcing as an integral profit-making component. By 1994 ValuJet was acquiring planes as fast as they could get their hands on them, most of which were older and in need of repairs. At the time of the crash of flight 592 the average age of ValuJet aircraft was 26.4 years old (Greising 1996). Since one of the cost cutting measures employed by ValuJet was contracting out maintenance duties, the older planes they purchased were sent to out-of-house contractors. Indeed, the only maintenance ValuJet did itself was routine inspections, and it was not equipped to do heavy maintenance. In all, ValuJet had contracts with 21 different certified maintenance facilities, including SabreTech (NTSB 1997).

In January of 1996 ValuJet purchased two McDonnell Douglas MS–82s from McDonnell Douglas Finance Corporation, and then in February 1996 purchased a third plane from them, a MD–83. ValuJet sent all three planes to Miami, where they were to be serviced by SabreTech.

One of the maintenance tasks requested of SabreTech by ValuJet was the inspection of oxygen generators on all three planes to determine if they had exceeded their allowable service life of 12 years. One of the planes had generators that were to expire in 1998 or later. However, the other two planes had generators that had already expired or were going to expire shortly. Thus, ValuJet contracted with SabreTech to remove the generators from these planes and replace them.

The guidelines for removing and disposing of oxygen generators are quite clear. The McDonnell Douglas manual, for instance, explicitly states that "if the generator has not been expended" workers are to "install safety cap over (the) primer" (NTSB 1997, p. 10). Furthermore, this manual states that the generators must be stored in a safe environment (i.e., noncombustible surface) where they are not exposed to high temperatures or possible damage until they are expended. Expenditure of the oxygen generators is done by securing them on a nonflammable surface in an area free of combustible substances. Once the chemical reaction has occurred, and the canister has cooled, it may be disposed of. Of the 144 canisters on the two planes, only six were expended (NTSB 1997).

In March of 1996 SabreTech crews began removing the old oxygen generators from the ValuJet planes and replacing them with new ones. According to the mechanics from SabreTech, almost all the generators that were removed were placed in cardboard boxes and then placed on racks in the hangar near the airplanes themselves (NTSB 1997). All the work at SabreTech was to have been completed by April 24, 1996, for the first plane, and April 30, 1996, for the second. This time line was established with an agreement between ValuJet and SabreTech, which explicitly stated that ValuJet was to be credited $2,500.00 per calendar day for each day the aircraft was delayed beyond the redelivery date (NTSB 1997). According to the mechanics at SabreTech, there was considerable pressure to complete work on the aircraft. They reportedly worked 12-hour shifts, 7 days a week to complete the task (NTSB 1997).

The NTSB investigation of the crash of ValuJet flight 592 revealed that the mechanic who signed the work card for ValuJet said that the canisters were placed in the cardboard boxes without packing material between them and without safety caps. He later testified that he assumed that they would not be shipped that way (NTSB 1997). However, many mechanics asked about acquiring safety caps for the generators. Since this was the first time SabreTech had performed this sort of task, the SabreTech company had no new safety caps available (NTSB 1997). Some mechanics inquired about placing the caps from the new generators on the old generators, but their inquiries were not followed up. The supervisor from SabreTech would later claim that no one who had worked with the oxygen generators had asked about safety caps (NTSB 1997).

The NTSB investigation revealed that the SabreTech inspector who signed off on the final inspection block said that he was aware that the generators needed safety caps, and that he had brought this to the attention of the lead mechanic. He further testified that he was told that both the SabreTech supervisor and the ValuJet technical representative were aware of the problem (NTSB 1997). According to this inspector, it was only after he was assured that the problem would be taken care of that he signed the card.

Two of the three ValuJet technical representatives assigned to SabreTech and a ValuJet quality assurance inspector claimed not to have seen the generator removal or the generators once they were removed (NTSB 1997). Further, they claimed that they were unaware of the lack of caps on the generators. However, one of the ValuJet technical representatives did see unexpended canisters lying on racks outside the planes without safety caps. He told the mechanics that they were dangerous and should be disposed of with the rest of SabreTech's hazardous waste (NTSB 1997).

The oxygen generators were eventually packed into five cardboard boxes and brought to SabreTech's receiving and shipping area for ValuJet. Three of the five boxes were delivered by one of the mechanics who had made earlier inquiries about the lack of safety caps. When asked if he had informed anyone in the receiving and shipping area about the specific contents of the boxes he said he had not (NTSB 1997). Unlike other facilities that do repair maintenance for airlines, SabreTech "had no formal procedure in place that required an individual leaving items in the shipping and receiving area to inform anyone in that area of what the items were, or that they were hazardous" (NTSB 1997, p. 118). To complicate matters, none of the SabreTech mechanics could recall seeing hazardous waste warnings on any of the boxes. After an extensive NTSB investigation, it is still unclear who brought the other two boxes to this area.

A few days later a SabreTech stores clerk asked the director of logistics if he could close up the boxes and prepare them for shipping. When the logistics director gave his approval, the stores clerk reorganized the contents of the boxes, placing the generators on their sides end-to-end and redistributing the number in each box. A few inches of bubble wrap was placed on the top of the generators in each box, and then they were sealed. After the boxes were sealed, they were marked "aircraft" parts (NTSB 1997).

On May 9 the shipping ticket for the five boxes was prepared by a SabreTech receiving clerk. The receiving clerk was given a piece of paper by the stock clerk and was told to write "Oxygen Canisters—Empty" on the shipping tickets (NTSB 1997, p. 19). When asked later if he knew the contents of the boxes, the receiving clerk said he did not, as the boxes were already sealed. ValuJet's Atlanta address was then placed on the boxes, and they were brought to the ValuJet loading ramp.

The ramp agents for ValuJet loaded the boxes on flight 592 headed for Atlanta, placing them in the forward cargo bin of the plane. None of the boxes were secured, and they were stacked on top of each other and around two spare airplane tires being shipped to Atlanta (NTSB 1997). One of the ramp agents said the contents of the boxes were "loose" and he heard "clinking" noises when he moved them (NTSB 1997).

THE CRASH OF FLIGHT 592

The NTSB used recorded radar data, cockpit voice recorder (CVR) comments and sounds, and flight data recorder (FDR) information to reconstruct the flight history of ValuJet flight 592. At 12:03 Flight 592 was cleared for takeoff. By 12:07 the plane was airborne and the pilot was instructed by air traffic controllers to turn left to begin the WINCO transition climb. Within three minutes of takeoff there was an unidentified sound that was recorded on the CVR, and the captain asked "what was that?" (NTSB 1997, p. 170). A few seconds later the captain remarked that they were experiencing some electrical problem. Five seconds later he said "we are losing everything" and within seconds he stated "we need, we need to go back to Miami" (NTSB 1997, p. 171). Shortly after, a male voice is heard on the CVR stating "we are on fire, we're on fire" (NTSB 1997, p. 171).

Amidst the shouting, the captain radioed and said they needed an immediate return. Air traffic control in Miami gave them a course and asked about the nature of the problem. The captain replied "fire" and the first officer said "uh, smoke in the cockp . . . smoke in the cabin" (NTSB 1997, p. 171). The plane then began to change direction, heading in a southerly direction. Within 40 seconds, the CVR recorded a flight attendant shouting "completely on fire!" (NTSB 1997, p. 171). This was the last contact with flight 592.

The plane crashed within 10 minutes of takeoff, about 17 miles northwest of Miami International Airport (NTSB 1997). The captain had turned the plane around but flames had engulfed the plane, causing it to crash nose down into the Florida Everglades. Subsequent tests with oxygen generators indicated that the heat generated by the fire was approximately 2,000° F within 10 to 15 minutes of ignition (NTSB 1997).

While ValuJet had a responsibility to oversee and regulate SabreTech's maintenance of its aircraft, the FAA had the ultimate responsibility of overseeing both ValuJet and SabreTech. However, we suggest that neither the FAA nor ValuJet fulfilled their responsibilities. ValuJet, for example, had outsourced its maintenance to the lowest possible bidder without ensuring the work was being done properly. It is clear that the FAA did not ensure that both ValuJet and SabreTech were following Federal Aviation Requirements (NTSB 1997).

To understand why the FAA did not adequately enforce federal regulations that may have prevented this accident, we suggest that their contradictory roles as regulators of airline safety and promoters of the airline industry lie at the core of the problem. The reasons why the FAA has such contradictory duties are rooted in its organizational development. However, the organizational development of the FAA is, we will argue, best understood within the broader historical contexts of laissez-faire economic philosophies.

THE FEDERAL AVIATION ADMINISTRATION, DEREGULATION, AND BENIGN TOLERANCE

The federal government's regulation of civil aviation began in 1926 with the passage of the Air Commerce Act, the intent of which was to help the infant airline industry reach its full commercial potential through increased safety standards enforced by a federal regulatory agency (Schiavo 1997). The training of pilots, air traffic rules and regulations, certification of aircraft, and the establishment of airways were all among the first responsibilities addressed by this act and were given to the Secretary of Commerce.

In 1938 the Civil Aeronautics Act created a new independent agency (the Civil Aeronautics Authority) to handle the responsibility of civil aviation standards. Then in 1940 President Franklin D. Roosevelt split the Civil Aeronautics Authority into two agencies, the Civil Aeronautics Board (CAB) and the Civil Aeronautics Administration (CAA). CAB was given the responsibility for safety rule-making, accident investigation, and economic regulation, while

CAA was responsible for air traffic control, pilot and aircraft certification, safety enforcement, and airway development. While both agencies were still under the Commerce Department, CAB functioned independently.

The Federal Aviation Act of 1958 transferred the CAA's responsibilities to the newly formed Federal Aviation Agency (which would be renamed the Federal Aviation Administration in 1967). The FAA was also given CAB's safety rule-making responsibilities. Thus, the newly formed FAA contained within a single agency the contradictory roles of promoting the airline industry, while at the same time overseeing safety regulations (Schiavo 1997).

In 1966 President Lyndon B. Johnson created the Department of Transportation (DOT), which combined all federal transportation responsibilities in order to integrate and facilitate national interests in the distribution and transportation of goods. The DOT would become the agency under which the FAA was placed. However, CAB's accident investigation responsibilities were placed under the auspices of the newly formed National Transportation Safety Board (NTSB). In short, the NTSB was given the responsibility of investigating accidents and making recommendations to the FAA, and the FAA, as a branch of the DOT, was given the responsibility of enforcing federal regulations within the airline industry.

Since the Airline Deregulation Act (ADA) of 1978 the airline industry has undergone several phases of growth and decline (Dempsy and Goetz 1992). However, rather than an increase in the competition between airlines, deregulation has resulted in increased consolidation and decreased competition. For example, during the first decade of deregulation, more than 150 carriers went into bankruptcy, and by the early 1990s only eight domestic air carriers accounted for nearly 95 percent of all the passenger industry (Dempsy and Goetz 1992).

Within this context, then, the FAA attempted to promote the growth of start-ups like ValuJet while also overseeing their compliance with Federal Aviation Regulations. While success rates for most start-ups were low, ValuJet seemed to be the exception to the rule. In many ways, ValuJet justified the laissez-faire philosophy of the ADA of 1978, and was touted as a model startup company in the age of deregulation. Given that only 3 of the over 250 airline companies had survived since 1978, the success of ValuJet was important to the FAA (particularly in its capacity of promoting the economic success of the airline industry in the wake of deregulation) and to the several political administrations that supported it (i.e., every presidential administration from Carter to Clinton).

In terms of safety, the FAA had attempted to coax ValuJet into federal compliance rather than imposing stiff penalties (Cary, Heges, and Sieder 1996). As former Inspector General Schiavo (1997) notes, the FAA had inspected ValuJet planes nearly 5,000 times in the three years it was in operation, and had never reported any significant problems or concerns. It has become clear that since the FAA had a vested interest in the *economic* success of the airline industry as a whole, and ValuJet in particular in the wake of deregulation, they did not adequately pursue ValuJet's violations (NTSB 1997; Schiavo 1997)

Some FAA inspectors, however, had serious concerns about ValuJet, even though the administration of the FAA did not. Internal reports and memos indicate that there were increasing problems that should have been addressed with regard to ValuJet's rapid growth, enormous profitability, and subsequently atrocious safety record (Schiavo 1997). However, according to Schiavo (1997), the FAA did not know what to do with ValuJet:

> the airline's safety record had deteriorated almost in direct proportion to its growth. ValuJet pilots made fifteen emergency landings in 1994, then were forced down fifty-seven times in 1995 . . . but that record would be surpassed within months with fifty-nine emergency landings from February through May of 1996 . . . an unscheduled landing almost *every other day*. (Schiavo 1997, p. 12, emphasis original) . . .

On May 2, just nine days before the ValuJet crash, the FAA produced a nine-page report on the safety records of the various new airlines. Ordered by Anthony Broderick, who was then the FAA's associate administrator of regulation and certification, the report was prepared by Bob Matthews, an analyst with the FAA's office of Accident Investigation (Fumento 1996). Matthews had two sets of data, one with SouthWest included in the new airline starts, and one without. Contrary to their claims, ValuJet's safety record was far from exceeding FAA standards. While the other start-ups had one accident annually, ValuJet averaged five (Fumento 1996). To make matters worse, ValuJet's accident rate was 14 times the major air carriers, and its serious accident rate was 32 times higher (Fumento 1996). Additionally, other incidents uncovered by the FAA before the crash of flight 592 included planes skidding off runways, planes landing with nearly empty fuel tanks, oil and fuel leaks that were left unfixed for long periods of time, and inexperienced pilots making errors of judgement. In an internal FAA report on ValuJet, there were nearly 100 safety-related problems (Stern 1996). However, the FAA did not officially recommend closing ValuJet down until after the crash of flight 592.

There is little question that the rapid growth of ValuJet came at the cost of safety, particularly in the area of maintenance (Schiavo 1997). One of the most contested areas in the debate surrounding the effects of the ADA of 1978 is that of maintenance and safety (Oster, Strong, and Zorn 1992). Since deregulation, some airline companies have reduced their maintenance programs to the FAA's minimum standards—or below—in order to increase short-term profits and cash flow (Oster et al. 1992). This, however, is contrary to the initial arguments by deregulation supporters that the airline industry would not only be prevented from neglecting maintenance standards by the FAA, and contrary to the argument that the economic consequences of unsafe practices in terms of lost revenues would deter such practices (Brown 1987). As Oster et al. (1992) argue, the FAA has generally been ill-equipped to ensure that maintenance programs are being adequately followed, as evidenced by the Eastern Airlines maintenance record falsification case of 1991.[3]

While ValuJet's failure to comply with safety regulations and the FAA's unwillingness or inability to enforce them are troubling enough, it is evident that the NTSB had made safety recommendations to the FAA long before the crash of flight 592 that could have prevented the accident. For example, in 1981 the NTSB had recommended that the FAA reevaluate the classification of class D cargo holds. The first recommendation (A–81–012) from the NTSB was that the FAA reevaluate the class D certification of the Lockheed L–1011, with the suggestion that it be changed to class C, which requires extinguishing equipment or changing the liner material to insure fire containment. The second recommendation (A–81–013) was to reevaluate class D cargo holds over 500 cubic feet to ensure that any fires would die from oxygen starvation and that the rest of the plane was properly protected. This recommendation came after a plane operated by Saudi Arabian Airlines in 1980 caught fire shortly after departure. The plane landed successfully, but all 301 occupants died. The fire on the Saudi Arabian Airlines plane started in the class D cargo hold. The FAA responded by stating that the NTSB recommendations should be addressed by making sure that class D cargo liners be made of fire-resistant materials better than the ones that were being used at the time.

In 1988 American Airlines flight 132 experienced a fire in its class D cargo hold en route to Nashville Metropolitan Airport. As the plane was on its final approach, smoke began to enter the passenger cabin. The fire was not contained in the cargo hold, and could be felt on the floor of the passenger cabin. Fortunately, the plane landed safely, and all passengers were evacuated. However, contrary to FAA claims, the fire in the class D cargo hold on this plane did not extinguish itself, and was not contained to the cargo hold. The cause of the fire was a hydrogen peroxide solution (an oxidizer) and a sodium orthosilicate-based mixture that had been shipped in the class D cargo hold. Neither of the chemicals was properly packaged, nor were they identified as hazardous materials.

After investigating this accident, the NTSB urged the FAA to require smoke detectors in all class D cargo compartments, and to require fire extinguishment systems for them. Additionally, the NTSB asked the FAA to evaluate the possibility of prohibiting the transportation of oxidizers in cargo compartments without smoke detectors or extinguishing systems. After several exchanges of correspondence, the FAA informed the NTSB that its cost/benefit analysis revealed the $350 million pricetag attached to this recommendation was not feasible. The FAA took the position that it was not going to force the airline industry to make these improvements because it felt they were not cost effective in terms of the amount of money required to possibly prevent a small number of accidents.

NOTES

1. The debate on the definition of "crime" has a long and contentious history. Some have argued that crime should be defined as a violation of a state's criminal laws or through the process of criminal justice (Adler and Adler 1933; Tappan 1947). Others like Sutherland (1945) have suggested that crimes can also

be considered violations of regulatory or administrative law, for example, consumer protection laws. Still others have argued that the definition of crime need not be related to any form of official state definition, but rather based on basic notions of human rights, conduct norms, or analogous social injury (Michalowski 1985; Schwendinger and Schwendinger 1970; Sellin 1938). We adopt the latter positions (as do most state crime scholars, e.g., Barak 1991; Tunnell 1993; and Friedrichs 1996a, 1998) which may or may not rely on the state's definition of an act or omission as a crime, particularly in reference to state or corporate crime.

2. Vaughan's (1996) monumental work challenges the usefulness of an exclusively criminological perspective on the *Challenger* explosion, though her theoretical explanation closely parallels some versions of organizational crime theory.

3. In 1991 Eastern Airlines was fined $3.5 million by the FAA for maintenance record falsification in their attempt to conceal failure to do required maintenance.

REFERENCES

Adler, Michael J. and Mortimer Adler. 1933. *Crime, Law and Social Science.* New York: Harcourt, Brace.

Aulette, Judy R. and Raymond J. Michalowski. 1993. "Fire in Hamlet: A Case Study of State-Corporate Crime." Pp. 171–206 in *Political Crime in Contemporary America,* edited by Kenneth Tunnell. New York: Garland.

Barak, Gregg. 1991. *Crimes by the Capitalist State.* Albany: State University of New York Press.

———. 1993. "Crime, Criminology, and Human Rights: Toward an Understanding of State Criminality." Pp. 207–230 in *Political Crime in Contemporary America,* edited by Kenneth Tunnell. New York: Garland.

Bohm, Robert. 1993. "Social Relationship That Arguably Should be Criminal Although They Are Not: On the Political Economy of Crime." Pp. 3–30 in *Political Crime in Contemporary America,* edited by Kenneth Tunnell. New York: Garland.

Borkin, Joseph. 1978. *The Crime and Punishment of I.G. Farben.* New York: Free Press.

Brown, Anthony. 1987. *The Politics of Airline Deregulation.* Knoxville: University of Tennessee Press.

Cary, Peter, Stephen Hedges, and Jill Sieder. 1996. "A Start-up's Struggles." *U.S. News and World Report,* June 24:50.

Caulfield, Susan and Nancy Wonders. 1993. "Personal AND Political: Violence against Women and the Role of the State." Pp. 79–100 in *Political Crime in Contemporary America.* edited by Kenneth Tunnell. New York: Garland.

Clinard, Marshall and Richard Quinney. 1973. *Criminal Behavior Systems: A Typology.* New York: Holt, Rinehart, and Winston.

Dempsy, Paul and Andrew Goetz. 1992. *Airline Deregulation and Laissez-Faire Mythology.* Westport, CT: Quorum.

Friedrichs, David. 1996a. *Trusted Criminals: White Collar Crime in Contemporary Society.* New York: Wadsworth.

———. 1996b. "Governmental Crime, Hitler, and White Collar Crime: A Problematic Relationship." *Caribbean Journal of Criminology and Social Psychology* 1:44–63.

———. 1998. *State Crime.* (Volumes I and II). Aldershot, U.K.: Dartmouth.

Fumento, Michael. 1996. "Flight From Reality." *The New Republic,* October 20.

Green, Gary. 1990. *Occupational Crime.* Chicago: Nelson-Hall.

Greising, David. 1996. "Managing Tragedy at ValuJet." *Business Week,* June 3:40.

Hamm, Mark. 1991. "The Abandoned Ones: A History of the Oakdale and Atlanta Prison Riots." Pp. 111–129 in *Crimes by the Capitalist State,* edited by Gregg Barak. Albany: State University of New York Press.

Hazlehurst, Kayleen. 1991. "Passion and Policy: Aboriginal Deaths in Custody in Australia 1980–1989." Pp. 21–48 in *Crimes by the Capitalist State,* edited by Gregg Barak. Albany: State University of New York Press.

Hosenball, Mark and Anne Underwood. 1996. "Seeing No Evil: Did Industry Regulators Take Too Long to Scrutinize Atlanta's ValuJet?" *Newsweek,* May 27:46.

Kauzlarich, David. 1995. "A Criminology of the Nuclear State." *Humanity and Society* 19:37–57.

Kauzlarich, David and Ronald C. Kramer. 1993. "State-Corporate Crime in the U.S. Nuclear Weapons Production Complex." *The Journal of Human Justice* 5:4–28.

———. 1998. *Crimes of the American Nuclear State.* Boston: Northeastern University Press.

Kramer, Ronald C. 1992. "The Space Shuttle *Challenger* Explosion: A Case Study of State-Corporate Crime." Pp. 214–243 in *White Collar Crime Reconsidered,* edited by Kip Schlegel and David Weisburd. Boston: Northeastern University Press.

Kramer, Ronald C. and Raymond Michalowski. 1990. "State-Corporate Crime." Paper presented at the Annual Society of Criminology Meeting, Baltimore, MD.

———. 1991. "State-Corporate Crime: Case Studies in Organizational Deviance." Unpublished manuscript.

Levinson, Marc, Anne Underwood, and Bill Turque. 1996."A New Day at the FAA?" *Newsweek,* July 1:46.

Michalowski, Raymond. 1985. *Order, Law, and Crime.* New York: Random House.

National Transportation Safety Board. 1997. *Aircraft Accident Report: In-Flight Fire and Impact with Terrain, ValuJet Airlines Flight 592.* Washington, DC: U.S. GPO.

Oster, Clinton, John Strong, and Kurt Zorn. 1992. *Why Airplanes Crash: Aviation Safety in a Changing World.* New York: Oxford.

Ross, Jeffrey Ian. 1995. *Controlling State Crime.* New York: Garland.

Schiavo, Mary. 1997. *Flying Blind, Flying Safe.* New York: Avon Books.

Schwendinger, Herman and Julia Schwendinger. 1970. "Defenders of Order or Guardians of Human Rights?" *Issues in Criminology,* 5(2):123–157.

Sellin, Thorsten. 1938. *Culture, Conflict and Crime.* New York: Social Science Research Counsel.

Stern, Willy. 1996. "Has the FAA Been Coming Clean?" *Business Week,* June 17:37.

Sutherland, Edwin. 1945. "Is 'White Collar Crime' Crime?" *American Sociological Review* 10:132–139.

Tappan, Paul. 1947. "Who is the Criminal?" *American Sociological Review* 12:96–102.

Tunnell, Kenneth. 1993a. *Political Crime in Contemporary America.* New York: Garland.

———. 1993b. "Political Crime and Pedagogy: A Content Analysis of Criminology and Criminal Justice Texts." *The Journal of Criminal Justice Education* 4:101–114.

Vaughan, Diane. 1996. *The Challenger Launch Decision: Risky Technology, Culture, and Deviance at NASA.* Chicago: University of Chicago Press.

PART IX

Structure of the Deviant Act

Although the structure of deviant associations is revealing, in Part IX we investigate the characteristics of the acts of deviance themselves. Deviant acts involve one or more persons aiming to accomplish a particular deviant goal. These vary widely in character from those enduring over a period of months to the more fleeting encounters that last only a few minutes, from those conducted alone to those requiring the participation of several or many people, and from those where the participants are face-to-face to those where they are physically separated. At the same time, acts of deviance can be looked at in terms of what they have in common. All deviant acts consist of purposeful behavior intended to accomplish a gratifying end, require the coordination of participants (if there are more than one), and depend on individuals reacting flexibly to unexpected events that may arise in this relatively unstructured and unregulated arena. Like the relations among deviants, deviant acts fall along a continuum of sophistication and organizational complexity. Following Best and Luckenbill's (1981) typology, we arrange them here according to the minimum number of their participants and the intricacy of the relations among these participants.

Some deviant acts can be accomplished by a lone *individual,* without recourse to the assistance or presence of other people. This does not mean that others cannot accompany the deviant, either before or during the deviant act, or even that two deviants cannot commit acts of individual deviance together. Rather, the defining characteristic of individual deviance is that it can be committed by one person, to that person, on that person, alone. A teenager's suicide, a drug addiction, a skid row transient's alcoholism, and a self-induced abortion are all examples of individual behavioral deviance. Nonbehavioral forms of individual deviance include obesity, minority group status, a physical handicap, and a deviant belief system (such as alternative religious or political beliefs). Lowery and Wetli's earlier selection on sexual asphyxia discusses such a solitary deviant practice, where people diminish the flow of oxygen to their brain during auto-eroticism in order to enhance their sexual climaxes.

A second type of deviant act involves the *cooperation* of at least two voluntary participants. This cooperation usually involves the transfer of illicit goods, such as pornography, arms, or drugs, or the provision of deviant services, such as those in the sexual or medical realm. Cooperative deviant acts may involve the exchange of money. Where this is absent, participants usually trade reciprocal acts. They both come to the interaction wanting to give and get something. In deviant sales, one participant supplies an illicit good or service in exchange for money. One or more of the participants in such acts may be earning a living through this means.

Tewksbury's selection on cruising for sex in public places offers an excellent illustration of a type of cooperative deviant activity that does not involve the transfer of money, where men exchange sexual services. Flowers' selection on telephone sex operators shows how these voluntary exchanges are changed when money comes into the picture.

The final type of deviant act is one of *conflict* between the involved parties. A perpetrator forces the interaction on the unwilling other, or an act entered into through cooperation turns out with one party "setting-up" the other. In either case, the core relationship between the interactants is one of hostility, with one person getting the more favorable outcome. Conflictual acts may be carried out through secrecy, trickery, or physical force, but they end up with one person giving up goods or services to the other, involuntarily or without adequate compensation. Conflictful acts may be highly volatile in character, as victims may complain to the authorities or enlist the aid of outside parties if they have the chance. To be successful, therefore, perpetrators must control not only their victims' activities, but also victims' perception of what is going on. Such acts can range from kidnapping and blackmail, to theft, fraud, arson, and assault. Our first reading by Martin and Hummer, on fraternity rape, deals with simple coercion. The Liederbach selection on deviance in medicine considers various types of fraud by these white collar professionals.

COOPERATION

33

Cruising for Sex in Public Places

RICHARD TEWKSBURY

Tewksbury offers an updated version of a classic study of homosexual cruising for impersonal sex partners in public places in his look at the way men negotiate the sensitive demands of this high-risk activity. He discusses the way the participants, a fringe part of the gay scene who are mostly heterosexually-identified, locate the promising places for finding sex, and try to "score" without getting beaten up, arrested, or exposed. Interested participants follow a sequence of mostly non-verbal cues where they follow a scripted set of behaviors designed to let others know how to proceed. These overtures are subtle enough to remain inconspicuous to families and straights that might be using the area, while signaling to people in the know how to proceed with the interaction. Tewksbury outlines the hidden structure and language of these men's hidden erotic worlds.

To date, studies of interpersonal, especially sexual, attraction have focused almost entirely on heterosexual relationships (but see Ross and Paul 1992). Although these relationships are important to understand, such studies are too narrow for practical use in contemporary society. Understanding how and why men may be attracted to other men, or women to other women, can lead to important conclusions regarding social structures, public health, marketing, legal implications, and daily work and leisure activities.

The mere fact that some sexual settings are homosocial necessarily leads to expectations for varying sexual scripts based on differing values, desires, and expectations. Men are reputed, and shown in the research literature, to be more likely than women to have engaged in casual sex (Herold and Mewhinney 1993) and to be more willing to accept a sexual invitation from an unknown other (Clark and Hatfield 1989; Clark 1990). With this in mind, an examination of the means by which men who have sex with men (MSMs) pursue and carry through with casual sexual encounters via cruising—seeking sexual partners, often in public places—becomes an important and socially relevant topic of research.

Copyright 1996 from *Deviant Behavior*, Vol. 17, No. 1. Reproduced by permission of Taylor & Francis, Inc., http://www.routledge-ny.com.

THE STUDY OF SEX BETWEEN MEN

The study of MSMs is not necessarily a study of gay culture, but simply the study of . . . men who happen to engage in sex with other men (whether or not they are sexually active with women as well). There is a well developed body of literature that has documented that men of varying sexual identities . . . do engage in sex with other men (Kinsey, Pomeroy, and Martin 1948; Humphreys 1970; Sundholm 1973; Corzine and Kirby 1977; Delph 1978; Donnelly 1981; Weatherford 1986; Gray 1988; Desroches 1990; Earl 1990; Tewksbury 1990; Doll et al. 1992).

Although we know that large numbers of men do engage—at least at some point in their lives—in sex with other men, we have only limited knowledge regarding how men initiate such encounters. What research we do have available typically focuses on one specific type of sexual arena; the present research adds to this aggregation by explicating the processes and perceptions of participants in a previously overlooked sexual arena—the urban public park.

The present research also expands our current understandings by drawing on a research method not yet fully implemented in investigations of public sexual arenas: formal, in-depth interviews. Earlier researchers have examined male-male, public, and anonymous sexual encounters but have typically approached such topics from a detached position. Humphreys's (1970) classic work stands as one exception among these pieces; however, Humphreys's means of obtaining interview data on involved men was covert and as such did not directly address men's experiences in casual, same-sex sexual encounters. Other important exceptions to the norm of detached observation and analysis are Lee's (1979) and Kamel's (1983) reflections on gay sadism/masochism (S&M) and Styles's (1979) and Brodsky's (1993) analyses of social organization in gay bathhouses and sex clubs. Most common among the other approaches to the topic have been the utilization of law enforcement surveillance techniques (Gray 1988; Desroches 1990; Maynard 1994), observations supplemented with informal interviews (Ponte 1974; Weinberg and Williams 1975; Corzine and Kirby 1977), and simple observation (Delph 1978; Tewksbury 1990).

The discussion that follows relies on indepth interviews with eleven men who have had sex with men in public, anonymous encounters. . . . The following discussion yields an advanced understanding of the behaviors and strategies men employ while cruising in public places. . . .

CRUISING FOR SEX PARTNERS

In order for an individual to participate in public cruising activities it is, of course, first necessary for him to know that such activities occur, and to know where, when, and how such activities occur. Within urban gay communities, the fact that cruising and public sex occur is relatively common knowledge.

Locations where cruising takes place are basic kernels of subcultural knowledge, often including specifics about particular "types" of men one may expect to find in particular settings (Lee 1979; Brodsky 1993). However, in the rare instance that a man does not know where or when he may find cruising activity, there are annually published guides to public sex locations for most American urban areas. Therefore, the where issue is addressed, leaving a man to answer only when and how.

Both sexual scripting (Gagnon and Simon 1973; Simon and Gagnon 1986) and imaging (Kamel 1983) are important means by which individuals manage presentations of self, which in turn shape and direct interactional (especially sexual) possibilities. . . . Depending on the imaging practices and scripts pursued by setting participants, the encounters among men vary in form, degree, intensity, and success.

The playing out of sexual scripts and imaging practices occur everywhere in society, but [is] most common (and obvious) in subcultural locations known for cruising and sexual activity. The process of cruising has a long history in gay male communities. The perpetuation of such behavior is so well known, and so commonly practiced, that one theorist has been led to critique cruising as a "ritual" (Pollack 1993).

Cruising can be a dangerous activity. Media and anecdotal accounts abound about violence and blackmail arising from public sex and the search for it. Consequently, many MSMs seek out protected environments for casual . . . sex, places where access can be controlled and victimization potential limited. Such controlled environments—sex clubs, baths, and some gay bars—not only function to provide a sense of protection from invading cultural outsiders, but also allow for transformations of sexual and sex-seeking activities within the controlled boundaries. . . . [T]he virtual absence of access control and less authoritative means of enforcing situational norms [in public parks] means the structure and process of cruising public places is significantly different from the sex-seeking behaviors of other, more restricted, cruising locales. Understanding the men who cruise for sex with other men in public places, and how the ritual is performed, are the next topics of this discussion.

LEARNING THE ROPES

Once a man seeking to locate other men interested in anonymous sexual encounters identifies a location, all that he needs to do is go to the location and, as Steve succinctly explained, "just start walking, it'll be there. At least somebody there is waiting to service your every need." In other words, even the naive, inexperienced cruiser can expect to engage a willing sexual partner. Where the individual encounters challenges and has a need for subcultural knowledge is in the area of facilitating his likelihood for achieving a successful and safe encounter. As Lee (1979) has elaborated, concerns about finding structured and protected settings and establishing means to negotiate encounters are central to the experience of sexual encounters with anonymous or casual-acquaintance others. These are abilities and skills that must be learned.

Although public park cruising locales are widely known within the subculture of an urban area's MSMs, such locations cannot be approached as if everyone found there [were] interested in sex with everyone and anyone else. Obviously, not all individuals or even everyone in public parks are in search of sexual encounters. Despite the apparent openness and unrestricted nature of the sexual activity, there are norms that regulate activities. Many of these norms may appear to be simple courtesies, maintaining a sense of civility within the setting. Additionally, and more importantly, norms structure activities and foster avenues by which men may screen potential partners for safety and enter into familiar negotiation patterns. As Kirk, a well-practiced cruiser, explains,

> You don't walk straight up to someone and put your hand in their crotch, that's for sure. That's a little forward. You don't assume that just because the other person's there to have sex, that they want to have sex with you.

It should be noted, however, that many men do report norm-violating experiences including things such as having another simply walk up and place a hand on their crotch. Such instances, where sexual receptiveness is assumed and contact is immediately initiated, do occur. However, typically men methodically seek out and negotiate sexual interactions, and only then perhaps enter into a sexual exchange. Regardless of whether an encounter is methodically negotiated or a surprise contact, both men are allowed to withdraw quickly, easily, and without negative consequence, at any time (Weinberg and Williams 1975; Brown 1976; Lee 1979; Tewksbury 1990; Brodsky 1993). Simply because sexual contact is initiated does not mean that it will necessarily result in both men's (or either individual's) ejaculation. When a man wishes to remove himself from a sexual interaction, he does just that. Replacing one's clothing and walking away are acceptable, although not always desirable, to one's anonymous sex partners.

How does one learn the norms of a public sex arena? Explicit instruction from a friend who is knowledgeable may be one way for some men to learn how to cruise. Or, simply being present and observing the actions and interactions of others may be the most fruitful method. Or, as has been the case for some, learning may occur as a result of violating the unknown norms and being subsequently sanctioned. To successfully learn via observation, and to successfully complete a sexual encounter, the trick is, as Vince puts it, to "look inconspicuous."

Meticulous scrutiny of co-present others is generally believed not only to be the key to locating desirable, willing sexual partners, but also to serve as a screening device to assess the situation for risk. Such a practice is by no means unique to public sex arenas. What may be unique here, though, is that the use of such an activity serves both to facilitate seeking and to assess the setting for possible threats to safety. If such a canvassing reveals the presence of men of dubious appearance, activities can be restricted, or the individual can safely retreat from the perceived threat.

CHARACTERISTICS OF MEN FOUND IN PUBLIC SEX ARENAS

The men found in public cruising areas are, according to Steve, "just your average build, average body, average hair, average, average. Very average-looking." In short, the men in public sex arenas run the full spectrum from highly unattractive to attractive, old to young, poor to wealthy, short to tall, and so forth. Stating this position more completely is Matt, who says cruising men are not stereotypical gay men, but rather:

> You're finding trolls, you're finding young adults, you're finding married men, you're finding truckers, you're finding the whole spectrum. I mean, that's who you would find in any cruising spots, I guess. . . . These are often people that don't want to admit their homosexuality yet.

Standing on the edge of social acceptability, and on the edge of being "out," the men who cruise in public parks (and in other public sex arenas) do not fit common stereotypes of gay men. Neither do men in public cruising areas adorn themselves in "costumes" common to other gay settings (Lee, 1979; Kamel, 1983; Brodsky, 1993). Cruising men are not necessarily feminine, highly fashionable, flamboyant, or hypermasculine. If anything, it can be expected that cruising men fit more closely with stereotypes of "traditional masculine" presentations of self. This is clearly seen in Jack's definition of men in the park:

> They are not your average queen. They're not the little faggot down the street with the loose wrist and the lisp and the high heels that walks like they're stepping on eggshells. You will very seldom find that person in the park, because the park intones to them more masculinity.

For some, the more masculine nature of cruising for sex outdoors might suggest a parallel with ideals of heterosexuality. This may stand as one reason, together with the setting's norms of anonymity and quick sexual consumption, that all men interviewed believed a significant percentage of those who cruise in public are self-identified heterosexuals. Such "men who have sex with men" are more likely to venture into the more easily permeable setting of the public park than they are to cross the boundaries that maintain more recognizable gay settings (such as gay bars and baths).

This evidence suggests, then, that cruising public parks is something that is looked upon as, at best, marginal to the gay community. Although a sizable minority (or, perhaps a slight majority) of gay men may at one time or another seek out anonymous sex in a public sex arena, this is not an activity that is commonly discussed openly and honestly among gay men. Rather, to be known as a man who cruises in the park is to be, as Albert claims from personal experience, severely stigmatized. In his experiences, Albert says he has learned, "Oh God, everybody that goes to the park, if you are seen at the park you are automatically tagged a scuzz. You are a real slut.". . .

WHEN IS CRUISING MOST PRODUCTIVE?

Knowing that cruising in public sex arenas is not a highly respected or respectable activity, men clearly guide the temporal aspects of cruising by a combination of utilitarian and self-protective interests. The utilitarian interests are seen in efforts to be present when other, similarly interested men will also be present. The self-protective interests are the efforts not to be present when "outsiders" are present; in this way, men seek to maximize the effects of screening procedures and to provide opportunities for sexual encounters that are not likely to be interrupted by the "unwise." In practice, most cruising activity in public parks occurs at night. Although men do visit the park during daylight hours, and may meet and leave with other men, actual consummation of relations most often is an after dark activity. This is a practical practice, for, as Kirk says, "If you're intent on having sex in the park, yes, it is at night. . . . That's your cover; people don't see you." Darkness facilitates the anonymity of the setting and thus the sexual activity.

Minimizing the likelihood of intrusions (including arrest) means that cruising rarely takes place during hours when large numbers of "straight" people are using the park. Often referred to as "family time," these hours (afternoons, early evenings, weekend daylight hours) are perceived as both too dangerous for sexual activity and simply inappropriate. Marc explained this saying, "A bad time is when, well, family time. . . . It's our criteria. Family time is kind of an etiquette, you just don't do it. There is etiquette out there."

Even during the hours when supposedly "polite" men do not have sex where they may be detected, cruising does occur. Such hours host sex between men, but in more secluded areas, places where "straight" people are unlikely to venture. Daytime sexual relations, when they occur, are most likely to be consummated deep in the woods, or in locations removed from the easily accessed regions of the park. In other words, especially when running increased risks of detection, active attempts are made to "hide" sexual activities. Delph (1978), discussing how men transform innocuous public settings into erotic oases, described prime locations as those having distinct boundaries and structural features that help acknowledge the approach of others. Out-of-sight, deep-in-the-woods locations offer these same attractions to men who transform public parks into erotic oases.

Therefore, both time of day and season (in most climates) influence the structure and amount of cruising that occurs in public parks. In the Midwest, late fall through early spring is a very slow time for cruising; weather turns cold, leaves fall from trees, and the setting loses its veils of secrecy. However, even on cold, sunlit days, at least a few men are likely to be cruising in the park. The setting, both physically and temporally, may facilitate or complicate cruising, but it does not strictly govern the who, when, where, or how.

CONTACTING AND CONTRACTING WITH ANONYMOUS SEXUAL PARTNERS

As outlined above, cruising for anonymous sexual partners in public sex arenas is a subcultural phenomenon that requires knowledge of locations, comprehension of norms, and often a period of time during which skills are acquired and refined. However, what remains to be discussed is the crux of the matter: the process by which men actually identify mutually interested others and the ways they communicate and negotiate sexual interactions. The "information game" (Goffman 1959) is a process by which individuals carefully seek to negotiate a mutual understanding of the situation: involved others and mutual needs, wants, and expectations. It is this process that cruising men recreate each and every time they return to the park and pursue a new sexual partner.

What is perhaps the most notable characteristic of interactions in pursuit of sexual relations in public sex arenas is the lack of verbal communications. As seen above, cover of darkness, bushes, and other obstacles to visibility are desirable elements of an erotic oasis, in part because they provide means to shield identities, as well as activities. Nonverbal communications are also, at least partially, a shield against identity disclosures. When one speaks, one conveys more information than the mere content of one's words. If the verbal aspects of communication are removed, so too may some of the additional identity elements be removed from an individual's interactions. Consequently, the great majority of cruising activities found in public sex arenas are conducted under the veil of silence.

Because verbal conversations are severely limited, and due to the subcultural stigmas associated with cruising public parks, men interested in other men they find in the park have restricted means to learn of others' sexual and social reputations. Whereas in other settings men may ask their associates about particular others, this is not generally possible in the arena of the public park. Additionally, because of both the very fluid nature of the park's patrons and their preference for after-dark cruising, identifying others can often be a very difficult task. Therefore, there may be only minimal assurances that a sexual partner is not prone to violence or to exposing others to social and physical dangers.

On those rare occasions when verbal conversation is used to establish contact, it is brief and presented in ambiguous fashion. One man may greet another, or make a simple request (ask for the time, a match, or directions to a landmark). Such opening comments allow for a quick appraisal of another man, in a very low-risk form of contact. Once a man is deduced as willing to speak, the conversation starter will ask some type of leading, or double-meaning, question. This is the way that Adam makes initial contact with men in the park. Adam says he likes to feel as if he knows *something* about the men with whom he has anonymous sexual encounters, so he always asks others a few questions.

> I might say something about any subject that I knew would be of common interest . . . or, I'd ask if you were driving through enjoying the

day, or looking? Everybody knows what that means! If you're driving through enjoying the day, or you're not interested, you'll tell me. That's a comfortable way out, and not being rude. If you are interested, you're going to tell me that you are—you are looking.

As explained earlier, withdrawing from an unwanted interaction is acceptable, and provided for, in men's homosocial, sexual environments. Ambiguous questions not only facilitate the negotiation of sexual encounters but can also provide for easy withdrawal from undesirable potential encounters.

The way that men communicate their interest in other men, and contact and contract with others, is via five primary modes of nonverbal communication: eye contact, use of personal space, body language, subtle forms of touching (of both themselves and others), and movement through the park in pursuit or in tandem with other men. Non-verbal communication is the rule in erotic oases (Delph 1978; Donnelly 1981; Weatherford 1986; Tewksbury 1990). The "language" of such a subcultural location is well summarized by Ted, who explains the way men who cruise public parks communicate, saying,

> It has its own language, and you don't—well, I'm sure I don't know all of it. But what I have come to learn and be able to recognize is that there is a subliminal language to it that is not words. It's certain actions, certain movement. . . . This is something I've never really put words to until now.

To understand the language is to use it while not needing to think consciously about it. It is only when one develops a fluency in the "subliminal language" of the public sex arena that successful cruising is likely to occur. The way men communicate while cruising is indeed a language in its own right, just as is any patterned form of communication that requires study, skill, and practice. . . .

Eye Contact

With verbal communication essentially absent, men need to have means of getting the attention of others, as well as a means of acknowledging the call for attention of another. This is the principal function of eye contact. As Goffman (1977:309) believed, "the male's assessing act—his ogling—constitutes the first move in the courtship process." Eye contact is the main means by which contact is made. Contact is not simply looking at a man, but looking consistently into his eyes and holding his gaze. The "prolonged" look, directly into the eyes of another, which might be considered deviant in ordinary, daily interactions, is the means by which men greet one another and quickly determine whether sufficient mutual interest exists to continue with increasingly intimate contact.

. . . Duane, who knows that police regularly visit the parks undercover, says that when he is cruising in the park,

> Eye contact for me is ultimate. By being able to catch a person's eye and the way that—well, I can tell whether they're really comfortable or not.

Even cops aren't comfortable enough to hold your gaze, you know—I mean 'cause they are just not to the point of doing it. That's one of the biggest clues for me.

Another way of looking at the functions of eye contact while cruising is offered by Matt, a self-described "long experienced" public sex arena cruiser: "Eye contact is the way you invite someone to participate with you . . . It's not just looking at someone, but like *really* looking at someone, and them really looking at you."

Eye contact is the contact point, the initial way to investigate another man's subcultural experience, and perhaps motivations. Eye contact can open the door to continued, increasingly focused, sexually directed interactions. The eyes serve as the filter for determining which others in the setting may be appropriate, interested, and suitable partners, with whom one may wish to move into the initial contracting stage. As amount and intensity of eye contact increase, so too do interactants' personal investments in continued interactions (Iizuka 1992).

Use of Personal Space and Body Language

There are two general, and related, means by which increasingly focused interactions are pursued. Communicating an interest in contracting is put forth either through one man entering another man's "personal space" (see Hall 1959) or through body language signaling a sexual interest.

Body language—positioning oneself within the setting, posing in the line of another's vision, and contorting the body so as to present or emphasize certain parts of it—can communicate meaningful messages. Many men maintain that even though they don't speak with others, "body language says a lot." Or, more specifically, as Ted explains it,

> There are several forms of body language. If a person is standing facing toward you and they're moving about in place. . . . Fondling himself, making it apparent that they're interested in some kind of sexual contact. Maybe by fondling themselves directly at you, so that is something that they perceive is what you're looking for. Their intention is pretty clear and very tight. Then that followed with a smile.

Body language often involves motions and signals. As attention is maintained between men, these signals frequently involve touching oneself. Touching, stroking, and massaging parts of one's own body are taken to communicate interests (often specific sexual interests). Rubbing one's chest or lips, or stroking the inside of one's thighs or, most obviously, stroking, massaging, or caressing one's genitals (usually, but not always, through one's clothing) communicates sexual interest, and for most men is taken as an invitation for sexual contact. Body rubbing is done in conjunction with eye contact. To create the most direct message possible, a man moves, rubs, or serially poses while maintaining what he hopes is a mutually held gaze.

Body language also means presenting one's body in what is self-perceived as an attractive manner. Often, especially during warm weather, careful attention is given to the apparel a man wears to the public sex arena. Clothing not only functions as costume, but ideally for the cruising man also provides easy access to critical sexual body parts. Ease and speed of removal and replacement are important. More important, though, is the way that clothing accentuates what a man perceives as his appearance strengths. On occasion men will be seen in cruising areas wearing very revealing clothing (e.g., bathing suits, leather accessories, workout clothing). The intention of men so attired is presumably obvious to others, both those cruising and not. Among many men in public sex arenas, "there's this whole mentality of putting yourself on display," says Duane. Being on display can be dangerous, though; this is why many gay men prefer cruising in controlled access environments (e.g., bars, baths, and clubs) where costuming can be beneficial to one's displays but only minimally dangerous (Brodsky 1993).

Touching

In addition to the presentation of oneself, communicating a sexual interest in another man involves personal space invasions and perhaps subtle, "accidental" forms of contact. When one man allows himself to be touched in any way by another man and does not indicate a dislike or unwillingness to have it continue or occur again, an agreement to progress in the form and intensity of touching is presumed. Vince explained that when he cruises for a sexual partner in the park, after making eye contact with a man he next moves to

> invasion of personal space. Another thing after that is accidentally brushing up against him and neither one of you moves away. That's important, 'cause that dude not moving shows a willingness to be touched.

Touching, posing, placing oneself in another's line of vision, gazing at and receiving a return gaze: these are the tools of communication men in public sex arenas use to replace verbal exchanges when contacting and contracting for exchanges.

Collateral Movement

Intervening between the initial eye contact and an eventual sexual contact are a period of time and a series of interactions in which men pursue each other (both literally and figuratively). The pursuit involves men following each other through the park, whether on foot or in their automobiles, testing one another's resolve and commitment to an exchange. For many men, this phase takes on the quality of a game, or becomes a literal "hunt" and conquest series of interactions. Jack, who believes himself to be among the most skillful and experienced public park cruisers in his city, related the following description of "the chase":

> It's more like playing a game. Basically, that's what it is with a lot of them, it's a big game. You see somebody, you walk by, they'll take off and you follow them from one end to the other, chasing each other. If you give up and turn around, then they'll start chasing after you. That will happen as much as seven or ten times; it's a two-way game, until the connection is made.

The rules of the "chase" game are fairly simple. If you make initial contact with someone that you are interested in, you either follow him or position yourself so that he can easily follow you. The chase is not supposed to be simply one man following another to a location where they have sex, however. Rather, the chase is a test of each man's dedication to the process and the possibility of a sexual encounter with his chosen other.

Many times the chase lasts for extended periods of time, perhaps two hours or more. During the course of this time, men report, they are consistently strategizing ways to read the intentions of the other and to seek ways to draw the other man into making a more assertive advance. Adam, a friend who often goes to the park with Jack, but who always cruises alone, describes how his mind is constantly at work strategizing while playing the "chase game." According to Adam, after making an initial move to follow a man with whom he has already made contact, his mind turns to:

> The next move is his, what will he do? Will he go back the other way? Will he at that point take the initiative and come up to me, or will he pass me and go down the path ten or fifteen feet and stop? If he does that, it's quite obvious he knows how to play the game. He's interested. He knows that second stop on his part has told me something. . . . Then I know. He already knows, because I've already stopped. I was first following him, and I stopped to give him a chance to follow me. If he's not interested, he'll do something else. He'll go the other way or do something that will tell me that. It's all just a game!

In the game, the role with the greater degree of both control and prestige is the role of the pursued. To be followed is to be complimented. Men who are sought after may, in fact, experience situations when more than one individual follows them. In such a situation the matter is complicated when only one of the others involved in the game is desirable to the sought-after man.

Throughout the chase game, communications remain on the non-verbal level, and they may often include heavy doses of eye contact. Whenever a man is being followed, he needs to be aware constantly of whether his pursuer is still in pursuit. This means both visually checking behind him and carefully listening for the other man.

Obviously, to be successful at the chase game requires skills, yet not every man in the park possesses these skills. They can be learned but frequently may require numerous futile (and therefore discouraging) forays. However, with time, dedication, and careful observation (if not direct instruction from a friend) . . . a man can become a skillful player of the game.

CONCLUSION

The processes employed by men who have anonymous sexual encounters with men in public places are culturally created and reinforced phenomena. The idea that men seek out and consummate sexual relationships with anonymous others, while perhaps shocking or disturbing to some, actually carries many similarities to the processes by which men and women seek out both sexual and long-term relationship partners.

Regardless of the gender of those involved, partner-seeking activities require that an individual be aware of the locations where potential partners can be located, have a sense of when others are likely to be in these locations, understand the basic scripts for normative interactional patterns, and be at least somewhat fluent in the language and dialects of cruising. Where differences become apparent is in the fact that gender-based roles are altered, and only men are present. Rather than having attraction based on gender differences, these men must establish differences in roles, expectations, and desires among potential partners. Communication among and between those seeking partners must be clear and complete.

However, because men who cruise for male sexual partners in public settings must be ever attentive to the possibilities (and likely negative consequences) that other men present may not be seeking sex, common communication modes present obstacles. Therefore, the process of communication is tracked into a (typically) prolonged, carefully navigated cruise through double-entendres and ambiguous non-verbal statements. Communication modes include body language, movement throughout the setting, eye contact, and subtle forms of gestures and touch. In essence, men seeking male sex partners in anonymous public settings employ some of the traditionally feminine means of communication. The homosocial nature of the interactional environment necessitates reconfiguration of gendered interactions.

The language of cruising serves as a gatekeeping mechanism for the subcultural setting of the public sex arena. As an open, yet carefully guarded, subcultural setting, a closed set of discreditable (Goffman 1963) men are provided an environment that meets their needs and desires. Although open, the setting does provide some degree of territorial bounding as well as a potential "cover" for men's presence.

REFERENCES

Brodsky, Joel I. 1993. "The Mineshaft: A Retrospective Ethnography." *Journal of Homosexuality* 24:233–251.

Brown, Rita Mae. 1976. "Strangers in Paradise." *Body Politic* (no volume):23.

Clark, Russell III. 1990. "The Impact of AIDS on Gender Differences in Willingness to Engage in Casual Sex." *Journal of Applied Social Psychology* 20:771–782.

Clark, Russell III, and Elaine Hatfield. 1989. "Gender Differences in Receptivity to Sexual Offers." *Journal of Psychology and Human Sexuality* 2:39–55.

Corzine, Jay, and Richard Kirby. 1977. "Cruising the Truckers: Sexual Encounters in a Highway Rest Area." *Urban Life* 6:171–192.

Delph, Edward. 1978. *The Silent Community: Public Homosexual Encounters.* Beverly Hills, CA: Sage.

Desroches, Frederick. 1990. "Tearoom Trade: A Research Update." *Qualitative Sociology* 13:39–61.

Doll, Lynda, Lyle Peterson, Carol White, Eric Johnson, John Ward, and the Blood Donor Study Group. 1992. "Homosexually and Non-homosexually Identified Men Who Have Sex With Men: A Behavioral Comparison." *The Journal of Sex Research* 29:1–14.

Donnelly, Peter. 1981. "Running the Gauntlet: The Moral Order of Pornographic Movie Theaters." *Urban Life* 10:239–264.

Earl, William. 1990. "Married Men and Safe Sex Activity: A Field Study on HIV Risk Among Men Who Do Not Identify as Gay or Bisexual." *Journal of Sex and Marital Therapy* 16:251–257.

Gagnon, John, and William Simon. 1973. *Sexual Conduct: The Social Sources of Human Sexuality*. Chicago: Aldine.

Goffman, Erving. 1959. *The Presentation of Self in Everyday Life*. Garden City, NY: Doubleday.

———. 1963. *Stigma: Notes on the Management of Spoiled Identity*. Englewood Cliffs, NJ: Prentice-Hall.

———. 1977. "The Arrangement Between the Sexes." *Theory and Society* 4:301–331.

Gray, Jan. 1988. *The Tearoom Revisited: A Study of Impersonal Homosexual Encounters in Public Setting*. Unpublished PhD dissertation, The Ohio State University, Columbus.

Hall, Edward. 1959. *The Silent Language*. Garden City, NY: Doubleday.

Herold, Edward, and Dawn-Marie Mewhinney. 1993. "Gender Differences in Casual Sex and AIDS Prevention: A Survey of Dating Bars." *The Journal of Sex Research* 30:36–42.

Humphreys, Laud. 1970. *Tearoom Trade: Impersonal Sex in Public Places*. Chicago: Aldine.

Iizuka, Yuichi. 1992. "Eye Contact in Dating Couples and Unacquainted Couples." *Perceptual and Motor Skills* 75:457–461.

Kamel, G.W. Levi. 1983. *Downtown Street Hustlers: The Role of Dramaturgical Imaging Practices in the Social Construction of Male Prostitution*. Unpublished Ph.D. dissertation, University of California, San Diego.

Kinsey, Alfred, Wardell Pomeroy, and Clyde Martin. 1948. *Sexual Behavior in the Human Male*. Philadelphia: W.B. Saunders.

Lee, John Alan. 1979. "The Social Organization of Sexual Risk." *Alternative Lifestyles* 2:69–100.

Maynard, Steven. 1994. "Through a Hole in the Lavatory Wall: Homosexual Subcultures, Police Surveillance, and the Dialectics of Discovery, Toronto, 1890–1930." *Journal of the History of Sexuality* 5:207–242.

Pollack, Michael. 1993. "Homosexual Rituals and Safer Sex." *Journal of Homosexuality* 25:307–317.

Ponte, Meredith. 1974. "Life in a Parking Lot: An Ethnography of a Homosexual Drive-In." In *Deviance: Field Studies and Self Disclosures,* edited by J. Jacobs. Palo Alto, CA: National Press Books.

Ross, Michael and Jay Paul. 1992. "Beyond Gender: The Basis of Sexual Attraction in Bisexual Men and Women." *Psychological Reports* 71:1283–1290.

Simon, William, and John Gagnon. 1986. "Sexual Scripts: Permanence and Change." *Archives of Sexual Behavior* 15:97–120.

Styles, Joseph. 1979. "Outsider/Insider: Researching Gay Baths." *Urban Life* 8:135–152.

Sundholm, Charles. 1973. "The Pornographic Arcade: Ethnographic Notes on Moral Men in Immoral Places." *Urban Life and Culture*. 2:85–104.

Tewksbury, Richard. 1990. "Patrons of Porn: Research Notes on the Clientele of Adult Bookstores." *Deviant Behavior* 11:259–271.

Weatherford, Jack McIver. 1986. *Porn Row*. New York: Arbor House.

Weinberg, Martin, and Colin Williams. 1975. "Gay Baths and the Social Organization of Impersonal Sex." *Social Problems* 23:124–136.

34

The Manufacture of Fantasy

AMY FLOWERS

Flowers draws on her own experiences as a phone sex worker to discuss the intricate ways that women who work in the phone sex trade use to sell their services. Occasionally referring to each other as "phone sex whores," these young women (and occasionally men) work in a room of phone banks producing fraudulent images of themselves and their behavior designed to entice customers to call them on the phone and remain there as long as possible. Flowers found that the exchange of information, although superficial and scripted to some customers, often becomes highly involved for these phone operators, drawing them into the fantasies they are constructing and selling to others. Operators sometimes find themselves embroiled in the very needs they are paid to create, needing to be wanted by the repeat, regular customers, who they occasionally are tempted to meet outside of work, in violation of the company's policies. Flowers describes the complex way operators attempt to manipulate these interactions, maintain control, and avoid getting emotionally entangled. Phone sex sellers personalize their presentations, investing a deeper part of themselves in the fantasies they are creating.

In the ten years since its inception, the phone sex industry has become a billion dollar business and is continuing to grow despite widespread varied efforts to curb its availability. Psychic phone lines, cyber chat rooms, and many other services are emerging to take advantage of the desire for disembodied intimacy. The users and providers of these services are not strange or extraordinary; they are founding members of modern society.

You or I might never call a phone sex line, but apparently others do. Pacific Bell, in the first twelve months after it began contracting phone sex exchanges local to the New York City area, reaped $13.5 million in revenues (*Time* 1987). In 1994, the four international exchange carries (Telesphere, AT&T, US Sprint, and MCI) collected $900 million in profits from international 900 number calls alone, a market that is increasing because of its relative freedom from Federal Communications Commission (FCC) regulations of obscene content and Federal Trade Commission (FTC) restrictions on business practices. These international profits are in addition to even greater domestic profits from the 900 and 976 lines within the United States (*Economist* 1994)....

From *The Fantasy Factor: An Insider's View of the Phone Sex Industry* by Amy Flowers. Copyright © 1998 by Amy Flowers. Reprinted with permission of the University of Pennsylvania Press.

INSIDE THE FANTASY FACTORY

The "phone-actress" wanted ads in a local underground newspaper were abundant, and I applied to several. My first phone sex job consisted of reading scripts in a recording studio. Most of the scripts were sexually explicit "fantasy" material; for example, "male schoolteacher disciplines flirtatious female student." Usually the scripts relied on social stratification themes; doctor-patient, teacher-student, officer-citizen motifs were most common. "Fresh" scripts were sought after but were rigidly held to a three-minute, social-stratification, gradual-build-to-climax structure. . . .

I, like the other actresses, was paid by the completed script, which often took hours of rehearsals and retakes. Sometimes an acceptable take was never recorded, in which case the actress went unpaid. Ostensibly we made $40.00 for every "completed" three-minute tape. I never saw anyone complete more than one script per four-hour day, however, though many takes of three to five scripts were read and recorded by each actress. . . .

The phone room had six rows of six cubicles for a total of thirty-six stations, although several were nonoperational. Each station had two single-line telephones (well-used office models older than their surroundings and clearly leased or purchased used), and an equally worn list of birth years with their corresponding ages and an approximate year of high-school graduation dates. A nearby operator offered me some rubbing alcohol to disinfect my phones. She also explained that I should first ask the caller's age and when he had graduated from high school in order to screen underage callers. Within moments my phone began to ring.

For the most part it was . . . easy work. The lonely callers liked me, as did the straight-sex callers, and a cross-dresser called back, paying additional charges, just to say thanks. However, it was also clear that the fetish and domination callers were hanging up on me. At one point, Nancy announced, somewhat angrily, that she had been going easy on me because I was new, but that she would have to start sending me more difficult callers. Obviously there was knowledge and skill to acquire, and some operators were more successful than others at getting particular types of callers to linger. . . .

My naiveté about the content of calls I would receive is surprising in retrospect, but I soon discovered that the scripts I had been reading were remarkably pedestrian. The phone sex repertoire includes sadomasochism, fetishism, and many other forms of sexuality I had never explored personally or professionally. I soon learned that phone sex requires real skill and creativity. Imagine, for example, talking for ten minutes or longer to someone with a foot fetish. Removing high-heeled shoes and stockings and then describing the color of nail polish and the shape of my toes took a scant two minutes. As soon as I would begin to slow down or stammer for lack of material, the caller would hang up, so I listened to other operators and asked for their advice in order to learn the details and fillers used for various kinds of callers.

The biggest challenge of phone sex was not the unexpected commonality of fetishism but the interaction itself. Its interactive nature makes phone sex akin to prostitution, and operators often teased each other about being "phone whores." New operators were often troubled by the intimacy they felt with some callers. One experienced operator, who was open about also working as a prostitute, was especially keen on expressing the similarities to anyone who tried to detach herself. "Oooh, baby, that feels so good," she would mimic before adding, "You ain't nothing but a phone-whore."

In obtaining a job as a phone sex operator, I did not intend to deceive anyone. I did not intend to write about my experience for the first two months out of the four I worked as an operator. In fact, although I took notes, it was for an entirely different purpose: they were meant as a personal journal and an inspiration for fictional stories based on interactions with my more imaginative callers. As my intentions changed, I could have begun again. I could have gone to the phone sex management, explained my new objective, and hoped to continue working and observing under these new conditions. However, there were practical concerns. Pornographic businesses, known for their illegal activities (See 1974), are not likely to accept a researcher hanging around, and I had already witnessed illegal activity. In my experience managers were downright oppressive toward workers, insisting on a great deal of control over workers' personal interaction and conversation, especially on such matters as salary comparison, tardiness, and insolence toward supervisors. Since I had no intention to organize workers or report illegal practices, I did not wish to give management any reason to think I might.

My position as a privileged, white, middle-class, educated, noncommitted worker was obvious from my speech, appearance, and demeanor, and I made no pretense of hiding that at work. In reality, there was no need to announce that I was a graduate student. It would have been self-aggrandizing. Describing myself as a Ph.D. candidate would have been seen as putting on airs. I also avoided using my position as an educated, marginal worker to explain other workers' experiences to them, to mediate or counsel, or to develop more effective training methods for the phone sex industry.

I did, when asked, reveal to coworkers that I was a graduate student, but this information was generally not well received. It put distance where there had been camaraderie. They usually confessed, in a tone of inferiority, that they were not very good at school. I often responded with stories of harassment and agony, which seemed to increase their comfort level greatly.

The benefits of these modest "deceptions" were monumental. I gained entry not only to the setting but also to the substance of the material by working as a phone sex operator. I was neither completely an observer nor completely a participant. Far from being problematic, this situation was ideal for the purpose of exploration. After all, all participants observe, and all observers participate. . . .

THE SCRIPT

The script of a phone sex conversation exists like predawn Disneyland—awaiting the noise and animation that only live consumers can supply. The consumers of phone sex enjoy an illusion of great freedom and adventure, but their actions are considerably constrained. The customers choose the ride—the Mad Hatter or Alice in Wonderland—but once their choice is made, they can anticipate a predictable sequence of events. Their experience can be influenced by variations in weather or mood, but for the most part the consumers can enjoy both the fiction of adventure and an experience that is guaranteed safe and dependable.

With few exceptions, phone sex fantasies follow standard scripts. A foot fetishist, for example, wants to hear that toes are painted, skin is supple and smooth. The polish might be pale pink on dark Nigerian toes; sunlit orange on lightly tanned California feet; dark red on pale pink feet fresh out of black satin pumps. The content varies in the details, but the form is constant. What the consumer experiences as spontaneity is actually the product of careful planning and design.

Phone sex, like any interaction, is constructed within many preexisting contexts, and it is assembled from an especially grimy mix of belief, cynicism, pretense, and sincerity. This mix produces a tenuous facade. The caller wants to believe in the fantasy, and he often behaves imprudently, establishing a willingness to act on his belief. The operators themselves are often caught acting on their beliefs in details of the fantasy, despite great familiarity with their own fabrications. Elements of belief and cynicism are necessary.

Each fantasy relies on both participants' commitment to the story line. If either party alters it in an inconsistent or inappropriate way, the constructed reality is easily destroyed. Such an interruption is very rare, except when an inexperienced operator blunders and mistakenly breaks the rules and disrupts the process. Even if the caller or the operator wants to end an encounter, they usually stay within the context of the fantasy to its conclusion. To get rid of callers they find tiresome, operators fake car and plane crashes, terminal diseases, and impending engagements. I also heard stories about steady callers who had "moved away" from operators to new cities, relationships, or prison.

Real-life situations are commonly introduced into the fantasies, and the fantasies have elements that transcend their unreality and influence real events. The process that transforms fantasy into reality resides in the structure of the phone conversations. The formal requirements and rules of civility oblige participants to use certain known falsehoods as if they were true. This process, widely used in the conduct of disembodied intimacy, often leads to a disillusionment with reality and an increasing craving for the comforts of fantasy.

CRAVING BELIEF

Many of the operators expressed surprise that callers actually believe the details of the fantasies and seem to be engrossed at a level of belief beyond fantasy. New operators, particularly, expected callers to be able to enjoy the imagery without actual belief. The empirical evidence, however, of a few callers acting on their beliefs gradually convinces them that callers are taking the details seriously.

"You'll tell them 'I'm a blonde' and they hang up. They don't want a blonde, they want a brunette. Or they want a redhead. But the picture has got to be right or they hang up." Another operator said, "They really think that we're sitting there having sex with dildos and vibrators. They get so caught up in their fantasy that they lose touch with reality." Still another said, "I didn't want them to hang up, you know, so, I'd say, 'Go get me a piece of paper.' And they would."

All the operators I interviewed believed that some callers move beyond fantasy. Peter often enjoyed manipulating callers for both fun and profit.

> I [have] made them go and get things and they have come back with it, which I always thought was the funniest thing. Why don't they just lie? Just say you have the brush in your hand. You don't have to, but they would do that. They would do things, you know, like sing. I'd have them sing to me, or spank themselves. They would do that, put vibrators up their butt. You'd hear them crying on the phone from the things you have them doing. Really crying in pain, you know? "And does it hurt?" "Yeah." "Well, that's good, too bad." Poor guy is crying, but he likes it. So, I'll go ahead and do this. That's what the guy wants me to do. But it's . . . it's funny.

When Peter asks "Why doesn't he just lie?" he never questions whether the caller *did* lie. How does he know that the caller got the hairbrush he was ordered to find? He assumes that if the caller uses expensive time to search for an object, he must be physically seeking the object. He assumes that it is in the caller's interest to use as much of his paid time as possible in actual conversation, and that the time he spends in nonconversational activity is wasted time. Furthermore, Peter assumes that the caller's enjoyment of pauses is very different from his own enjoyment of the very same pauses. Perhaps the caller is as happy as Peter to have a short break from the conversation while still maintaining contact.

The caller offers his belief first, simply by placing the call, and acting first leaves the caller vulnerable to being manipulated by his commitment. For example, if a caller reveals a penchant for elaborate lingerie, the operator might use this information by repeatedly asking him to hold while she fixes the seams of her stockings. The operator sees herself getting away with something, because in her estimation she is. She thinks that she is cheating the caller of conversation and stealing a few moments of privacy in the midst of the encounter.

It is quite likely however, that the caller does not feel his time is being wasted and continues to enjoy the fantasy even during the pauses. A skilled operator can maintain a fantasy even while she flips through the pages of a magazine, whispers to a coworker, or runs for a Coke, and she considers the caller a fool to spend his time and money while she does these mundane things. But these are not the actions the caller visualizes. He sees the adjustment of stockings, the high heels, the pedicured feet. He can enjoy an interlude as part of the interaction, because, presumably, he stays within the realm of his fantasy. From his perspective, the rules of his fantasy have never been violated.

Mimi noticed that the callers themselves did most of the work, creating their own fantasies from the raw material she provides.

> It's funny how many men get into it, and it's really funny to me how easy it is to get into a man's psyche and make him think whatever you want him to think. From my voice he'll create this entire picture and fall in love with it, and will keep calling and spend thirty-five to forty dollars a night talking to a woman about whatever he wants, sex or whatever, and then calling back the next night. I think it's interesting—not funny, but interesting. . . .

Many operators said, enthusiastically, that on the phone an operator has the freedom to "be" whatever she wants. The woman who hired me as an operator listed this freedom as an extra benefit that went with the job. And it is true: the creation of the persona is free of physical constraints. Phone sex allows an operator the freedom to be slender-waisted, buxom, and blonde. Yet the phone sex operator is in another sense more constrained and restricted than ever. Phone sex characters have less variation and less individuality than real life human beings. The characteristics of her persona and the circumstances of her stories are pre-fabricated pornographic images much older than the operator herself. Her on-line character is like a Jungian archetype, emerging not from any individual but from a collective, primitive, human consciousness.

A dominatrix, for example, is tall, strong, and buxom; her hair may be long or short, but it is always straight and black, in keeping with her harsh, direct temperament. A submissive, in contrast, is small and thin, delicate, refined, dependent. Her hair is more likely brown or blonde, and must be long enough to pull. Either fantasy girl can be found in dark, dank environments, but neither is likely to be found outdoors, swimming on a deserted beach. Despite the possibilities for detailed variety within general themes, the major components are rigidly set. When the operator dons a particular mask, her self shows through, but only appropriate features are emphasized, while any inappropriate characteristic is altered, hidden, or downplayed. The operator interprets her self in light of her persona.

By looking at herself through the lens of fantasy, an operator comes to see her inner nature as fitting the fantasy. Many operators I interviewed and worked with sexualized their personal histories and described them as leading directly to, and culminating in, the occupation of phone sex operator. By describing themselves as a "phone person," "a natural-born phone whore," or "always lost in my own fantasies," operators incorporate their working identity into their private conception of self.

Some operators noticed how others confused their real selves with their on-line characters. This confusion was in the form of conceit—they expected the social privileges that accompany the attractiveness and social position of their fantasy characters. Charlene, discussing her own on-line character, said "I exaggerate a little bit." Then she added:

> I don't take it too seriously. Some people do. I have seen some operators, they talk about themselves, and they believe that things are as wonderful as they are saying. The way they describe themselves, and they really think that's how they are. And that's as amazing as the people at the other end of the phone.

Goffman, using the same metaphor of the mask, has suggested that the mask may overtake the self and become more real than the actor who wears it (1959, 1961). In the case of the phone sex operators, other entities—from supervisors to International Exchange Carriers—have as great an interest in their personas as in the actors themselves. If an operator is overtaken by this persona, however, she can be alienated from herself, seeing herself through the eyes of the persona as a pornographer might, and judging herself by those uncompromising standards.

Callers also present a persona on the phone which they reveal in several ways. Callers sometimes admit to an operator that they are shorter, heavier, older, poorer, or balder than they had originally described, and as they reveal their true selves they often refer to the image they had originally created. One called explained to me that he often started out a phone sex call by describing himself as a "*Playgirl* centerfold," and then as the conversation progressed, he would "check out" the operator to assess her compassion. Other callers revealed using a pseudonym ("My real name is Jim") or acknowledged an exaggeration.

Callers, like operators, present a persona on the phone. And operators, in return, have fantasies they expect the callers to fulfill. Callers may blur the line between the fantasy they initially request and the reality that takes over the call, but operators rarely acknowledge their own fantasy, their own confusion, or the gullibility that results. For example, operators often believe that they can tell the difference between the real perverts, who truly believe that an operator exists as described, and the fakes, who know that the call is make-believe.

Some operators' fantasy beliefs seemed every bit as potent as the ones we commonly call "real." Valerie, a seasoned expert in many avenues of the sex business, was far from naive, yet she demonstrated a gullibility and vulnerability common to many operators.

> I got involved with someone who was in the movie business who is very, very wealthy. But he would never meet me. He never said anything, like, to lead me on, and say, "Well, we'll meet." It's a very complicated story, but he started calling, and we just talked, and it wasn't phone sex anymore. Well, I found out that he lives with someone. He just lied and

said he was on and off with somebody, you know? He lied about everything. I found out he had other girls that he was talking to, getting free phone sex from girls he picked up off of 900 lines. I knew I wasn't the only one. So that's just such a complicated story. I'm just getting over it now. . . .

Valerie is a part-time topless dancer who often arranged, clandestinely, private phone sex relationships with her callers; they would call her at home and mail their payments to her post office box, thereby skipping the phone sex company that had introduced them. A caller developing such a relationship with Valerie might assume that he could lie to her about details such as whether or not he lives with someone. A caller might imagine that his lifestyle is as much a part of his fantasy as his "ten inch paw." Despite the fact that Valerie claims she draws the line at fraudulent credit depictions, the examples she gives of harmful lies that have hurt her are concerned with lies of the heart. She has been hurt by the pretense of love and her imaginary lover's infidelity, not by false credit card numbers. . . .

Desire includes the need to feel desirable. "[Sex] involves not only desiring another but also . . . involves a desire that one's partner be aroused by the recognition of one's desire. . . . Desire is therefore not merely the perception of a preexisting embodiment of the other, but ideally a contribution to his further embodiment which in turn enhances the original subject's sense of himself" (Nagel, quoted in Sable 1980: 84–85). The rarity of human contact in modern society and the ideology of male sexuality that must affirm itself through the "enhancement" offered by women amplifies this need for reciprocity and creates an industry to fill it. Several phone sex advertisements proclaim that the operators on their line work "strictly for pleasure," are "unpaid," or are working "just for fun." These advertisements are not seen by the operators as lies, but as part of the fantasy required by the callers.

A caller uses a phone sex fantasy to experience his own desirability. Thus, through phone sex, pornographers are able to sell the reciprocity of desire without posing the risks of discovery and disease associated with more "traditional" forms of prostitution or the emotional risks of real intimacy. The operators are paid for their contribution to the interaction, though apparently not enough to negate their own needs for reciprocity in interaction.

REFERENCES

The Economist. 1994. "Heavy Breathing." 332, 7874: 64(1).

Goffman, Erving. 1959. The Presentation of Self in Everyday Life. NY: Overlook Press.

———. 1961. Encounter Indianapolis: Bobbs Merrill.

See, Carolyn. 1974. Blue Money: Pornography and the Pornographers. NY: David McKay.

Sable, Alan (ed.). 1980. The Philosophy of Sex: Contemporary Readings. Totowa, NJ: Littlefield, Adams.

Time Magazine. 1987. "Reach Out and Touch Someone." 130 (December 21): 58.

CONFLICT

35

Fraternities and Rape on Campus

PATRICIA YANCEY MARTIN AND ROBERT A. HUMMER

In this selection Martin and Hummer consider the situation and activities of college fraternities and the way they promote rape. This compelling article examines the culture of masculinity that flourishes within college fraternities and the way men engage in bonding and status-building by degrading women and making them objects of sexual conquest. Martin and Hummer offer insight into the way masculine culture binds male participants to potentially abusive behavior through its hierarchical status system, as well as into the way women accept the values of this culture and passively follow them to their own exploitation. They also specifically describe the strategies that fraternity men use to accomplish their rapes, from using alcohol as a weapon to employing violence. They conclude that basic features of society and fraternities foster the continuation of such rampant behavior.

Rapes are perpetrated on dates, at parties, in chance encounters, and in specially planned circumstances. That group structure and processes, rather than individual values or characteristics, are the impetus for many rape episodes was documented by Blanchard (1959) 30 years ago (also see Geis 1971), yet sociologists have failed to pursue this theme (for an exception, see Chancer 1987). A recent review of research (Muehlenhard and Linton 1987) on sexual violence, or rape, devotes only a few pages to the situational contexts of rape events, and these are conceptualized as potential risk factors for individuals rather than qualities of rape-prone social contexts.

Many rapes, far more than come to the public's attention, occur in fraternity houses on college and university campuses, yet little research has analyzed fraternities at American colleges and universities as rape-prone contexts

---

From "Fraternities and Rape on Campus," Patricia Y. Martin and Robert A. Hummer. *Gender & Society,* Vol. 3, No. 4, 1989. Reprinted by permission of Sage Publications, Inc.

(cf. Ehrhart and Sandler 1985). Most of the research on fraternities reports on samples of individual fraternity men. One group of studies compares the values, attitudes, perceptions, family socioeconomic status, psychological traits (aggressiveness, dependence), and so on, of fraternity and nonfraternity men (Bohrnstedt 1969; Fox, Hodge, and Ward 1987; Kanin 1967; Lemire 1979; Miller 1973). A second group attempts to identify the effects of fraternity membership over time on the values, attitudes, beliefs, or moral precepts of members (Hughes and Winston 1987; Marlowe and Auvenshine 1982; Miller 1973; Wilder, Hoyt, Doren, Hauck, and Zettle 1978; Wilder, Hoyt, Surbeck, Wilder, and Carney 1986). With minor exceptions, little research addresses the group and organizational context of fraternities or the social construction of fraternity life (for exceptions, see Letchworth 1969; Longino and Kart 1973; Smith 1964).

Gary Tash, writing as an alumnus and trial attorney in his fraternity's magazine, claims that over 90 percent of all gang rapes on college campuses involve fraternity men (1988, p. 2). Tash provides no evidence to substantiate this claim, but students of violence against women have been concerned with fraternity men's frequently reported involvement in rape episodes (Adams and Abarbanel 1988). Ehrhart and Sandler (1985) identify over 50 cases of gang rapes on campus perpetrated by fraternity men, and their analysis points to many of the conditions that we discuss here. Their analysis is unique in focusing on conditions in fraternities that make gang rapes of women by fraternity men both feasible and probable. They identify excessive alcohol use, isolation from external monitoring, treatment of women as prey, use of pornography, approval of violence, and excessive concern with competition as precipitating conditions to gang rape (also see Merton 1985; Roark 1987).

The study reported here confirmed and complemented these findings by focusing on both conditions and processes. We examined dynamics associated with the social construction of fraternity life, with a focus on processes that foster the use of coercion, including rape, in fraternity men's relations with women. Our examination of men's social fraternities on college and university campuses as groups and organizations led us to conclude that fraternities are a physical and sociocultural context that encourages the sexual coercion of women. We make no claims that all fraternities are "bad" or that all fraternity men are rapists. Our observations indicated, however, that rape is especially probable in fraternities because of the kinds of organizations they are, the kinds of members they have, the practices their members engage in, and a virtual absence of university or community oversight. Analyses that lay blame for rapes by fraternity men of "peer pressure" are, we feel, overly simplistic (cf. Burkhart 1989; Walsh 1989). We suggest, rather, that fraternities create a sociocultural context in which the use of coercion in sexual relations with women is normative and in which the mechanisms to keep this pattern of behavior in check are minimal at best and absent at worst. We conclude that unless fraternities change in fundamental ways, little improvement can be expected.

METHODOLOGY

Our goal was to analyze the group and organizational practices and conditions that create in fraternities an abusive social context for women. We developed a conceptual framework from an initial case study of an alleged gang rape at Florida State University that involved four fraternity men and an 18-year-old coed. The group rape took place on the third floor of a fraternity house and ended with the "dumping" of the woman in the hallway of a neighboring fraternity house. According to newspaper accounts, the victim's blood-alcohol concentration, when she was discovered, was .349 percent, more than three times the legal limit for automobile driving and an almost lethal amount. One law enforcement officer reported that sexual intercourse occurred during the time the victim was unconscious: "She was in a life-threatening situation" (*Tallahassee Democrat*, 1988b). When the victim was found, she was comatose and had suffered multiple scratches and abrasions. Crude words and a fraternity symbol had been written on her thighs (*Tampa Tribune*, 1988). When law enforcement officials tried to investigate the case, fraternity members refused to cooperate. This led, eventually, to a five-year ban of the fraternity from campus by the university and by the fraternity's national organization.

In trying to understand how such an event could have occurred, and how a group of over 150 members (exact figures are unknown because the fraternity refused to provide a membership roster) could hold rank, deny knowledge of the event, and allegedly lie to a grand jury, we analyzed newspaper articles about the case and conducted open-ended interviews with a variety of respondents about the case and about fraternities, rapes, alcohol use, gender relations, and sexual activities on campus. Our data included over 100 newspaper articles on the initial gang rape case; open-ended interviews with Greek (social fraternity and sorority) and non-Greek (independent) students (N = 20); university administrators (N = 8, five men, three women); and alumni advisers to Greek organizations (N = 6). Open-ended interviews were held also with judges, public and private defense attorneys, victim advocates, and state prosecutors regarding the processing of sexual assault cases. Data were analyzed using the grounded theory method (Glaser 1978; Martin and Turner 1986). In the following analysis, concepts generated from the data analysis are integrated with the literature on men's social fraternities, sexual coercion, and related issues.

FRATERNITIES AND THE SOCIAL CONSTRUCTION OF MEN AND MASCULINITY

Our research indicated that fraternities are vitally concerned—more than with anything else—with masculinity (cf. Kanin 1967). They work hard to create a macho image and context and try to avoid any suggestion of "wimpishness," effeminacy, and homosexuality. Valued members display, or are willing to go along with, a narrow conception of masculinity that stresses competition, athleticism, dominance, winning, conflict, wealth, material possessions, willingness to drink alcohol, and sexual prowess vis-à-vis women.

Valued Qualities of Members

When fraternity members talked about the kind of pledges they prefer, a litany of stereotypical and narrowly masculine attributes and behaviors was recited and feminine or woman-associated qualities and behaviors were expressly denounced (cf. Merton 1985). Fraternities seek men who are "athletic," "big guys," good in intramural competition, "who can talk college sports." Males "who are willing to drink alcohol," "who drink socially," or "who can hold their liquor" are sought. Alcohol and activities associated with the recreational use of alcohol are cornerstones of fraternity social life. Nondrinkers are viewed with skepticism and rarely selected for membership.[1]

Fraternities try to avoid "geeks," nerds, and men said to give the fraternity a "wimpy" or "gay" reputation. Art, music, and humanities majors, majors in traditional women's fields (nursing, home economics, social work, education), men with long hair, and those whose appearance or dress violate current norms are rejected. Clean-cut, handsome men who dress well (are clean, neat, conforming, fashionable) are preferred. One sorority woman commented that "the top ranking fraternities have the best looking guys."

One fraternity man, a senior, said his fraternity recruited "some big guys, very athletic" over a two-year period to help overcome its image of wimpiness. His fraternity had won the interfraternity competition for highest grade-point average several years running but was looked down on as "wimpy, dancy, even gay." With their bigger, more athletic recruits, "our reputation improved; we're a much more recognized fraternity now." Thus a fraternity's reputation and status depends on members' possession of stereotypically masculine qualities. Good grades, campus leadership, and community service are "nice" but masculinity dominance—for example, in athletic events, physical size of members, athleticism of members—counts most.

Certain social skills are valued. Men are sought who "have good personalities," are friendly, and "have the ability to relate to girls" (cf. Longino and Kart 1973). One fraternity man, a junior, said: "We watch a guy [a potential pledge] talk to women . . . we want guys who can relate to girls." Assessing a pledge's ability to talk to women is, in part, a preoccupation with homosexuality and a conscious avoidance of men who seem to have effeminate manners or qualities. If a member is suspected of being gay, he is ostracized and informally drummed out of the fraternity. A fraternity with a reputation as wimpy or tolerant of gays is ridiculed and shunned by other fraternities. Militant heterosexuality is frequently used by men as a strategy to keep each other in line (Kimmel 1987).

Financial affluence or wealth, a male-associated value in American culture, is highly valued by fraternities. In accounting for why the fraternity involved in the gang rape that precipitated our research project had been recognized recently as "the best fraternity chapter in the United States," a university official said: "They were good-looking, a big fraternity, had lots of BMWs [expensive, German-made automobiles]." After the rape, newspaper stories described the fraternity members' affluence, noting the high number of members who owned expensive cars (*St. Petersburg Times,* 1988).

The Status and Norms of Pledgeship

A pledge (sometimes called an associate member) is a new recruit who occupies a trial membership status for a specific period of time. The pledge period (typically ranging from 10 to 15 weeks) gives fraternity brothers an opportunity to assess and socialize new recruits. Pledges evaluate the fraternity also and decide if they want to become brothers. The socialization experience is structured partly through assignment of a Big Brother to each pledge. Big Brothers are expected to teach pledges how to become a brother and to support them as they progress through the trial membership period. Some pledges are repelled by the pledging experience, which can entail physical abuse; harsh discipline; and demands to be subordinate, follow orders, and engage in demeaning routines and activities, similar to those used by the military to "make men out of boys" during boot camp.

Characteristics of the pledge experience are rationalized by fraternity members as necessary to help pledges unite into a group, rely on each other, and join together against outsiders. The process is highly masculinist in execution as well as conception. A willingness to submit to authority, follow orders, and do as one is told is viewed as a sign of loyalty, togetherness, and unity. Fraternity pledges who find the pledge process offensive often drop out. Some do this by openly quitting, which can subject them to ridicule by brothers and other pledges, or they may deliberately fail to make the grades necessary for initiation or transfer schools and decline to reaffiliate with the fraternity on the new campus. One fraternity pledge who quit the fraternity he had pledged described an experience during pledgeship as follows:

> This one guy was always picking on me. No matter what I did, I was wrong. One night after dinner, he and two other guys called me and two other pledges into the chapter room. He said, "Here, X, hold this 25 pound bag of ice at arms' length 'til I tell you to stop." I did it even though my arms and hands were killing me. When I asked if I could stop, he grabbed me around the throat and lifted me off the floor. I thought he would choke me to death. He cussed me and called me all kinds of names. He took one of my fingers and twisted it until it nearly broke. . . . I stayed in the fraternity for a few more days, but then I decided to quit. I hated it. Those guys are sick. They like seeing you suffer.

Fraternities' emphasis on toughness, withstanding pain and humiliation, obedience to superiors, and using physical force to obtain compliance contributes to an interpersonal style that de-emphasizes caring and sensitivity but fosters intragroup trust and loyalty. If the least macho or most critical pledges drop out, those who remain may be more receptive to, and influenced by, masculinist values and practices that encourage the use of force in sexual relations with women and the covering up of such behavior (cf. Kanin 1967).

Norms and Dynamics of Brotherhood

Brother is the status occupied by fraternity men to indicate their relations to each other and their membership in a particular fraternity organization or group. Brother is a male-specific status; only males can become brothers, although women can become "Little Sisters," a form of pseudomembership. "Becoming a brother" is a rite of passage that follows the consistent and often lengthy display by pledges of appropriately masculine qualities and behaviors. Brothers have a quasi-familial relationship with each other, are normatively said to share bonds of closeness and support, and are sharply set off from nonmembers. Brotherhood is a loosely defined term used to represent the bonds that develop among fraternity members and the obligations and expectations incumbent upon them (cf. Marlowe and Auvenshine [1982] on fraternities' failure to encourage "moral development" in freshman pledges).

Some of our respondents talked about brotherhood in almost reverential terms, viewing it as the most valuable benefit of fraternity membership. One senior, a business-school major who had been affiliated with a fairly high-status fraternity throughout four years on campus, said:

> Brotherhood spurs friendship for life, which I consider its best aspect, although I didn't see it that way when I joined. Brotherhood bonds and unites. It instills values of caring about one another, caring about community, caring about ourselves. The values and bonds [of brotherhood] continually develop over the four years [in college] while normal friendships come and go.

Despite this idealization, most aspects of fraternity practice and conception are more mundane. Brotherhood often plays itself out as an overriding concern with masculinity and, by extension, femininity. As a consequence, fraternities comprise collectivities of highly masculinized men with attitudinal qualities and behavior norms that predispose them to sexual coercion of women (cf. Kanin 1967; Merton 1985; Rapport and Burkhart 1983). The norms of masculinity are complemented by conceptions of women and femininity that are equally distorted and stereotyped and that may enhance the probability of women's exploitation (cf. Ehrhart and Sandler 1985; Sanday 1981, 1986).

Practices of Brotherhood

Practices associated with fraternity brotherhood that contribute to the sexual coercion of women include a preoccupation with loyalty, group protection and secrecy, use of alcohol as a weapon, involvement in violence and physical force, and an emphasis on competition and superiority.

Loyalty, Group Protection, and Secrecy Loyalty is a fraternity preoccupation. Members are reminded constantly to be loyal to the fraternity and to their brothers. Among other ways, loyalty is played out in the practices of group protection and secrecy. The fraternity must be shielded from criticism. Members are admonished to avoid getting the fraternity in trouble and to bring all problems "to the chapter" (local branch of a national social fraternity) rather than to outsiders. Fraternities try to protect themselves from close scrutiny and criticism by the Interfraternity Council (a quasi-governing body composed of representatives from all social fraternities on campus), their fraternity's national office, university officials, law enforcement, the media, and the public. Protection of the fraternity often takes precedence over what is procedurally, ethically, or legally correct. Numerous examples were related to us of fraternity brothers' lying to outsiders to "protect the fraternity."

Group protection was observed in the alleged gang rape case with which we began our study. Except for one brother, a rapist who turned state's evidence, the entire remaining fraternity membership was accused by university and criminal justice officials of lying to protect the fraternity. Members consistently failed to cooperate even though the alleged crimes were felonies, involved only four men (two of whom were not even members of the local chapter), and the victim of the crime nearly died. According to a grand jury's findings, fraternity officers repeatedly broke appointments with law enforcement officials, refused to provide police with a list of members, and refused to cooperate with police and prosecutors investigating the case (*Florida Flambeau*, 1988).

Secrecy is a priority value and practice in fraternities, partly because full-fledged membership is premised on it (for confirmation, see Ehrhart and Sandler 1985; Longino and Kart 1973; Roark 1987). Secrecy is also a boundary-maintaining mechanism, demarcating in-group from out-group, us from them. Secret rituals, handshakes, and mottoes are revealed to pledge brothers as they are initiated into full brotherhood. Since only brothers are supposed to know a fraternity's secrets, such knowledge affirms membership in the fraternity and separates a brother from others. Extending secrecy tactics from protection of private knowledge to protection of the fraternity from criticism is a predictable development. Our interviews indicated that individual members knew the difference between right and wrong, but fraternity norms that emphasize loyalty, group protection, and secrecy often overrode standards of ethical correctness.

Alcohol as Weapon Alcohol use by fraternity men is normative. They use it on weekdays to relax after class and on weekends to "get drunk," "get crazy," and "get laid." The use of alcohol to obtain sex from women is pervasive—in other words, it is used as a weapon against sexual reluctance. According to several fraternity men whom we interviewed, alcohol is the major tool used to gain sexual mastery over women (cf. Adams and Abarbanel 1988; Ehrhart and Sandler 1985). One fraternity man, a 21-year-old senior, described alcohol use to gain sex as follows: "There are girls that you know will fuck, then some you have to put some effort into it. . . . You have to buy them drinks or find out if she's drunk enough."

A similar strategy is used collectively. A fraternity man said that at parties with Little Sisters: "We provide them with 'hunch punch' and things get wild. We get them drunk and most of the guys end up with one." " 'Hunch punch,' " he said, "is a girls' drink made up of overproof alcohol and powdered Kool-Aid, no water or anything, just ice. It's very strong. Two cups will do a number on a female." He had plans in the next academic term to surreptitiously give hunch punch to women in a "prim and proper" sorority because "having sex with prim and proper sorority girls is definitely a goal." These women are a challenge because they "won't openly consume alcohol and won't get openly drunk as hell." Their sororities have "standards committees" that forbid heavy drinking and easy sex.

In the gang rape case, our sources said that many fraternity men on campus believed the victim had a drinking problem and was thus an "easy make." According to newspaper accounts, she had been drinking alcohol on the evening she was raped; the lead assailant is alleged to have given her a bottle of wine after she arrived at his fraternity house. Portions of the rape occurred in a shower, and the victim was reportedly so drunk that her assailants had difficulty holding her in a standing position (*Tallahassee Democrat*, 1988a). While raping her, her assailants repeatedly told her they were members of another fraternity under the apparent belief that she was too drunk to know the difference. Of course, if she was too drunk to know who they were, she was too drunk to consent to sex (cf. Allgeier 1986; Tash 1988).

One respondent told us that gang rapes are wrong and can get one expelled, but he seemed to see nothing wrong in sexual coercion one-on-one. He seemed unaware that the use of alcohol to obtain sex from a woman is grounds for a claim that a rape occurred (cf. Tash 1988). Few women on campus (who also may not know these grounds) report date rapes, however; so the odds of detection and punishment are slim for fraternity men who use alcohol for "seduction" purposes (cf. Byington and Keeter 1988; Merton 1985).

Violence and Physical Force Fraternity men have a history of violence (Ehrhart and Sandler 1985; Roark 1987). Their record of hazing, fighting, property destruction, and rape has caused them problems with insurance companies (Bradford 1986; Pressley 1987). Two university officials told us that fraternities "are the third riskiest property to insure behind toxic waste dumps and amusement parks." Fraternities are increasingly defendants in legal actions brought by pledges subjected to hazing (Meyer 1986; Pressley 1987) and by women who were raped by one or more members. In a recent alleged gang rape incident at another Florida university, prosecutors failed to file charges but the victim filed a civil suit against the fraternity nevertheless (*Tallahassee Democrat*, 1989).

Competition and Superiority Interfraternity rivalry fosters in-group identification and out-group hostility. Fraternities stress pride of membership and superiority over other fraternities as major goals. Interfraternity rivalries take many forms, including competition for desirable pledges, size of pledge class, size of membership, size and appearance of fraternity house, superiority in intramural sports, highest grade-point averages, giving the best parties, gaining the best or most campus leadership roles, and, of great importance, attracting and displaying "good looking women." Rivalry is particularly intense over members, intramural sports, and women (cf. Messner 1989).

FRATERNITIES' COMMODIFICATION OF WOMEN

In claiming that women are treated by fraternities as commodities, we mean that fraternities knowingly, and intentionally, *use* women for their benefit. Fraternities use women as bait for new members, as servers of brothers' needs, and as sexual prey.

Women as Bait Fashionably attractive women help a fraternity attract new members. As one fraternity man, a junior, said, "They are good bait." Beautiful, sociable women are believed to impress the right kind of pledges and give the impression that the fraternity can deliver this type of woman to its members. Photographs of shapely, attractive coeds are printed in fraternity brochures and videotapes that are distributed and shown to potential pledges. The women pictured are often dressed in bikinis, at the beach, and are pictured hugging the brothers of the fraternity. One university official says such recruitment materials give the message: "Hey, they're here for you, you can have whatever you want," and, "We have the best looking women. Join us and you can have them too." Another commented: "Something's wrong when males join an all-male organization as the best place to meet women. It's so illogical."

Fraternities compete in promising access to beautiful women. One fraternity man, a senior, commented that "the attraction of girls [i.e., a fraternity's success in attracting women] is a big status symbol for fraternities." One university official commented that the use of women as a recruiting tool is so well entrenched that fraternities that might be willing to forgo it say they cannot afford to unless other fraternities do so as well. One fraternity man said, "Look, if we don't have Little Sisters, the fraternities that do will get all the good pledges." Another said, "We won't have as good a rush [the period during which new members are assessed and selected] if we don't have these women around."

In displaying good-looking, attractive, skimpily dressed, nubile women to potential members, fraternities implicitly, and sometimes explicitly, promise sexual access to women. One fraternity man commented that "part of what being in a fraternity is all about is the sex" and explained how his fraternity uses Little Sisters to recruit new members:

> We'll tell the sweetheart [the fraternity's term for Little Sister], "You're gorgeous; you can get him." We'll tell her to fake a scam and she'll go

hang all over him during a rush party, kiss him, and he thinks he's done wonderful and wants to join. The girls think it's great too. It's flattering for them.

Women as Servers The use of women as servers is exemplified in the Little Sister program. Little Sisters are undergraduate women who are rushed and selected in a manner parallel to the recruitment of fraternity men. They are affiliated with the fraternity in a formal but unofficial way and are able, indeed required, to wear the fraternity's Greek letters. Little Sisters are not full-fledged fraternity members, however; and fraternity national offices and most universities do not register or regulate them. Each fraternity has an officer called Little Sister Chairman who oversees their organization and activities. The Little Sisters elect officers among themselves, pay monthly dues to the fraternity, and have well-defined roles. Their dues are used to pay for the fraternity's social events, and Little Sisters are expected to attend and hostess fraternity parties and hang around the house to make it a "nice place to be." One fraternity man, a senior, described Little Sisters this way: "They are very social girls, willing to join in, be affiliated with the group, devoted to the fraternity." Another member, a sophomore, said: "Their sole purpose is social—attend parties, attract new members, and 'take care' of the guys."

Our observations and interviews suggested that women selected by fraternities as Little Sisters are physically attractive, possess good social skills, and are willing to devote time and energy to the fraternity and its members. One undergraduate woman gave the following job description for Little Sisters to a campus newspaper:

> It's not just making appearances at all the parties but entails many more responsibilities. You're going to be expected to go to all the intramural games to cheer the brothers on, support and encourage the pledges, and just be around to bring some extra life to the house. [As a Little Sister] you have to agree to take on a new responsibility other than studying to maintain your grades and managing to keep your checkbook from bouncing. You have to make time to be a part of the fraternity and support the brothers in all they do. (*The Tomahawk*, 1988)

The title of Little Sister reflects women's subordinate status; fraternity men in a parallel role are called Big Brothers. Big Brothers assist a sorority primarily with the physical work of sorority rushes, which, compared to fraternity rushes, are more formal, structured, and intensive. Sorority rushes take place in the daytime and fraternity rushes at night so fraternity men are free to help. According to one fraternity member, Little Sister status is a benefit to women because it gives them a social outlet and "the protection of the brothers." The gender-stereotypic conceptions and obligations of these Little Sister and Big Brother statuses indicate that fraternities and sororities promote a gender hierarchy on campus that fosters subordination and dependence in women, thus encouraging sexual exploitation and the belief that it is acceptable.

Women as Sexual Prey Little Sisters are a sexual utility. Many Little Sisters do not belong to sororities and lack peer support for refraining from unwanted sexual relations. One fraternity man (whose fraternity has 65 members and 85 Little Sisters) told us they had recruited "wholesale" in the prior year to "get lots of new women." The structural access to women that the Little Sister program provides and the absence of normative supports for refusing fraternity members' sexual advances may make women in this program particularly susceptible to coerced sexual encounters with fraternity men.

Access to women for sexual gratification is a presumed benefit of fraternity membership, promised in recruitment materials and strategies and through brothers' conversations with new recruits. One fraternity man said: "We always tell the guys that you get sex all the time, there's always new girls.... After I became a Greek, I found out I could be with females at will." A university official told us that, based on his observations, "no one [i.e., fraternity men] on this campus wants to have 'relationships.' They just want to have fun [i.e., sex]." Fraternity men plan and execute strategies aimed at obtaining sexual gratification, and this occurs at both individual and collective levels.

Individual strategies include getting a woman drunk and spending a great deal of money on her. As for collective strategies, most of our undergraduate interviewees agreed that fraternity parties often culminate in sex and that this outcome is planned. One fraternity man said fraternity parties often involve sex and nudity and can "turn into orgies." Orgies may be planned in advance, such as the Bowery Ball party held by one fraternity. A former fraternity member said of this party:

> The entire idea behind this is sex. Both men and women come to the party wearing little or nothing. There are pornographic pinups on the walls and usually porno movies playing on the TV. The music carries sexual overtones.... They just get schnockered [drunk] and, in most cases, they also get laid.

When asked about the women who come to such a party, he said: "Some Little Sisters just won't go.... The girls who do are looking for a good time, girls who don't know what it is, things like that."

Other respondents denied that fraternity parties are orgies but said that sex is always talked about among the brothers and they all know "who each other is doing it with." One member said that most of the time, guys have sex with their girlfriends "but with socials, girlfriends aren't allowed to come and it's their [members'] big chance [to have sex with other women]." The use of alcohol to help them get women into bed is a routine strategy at fraternity parties.

CONCLUSIONS

In general, our research indicated that the organization and membership of fraternities contribute heavily to coercive and often violent sex. Fraternity houses are occupied by same-sex (all men) and same-age (late teens, early twenties) peers whose maturity and judgment is often less than ideal. Yet fraternity houses are private dwellings that are mostly off-limits to, and away from scrutiny of, university and community representatives, with the result that fraternity house events seldom come to the attention of outsiders. Practices associated with the social construction of fraternity brotherhood emphasize a macho conception of men and masculinity, a narrow, stereotyped conception of women and femininity, and the treatment of women as commodities. Other practices contributing to coercive sexual relations and the cover-up of rapes include excessive alcohol use, competitiveness, and normative support for deviance and secrecy (cf. Bogal-Allbritten and Allbritten 1985; Kanin 1967).

Some fraternity practices exacerbate others. Brotherhood norms require "sticking together" regardless of right or wrong; thus rape episodes are unlikely to be stopped or reported to outsiders, even when witnesses disapprove. The ability to use alcohol without scrutiny by authorities and alcohol's frequent association with violence, including sexual coercion, facilitates rape in fraternity houses. Fraternity norms that emphasize the value of maleness and masculinity over femaleness and femininity and that elevate the status of men and lower the status of women in members' eyes undermine perceptions and treatment of women as persons who deserve consideration and care (cf. Ehrhart and Sandler 1985; Merton 1985).

Androgynous men and men with a broad range of interests and attributes are lost to fraternities through their recruitment practices. Masculinity of a narrow and stereotypical type helps create attitudes, norms, and practices that predispose fraternity men to coerce women sexually, both individually and collectively (Allgeier 1986; Hood 1989; Sanday 1981, 1986). Male athletes on campus may be similarly disposed for the same reasons (Kirshenbaum 1989; Telander and Sullivan 1989).

Research into the social contexts in which rape crimes occur and the social constructions associated with these contexts illumine rape dynamics on campus. Blanchard (1959) found that group rapes almost always have a leader who pushes others into the crime. He also found that the leader's latent homosexuality, desire to show off to his peers, or fear of failing to prove himself a man are frequently an impetus. Fraternity norms and practices contribute to the approval and use of sexual coercion as an accepted tactic in relations with women. Alcohol-induced compliance is normative, whereas, presumably, use of a knife, gun, or threat of bodily harm would not be because the woman who "drinks too much" is viewed as "causing her own rape" (cf. Ehrhart and Sandler 1985).

Our research led us to conclude that fraternity norms and practices influence members to view the sexual coercion of women, which is a felony crime, as sport, a contest, or a game (cf. Sato 1988). This sport is played not between men and women but between men and men. Women are the pawns or prey in the interfraternity rivalry game; they prove that a fraternity is successful or prestigious. The use of women in this way encourages fraternity men to see women as objects and sexual coercion as sport. Today's societal norms support young women's right to engage in sex at their discretion, and coercion is unnecessary in a mutually desired encounter. However, nubile young women say they prefer to be "in a relationship" to have sex while young men say they prefer to "get laid" without a commitment (Muehlenhard and Linton 1987). These differences may reflect, in part, American puritanism and men's fears of sexual intimacy or perhaps intimacy of any kind. In a fraternity context, getting sex without giving emotionally demonstrates "cool" masculinity. More important, it poses no threat to the bonding and loyalty of the fraternity brotherhood (cf. Farr 1988). Drinking large quantities of alcohol before having sex suggests that "scoring" rather than intrinsic sexual pleasure is a primary concern of fraternity men.

Unless fraternities' composition, goals, structures, and practices change in fundamental ways, women on campus will continue to be sexual prey for fraternity men. As all-male enclaves dedicated to opposing faculty and administration and to cementing in-group ties, fraternity members eschew any hint of homosexuality. Their version of masculinity transforms women, and men with womanly characteristics, into the out-group. "Womanly men" are ostracized; feminine women are used to demonstrate members' masculinity. Encouraging renewed emphasis on their founding values (Longino and Kart 1973), service orientation and activities (Lemire 1979), or members' moral development (Marlowe and Auvenshine 1982) will have little effect on fraternities' treatment of women. A case for or against fraternities cannot be made by studying individual members. The fraternity qua group and organization is at issue. Located on campus along with many vulnerable women, embedded in a sexist society, and caught up in masculinist goals, practices, and values, fraternities' violation of women—including forcible rape—should come as no surprise.

NOTE

1. Recent bans by some universities on open-keg parties at fraternity houses have resulted in heavy drinking before coming to a party and an increase in drunkenness among those who attend. This may aggravate, rather than improve, the treatment of women by fraternity men at parties.

REFERENCES

Adams, Aileen, and Gail Abarbanel. 1988. *Sexual Assault on Campus: What Colleges Can Do*. Santa Monica, CA: Rape Treatment Center.

Allgeier, Elizabeth. 1986. "Coercive Versus Consensual Sexual Interactions." G. Stanley Hall Lecture to American Psychological Association Annual Meeting, Washington, DC, August.

Blanchard, W. H. 1959. "The Group Process in Gang Rape." *Journal of Social Psychology* 49: 259–66.

Bogal-Allbritten, Rosemarie B., and William L. Allbritten. 1985. "The Hidden Victims: Courtship Violence Among College Students." *Journal of College Student Personnel* 43: 201–4.

Bohrnstedt, George W. 1969. "Conservatism, Authoritarianism and Religiosity of Fraternity Pledges." *Journal of College Student Personnel* 27: 36–43.

Bradford, Michael. 1986. "Tight Market Dries Up Nightlife at University." *Business Insurance* (March 2): 2, 6.

Burkhart, Barry. 1989. Comments in Seminar on Acquaintance/Date Rape Prevention: A National Video Teleconference, February 2.

Byington, Diane B., and Karen W. Keeter. 1988. "Assessing Needs of Sexual Assault Victims on a University Campus." In *Student Services: Responding to Issues and Challenges* (pp. 23–31). Chapel Hill: University of North Carolina Press.

Chancer, Lynn S. 1987. "New Bedford, Massachusetts, March 6, 1983–March 22, 1984: The 'Before and After' of a Group Rape." *Gender & Society* 1: 239–60.

Ehrhart, Julie K., and Bernice R. Sandler. 1985. *Campus Gang Rape: Party Games?* Washington, DC: Association of American Colleges.

Farr, K. A. 1988. "Dominance Bonding through the Good Old Boys Sociability Network." *Sex Roles* 18: 259–77.

Florida Flambeau. 1988. "Pike Members Indicted in Rape." (May 19): 1, 5.

Fox, Elaine, Charles Hodge, and Walter Ward. 1987. "A Comparison of Attitudes Held by Black and White Fraternity Members." *Journal of Negro Education* 56: 521–34.

Geis, Gilbert. 1971. "Group Sexual Assaults." *Medical Aspects of Human Sexuality* 5: 101–13.

Glaser, Barney G. 1978. *Theoretical Sensitivity: Advances in the Methodology of Grounded Theory.* Mill Valley, CA: Sociology Press.

Hood, Jane. 1989. "Why Our Society Is Rape-Prone." *New York Times,* May 16.

Hughes, Michael J., and Roger B. Winston, Jr. 1987. "Effects of Fraternity Membership on Interpersonal Values." *Journal of College Student Personnel* 45: 405–11.

Kanin, Eugene J. 1967. "Reference Groups and Sex Conduct Norm Violations." *The Sociological Quarterly* 8: 495–504.

Kimmel, Michael, ed. 1987. *Changing Men: New Directions in Research on Men and Masculinity.* Newbury Park, CA: Sage.

Kirshenbaum, Jerry. 1989. "Special Report, an American Disgrace: A Violent and Unprecedented Lawlessness Has Arisen Among College Athletes in all Parts of the Country." *Sports Illustrated* (February 27): 16–19.

Lemire, David. 1979. "One Investigation of the Stereotypes Associated with Fraternities and Sororities." *Journal of College Student Personnel* 37: 54–57.

Letchworth, G. E. 1969. "Fraternities Now and in the Future." *Journal of College Student Personnel* 10: 118–22.

Longino, Charles F., Jr., and Cary S. Kart. 1973. "The College Fraternity: An Assessment of Theory and Research." *Journal of College Student Personnel* 31: 118–25.

Marlowe, Anne F., and Dwight C. Auvenshine. 1982. "Greek Membership: Its Impact on the Moral Development of College Freshmen." *Journal of College Student Personnel* 40: 53–57.

Martin, Patricia Yancey, and Barry A. Turner. 1986. "Grounded Theory and Organizational Research." *Journal of Applied Behavioral Science* 22: 141–57.

Merton, Andrew. 1985. "On Competition and Class: Return to Brother-hood." *Ms.* (September): 60–65, 121–22.

Messner, Michael. 1989. "Masculinities and Athletic Careers." *Gender & Society* 3: 71–88.

Meyer, T. J. 1986. "Fight Against Hazing Rituals Rages on Campuses." *Chronicle of Higher Education* (March 12): 34–36.

Miller, Leonard D. 1973. "Distinctive Characteristics of Fraternity Members." *Journal of College Student Personnel* 31: 126–28.

Muehlenhard, Charlene L., and Melaney A. Linton. 1987. "Date Rape and Sexual Aggression in Dating Situations: Incidence and Risk Factors." *Journal of Counseling Psychology* 34: 186–96.

Pressley, Sue Anne. 1987. "Fraternity Hell Night Still Endures." *Washington Post* (August 11): B1.

Rapport, Karen, and Barry R. Burkhart. 1983. "Personality and Attitudinal Characteristics of Sexually Coercive College Males." *Journal of Abnormal Psychology* 93: 216–21.

Roark, Mary L. 1987. "Preventing Violence on College Campuses." *Journal of Counseling and Development* 65: 367–70.

St. Petersburg Times. 1988. "A Greek Tragedy." (May 29): 1F, 6F.

Sanday, Peggy Reeves. 1981. "The Socio-Cultural Context of Rape: A Cross-Cultural Study." *Journal of Social Issues* 37: 5–27.

———. 1986. "Rape and the Silencing of the Feminine." In *Rape,* edited by S. Tomaselli and R. Porter (pp. 84–101). Oxford: Basil Blackwell.

Sato, Ikuya. 1988. "Play Theory of Delinquency: Toward a General Theory of 'Action.'" *Symbolic Interaction* 11: 191–212.

Smith, T. 1964. "Emergence and Maintenance of Fraternal Solidarity." *Pacific Sociological Review* 7: 29–37.

Tallahassee Democrat. 1988a. "FSU Fraternity Brothers Charged." (April 27): 1A, 12A.

———. 1988b. "FSU Interviewing Students About Alleged Rape." (April 24): 1D.

———. 1989. "Woman Sues Stetson in Alleged Rape." (March 19): 3B.

Tampa Tribune. 1988. "Fraternity Brothers Charged in Sexual Assault of FSU Coed." (April 27): 6B.

Tash, Gary B. 1988. "Date Rape." *The Emerald of Sigma Pi Fraternity* 75(4): 1–2.

Telander, Rick, and Robert Sullivan. 1989. "Special Report, You Reap What You Sow." *Sports Illustrated* (February 27): 20–34.

The Tomahawk. 1988. "A Look Back at Rush, A Mixture of Hard Work and Fun" (April/May): 3D.

Walsh, Claire. 1989. Comments in Seminar on Acquaintance/Date Rape Prevention: A National Video Teleconference, February 2.

Wilder, David H., Arlyne E. Hoyt, Dennis M. Doren, William E. Hauck, and Robert D. Zettle. 1978. "The Impact of Fraternity and Sorority Membership on Values and Attitudes." *Journal of College Student Personnel* 36: 445–49.

Wilder, David H., Arlyne E. Hoyt, Beth Shuster Surbeck, Janet C. Wilder, and Patricia Imperatrice Carney. 1986. "Greek Affiliation and Attitude Change in College Students." *Journal of College Student Personnel* 44: 510–19.

36

Opportunity and Crime in the Medical Professions

JOHN LIEDERBACH

Malfeasance and misconduct are rampant in the medical professions, as we constantly read in the newspaper and as Liederbach discusses in this selection on the deviance of doctors. We are less likely to suspect doctors of wrong-doing, Liederbach suggests, due to their high status and role as altruistic healers, but they are nonetheless as subject to the temptations of fraud and abuse as any other individuals. Liederbach describes such common practices as fee-splitting; self-referrals, prescription violations, unnecessary treatments, and sexual misconduct that have led insurers and health maintenance organizations (HMOs) to tighten up their scrutiny of the medical industry. Regulation by the government is less conscientious, it appears, and the presence of overpayment, fraud, and abuse of the Medicaid and Medicare systems are more widespread. Liederbach notes that occupational crimes such as these, where people engage in deviance for their own benefit not that of their organizations, will occur wherever opportunities can be found.

Over time, health care has grown into a trillion dollar a year enterprise. The delivery of patient services involves not only physicians, but also large-scale insurance companies, government-financed benefit programs, and Health Maintenance Organizations (HMOs). Estimates of the cost of health-care fraud range from fifty to eighty billion dollars annually (Witkin, Friedman, and Doran 1992). In the Medicare program alone, some seventeen billion dollars a year is lost (Shogren 1995). The financial cost of medical crime has led one observer to characterize the situation as "white-collar wilding" (Witkin, Friedman, and Doran 1992).

The consequences of medical crime are not merely financial. Unnecessary medical procedures, negligent care, prescription violations, and the sexual abuse of patients exact an enormous physical toll as well. Each year some 400,000 patients become victims of negligent mistakes or misdiagnoses. One Harvard researcher estimates that 180,000 patients die every year, due at least in part to negligent care (Harvard Medical Practices Study 1990). Up to two

---

From "Opportunity and Crime in the Medical Professions" by John Liederbach, from *Crimes of Privilege*, edited by Neal Shover and John Paul Wright, copyright © 2000 by Oxford University Press, Inc. Used by permission of Oxford University Press, Inc.

million patients are needlessly subjected to physical risks through unnecessary operations each year; the resulting price tag approaches four billion dollars (Jesilow, Pontell, and Geis 1993).

THE "PROTECTIVE CLOAK": STATUS, ALTRUISM, AND AUTONOMY

Physicians are recognized as a special group in society—a privileged caste able to decipher puzzling ailments and able to fix broken-down bodies. The privilege is hard won through years of education and exhaustive training. The physician's honored rank, however, sponsors opportunities for doctors to commit crimes within their profession. Attributes synonymous with medical practice, such as high social status, trustworthiness, and professional autonomy, have provided doctors with what some have termed a "protective cloak" that has shielded doctors from scrutiny and legal accountability (Jesilow, Pontell, and Geis 1993; Parsons 1951).

One element of the "cloak," high social status, has helped to afford doctors the protections necessary to commit medical crimes. Doctors' traditional high status derives from two related elements, namely, lucrative salaries and occupational prestige. Physicians remain one of the most highly compensated occupational groups, with median annual incomes exceeding $120,000 (Ruffenach 1988). Aided by the prestige that typically accompanies high salaries in American culture, physicians have been able to retain an elite social position.

Historically, there has also been a general reluctance in American society to use the criminal law against high-status offenders, and criminologists have long recognized the important role that status plays in shaping the criminal opportunities afforded to professional groups. Professionals possess the financial and political wherewithal to influence the manner in which criminal statutes are written and enforced, and they are more apt to "escape arrest and conviction . . . than those who lack such power" (Sutherland 1949). While scholars debate whether this reluctance stems from public apathy and/or ignorance concerning the costs connected to elite crimes (Cullen, Maakestad and Cavender 1987; Evans, Cullen, Dubeck 1993; Wilson 1975), the typically lenient sanctions currently imposed on doctors who pillage government benefit programs, provide negligent care, or otherwise physically abuse their patients points to a historical reluctance to treat as criminal even the most egregious forms of physician malfeasance (Jesilow, Pontell, and Geis 1993; Rosoff, Pontell, and Tillman 1998; Tillman and Pontell 1992; Wolfe et al. 1998).

A second protective element is the altruistic and trustworthy image projected by doctors. This image is cemented in the physician's code of ethics. The oath serves to define doctors as selfless professionals who perform an invaluable service without regard to personal financial gain (Jesilow, Pontell, and Geis 1993). The image structures criminal opportunities in several ways: the image creates an assumption of good will on the part of doctors that makes charges of intentional wrongdoing difficult to justify. Prosecutors may find it too challenging to prove intentionally fraudulent or harmful behavior in cases against highly respected and trusted doctors. Also, the physician's altruistic image has traditionally engendered a certain level of trust from patients (McKinlay and Stoeckle 1988; Stoeckle 1989). Trusting patients who are victims of fraudulent medical schemes or negligent care may fail to hold doctors accountable for their crimes. One observer has defined the impact of these factors more generally as a "pattern of deference" to doctors-a prevailing unwillingness to question their presumed trustworthiness (Bucy 1989).

Third, doctors have been relatively immune from legal scrutiny because of the medical professions' historical preference for self-regulation. State medical review boards, whose members are predominantly physicians themselves. are supposed to provide a "first line of defense" against doctors who violate legal or ethical codes (Wolfe et al. 1998). These boards can revoke medical licenses or otherwise discipline doctors who fail to meet professional or legal standards. Doctors argue that self-regulation and autonomy characterize any profession— that is, doctors alone possess the specialized expertise and unique qualifications to judge the actions of other physicians. The medical community regards the imposition of civil and/or criminal penalties as both unwarranted and unnecessary, especially in cases that involve errors in clinical judgement (Abramovsky 1995). The profession's reliance on self-regulation, however, may facilitate criminal opportunities by shielding its members from more effective punishments. State medical boards, for example, have continually failed to identify doctors who are chronically incompetent, and often punish them with "slaps on the wrist" (Wolfe et al. 1998). The case of one New York doctor illustrates the dangers of relying solely on professional controls:

> During the mid-1980s [Dr. Benjamin] was investigated by the Department of Health in connection with numerous medical irregularities. In 1986, after a medical review board convicted him on 38 counts of gross negligence and incompetence, the New York State Health Commissioner asked the Board to revoke his license. . . . The doctor's punishment was reduced, to a three-month suspension and three years probation. In June 1993 . . . the Department revoked Dr. Benjamin's license for five botched abortions performed in one year. However, Dr. Benjamin was allowed to continue performing abortions pending appeal. Less than a month later (patient) Gaudalupe Negron met her death from another of Dr. Benjamin's botched procedures (Abramovsky 1995).

The lax enforcement typically provided by state medical boards has created an inviting opportunity structure for doctors to commit fraud and abuse within their profession. The problems related to professional control are exacerbated by the well-documented "code of silence" that exists among medical professionals (Rosoff, Pontell, and Tillman 1998). Doctors are often hesitant to report fraud and abuse for fear of professional recriminations (Karlin 1995; Levy 1995). Still, some in the medical community recognize the extent of the problem: "The profession has done a lousy job of policing its own," acknowledges Arthur Caplan, chairman of the Center for Bioethics at the University of Pennsylvania (Grey 1995).

SELECTED MEDICAL OFFENSES

Medical "Kickbacks": Fee Splitting and Self-Referrals

"Kickbacks" are generally defined as payments from one party to another in exchange for referred business or other income-producing deals. Their acceptance by doctors is considered unethical, and in most cases illegal, because they create a conflict of interest between the physicians' commitment to quality patient care and their own financial self-interest. Doctors who are primarily concerned with financial gain compromise their loyalty to patients, as well as their independent professional judgement. Two well-recognized types of medical kickbacks include fee splitting and self-referrals.

Fee splitting occurs when one physician (usually a general practitioner) receives payment from a surgeon or other specialists in exchange for patient referrals. Fee splitting artificially inflates medical costs, provides incentives for unnecessary tests and specialized treatment, and can also endanger the quality of patient care (Stevens 1971). As Sutherland (1949) explained, the fee-splitting doctor "tends to send his patients to the surgeon who will split the largest fee rather than to the surgeon who will do the best work." Early observers regarded fee splitting as an "almost universal" practice, and estimated that 50–90% of physicians split fees (MacEachern 1948; Williams 1948). Despite the advent of more secure payment sources provided by the spread of health insurance coverage in the post-World War II years, congressional investigations in the 1970s continued to recognize fee splitting as a problem (Rodwin 1992; U.S. Congress 1976). While it remains difficult to determine whether the prevalence of fee splitting has increased or decreased over time, it clearly has persisted (Rodwin 1992).

Alternatives to fee splitting have developed more recently, including self-referrals. Self-referrals involve sending patients to specialized medical facilities in which the physician has a financial interest. Between 50,000 and 75,000 doctors have a financial stake in ancillary medical services (quoted in Rosoff, Pontell, and Tillman 1998). Recent research identifies the problems associated with self-referrals, including higher utilization costs and unnecessary services (Hillman et al. 1990; Mitchell and Scott 1992; Rodwin 1992). Self-referring doctors refer patients for laboratory testing at a 45% higher rate than noninvesting physicians (U.S. Department of Health and Human Services 1989). Physicians' utilization of clinical laboratories, diagnostic imaging centers, and rehabilitation facilities was found to be significantly higher when physicians owned these facilities (Hillman et al. 1990).

The medical profession's traditional response to financial conflicts of interest has been less than overwhelming. While most states had declared fee splitting illegal by the mid-1950s, the medical profession did not explicitly address financial conflicts in ethical codes until the 1980s (Rodwin 1992). One newspaper characterized the profession's response with embarrassing clarity: "Fee splitting has been like a venereal disease . . . it exist(s), but nice people do not talk about it" (quoted in Rodwin 1992). The legal and professional response to self-referrals has been more ambivalent. Despite recent legislative attempts to prohibit self-referrals, the practice remains legal. Likewise, professional standards have not been effective in curtailing abuses:

> Unlike other professionals who are the subject to extensive conflict of interest regulation . . . physicians have addressed these issues largely on their own, and have been subject to minimal regulation by state and federal laws or even professional codes . . . The American Medical Association addresses these issues primarily by relying on professional norms, individual discretion and subjective standards . . . (the profession) lacks an effective means to hold physicians accountable (Rodwin 1992, 734).

Prescription Violations

Only doctors possess the education and specialized expertise required to safely prescribe dangerous and often addictive drugs, including narcotics, amphetamines, tranquilizers, and other controlled substances. The privilege is entrusted with the legal responsibility to limit access to these drugs on the basis of medical need. While the vast majority of physicians uphold these responsibilities, an alarming number of doctors violate this trust. Between 1988 and 1996, 1521 doctors were disciplined for misprescribing or over prescribing drugs (Wolfe et al. 1998). Numerous doctors were caught selling blank prescriptions to known addicts. One physician dispensed expired drugs from old, unlabeled spice jars. Another doctor prescribed dangerous weight loss pills to a patient for four years without even examining her—eventually resulting in the patient having a stroke (Wolfe et al. 1998). Some prescription violations are coupled with fraudulent billings schemes designed to maximize profits from illegal prescriptions. Jesilow, Pontell, and Geis (1993) relate one of the most appalling cases:

> In Los Angeles, one investigator reported a Medicaid doctor who saw so many patients daily that red, blue, and yellow lines had been painted on his office floor to expedite traffic. Each color represented a different kind of pill (22).

Similar to the case of medical kickbacks, the medical profession has also largely failed to adequately discipline doctors who violate prescription laws. At least 69% of the doctors cited between 1988 and 1996 were not even temporarily suspended from practicing medicine (Wolfe et al. 1998).

Unnecessary Treatments

Doctors who intentionally subject patients to medically unnecessary treatments violate the law in two ways. First, unnecessary treatments are fraudulent because they result in compensation that is deceptively gained. More important, unnecessary procedures that are invasive, such as surgery, may be considered a form of assault because they needlessly expose patients to physical risks (Lanza-Kaduce 1980). The highly publicized case of one California ophthalmologist dearly illustrates how unnecessary treatments can result in serious physical harm to patients. The doctor preformed unneeded cataract surgery on patients solely to collect a $584 operation fee. In the process, the doctor admitted needlessly "blinding a lot of people" (Pontell, Geis, and Jesilow 1984).

Determining the prevalence of "unnecessary" treatments is often difficult given the inherent uncertainties involved in diagnosing and treating patients. Given (1997) has outlined several methods used by researchers to estimate the extent of unnecessary surgeries: (1) geographical variations in surgical rates, (2) studies of second surgical opinions, (3) variations in surgical rates between payment plans, and (4) expert opinions based on predetermined criteria. Although these studies present mixed findings, wide geographical variations in surgical rates can be used to suggest that unnecessary surgeries occur with some frequency. For example, hysterectomies are performed at an 80% higher rate in Southern states, and 1000% rate variations in pacemaker operations have been identified in Massachusetts (quoted in Green 1997). Similarly, one government-sponsored study found a 120% higher surgical rate for patients enrolled in fee-for-service plans versus patients enrolled in HMOs (U.S. Department of Health, Education, and Welfare 1971).

Sexual Misconduct

Sexual misconduct by doctors can take a variety of forms (Jacobs 1994). Doctors may engage in sexual misconduct in exchange for professional services. Alternatively, doctors may allow relationships with patients to escalate beyond what is ethically acceptable. Finally, doctors may sexually assault patients while they are under the control of anesthesia or otherwise incapable of consenting to a sexual act (Green 1997). These offenses are especially abhorrent, because they represent an abuse of power by the doctor in situations where the patient is particularly vulnerable. From 1987 to 1996, 393 doctors were disciplined by state medical boards for sexual misconduct with patients. At least 34% of those doctors were not forced to even temporarily stop practicing (Wolfe et al. 1998).

MEDICAID FRAUD AND ABUSE

Medicaid began in 1965 as one of the "Great Society" programs initiated during the Lyndon B. Johnson administration. The program extended health coverage to needy Americans who could not otherwise afford it. Perhaps overshadowed by the nobility of this goal, the program's costs were not considered a primary concern (Jesilow, Pontell, and Geis 1993). Since its inception, fraud and abuse has been endemic to the program. Jesilow, Pontell, and Geis (1993), in their exposé on Medicaid crime, identify several reasons why the introduction of Medicaid has surreptitiously expanded the scope of the medical crime problem.

The medical profession opposed the initial Medicaid legislation. Doctors perceived Medicaid as a threat to their professional autonomy, because the program dictated the price of their professional medical services (Jesilow, Pontell, and Geis 1993). The Medicaid program introduced an unwelcome influence—the government—into decisions that were traditionally left to the independent professional discretion of doctors. This intrusion into the professional autonomy of doctors created widespread dissatisfaction within the medical community, and served to facilitate fraud and abuse within the program in at least two important ways (Jesilow, Pontell, and Geis 1993). First, the medical community's initial opposition led to certain flaws in the program's design that created easy opportunities to violate the law. For example, the original Medicaid legislation did not include provisions for punishing doctors who violated program rules (Jesilow, Pontell, and Geis 1993). The omission was not an accident, but an attempt to placate doctors, without whose cooperation the program could not be launched. As a result, the program lacked the effective sanctions necessary to police an additional design flaw, the program's fee-for-service payment structure. The fee-for-service plan reimbursed doctors a fixed amount for each procedure, but the doctor could earn additional income by double-billing for the same patient, billing for more expensive procedures than those performed, or even charging the program for services that were never done (Jesilow, Pontell, and Geis 1993). A significant number of Medicaid providers could not resist the combination of easy opportunities and lenient sanctions.

Second, the government's intrusion on the professional autonomy of doctors gave rise to an aggressively defiant attitude among some practitioners against the rules that governed Medicaid work. As Jesilow, Pontell, and Geis (1993) explain, this militant defiance "dramatically reduced one of the most powerful deterrents to crime, especially for middle and upper class perpetrators: the sense of guilt and the force of conscience associated with depredations against known human victims." Because they believed Medicaid to be an illegitimate intrusion on their professional autonomy, some Medicaid violators "redefined" their criminal behavior in a positive light. As a result, many of these doctors did not view their actions as wrong (Jesilow, Pontell, and Geis 1993).

This combination of opportunities, motivations, and lax sanctions has produced a dizzying array of violations related to the Medicaid program. Perhaps the most striking example of Medicaid fraud, however, is the discovery of Medicaid "mills" by fraud investigators (Jesilow, Pontell, and Geis 1993).

> Located in dilapidated areas, often in storefronts, and catering almost exclusively to patients on Medicaid rolls, the mills resemble clinics in that doctors with different specialties are gathered under one roof. But the mill's providers often rent space in the building and bill Medicaid individually.... Criminal activities flourished in the Medicaid mills. Some employed "hawkers" to round up customers. Several catered to drug traffic. Various government agents, all claiming to be suffering from

nothing more than the common cold, had been seen by eighty-five doctors in Medicaid mills. They underwent eighteen electrocardiograms, eight tuberculosis tests, four allergy tests, as well as a hearing, glaucoma, and electroencephalogram tests (50).

The inception of the Medicaid program posed one of he first significant challenges to the professional autonomy of physicians. Although this challenge did serve to extend health benefits to many of the nation's most indigent citizens, the program also altered the traditional opportunity structure for medical crime, and created new and unique avenues for physicians to commit offenses relating to their medical practice. . . .

One indication of increasing physician vulnerability is the recent spate of criminal prosecutions against doctors who have killed or maimed patients through negligent or reckless medical care. Because of the protections traditionally afforded to doctors, errors in clinical judgment—even if they resulted in the death of patient—have traditionally been sanctioned exclusively through civil actions or peer-oriented sanctions. The filing of *violent* criminal charges, such as manslaughter, assault, reckless homicide, or murder, against doctors who victimize patients in the course of their medical practice had been an exceedingly rare occurrence (Abramovsky 1995; Crane 1994; Green 1997). However, at least seven doctors have been criminally prosecuted for their violent victimization of patients over the last ten years (Liederbach, Cullen, Sundt, and Geis 1998). The case against Dr. Joseph Verbrugge is indicative not only because of the gravity of his mistakes, but also because he is believed to be the first doctor to stand trial in Colorado accused of a violent crime related to a medical procedure:

> Verbrugge was charged with reckless manslaughter in connection with the death of eight-year-old Richard Leonard during ear surgery. The normally routine procedure went awry when the boy's heart rate jumped significantly after Verbrugge administered the anesthetic. During the surgery, the patient's breathing became irregular and his temperature soared to 107 degrees. Prosecutors contended that Verbrugge failed to react to those danger signs because he had fallen asleep during the operation. The patient died after three hours in surgery. His reaction to the anesthesia had increased the level of carbon monoxide in his blood to four times the normal level (Liederbach, Cullen, Sundt, and Geis 1998).

It remains to be seen how prevalent criminal prosecutions against doctors for medical "mistakes" will become. However, the recent cluster of these cases may suggest the beginning of a trend toward increasing physician vulnerability. These cases may be a signpost indicating that the traditional protections afforded to doctors have eroded. In particular, as long-term doctor-patient relationships wane and as HMOs increasingly influence the delivery of medical services, patient trust in doctors—especially when things go badly in a medical procedure—will decline, and the ability to see reckless doctors as *criminals* may increase (Liederbach, Cullen, Sundt, and Geis 1998).

REFERENCES

Abramovsky, A. 1995. "Depraved Indifference and the Incompetent Doctor." *New York Law Journal* November 8, pp. 3–10.

Bucy, P. H. 1989. "Fraud by Fright: White Collar Crime by Health Care Providers." *North Carolina Law Review* 67: 855–937.

Crane, M. 1994. "Could Clinical Mistakes Land You in Jail? The Case of Gerald Einaugler." *Medical Economics* 71: 46–52.

Cullen, F. T., W. J. Maakestad, and G. Cavender. 1987. *Corporate Crime under Attack: The Ford Pinto Case and Beyond.* Cincinnati: Anderson.

Evans, D. T., F. T. Cullen, and P. J. Dubeck. 1993. "Public Perceptions of White Collar Crime." Pp. 85–114 in *Understanding Corporate Criminality,* edited by M. B. Blankenship. New York: Garland.

Green, G. S. 1997. *Occupational Crime,* 2nd ed. Chicago: Nelson Hall.

Grey, B. 1995. "Medical Scandal." *Baltimore Sun,* August 21.

Hillman, B. J., V. A. Joseph, M. R. Mabry, J. H. Sunshine, S. D. Kennedy, and M. Noether. 1990. "Frequency and Costs of Diagnostic Imaging in Office Practice—A Comparison of Self-Referring and Radiologist Referring Physicians." *New England Journal of Medicine* 323: 1604–8.

Himmelstein, D. V. 1996. "US Health Reform: Unkindest Cuts." *The Nation* January 22.

Jacobs, S. 1994. "Social Control of Sexual Assault By Physicians and Lawyers within the Professional Relationship: Criminal and Disciplinary Actions." *American Journal of Criminal Justice* 19(1): 43–60.

Jesilow, P., H. N. Pontell, and G. Geis. 1993. *Prescription for Profit: How Doctors Defraud Medicaid.* Berkeley: University of California Press.

Karlin, R. 1995. "Selective Silence Is under Scrutiny." *The Time Union* (Albany, NY), August 28, p. B1.

Lanza-Kaduce, L. 1980. "Deviance among Professionals: The Case of Unnecessary Surgery." *Deviant Behavior* 1: 333–59.

Levy, D. 1995. "Physicians Can Run, Hide from Deadly Errors." *USA Today,* September 11, p. 1D.

Liederbach, J., F. T. Cullen, J. Sundt, and G. Geis. 1998. "The Criminalization of Physician Violence: Social Control in Transformation?" Paper presented at the annual meeting of the American Society of Criminology, Washington, D.C.

MacEachern, M. 1948. "College Continues Militant Stance against Fee-Splitting and Rebates." *Bulletin of the American College of Surgeons* 33: 65–67.

McKinlay, J. B., and J. D. Stoeckle. 1988. "Corporatization and the Social Transformation of Medicine." *International Journal of Health Services* 18: 191–200.

Mitchell, J. M., and E. Scott. 1992. "New Evidence on the Prevalence and Scope of Physician Joint Ventures." *Journal of the American Medical Association* 268: 80–84.

Parsons, T. 1951. *The Social System.* Glencoe, IL: Free Press.

"Patients, Doctors, Lawyers: Medical Injury, Malpractice Litigation, and Patient Compensation." 1990. *Harvard Medical Practices Study.* Boston: President and Fellows of Harvard University.

Pontell, H. N., G. Geis, and P. D. Jesilow. 1984. "Practitioner Fraud and Abuse in Government Medical Benefit Programs." Washington, D.C.: U.S. Department of Justice.

Rodwin, M. A. 1992. "The Organized American Medical Profession's Response to Financial Conflicts of Interest: 1890–1992." *Milbank Quarterly* 70(4): 703–41.

Rosoff, S. M., H. N. Pontell, and R. Tillman. 1998. *Profit without Honor: White Collar Crime and the Looting of America.* Upper Saddle River, NJ: Prentice Hall.

Ruffenach, G. 1988. "No Need to Worry, Doctors Do Just Fine." *Wall Street Journal,* October 10.

Sherman, W. 1995. "Girl Waited 6 Months for Spine Surgery." *New York Post,* September 19.

Shogren, G. 1995. "Rampant Fraud Complicates Medicare Cures." *Los Angeles Times,* October 8, p. 1.

Stevens, R. 1971. *American Medicine and the Public Interest.* New Haven: Yale University Press.

Stoeckle, J. D. 1989. "Reflections on Modern Doctoring." *Milbank Quarterly* 66: 76–89.

Sutherland, E. H. 1949. *White-Collar Crime,* New Haven: Yale University Press.

Tillman, R., and H. N. Pontell. 1992. "Is Justice Color Blind?: Punishing Medicaid Provider Fraud." *Criminology* 30(4): 547–74.

U.S. Congesss. Senate Subcommittee on Long-Term Care, Special Committee on Aging. 1976. *Fraud and Abuse among Practitioners Participating in the Medicaid Program.* Washington, D.C.

U.S. Department of Health, Education, and Welfare.1971. *The Federal Employees Health Benefit Program—Enrollment and Utilization of Health Services 1961–1968.* Washington, D.C.: U.S. Government Printing Office.

U.S. Department of Health and Human Services. 1989. *Financial Arraignments between Physicians and Health Care Businesses.* Washington, D.C.: U.S. Government Printing Office.

Williams, G. 1948. "The Truth About Fee-Splitting." *Modern Hospital* 70: 43–48 (reprinted in Reader's Digest, July 1948).

Wilson, J. Q. 1975. *Thinking about Crime.* New York: Basic Books.

Witkin, G., D. Friedman, and G. Doran. 1992. "Health Care Fraud." *US News and World Report,* February 24, pp. 34–43.

Wolfe, S., K. M. Franklin, P. McCarthy, P. Bame, and B. M. Adler. 1998. *16,638 Questionable Doctors Disciplined by State and Federal Governments.* Washington, D.C.: Public Citizens Health Research Group.

PART X

Phases of the Deviant Career

One of the fascinating things about people's involvement in deviance is that it evolves, yielding a shifting and changing experience. Doing something for the first time is very different from doing it for the hundredth time. It is fruitful, then, to consider involvement in deviance from a career perspective, to see what the nature of deviance is and how it develops over the course of people's involvement with it. Sociologists have documented various stages of people's participation in such things as drug use, drug dealing, fencing, carrying out a professional hit, engaging in prostitution, and shoplifting. Although these activities are very different in character, they have structural similarities in the way people experience them according to the stage of people's involvement. In fact, the career analogy has been applied fruitfully to the study of deviance because people go through many of the same cycles of entry, upward mobility, achieving career peaks, aging in the career, burning out, and getting out of deviance as they do in legitimate work. The career phases most commonly analyzed are those at the beginning and the end, when participants are involved in the transition between deviance and the conventional world, but we will also look at some features of intermediary career deviance. In examining deviant careers we note the limitation of the comparison to enterprises in the legitimate realm; while legitimate work has several varieties of structure, the patterns for deviant careers are more flexible and varied (Luckenbill and Best 1981). Entry can take many shapes and lengths of time. Behavioral shifts once in deviance can be lateral and downward as well as upward, precipitous as well as gradual and controlled, repetitive as well as dissimilar, and involve continuity or complete shift into other venues. Exits are problematic, varying in degree of coercion or voluntariness, being temporary or lasting, and involve anything from going out on top to slinking away in debt and disgrace. Let us consider these aspects of the deviant career one-by-one.

Sociologists have been most fascinated by the process through which people *enter deviance*. Although some people venture into deviance on their own, the vast majority do it with the encouragement and assistance of others, often joining cooperative deviant enterprises. The turning points that mark significant phases in their transitions have been explored, as well as their changing self-identities. Most commonly, people who become involved in deviance do so through a process of shifting their circle of friends. They drift into new peer groups as they are drifting into deviance, or their whole peer group drifts into deviance together as the members enter a new phase of the lifecycle.

Managing deviance holds different challenges. Participants must navigate their deviance, their relationships within deviant communities, and their safety from agents of social control, and must evolve a personal style for their deviance. They must balance their deviance with the nondeviant aspects of their lives, such as their relationships with family members, people in the community, and those on whom they rely to meet their legitimate needs. Yip's selection on gay male Christian couples illustrates some of the creative ways that homosexuals find to maintain their relational commitment in a social environment where their union is sanctioned by neither church nor state.

Although people tend to think that getting into deviance represents the more difficult end of the career span, sociologists have found that there can be greater problems associated with *exiting deviance*. A variety of factors exist that both "push" people out of the deviant life and "pull" them back into the conventional world. People change during their involvement with deviance and the easily obtained money they often find there; returning to a more restricted base of funds is not always easy. They also become accustomed to the free-wheeling lifestyle and open value system associated with a deviant community, where conventional norms are disdained. Reentering the straight world with its morality may chafe. They may have difficulty earning a living if they have been involved in occupational deviance, where they were making money through illicit means. They may have difficulty putting together a resume that accounts for their gap in legitimate employment and finding someone who will hire them. They may have difficulty adhering to the structure of the 9-to-5 straight world. They may have difficulty finding legitimate skills through which they can support themselves. Yet most people do not want to spend their whole lives engaged in deviance. We discuss the process by which people burn out of deviance in our selection on the exiting patterns of upper level drug traffickers.

Parallel to exiting deviance is the process of finding something to move into that can hold an individual to conventional society and the legitimate economy. Brown highlights an alternate route out of deviance that has recently become more common in society: capitalizing on one's former deviance in his discussion of the professional ex- role. This article moves beyond our discussion of wanting to get out of deviance to actually making the move and sticking with it.

ENTERING DEVIANCE

37

Joining a Gang

MARTÍN SÁNCHEZ JANKOWSKI

Now it is thought to be the mark of a man of practical wisdom to be able to deliberate well about what is good and expedient for himself, not in some particular respect, e.g. about what sorts of thing conduce to health or to strength, but about what sorts of thing conduce to the good life in general.

ARISTOTLE
THE NICOMACHEAN ETHICS

We are looking for a few good men.
U.S. MARINE CORPS RECRUITING POSTER

Jankowski details and contrasts the many complex factors that induce individuals to enter urban street gangs, highly cohesive and illicit groups. These range from the physical to the economic and social. But joining a gang is not a one-way decision, as not every individual is acceptable as a member. Gangs have organizational needs that must be met through the continuous recruitment of new people, individuals who will mesh with existing members and be reliable in uncertain or dangerous situations. They may choose to entice new members with their attractiveness, to pressure them into joining through a sense of duty, or coerce them through threats. Jankowski also addresses why some individuals do not join gangs. This selection serves as an interesting counterpoint to the other discussions of gangs in the readings on women gang members and international organized crime.

I [have] argued that one of the most important features of gang members was their defiant individualist character. I explained the development of defiant individualism by locating its origins in the material conditions—the competition and conflict over resource scarcity—of the low-income neighborhoods

---

From Martín Sánchez Jankowski, *Islands in the Street: Gangs and American Urban Society* (Berkeley: University of California Press, 1991). Copyright © 1991 The Regents of the University of California. Reprinted by permission of the publisher.

of most large American cities. These conditions exist for everyone who lives in such neighborhoods, yet not every young person joins a gang. Although I have found that nearly all those who belong to gangs do exhibit defiant individualist traits to some degree, not all those who possess such traits join gangs. This chapter explores who joins a gang and why in more detail.

Many studies offer an answer to why a person joins a gang, or why a group of individuals start a gang. These studies can be divided into four groupings. First, there are those that hold the "natural association" point of view. These studies argue that people join gangs as a result of the natural act of associating with each other.[1] Their contention is that a group of boys, interrelating with each other, decide to formalize their relationship in an attempt to reduce the fear and anxiety associated with their socially disorganized neighborhoods. The individual's impetus to join is the result of his desire to defend against conflict and create order out of the condition of social disorganization.

The second group of studies explains gang formation in terms of "the subculture of blocked opportunities": gangs begin because young males experience persistent problems in gaining employment and/or status. As a result, members of poor communities who experience the strain of these blocked opportunities attempt to compensate for socioeconomic deprivation by joining a gang and establishing a subculture that can be kept separate from the culture of the wider society.[2]

The third group of studies focuses on "problems in identity construction." Within this broad group, some suggest that individuals join gangs as part of the developmental process of building a personal identity or as the result of a breakdown in the process.[3] Others argue that some individuals from low-income families have been blocked from achieving social status through conventional means and join gangs to gain status and self-worth, to rebuild a wounded identity.[4]

A recent work by Jack Katz has both creatively extended the status model and advanced the premise that sensuality is the central element leading to the commission of illegal acts. In Katz's "expressive" model, joining a gang, and being what he labels a "badass," involves a process whereby an individual manages (through transcendence) the gulf that exists between a sense of self located within the local world (the here) and a reality associated with the world outside (the there). Katz argues that the central elements in various forms of deviance, including becoming involved in a gang and gang violence, are the moral emotions of humiliation, righteousness, arrogance, ridicule, cynicism, defilement, and vengeance. "In each," he says, "the attraction that proves to be most fundamentally compelling is that of overcoming a personal challenge to moral—not material—existence."[5]

Most of these theories suffer from three flaws. First, they link joining a gang to delinquency, thereby combining two separate issues. Second, they use single-variable explanations. Third, and most important, they fail to treat joining a gang as the product of a rational decision to maximize self-interest, one in which both the individual and the organized gang play a role. This is especially true of Katz's approach, for two reasons. First, on the personal level, it underestimates the impact of material and status conditions in establishing the situations in which sensual needs/drives (emotions) present themselves, and overestimates/exaggerates the "seductive" impact of crime in satisfying these needs. Second, it does not consider the impact of organizational dynamics on the thought and action of gang members.

In contrast, the data presented here will indicate that gangs are composed of individuals who join for a variety of reasons. In addition, while the individual uses his own calculus to decide whether or not to join a gang, this is not the only deciding factor. The other deciding factor is whether the gang wants him in the organization. Like the individual's decision to join, the gang's decision to permit membership is based on a variety of factors. It is thus important to understand that who becomes a gang member depends on two decision-making processes: that of the individual and that of the gang.

THE INDIVIDUAL AND THE DECISION TO BECOME A MEMBER

Before proceeding, it is important to dismiss a number of the propositions that have often been advanced. The first is that young boys join gangs because they are from broken homes where the father is not present and they seek gang membership in order to identify with other males—that is, they have had no male authority figures with whom to identify. In the ten years of this study, I found that there were as many gang members from homes where the nuclear family was intact as there were from families where the father was absent.[6]

The second proposition given for why individuals join gangs is related to the first: it suggests that broken homes and/or bad home environments force them to look to the gang as a substitute family. Those who offer this explanation often quote gang members' statements such as "We are like a family" or "We are just like brothers" as indications of this motive. However, I found as many members who claimed close relationships with their families as those who denied them.

The third reason offered is that individuals who drop out of school have fewer skills for getting jobs, leaving them with nothing to do but join a gang. While I did find a larger number of members who had dropped out of school, the number was only slightly higher than those who had finished school.

The fourth reason suggested, disconfirmed by my data, is a modern version of the "Pied Piper" effect: the claim that young kids join gangs because they are socialized by older kids to aspire to gang membership and, being young and impressionable, are easily persuaded. I found on the contrary that individuals were as likely to join when they were older (mid to late teens) as when they were younger (nine to fifteen). I also found significantly more who joined when they were young who did so for reasons other than being socialized to think it was "cool" to belong to a gang. In brief, I found no evidence for this proposition.

What I did find was that individuals who live in low-income neighborhoods join gangs for a variety of reasons, basing their decisions on a rational calculation of what is best for them at that particular time. Furthermore, I found that they use the same calculus (not necessarily the same reasons) in deciding whether to stay in the gang, or, if they happen to leave it, whether to rejoin.

Reasons for Deciding to Join a Gang

Most people in the low-income inner cities of America face a situation in which a gang already exists in their area. Therefore the most salient question facing them is not whether to start a gang or not, but rather whether to join an existing one. Many of the reasons for starting a new gang are related to issues having to do with organizational development and decline—that is, with the existing gang's ability to provide the expected services, which include those that individuals considered in deciding to join. . . . This section deals primarily, although not exclusively, with the question of what influences individuals to join an existing gang. However, many of these are the same influences that encourage individuals to start a new gang.

Material Incentives

Those who had joined a gang most often gave as their reason the belief that it would provide them with an environment that would increase their chances of securing money. Defiant individualists constantly calculate the costs and benefits associated with their efforts to improve their financial well-being (which is usually not good). Therefore, on the one hand, they believe that if they engage in economic ventures on their own, they will, if successful, earn more per venture than if they acted as part of a gang. However, there is also the belief that if one participates in economic ventures with a gang, it is likely that the amount earned will be more regular, although perhaps less per venture. The comments of Slump, a sixteen-year-old member of a gang in the Los Angeles area, represent this belief:

> Well, I really didn't want to join the gang when I was a little younger because I had this idea that I could make more money if I would do some gigs [various illegal economic ventures] on my own. Now I don't know, I mean, I wasn't wrong. I could make more money on my own, but there are more things happening with the gang, so it's a little more even in terms of when the money comes in. . . . Let's just say there is more possibilities for a more steady amount of income if you need it.

It was also believed that less individual effort would be required in the various economic ventures in a gang because more people would be involved. In addition, some thought that being in a gang would reduce the *risk* (of personal injury) associated with their business ventures. They were aware that if larger numbers of people had knowledge of a crime, this would increase the risk that if someone were caught, others, including themselves, would be implicated. However, they countered this consideration with the belief that they faced less risk of being physically harmed when they were part of a group action. The comments of Corner, a seventeen-year-old resident of a poor Manhattan neighborhood, represent this consideration. During the interview, he was twice approached about joining the local gang. He said:

> I think I am going to join the club [gang] this time. I don't know, man, I got some things to decide, but I think I will. . . . Before I didn't want to join because when I did a job, I didn't want to share it with the whole group—hell, I was never able to make that much to share. . . . I would never have got enough money, and with all those dudes [other members of the gang] knowing who did the job, you can bet the police would find out. . . . Well, now my thinking is changed a bit 'cause there's more people involved and that'll keep me safer. [He joined the gang two weeks later.]

Others decided to join the gang for financial security. They viewed the gang as an organization that could provide them or their families with money in times of emergency. It represented the combination of a bank and a social security system, the equivalent of what the political machine had been to many new immigrant groups in American cities.[7] To these individuals, it provided both psychological and financial security in an economic environment of scarcity and intense competition. This was particularly true of those who were fifteen and younger. Many in this age group often find themselves in a precarious position. They are in need of money, and although social services are available to help during times of economic hardship, they often lack legal means of access to these resources. For these individuals, the gang can provide an alternative source of aid. The comments of Street Dog and Tomahawk represent these views. Street Dog was a fifteen-year-old Puerto Rican who had been in a New York gang for two years:

> Hey, the club [the gang] has been there when I needed help. There were times when there just wasn't enough food for me to get filled up with. My family was hard up and they couldn't manage all of their bills and such, so there was some lean meals! Well, I just needed some money to help for awhile, till I got some money or my family was better off. They [the gang] was there to help. I could see that [they would help] before I joined, that's why I joined. They are there when you need them and they'll continue to be.

Tomahawk was a fifteen-year-old Irishman who had been in a gang for one year:

> Before I joined the gang, I could see that you could count on your boys to help in times of need and that meant a lot to me. And when I needed money, sure enough they gave it to me. Nobody else would have given it to me; my parents didn't have it, and there was no other place to go. The gang was just like they said they would be, and they'll continue to be there when I need them.

Finally, many view the gang as providing an opportunity for future gratification. They expect that through belonging to a gang, they will be able to make contact with individuals who may eventually help them financially. Some look to meet people who have contacts in organized crime in the hope of entering that field in the future. Some hope to meet businessmen involved in the illegal market who will provide them with money to start their own illegal businesses. Still others think that gang membership will enable them to meet individuals who will later do them favors (with financial implications) of the kind fraternity brothers or Masons sometimes do for each other. Irish gang members in New York and Boston especially tend to believe this.

Recreation

The gang provides individuals with entertainment, much as a fraternity does for college students or the Moose and Elk clubs do for their members. Many individuals said they joined the gang because it was the primary social institution of their neighborhood—that is, it was where most (not necessarily the biggest) social events occurred. Gangs usually, though not always, have some type of clubhouse. The exact nature of the clubhouse varies according to how much money the gang has to support it, but every clubhouse offers some form of entertainment. In the case of some gangs with a good deal of money, the clubhouse includes a bar, which sells its members drinks at cost. In addition, some clubhouses have pinball machines, soccer-game machines, pool tables, Ping-Pong tables, card tables, and in some cases a few slot machines. The clubhouse acts as an incentive, much like the lodge houses of other social clubs.[8]

The gang can also be a promoter of social events in the community, such as a big party or dance. Often the gang, like a fraternity, is thought of as the organization to join to maximize opportunities to have fun. Many who joined said they did so because the gang provided them with a good opportunity to meet women. Young women frequently form an auxiliary unit to the gang, which usually adopts a version of the male gang's name (e.g., "Lady Jets"). The women who join this auxiliary do so for similar reasons—that is, opportunities to meet men and participate in social events.[9]

The gang is also a source of drugs and alcohol. Here, most gangs walk a fine line. They provide some drugs for purposes of recreation, but because they also ban addicts from the organization, they also attempt to monitor members' use of some drugs.[10]

The comments of Fox and Happy highlight these views of the gang as a source of recreation.[11] Fox was a twenty-three-year-old from New York and had been in a gang for seven years:

> Like I been telling you, I joined originally because all the action was happening with the Bats [gang's name]. I mean, all the foxy ladies were going to their parties and hanging with them. Plus their parties were great.
>
> They had good music and the herb [marijuana] was so smooth. . . . Man, it was a great source of dope and women. Hell, they were the kings of the community so I wanted to get in on some of the action.

Happy was a twenty-eight-year-old from Los Angeles, who had been a gang member for eight years:

> I joined because at the time, Jones Park [gang's name] had the best clubhouse. They had pool tables and pinball machines that you could use for free. Now they added a video game which you only have to pay like five cents for to play. You could do a lot in the club, so I thought it was a good thing to try it for awhile [join the gang], and it was a good thing.

A Place of Refuge and Camouflage

Some individuals join a gang because it provides them with a protective group identity. They see the gang as offering them anonymity, which may relieve the stresses associated with having to be personally accountable for all their actions in an intensely competitive environment. The statements of Junior J. and Black Top are representative of this belief. Junior J. was a seventeen-year-old who had been approached about becoming a gang member in one of New York's neighborhoods:

> I been thinking about joining the gang because the gang gives you a cover, you know what I mean? Like when me or anybody does a business deal and we're members of the gang, it's difficult to track us down 'cause people will say, oh, it was just one of those guys in the gang. You get my point? The gang is going to provide me with some cover.

Black Top was a seventeen-year-old member of a Jamaican gang in New York:

> Man, I been dealing me something awful. I been doing well, but I also attracted me some adversaries. And these adversaries have been getting close to me. So joining the brothers [the gang] lets me blend into the group. It lets me hide for awhile, it gives me refuge until the heat goes away.

Physical Protection

Individuals also join gangs because they believe the gang can provide them with personal protection from the predatory elements active in low-income neighborhoods. Nearly all the young men who join for this reason know what dangers exist for them in their low-income neighborhoods. These individuals are not the weakest of those who join the gang, for all have developed the savvy and skills to handle most threats. However, all are either tired of being on the alert or want to reduce the probability of danger to a level that allows them to devote more time to their effort to secure more money. Here are two representative comments of individuals who joined for this reason. Chico was a seventeen-year-old member of an Irish gang in New York:

> When I first started up with the Steel Flowers, I really didn't know much about them. But, to be honest, in the beginning I just joined because there were some people who were taking my school [lunch] money, and after I joined the gang, these guys laid off.

Cory was a sixteen-year-old member of a Los Angeles gang:

> Man I joined the Fultons because there are a lot of people out there who are trying to get you and if you don't got protection you in trouble sometimes. My homeboys gave me protection, so hey, they were the thing to do. . . . Now that I got some business things going I can concentrate on them and not worry so much. I don't always have to be looking over my shoulder.

A Time to Resist

Many older individuals (in their late teens or older) join gangs in an effort to resist living lives like their parents'. As Joan Moore, Ruth Horowitz, and others have pointed out, most gang members come from families whose parents are underemployed and/or employed in the secondary labor market in jobs that have little to recommend them.[12] These jobs are low-paying, have long hours, poor working conditions, and few opportunities for advancement; in brief, they are dead ends.[13] Most prospective gang members have lived through the pains of economic deprivation and the stresses that such an existence puts on a family. They desperately want to avoid following in their parents' path, which they believe is exactly what awaits them. For these individuals, the gang is a way to resist the jobs their parents held and, by extension, the life their parents led. Deciding to become a gang member is both a statement to society ("I'll not take these jobs passively") and an attempt to do whatever can be done to avoid such an outcome. At the very least, some of these individuals view being in a gang as a temporary reprieve from having to take such jobs, a postponement of the inevitable. The comments of Joey and D. D. are representative of this group. Joey was a nineteen-year-old member of an Irish gang in Boston:

> Hell, I joined because I really didn't see anything in the near future I wanted to do. I sure the hell didn't want to take that job my father got me. It was a shit job just like his. I said to myself, "Fuck this!" I'm only nineteen, I'm too young to start this shit. . . . I figured that the Black Rose [the gang] was into a lot of things and that maybe I could hit it big at something we're doing and get the hell out of this place.

D. D. was a twenty-year-old member of a Chicano gang in Los Angeles:

> I just joined the T-Men to kick back [relax, be carefree] for a while. My parents work real hard and they got little for it. I don't really want that kind of job, but that's what it looked like I would have to take. So I said, hey, I'll just kick back for a while and let that job wait for me. Hey, I just might make some money from our dealings and really be able to forget these jobs. . . . If I don't [make it, at least] I told the fuckers in Beverly Hills what I think of the jobs they left for us.

People who join as an act of resistance are often wrongly understood to have joined because they were having difficulty with their identity and the gang provided them with a new one. However, these individuals actually want a new identity less than they want better living conditions.

Commitment to Community

Some individuals join the gang because they see participation as a form of commitment to their community. These usually come from neighborhoods where gangs have existed for generations. Although the character of such gangs may have changed over the years, the fact remains that they have continued to exist. Many of these individuals have known people who have been in gangs, including family members—often a brother, but even, in considerable number of cases, a father and grandfather. The fact that their relatives have a history of gang involvement usually influences these individuals to see the gang as a part of the tradition of the community. They feel that their families and their community expect them to join, because community members see the gang as an aid to them and the individual who joins as meeting his neighborhood obligation. These attitudes are similar to attitudes in the larger society about one's obligation to serve in the armed forces. In a sense, this type of involvement represents a unique form of local patriotism. While this rationale for joining was present in a number of the gangs studied, it was most prevalent among Chicano and Irish gangs. The comments of Dolan and Pepe are representative of this line of thinking. Dolan was a sixteen-year-old member of an Irish gang in New York:

> I joined because the gang has been here for a long time and even though the name is different a lot of the fellas from the community have been involved in it over the years, including my dad. The gang has helped the community by protecting it against outsiders so people here have kind of depended on it. . . . I feel it's my obligation to the community to put in some time helping them out. This will help me to get help in the community if I need it some time.

Pepe was a seventeen-year-old member of a Chicano gang in the Los Angeles area:

> The Royal Dons [gang's name] have been here for a real long time. A lot of people from the community have been in it. I had lots of family in it so I guess I'll just have to carry on the tradition. A lot of people from outside this community wouldn't understand, but we have helped the community whenever they've asked us. We've been around to help. I felt it's kind of my duty to join 'cause everybody expects it. . . . No, the community doesn't mind that we do things to make some money and raise a little hell because they don't expect you to put in your time for nothing. Just like nobody expects guys in the military to put in their time for nothing.

In closing this section on why individuals join gangs, it is important to reemphasize that people choose to join for a variety of reasons, that these reasons are not exclusive of one another (some members have more than one), that gangs are composed of individuals whose reasons for joining include all those mentioned, that the decision to join is thought out, and that the individual believes this was best for his or her interests at the moment.

ORGANIZATIONAL RECRUITMENT

Deciding whether or not to join a gang is never an individual decision alone. Because gangs are well established in most of these neighborhoods, they are ultimately both the initiators of membership and the gatekeepers, deciding who will join and who will not.

Every gang that was studied had some type of recruitment strategy. A gang will frequently employ a number of strategies, depending on the circumstances in which recruitment is occurring. However, most gangs use one particular style of recruitment for what they consider a "normal" period and adopt other styles as specific situations present themselves. The three most prevalent styles of recruitment encountered were what I call the fraternity type, the obligation type, and the coercive type.

The Fraternity Type of Recruitment

In the fraternity type of recruitment, the gang adopts the posture of an organization that is "cool," "hip," the social thing to be in. Here the gang makes an effort to recruit by advertising through word of mouth that it is looking for members. Then many of the gangs either give a party or circulate information throughout the neighborhood, indicating when their next meeting will be held and that those interested in becoming members are invited. At this initial meeting, prospective members hear a short speech about the gang and its rules. They are also told about the gang's exploits and/or its most positive perks, such as the dances and parties it gives, the availability of dope, the women who are available, the clubhouse, and the various recreational machinery (pool table, video games, bar, etc.). In addition, the gang sometimes discusses, in the most general terms, its plans for creating revenues that will be shared among the general membership. Once this pitch is made, the decision rests with the individual. When one decides to join the gang, there is a trial period before one is considered a solid member of the group. This trial period is similar, but not identical, to the pledge period for fraternities. There are a number of precautions taken during this period to check the individual's worthiness to be in the group. If the individual is not known by members of the gang, he will need to be evaluated to see if he is an informant for one of the various law enforcement agencies (police, firearms and alcohol, drug enforcement). In addition, the individual will need to be assessed in terms of his ability to fight, his courage, and his commitment to help others in the gang.

Having the *will* to fight and defend other gang members or the "interest" of the gang is considered important, but what is looked upon as being an even more important asset for a prospective gang member is the *ability* to fight and to carry out group decisions. Many researchers have often misinterpreted this preference by gangs for those who can fight as an indication that gang members, and thus gangs as collectives, are primarily interested in establishing reputations as fighters.[14] They interpret this preoccupation as being based on adolescent drives for identity and the release of a great deal of aggression. However, what is most often missed are the functional aspects of fighting and its significance to a gang. The prospective member's ability to fight well is not looked upon by the organization simply as an additional symbol of status. Members of gangs want to know if a potential member can fight because if any of them are caught in a situation where they are required to fight, they want to feel confident that everyone can carry his or her own responsibility. In addition, gang members want to know if the potential gang member is disciplined enough to avoid getting scared and running, leaving them vulnerable. Often everyone's safety in a fight depends on the ability of every individual to fight efficiently. For example, on many occasions I observed a small group of one gang being attacked by an opposing gang. Gang fights are not like fights in the movies: there is no limit to the force anybody is prepared to use—it is, as one often hears, "for all the marbles." When gang members were attacked, they were often outnumbered and surrounded. The only way to protect themselves was to place themselves back to back and ward off the attackers until some type of help came (ironically, most often from the police). If someone cannot fight well and is overcome quickly, everyone's back will be exposed and everyone becomes vulnerable. Likewise, if someone decides to make a run for it, everyone's position is compromised. So assessing the potential member's ability to fight is not done simply to strengthen the gang's reputation as "the meanest fighters," but rather to strengthen the confidence of other gang members that the new member adds to the organization's general ability to protect and defend the collective's interests. The comments of Vase, an eighteen-year-old leader of a gang in New York, highlight this point:

> When I first started with the Silk Irons [gang's name], they checked me out to see if I could fight. After I passed their test, they told me that they didn't need anybody who would leave their butts uncovered. Now that I'm a leader I do the same thing. You see the guy over there? He wants to be in the Irons, but we don't know nothing about whether he can fight or if he got no heart [courage]. So we going to check out how good he is and whether he going to stand and fight. 'Cause if he ain't got good heart or skills [ability to fight], he could leave some of the brothers [gang members] real vulnerable and in a big mess. And if [he] do that, they going to get their asses messed up!

As mentioned earlier, in those cases where the gang has seen a prospective member fight enough to know he will be a valuable member, they simply admit him. However, if information is needed in order to decide whether the prospective gang member can fight, the gang leadership sets up a number of situations to test the individual. One favorite is to have one of the gang members pick a fight with the prospective member and observe the response. It is always assumed that the prospective member will fight; the question is, how well will he fight? The person selected to start the fight is usually one of the better fighters. This provides the group with comparative information by which to decide just how good the individual is in fighting.[15] Such fights are often so intense that there are numerous lacerations on the faces of both fighters. This test usually doubles as an initiation rite, although there are gangs who follow up this test phase with a separate initiation ritual where the individual is given a beating by all those gang members present. This beating is more often than not symbolic, in that the blows delivered to the new members are not done using full force. However, they still leave bruises.

Assessing whether a prospective gang member is trustworthy or not is likewise done by setting up a number of small tests. The gang members are concerned with whether the prospective member is an undercover agent for law enforcement. To help them establish this, they set up a number of criminal activities (usually of medium-level illegality) involving the individual(s); then they observe whether law enforcement proceeds to make arrests of the specific members involved. One gang set up a scam whereby it was scheduled to commit an armed robbery. When a number of the gang members were ready to make the robbery, the police came and arrested them—the consequence of a new member being a police informer. The person responsible was identified and punished. Testing the trustworthiness of new recruits proved to be an effective policy because later the gang was able to pursue a much more lucrative illegal venture without the fear of having a police informer in the organization.

Recruiting a certain number of new members who have already established reputations as good fighters does help the gang. The gang's ability to build and maintain a reputation for fighting reduces the number of times it will have to fight. If a gang has a reputation as a particularly tough group, it will not have as much trouble with rival gangs trying to assume control over its areas of interest. Thus, a reputation acts as an initial deterrent to rival groups. However, for the most part, the gang's concern with recruiting good fighters for the purpose of enhancing its reputation is secondary to its concern that members be able to fight well so that they can help each other.

Gangs who are selective about who they allow in also scrutinize whether the individual has any special talents that could be useful to the collective. Sometimes these special talents involve military skills, such as the ability to build incendiary bombs. Some New York gangs attempted to recruit people with carpentry and masonry skills so that they could help them renovate abandoned buildings.

Gangs that adopt a "fraternity recruiting style" are usually quite secure within their communities. They have a relatively large membership and have integrated themselves into the community well enough to have both legitimacy and status. In other words, the gang is an organization that is viewed by members of the community as legitimate. The comments of Mary, a 53-year-old garment worker who was a single parent in New York, indicate how some community residents feel about certain gangs:

> There are a lot of young people who want in the Bullets, but they don't let whoever wants to get in. Those guys are really selective about who

they want. Those who do get in are very helpful to the whole community. There are many times that they have helped the community . . . and the community appreciates that they have been here for us.

Gangs that use fraternity style recruitment have often become relatively prosperous. Having built up the economic resources of the group to a level that has benefited the general membership, they are reluctant to admit too many new members, fearing that increased numbers will not be accompanied by increases in revenues, resulting in less for the general membership. Hackman, a twenty-eight-year-old leader of a New York gang, represented this line of thought:

> Man, we don't let all the dudes who want to be let in. We can't do that, or I can't 'cause right now we're sitting good. We gots a good bank account and the whole gang is getting dividends. But if we let in a whole lot of other dudes, everybody will have to take a cut unless we come up with some more money, but that don't happen real fast. So you know the brothers ain't going to dig a cut, and if it happens, then they going to be on me and the rest of the leadership's ass and that ain't good for us.

The Obligation Type of Recruitment

The second recruiting technique used by gangs is what I call the "obligation type." In this form, the gang contacts as many young men from its community as it can and attempts to persuade them that it is their duty to join. These community pressures are real, and individuals need to calculate how to respond to them, because there are risks if one ignores them. In essence, the gang recruiter's pitch is that everyone who lives in this particular community has to give something back to it in order to indicate both appreciation of and solidarity with the community. In places where one particular gang has been in existence for a considerable amount of time (as long as a couple of generations), "upholding the tradition of the neighborhood" (not that of the gang) is the pitch used as the hook. The comments of Paul and Lorenzo are good examples. Paul was a nineteen-year-old member of an Irish gang in New York:

> Yeah, I joined this group of guys [the gang] because they have helped the community and a lot of us have taken some serious lumps [injuries] in doing that. . . . I think if a man has any sense of himself, he will help his community no matter what. Right now I'm talking to some guys about joining our gang and I tell them that they can make some money being in the gang, but the most important thing is they can help the community too. If any of them say that they don't want to get hurt or something like that, I tell'm that nobody wants to get hurt, but sometimes it happens. Then I tell them the bottom line, if you don't join and help the community, then outsiders will come and attack the people here and this community won't exist in a couple of years.

Lorenzo was a 22-year-old Chicano gang member from Los Angeles. Here he is talking to two prospective members:

> I don't need to talk to you dudes too much about this [joining a gang]. You know what the whole deal is, but I want you to know that your barrio [community] needs you just like they needed us and we delivered. We all get some battle scars [he shows them a scar from a bullet wound], but that's the price you pay to keep some honor for you and your barrio. We all have to give something back to our community.[16]

This recruiting pitch is primarily based on accountability to the community. It is most effective in communities where the residents have depended on the gang to help protect them from social predators. This is because gang recruiters can draw on the moral support that the gang receives from older residents.

Although the power of this recruiting pitch is accountability to the community, the recruiter can suggest other incentives as well. Three positive incentives generally are used. The first is that gang members are respected in the community. This means that the community will tolerate their illegal business dealings and help them whenever they are having difficulty with the police. As Cardboard, a sixteen-year-old member of a Dominican gang, commented:

> Hey, the dudes come by and they be putting all this shit about that I should do my part to protect the community, but I told them I'm not ready to join up. I tell you the truth, I did sometimes feel a little guilty, but I still didn't think it was for me. But now I tell you I been changing my mind a little. I thinking more about joining. . . . You see the dudes been telling me the community be helping you do your business, you understand? Hey, I been thinking, I got me a little business and if they right, this may be the final straw to get me, 'cause a little help from the community could be real helpful to me. [He joined the gang three weeks later.]

The second incentive is that some members of the community will help them find employment at a later time. (This happens more in Irish neighborhoods.) The comments of Andy, a seventeen-year-old Irish-American in Boston, illustrate this view:

> The community has been getting squeezed by some developers and there's been a lot of people who aren't from the community moving in, so that's why some of the Tigers [gang's name] have come by while we've been talking. They want to talk to me about joining. Just like they been saying, the community needs their help now and they need me. I really was torn because I thought there might be some kind violence used and I don't really want to get involved with that. But the other day when you weren't here, they talked to me and told me that I should remember that the community remembers when people help and they take care of their own. Well, they're right, the community does take care of its own. They help people get jobs all the time 'cause they got contacts at city hall and at the docks, so I been thinking I might join. [He joined three weeks later.]

The third incentive is access to women. Here the recruiter simply says that because the gang is a part of the community and is respected, women look up to gang members and want to be associated with them. So, the pitch continues, if an individual wants access to a lot of women, it will be available through the gang. The comments of Topper, a fifteen-year-old Chicano, illustrate the effectiveness of this pitch:

> Yeah, I was thinking of joining the Bangers [a gang]. These two homeboys [gang members] been coming to see me about joining for two months now. They've been telling me that my barrio really needs me and I should help my people. I really do want to help my barrio, but I never really made up my mind. But the other day they were telling me that the *mujeres* [women] really dig homeboys because they do help the community. So I was checking that out and you know what? They really do! So, I say, hey, I need to seriously check the Bangers out. [One week later he joined the gang.]

In addition to the three positive incentives used, there is a negative one. The gang recruiter can take the tack that if a prospective member decides not to join, he will not be respected as much in the community, or possibly even within his own family. This line of persuasion can be successful if other members of the prospective recruit's family have been in a gang and/or if there has been a high level of involvement in gangs throughout the community. The suggestion that people (including family) will be disappointed in him, or look down on his family, is an effective manipulative tool in the recruiting process in such cases. The comments of Texto, a fifteen-year-old Chicano, provide a good example:

> I didn't want to join the Pearls [gang's name] right now cause I didn't think it was best for me right now. Then a few of the Pearls came by to try to get me to join. They said all the stuff about helping your barrio, but I don't want to join now. I mean I do care about my barrio, but I just don't want to join now. But you heard them today ask me if my father wanted me to join. You know I got to think about this, I mean my dad was in this gang and I don't know. He says to me to do what you want, but I think he would be embarrassed with his friends if they heard I didn't want to join. I really don't want to embarrass my dad, I don't know what I'm going to do. [He joined the gang one month later.]

The "obligation method of recruitment" is similar to that employed by governments to secure recruits for their armed services, and it meets with only moderate results. Gangs using this method realized that while they would not be able to recruit all the individuals they made contact with, the obligation method (sometimes in combination with the coercive method) would enable them to recruit enough for the gang to continue operating.

This type of recruitment was found mostly, although not exclusively, in Irish and Chicano communities where the gang and community had been highly integrated. It is only effective in communities where a particular gang or a small number of gangs have been active for a considerable length of time.

The Coercive Type of Recruitment

A third type of recruitment involves various forms of coercion. Coercion is used as a recruitment method when gangs are confronted with the need to increase their membership quickly. There are a number of situations in which this occurs. One is when a gang has made a policy decision to expand its operations into another geographic area and needs troops to secure the area and keep it under control. The desire to build up membership is based on the gang's anticipation that there will be a struggle with a rival gang and that, if it is to be successful, it will be necessary to be numerically superior to the expected adversary.

Another situation involving gang expansion also encourages an intense recruitment effort that includes coercion. When a gang decides to expand into a geographic area that has not hitherto been controlled by another gang, and is not at the moment being fought for, it goes into the targeted area and vigorously recruits members in an effort to establish control. If individuals from this area are not receptive to the gang's efforts, then coercion is used to persuade some of them to join. The comment of Bolo, a seventeen-year-old leader of a New York gang, illustrates this position:

> Let me explain what just happened. Now you might be thinking, what are these dudes doing beating up on somebody they want to be in their gang? The answer is that we need people now, we can't be waiting till they make up their mind. They don't have to stay for a long time, but we need them now. . . . We don't like to recruit this way 'cause it ain't good for the long run, but this is necessary now because in order for us to expand our business in this area we got to get control, and in order to do that we got to have members who live in the neighborhood. We can't be building no structure to defend ourselves against the Wings [the rival gang in the area], or set up some communications in the area, or set up a connection with the community. We can't do shit unless we got a base and we ain't going to get any base without people. It's that simple.

A third situation where a gang feels a need to use a coercive recruiting strategy involves gangs who are defending themselves against a hostile attempt to take over a portion of their territory. Under such conditions, the gang defending its interests will need to bolster its ranks in order to fend off the threat. This will require that the embattled gang recruit rapidly. Often, a gang that normally uses the fraternity type of recruitment will be forced to abandon it for the more coercive type. The actions of these gangs can be compared to those of nation-states when they invoke universal conscription (certainly a form of coercion) during times when they are threatened and then abrogate it when they believe they have recruited a sufficient number to neutralize the threat, or, more usually, when a threat no longer exists. The comments of M. R. and Rider represent those who are recruited using coercion. M. R. was a nineteen-year-old ex-gang member from Los Angeles:[17]

> I really didn't want to be in any gang, but one day there was this big blowout [fight] a few blocks from here. A couple of O Streeters who were from another barrio came and shot up a number of the Dukes [local gang's name]. Then it was said that the O Streeters wanted to take over the area as theirs, so a group of the Dukes went around asking people to join for a while till everything got secure. They asked me, but I still didn't want to get involved because I really didn't want to get killed over something that I had no interest in. But they said they wanted me and if I didn't join and help they were going to mess me up. Then the next day a couple of them pushed me around pretty bad, and they did it much harder the following day. So I thought about it and then decided I'd join. Then after some gun fights things got secure again and they told me thanks and I left.

Rider was a sixteen-year-old member of an Irish gang from New York:

> Here one day I read in the paper there was fighting going on between a couple of gangs. I knew that one of the gangs was from a black section of the city. Then some of the Greenies [local Irish gang] came up to me and told me how some of the niggers from this gang were trying to start some drugs in the neighborhood. I didn't want the niggers coming in, but I had other business to tend to first. You know what I mean? So I said I thought they could handle it themselves, but then about three or four Greenies said that if I didn't go with them that I was going to be ground meat and so would members of my family. Well, I know they meant business because my sister said they followed her home from school and my brother said they threw stones at him on his way home. So, they asked me again and I said OK. . . . Then after we beat the niggers' asses, I quit. . . . Well, the truth is that I wanted to stay, but after the nigger business was over, they didn't want me. They just said that I was too crazy and wouldn't work out in the group.

This last interview highlights the gang's movement back to their prior form of recruitment after the threat was over. Rider wanted to stay in the gang but was asked to leave. Many of the members of the gang felt Rider was too crazy, too prone to vicious and outlandish acts, simply too unpredictable to trust. The gang admired his fighting ability, but he was the kind of person who caused too much trouble for the gang. As T.R., an eighteen-year-old leader of the gang:

> There's lots of things we liked about Rider. He sure could help us in any fight we'd get in, but he's just too crazy. You just couldn't tell what he'd do. If we kept him, he'd have the police on us all the time. He just had to go.

There is also a fourth situation in which coercion is used in recruiting. Sometimes a gang that has dominated a particular area has declined to such an extent that it can no longer control all its original area. In such situations, certain members of this gang often decide to start a new one. When this occurs, the newly constituted gang often uses coercive techniques to recruit members and establish authority over its defined territory. Take the comments of Rob and Loan Man, both of whom were leaders of two newly constituted gangs. Rob was a sixteen-year-old leader of the gang:

> There was the Rippers [old gang's name], but so many of their members went to jail that there really wasn't enough leadership people around. So a number of people decided to start a new gang. So then we went around the area to check who wanted to be in the gang. We only checked out those we really wanted. It was like pro football scouts, we were interested in all those that could help us now. Our biggest worry was getting members, so when some of the dudes said they didn't want to join, we had to put some heavy physical pressure on them; because if you don't get members, you don't have anything that you can build into a gang. . . . Later after we got established we didn't need to pressure people to get them to join.

Loan Man was a twenty-five-year-old member of a gang in New York:

> I got this idea to start a new gang because I thought the leaders we had were all fucked up. You know, they had shit for brains. They were ruining everything we built up and I wasn't going to go down with them and lose everything. So I talked to some others who didn't like what was going on and we decided to start a new club [gang] in the neighborhood we lived in. So we quit. . . . Well, we got new members from the community, one way or the other. . . . You know we had to use a little persuasive muscle to build our membership and let the community know we were able to take control and hold it, but after we did get control, then we only took brothers who wanted us [they used the fraternity type of recruiting].

In sum, the coercive method of recruitment is used most by gangs that find their existence threatened by competitor gangs. During such periods, the gang considers that its own needs must override the choice of the individual and coercion is used to induce individuals to join their group temporarily. . . .

WHO DOES NOT JOIN A GANG?

Who does not become involved with a gang, and why? There are two answers. First, individuals who see no personal advantages in participating in a gang do not become involved. These individuals can be separated into two distinct groups. The first is those who possess all the characteristics associated with defiant individualism, but have decided that participation in a gang is not to their advantage at the present. Most of these individuals are involved in a variety of economic ventures (usually illegal) that they hope will make them rich and do not perceive any advantage to becoming involved with the gang's activities, but the vast majority of them will become involved with a gang at some time in the future. The comment of Cover, a seventeen-year-old who lives in New York, illustrates this point:

> Right now, man, there ain't no reason for me to join the Black Widows. I got some good business going and I'm making some decent money. If the police don't mess me up, I can get some good cash flow going. So right now there ain't any incentive to join, you know what I mean? Now I ain't saying that I won't join sometime, 'cause they gots some good business going themselves. But right now I want to keep with what I'm doing and see where it goes. [He joined the gang one year later.]

There are also those who not only see no advantage to becoming involved in a gang, but also see significant disadvantages: the risks of being killed or imprisoned associated with gang life are too great in their view. They want to get out of the area they live in, but they have developed a strategy for doing so that does not involve the gang. Some of these people will seek socioeconomic mobility through sports, placing their hopes for a better life in their ability to become professional athletes. Ironically, however, there is probably less likelihood of their achieving such mobility through sports than through becoming involved in some illegal type of business venture.

Others from low-income neighborhoods who seek socioeconomic mobility but do not want to join a gang are those prepared to take the risk that investing their time and money in some type of formal training (formal education or the trades) will produce mobility for them. The comments of Phil, the eighteen-year-old son of a window washer in Los Angeles, represent this group:

> No, I don't want to join any gang. I know you can make money by being in, but frankly I don't want to take the risks of being killed or something. I mean some of the dudes in the gang make a whole lot of money, but they take some big risks too. I just don't want to do that. I want to get out of this neighborhood, so I'll just take my chances trying to get out by studying and trying to go to college. I know there are risks with that too. I mean even if you go to college don't mean you going to make a fortune. My cousin went to college and he started a business and it failed, so I know there is risks that I won't make doing it my way, but at least they don't include getting shot or going to prison.

The individuals in both these groups (those seeking mobility through sports and those who seek it through training) possess some of the characteristics associated with defiant individualism, but not the full defiant individualist character structure. Why some people from low-income areas have only a few of the characteristics of defiant individualism and others have them all has to do with a number of contingent factors related to exactly how each individual has experienced his or her social environment.

The second answer to the question of who does not become involved in gangs and why has to do with the fact that gangs do not want everyone who seeks membership. They will reject people if they are not good fighters, cannot be trusted, or are unpredictable (cannot be controlled), as well as if the gang itself already has too many members (in which case additional members would create difficulties in terms of social control and/or a burden in providing the services expected by the membership). Take the comment of Michael, the eighteen-year-old son of a street sweeper in New York:

> Sure, I wanted to join the Spears and I hung out with them, but they never invited me to formally join. You have to get a formal invitation to join, and they didn't give me one. They just told me that they had too many members right now, maybe sometime later.

WHAT HAPPENS TO GANG MEMBERS?

What is the trajectory of the individual who is involved in gangs? Thrasher and most of the subsequent studies on gangs believed that individuals who were in gangs simply matured out of them as they got older.[18] However, evidence from the present study suggests that the real story is more complicated. I found there are seven possible outcomes, some of which are not necessarily exclusive of each other. First, some people stay in the gang. As of this writing some individuals in their late thirties were still members of gangs. What will happen to them is open to question.

Second, some members will drop out of their gangs and pursue various illegal economic activities on their own.

Third, a number of gang members will move on to another type of organization or association. Many of the Irish will join Irish social clubs. Others will move to various branches of organized crime.

Fourth, there will be individuals who move from gangs and become involved in smaller groups like "crews" where they can receive a larger take of the money they have stolen than if they were in a gang.

Fifth, some will be imprisoned for a considerable part of their lives. While this will negate their involvement in the gangs of the streets, they will remain involved in the prison gangs.

Sixth, there will be those whose fate will be death as a consequence of a drug overdose, a violent confrontation, or the risks of lower-class life.[19]

Seventh, a large number will take the jobs and live the lives they were trying to avoid. While this may appear to be what Thrasher and others have previously reported, it is hardly accurate to think of it as "maturing out of the gang."

What do these future paths mean for the gang? Gangs are composed of individuals with defiant individualist characters who make decisions on the basis of what is good for them. On an individual level, this means that gang members will often come and go throughout long periods of their lives. Take the comment of Clip, a thirty-six-year-old member of a Los Angeles gang:

> I've been in gangs since I was fifteen. I joined and then quit and joined again. I did different things, I been married twice, but I come back to the

gang 'cause there is always a chance if you get some business going, you can make some big money and live in leisure. That's been my goal and always will be.

On the aggregate level this means that coming and going is merely an integral part of the organizational environment.

CONCLUSION

Who joins a gang and why have been central concerns of many studies having to do with gangs. One set of studies concentrates on delinquency, asking why individuals are inclined to engage in illegal acts. Another incorporates gangs into a larger analysis of community. Neither approach directly addresses the gang itself. If one begins not from delinquency or community but from the defiant individualist character of gang recruits, one sees that defiant individuals make rational decisions as to what is best for them. Although previous studies have argued that all individuals have the same reason for becoming involved with gangs, prospective members in fact have a variety of reasons for doing so. However, this [paper] also shows that the varying motives of defiant individuals do not suffice to explain who joins a gang and why. Gang involvement is also determined by the needs and desires of the organization. The answer to the question of who joins a gang and why depends on the complex interplay between the individual's decision concerning what is best for him and the organization's decision as to what is best for it.

NOTES

1. See Frederic Thrasher, *The Gang* (Chicago: University of Chicago Press, 1928); Gerald D. Suttles, *The Social Order of the Slum* (Chicago: University of Chicago Press, 1968); John Hagedorn, *People and Folks* (Chicago: Lakeview Press, 1988).

2. Of course, some of the studies cited here overlap these categories, and I have therefore placed them according to the major emphasis of the study. See Richard A. Cloward and Lloyd B. Ohlin, *Delinquency and Opportunity* (New York: Free Press, 1960); Hagedorn, *People and Folks;* Joan Moore, *Homeboys* (Philadelphia: Temple University Press, 1978); James F. Short, Jr., and Fred L. Strodtbeck, *Group Process and Gang Delinquency* (Chicago: University Press, 1965).

3. Here again it is important to restate that many of these studies overlap the categories I have created, but I have attempted to identify them by what seems to be their emphasis. See Herbert A. Block and Arthur Niederhoffer, *The Gang* (New York: Philosophical Library, 1958); Lewis Yablonsky, *The Violent Gang* (New York: Macmillan, 1966).

4. See the qualifying statement in fn. 2 and 3 above. See Ruth Horowitz, *Honor and the American Dream* (New Brunswick, N.J.: Rutgers University Press, 1983); Albert Cohen, *Delinquent Boys* (Glencoe, Ill.: Free Press, 1955); Walter B. Miller, "Lower Class Culture as a Generating Milieu of Gang Delinquency," *Journal of Social Issues* 14 (1958): 5–19; James Diago Vigil, *Barrio Gangs* (Austin: University of Texas Press, 1988).

5. See Jack Katz, *The Seduction of Crime: Moral and Sensual Attractions in Doing Evil* (New York: Basic Books, 1988), p. 9.

6. Although the present study is not a quantitative study, the finding reported here and the ones to follow are based on observation of, and conversations and formal interviews with, hundreds of gang members.

7. For a discussion of the political machine's role in providing psychological and financial support for poor immigrant groups, see Robert K. Merton, *Social Theory and Social Structure* (New York: Free Press, 1968), pp. 126–36. Also see William L. Riordan, *Plunkitt of Tammany Hall* (New York: Dutton, 1963).

8. There are numerous examples throughout the society of social clubs using the lodge or clubhouse as one of the incentives for gaining members. There are athletic clubs for the wealthy (like the University Club and the Downtown Athletic Club in New York), social clubs in ethnic neighborhoods, the Elks and Moose clubs, the clubs of various veterans' associations, and tennis, yacht, and racket ball clubs.

9. See Anne Campbell, *Girls in the Gang* (New York: Basil Blackwell, 1987).

10. For the use of drugs as recreational, see Vigil, *Barrio Gangs;* and Jeff Fagan, "Social Organization of Drug Use and Drug Dealing Among Urban Gangs," *Criminology* 27 (1986): 633–70, who reports varying degrees of drug use among various types of gangs. For studies that report the monitoring and/or prohibition of certain drugs by gangs, see Vigil, *Barrio Gangs,* on the prohibition of heroin use in Chicano gangs; and Thomas Mieczkowski, "Getting Up and Throwing Down: Heroin Street Life in Detroit," *Criminology* 24 (November 1986): 645–66.

11. See Thrasher, *The Gang,* pp. 84–96. He also discusses the gang as a source of recreation.

12. See Moore, *Homeboys,* ch. 2; Horowitz, *Honor and the American Dream,* ch. 8; Vigil, *Barrio Gangs;* Hagedorn, *People and Folks.*

13. For a discussion of these types of jobs, see Michael J. Piore, *Notes for a Theory of Labor Market Stratification.* Working Paper no. 95 (Cambridge, Mass.: Massachusetts Institute of Technology, 1972).

14. See Horowitz, *Honor and American Dream:* and Ruth Horowitz and Gary Schwartz, "Honor, Normative Ambiguity and Gang Violence," *American Sociological Review* 39 (April 1974): 238–51. There are many other studies that could have been cited here. These two are given merely as examples.

15. The testing of potential gang members as to their fighting ability was also observed by Vigil. See his *Barrio Gangs,* pp. 54–55.

16. This quotation was recorded longhand, not tape-recorded.

17. I first met M. R. when he was in one of the gangs that I was hanging around with. He subsequently left the gang, and I stayed in touch with him by talking to him when our paths crossed on the street. This quotation is from a long conversation that I had with him during one of our occasional encounters.

18. See Thrasher, *The Gang,* pp. 66–67. A great number of studies take a similar position. I shall mention but a few. See Suttles, *Social Order of the Slum,* and William F. Whyte, *Street Corner Society* (Chicago: University of Chicago Press, 1943). The work of Ruth Horowitz represents a modified exception to the other findings. While she does imply that gang members grow old and abandon their street lives, she also reports that involvement in gangs may last well into the thirties for individuals. See her *Honor and the American Dream,* pp. 177–97.

19. Some of these people have died from illnesses, food poisoning, and various accidents.

MANAGING DEVIANCE

38

Gay Male Christian Couples and Sexual Exclusivity

ANDREW K. T. YIP

Yip offers a fascinating look into another end of the homosexual spectrum from the cruising scene in his study of the sexual exclusivity of Christian gay men. While many portrayals of homosexuals sensationalize their behavior and emphasize promiscuity, disease, and stigma, this selection discusses the feelings of satisfaction gay couples have with each other. It sensitively contrasts the commitment and expectations that parallel and contrast homosexual and heterosexual relationships. Without the institutional reinforcement of the church and the state sanctioning their liaisons, Yip finds, gay men's commitments lack the "side bets" available to heterosexual marriages. The gay culture also accepts promiscuity to a greater degree than is found in the straight community, partly because of the unavailability of marriage. Yet, many gay men maintain their deviant relationships for years, despite the cultural and structural obstacles that beset them. The way they deal with sexual nonexclusivity is certainly relevant beyond the confines of homosexuality, as the larger heterosexual population engages in considerable extramarital sexuality, and divorce rates are high. Yip describes how some couples maintain commitment in the face of sexual nonexclusivity by being discreet in their affairs, while others stay together for precisely the opposite reason: they have a searing commitment to honesty. A range of ground rules and strategies are discussed that help participants navigate these difficult and unconventional relationships and encounters over the course of their lifetimes, keeping them together despite the obstacles challenging them from the gay and straight worlds.

The issue of sexual exclusivity constitutes one of the most researched areas in the studies on gay male couples. My review on the empirical literature in this field revealed that it is one of the five most prominent themes (Yip 1995). In a gay male partnership itself, sexual exclusivity is one of the

From *Sociology,* Vol. 31, No. 2 (1997) pp. 289–306. Reprinted by permission of Cambridge University Press.

major issues with which the partners have to grapple (Silverstein 1981). In his study on the factors leading to the dissolution of gay and lesbian partnerships, Kurdek (1991:269) reported that sexual non-exclusivity is ranked third, closely behind "partner's non-responsiveness" and "partner's personal problems."

Research evidence tends to suggest that the majority of gay male partnerships are sexually non-exclusive in nature. Many researchers argued that sexual exclusivity in gay male partnerships is the exception rather than the rule (Saghir and Robins 1973; Bell and Weinberg 1978; Mendola 1980; Peplau and Cochran 1981; Blumstein and Schwartz 1983; Peplau and Gordon 1983; McWhirter and Mattison 1984; Blasband and Peplau 1985; Buunk and Van Driel 1989; Davies et al. 1993).

Nevertheless, there exists more than one way of defining sexual exclusivity. Most researchers define sexual exclusivity in terms of the *behaviour* of the partners. A partnership is defined as sexually exclusive if both partners do not have sexual encounters outside the framework of the partnership throughout a particular period of time prior to the research. Researchers use different time periods in this connection, for instance, six months (Peplau 1981), a year (Harry and DeVall 1978) or the entire length of the partnership (Saghir and Robins 1973; Mendola 1980; Peplau and Cochran 1981; Blumstein and Schwartz 1983; McWhirter and Mattison 1984).

However, Hickson (1991) argued that there should be two classificatory systems in this respect—in terms of behaviour, as discussed, but also in terms of the partners' *expectation*. Based on the second criterion, a sexually exclusive partnership is defined as one in which sexual exclusivity is acknowledged by both partners. This acknowledgment can take the form of an explicit agreement or an implicit assumption either at the beginning of the partnership or it may be negotiated as the partnership develops. Therefore, a partnership with an expectation of sexual exclusivity but which is behaviourally otherwise over a stipulated period of time cannot be satisfactorily defined as sexually non-exclusive in general terms. In this case, the partnership is expectationally exclusive but behaviourally non-exclusive.

I support Hickson's (1991) argument. Unlike heterosexual couples who in general share the cultural ideal of sexual exclusivity and therefore the disapproval of non-exclusivity (Reiss et al. 1980; Lawson and Samson 1988; Greeley 1991), gay male couples do not have such a cultural assumption. They therefore have to resort to actively negotiating the constitutional arrangement of their partnership in this respect. Most gay male couples use a trial-and-error approach to construct relationship rules due to the lack of structural and cultural guidelines (Harry and DeVall 1978). Thus, the negotiation process that forms the expectation of their partnership must be taken into consideration alongside their actual behaviour when the issue of sexual exclusivity is examined.

The analysis of the different attitudes adopted by the couples in the area of sexual exclusivity also throws light on the complexity in the organisation of personal identity in postmodern society. All the respondents are currently in an advanced stage of their "moral career." They have already undergone what

Humphreys (1972:148) called "stigma redemption," a process through which gay Christians justify the acceptability of their sexuality—their discrediting attribute—with religious and moral rhetoric. Instead of succumbing to shame and guilt or attempting to rectify their discrediting attribute in the face of stigmatisation (Goffman 1963), they manage the stigma of being gay Christians by developing "accounts" (Lyman and Scott 1970) or "vocabularies of motives" (Mills 1940). This is done not only to defend their personal identity and the continuation of their social actions, but also to discredit and challenge the stigmatiser, namely the Church (for more details on this issue, see Yip 1997a; 1997b; 1997c).

Nevertheless, in spite of locating themselves within the social category of gay male Christians, the respondents do not speak with one voice with regards to the appropriate Christian moral framework within which their sexuality should be practised. As I will demonstrate later, the respondents hold differing perceptions of conventional Christian sexual ethics on intimate relationships which emphasise monogamy and "fidelity." Having justified the acceptability of their sexuality, they still need to negotiate a sexual lifestyle that incorporates harmoniously the religious and sexual aspects of their identity. This is but one of the primary issues that gay Christians have to negotiate in the construction of self identities and satisfactory partnerships. Their experiences illustrate the complexity of the organisation of personal identity in postmodern society.

This paper, therefore, aims to contribute to the acutely under-researched area of same-sex partnerships. It first demonstrates the necessity to define sexual exclusivity in terms of expectation and behaviour in the case of gay male couples. It then presents a typology comprising three categories of gay male Christian couples. Their justifications for their different sexual lifestyles are examined, particularly in relation to conventional Christian sexual ethics. It concludes that there is no significant difference between these couples in terms of their level of relationship satisfaction and commitment.

THE STUDY

Data collected through semi-structured interviews with thirty gay male Christian couples in Britain is presented here. The respondents were primarily recruited through personal contacts and three national gay Christian organisations. The organisations are: the Lesbian and Gay Christian Movement (LGCM), the organisation for lesbian and gay Catholics known as QUEST, and the Anglican Clergy Consultation (ACC). The ACC is a support group for Church of England priests and their partners.

For every couple, each partner was interviewed separately for approximately 70 minutes, followed by a couple interview of about 30 minutes. In view of the sensitive nature of the information pertaining to sexual exclusivity, I elicited all the data concerned in the individual rather than the couple interviews. Almost all the interviews took place in the respondents' homes. Fieldwork was carried out between June and September 1993.

The respondents' ages ranged from early 20s to late 70s, with an average of 42.9. They have been in partnerships between one and 33 years. The mean duration of the partnerships is 12 years and 3 months. Of these respondents, 63.3 per cent and 21.7 per cent respectively were affiliated to the Church of England and the Roman Catholic Church. The remaining respondents were affiliated to other denominations such as the Baptist, the Methodist and the United Reformed Church.

Except one Anglo-Chinese, one Japanese and one Afro-American, all the respondents were white, with the majority living in Greater London and the South-East. In the main, the respondents were highly educated. Half of the sample held a first degree, with another fourteen (23.3 per cent) possessing a postgraduate qualification. In addition, thirty-eight respondents (63.3 per cent) had some form of professional qualifications. In terms of occupation, fifteen of the fifty-two respondents in employment (28.2 per cent) were priests, followed by teachers/lecturers (eight, 15.4 per cent) and manager (five, 9.6 per cent).

TYPOLOGY OF GAY MALE CHRISTIAN COUPLES

Employing both classificatory systems, I constructed a typology which takes into consideration both expectation and behaviour throughout the entire duration of the partnership:

Category A: Couples who expect the partnership to be sexually exclusive, and are behaviourally so ($N=9$)

Category B: Couples who expect the partnership to be sexually exclusive, but are behaviourally non-exclusive ($N=8$)

Category C: Couples who expect the partnership to be sexually non-exclusive, and are behaviourally so ($N=13$)

Category D: Couples who expect the partnership to be sexually non-exclusive but are so far behaviourally exclusive ($N=0$)

The typology above shows four categories of couples. At the point of interview, no couples studied were in Category D. In terms of behaviour, nine (30 per cent) of the thirty couples are sexually exclusive (Category A). On the other hand, twenty-one couples (70 per cent) are sexually non-exclusive in terms of behaviour, with eight in Category B and thirteen in Category C. On the whole, twenty couples (66.7 per cent) have had an explicit agreement about sexual (non-)exclusivity since the beginning of their partnerships. The other ten couples (33.3 per cent) have an explicit assumption about it.

This typology reveals a scenario contradictory to Blasband and Peplau's (1985) argument that there is a consistency between partners' agreements about sexual exclusivity and their actual behaviour. All the thirteen couples in Category C who are sexually non-exclusive in terms of expectation are behaviourally so. However, only nine (52.9 per cent) of the seventeen couples in Category A and Category B who are sexually exclusive in terms of expectation are behaviourally so. The consistency between agreement and behaviour is found only in the former. Understandably, couples in Category C can achieve this consistency substantially more easily compared to couples in Category A and Category B.

Category A: Couples Who Are Expectationally and Behaviourally Exclusive

Research evidence suggests that gay male couples tend to demonstrate a trend towards sexual non-exclusivity over time. McWhirter and Mattison (1984), for instance, reported that all the 156 couples they studied whose partnerships had persisted for over five years adopted sexual non-exclusivity. Davies *et al.* (1993) also reported that 72.6 per cent of the partnerships that had lasted over five years which they studied were sexually non-exclusive. Harry and DeVall (1978) also found that most partnerships lasting longer than two years were non-exclusive.

These findings, however, are not confirmed by the nine couples in this category. Six of the partnerships are longer than five years. The shortest length is one year, and the longest twenty-seven years. The mean length of partnership is 9 years and 5 months.

Reasons for Sexual Exclusivity Three major reasons can be identified for these couples' adherence to the ideal of sexual exclusivity. First, they considered sexual exclusivity a symbol of *total* commitment between partners. Sexual non-exclusivity is therefore perceived to be a threat to this total commitment. Respondent No. 10B who has been in a partnership for 8 years and 2 months expresses this point:

> If there is a second or third person in the sharing of sex, I am not necessarily getting all the attention. . . . There are also the second or the third person's feelings and thoughts to take account of as well. So it then means that you could end up with a less-than-100-per-cent relationship because part of the relationship belongs to somebody else. It's detrimental to the relationship.

This personal commitment, clearly illustrated in the above comment, is also reported to be the main reason for sexual exclusivity in Blasband and Peplau's (1985) study. Seventy-nine per cent of the couples in sexually exclusive partnerships they studied cited this reason.

Sexual exclusivity is also perceived as a symbol of *complete* mutual satisfaction the partners provide for each other. Therefore, the inclusion of a third party into the sexual dimension of their partnership would be an indication of their failure in mutually providing satisfaction. Respondent No. 12B who is in a one-year-old partnership expresses this sentiment:

> I think it [sexual exclusivity] is really important. . . . I think that if [his partner] or I felt the need to look elsewhere for sexual gratification, it wouldn't be just pure sex. I think that is dangerous. I think that is the sign that there are cracks in our relationship or our way of communicating or our emotional relationship, that we need to look elsewhere for gratification. I do think it is very important.

Finally, some respondents explicitly attributed their belief in sexual exclusivity to their commitment to the conventional Christian sexual ethics for intimate relationships. They argued that the Christian ideal of faithfulness, monogamy and fidelity should apply to both same-sex and cross-sex partnerships. While rejecting the Church's conventional interpretation of the Bible on the issue of homosexuality (see Yip 1997a; 1997c), they nevertheless refer to the Bible for moral guidelines for their sexual behavioural patterns. Respondent No. 56A argues this point:

> Christians . . . really should stick with one partner. And I believe that quite strongly. Because I don't think there is any other theological model for sexual relationship that works in Christian terms. You can't make it fit what I think as the inside of scripture and tradition about what a sexual relationship is . . . I think faithfulness to one person actually is quite a profound thing. I don't think it is just a social convention. It's something that affects who we are and the people we grow into. It has something to do with the fact that we are made in God's image and that we are capable of faithfulness.

This view upholds the "ethic of sexual monogamy" which argues that conventional Christian sexual ethics apply to both same-sex and cross-sex partnerships (John 1993:2) This is further elaborated below:

> Christianity has something to do with this [his commitment to sexual exclusivity]. It has to do with valuing the other person. There is an element of self-sacrifice for the other person . . . of saying that there is every aspect of our relationship, it is not just about personal satisfaction. It is about what we can give to one another. That's what love is about. It's about putting another person first. Regarding sex as just about having fun and enjoying yourself actually removes the possibility of sex being about something that you give. It makes it something you take. *(Resp. No. 47B).*

These views stand in stark contrast to that of many respondents who considered the Bible inadequate in providing moral guidelines for same-sex partnerships, since it remains silent on this topic. To them, the behavioural script for intimate relationships propounded by the Bible is meant solely for heterosexual relationships. The ethic of sexual monogamy should therefore be rejected in the case of same-sex partnerships. This point will be developed later.

Category B: Couples Who Are Expectationally Exclusive but Behaviourally Non-Exclusive

This category comprises eight couples who began their partnerships with the expectation that it should be exclusive, but either one or both partners violated that expectation at a certain point of their partnerships. All but one of the respondents in this category have had outside sexual encounters, albeit with differing degrees of activeness and regularity.

What brings about the change in the constitutional arrangement in the sexual dimension of their partnerships? Non-exclusivity became a feature of these partnerships within six months to two years of their inception. Three major factors lead to sexual non-exclusivity, despite the initial expectation of exclusivity: natural progression; dissatisfaction with certain aspects of the partnership; and desire for sexual experimentation.

To most couples, the first outside sexual encounter took place unexpectedly. Having experienced it and realised that it did not necessarily lead to relationship breakdown, they were encouraged to continue. This precipitated a process of re-evaluation and renegotiation of their initial expectation. In certain cases, the boundary of the sexual dimension of their partnerships was re-defined.

On the whole, no respondents reported that they are compelled to sexual non-exclusivity because of their dissatisfaction with the frequency and quality of love-making within the partnership. This contradicts research findings that the decline of love-making between partners encourages outside sexual activity (e.g. Saghir and Robins 1973; Blumstein and Schwartz 1983).

Regulatory Mechanisms Most couples developed several regulatory mechanisms to manage their non-exclusive lifestyles. The chief objective of the employment of these mechanisms is to reap the benefits of both worlds—being able to experience the security of a primary partnership and having the opportunity to seek sexual gratification outside it.

Four mechanisms have been developed in this connection: (a) the establishment of explicit ground rules; (b) the concealment of information about outside sexual encounters; (c) the disclosure of information about outside sexual encounters; and (d) the prevention of casual sexual encounters from developing into ongoing affairs. The mechanisms are generally employed in tandem with each other to achieve the maximum effect.

> a. The establishment of ground rules clearly draws out the framework within which outside sexual encounters are considered acceptable to each partner. Two examples of ground rules are discussed:
>
>> I am sure he still has flings outside. <Are you upset about that?> No, not really. I would feel upset if he were with somebody in the house when I came home. He sees others outside or when I am away. That I don't mind. We have a sort of house rule. People don't come back if the other one is here because it is not very nice. *(Resp. No. 9A)*

The account above illustrates the ground rule for this couple—casual sex partners can only be brought home in the *absence* of the other partner. This is for the obvious reason of minimising embarrassment and jealousy. For another couple, the ground rule is just the opposite. Casual sex partners can only be brought home in the *presence* of the other partner:

> We had an arrangement . . . that we would not bring someone back here as individuals. But if we saw someone that we thought might be attracted to the other one, then we will have a threesome here. . . . That was the arrangement. No one was going to be brought back here without the other person being here . . . . So it wouldn't be done until it was discussed or arranged between the two of us. *(Resp. No. 50A)*

For this couple, the primary reason for this ground rule of engaging in threesomes is the maximisation of participation of both partners as a couple, and in return the minimisation of a sense of exclusion and jealousy.

Ground rules perform two functions—manifest and latent. They stipulate various terms and conditions such as the regularity of outside sexual activity, with whom, the type of sexual activity and the inclusion or exclusion of partner. This performs the manifest function of ensuring the smooth operation of such an activity. However, it also performs its latent function in reinforcing the *specialness* of the primary partner and the partnership. The compliance to the mutually-agreed ground rule is *a sign of trust* between partners. The violation of the ground rule, not the actual sexual encounter, is perceived as the betrayal of that trust.

> b. Concealment of information about outside sexual encounters is used when a respondent realises that such information would prove distressing to his partner. The concealment aims to minimise his partner's jealousy and a sense of insecurity. Information about outside sexual encounters can lead to three types of jealousy: possessive jealousy—one's property rights are violated; exclusive jealousy—being left out of partner's experience; and fearful jealousy—fear of losing partner (Mazur 1977; Silverstein 1981).

This is especially true for a couple in which one partner regularly seeks outside sexual encounters while his partner does not, although the latter is generally accepting of it:

> I don't think it [sexual non-exclusivity] would bother me because I know that there is no way that [his partner] would leave me for anybody else. If sex happens, it's sex. It's not a threat to me. . . . If he should have a sexual relationship with somebody else, it wouldn't bother me. Although I am not sure I necessarily want to hear the details of it. *(Resp. No. 55B)*

Realising his partner's position in this respect, as shown in the data above, respondent No. 55A resorts to concealing all the information about his outside sexual encounters.

> c. In stark contrast to the mechanism above, some couples had mutually agreed to freely disclose information about outside sexual encounters between themselves. This policy of honesty is to minimise a sense of exclusion a partner might feel in response to the encounter of his partner.

Mutually disclosing information about outside sexual encounters also indicates the equality they uphold in their partnership. In this connection, respondent No. 51A explains:

> Because he has done something similar [having outside sexual encounters] and told me, so I must tell him too. In this way it balances each other and so it doesn't create a problem for us.

It is clear from this account that the respondents perceived fidelity as a form of mutual honesty rather than sexual exclusivity itself. Therefore, sexual non-exclusivity only proves problematic to the partnership if information is deliberately concealed.

d. The prevention of the possible development of a casual sexual encounter or "fling" into an ongoing affair is greatly emphasized. This attitude is prevalent in respondents who strongly believed in the difference between *making love* and *having sex*. One makes love to one's primary partner with unreserved emotional attachment. But one has sex with a casual sex partner for mutual sexual gratification without or with little emotional attachment (Harry 1977; Silverstein 1981). Respondent No. 7B explains the distinction:

> A fling is purely lust. I can tolerate that. But I probably couldn't tolerate if it were lust and emotional. I think a relationship is about two things, sexuality and the emotional side.

This distinction is firmly held in order that one does not compromise the primacy of one's partnership with casual sexual encounters. This is to prevent the sexual gratification obtained outside the partnership from eclipsing the ultimate significance of the primary partnership.

It can be concluded that sexual non-exclusivity that sets in at a particular point of the partnership with the knowledge of both partners leads to a process of de-construction of the initial arrangement and the subsequent reconstruction of the new. In the process the establishment of several regulatory mechanisms for the maintenance of the new arrangement emerge. Most couples who have undergone the negotiation process reported that they were handling the situation successfully. This is mainly due to the fact that they have developed different regulatory mechanisms to manage the constitutional change in order to minimise the negative impact such change might exert on the partnership. One of the most widely-used mechanisms is the establishment of ground rules. It appears that it is the *violation* of such ground rules that threatens the stability of the partnership and not the actual outside sexual encounter itself.

Category C: Couples Who Are Expectationally and Behaviourally Non-Exclusive

The thirteen couples in this category began their partnerships with the expectation of non-exclusivity and have been so behaviourally. Three reasons were cited in support of this arrangement. The majority of respondents drew a clear line of demarcation between sex within and outside of the primary partnership

(Silverstein 1981). With their primary partners, they make love. With casual sex partners, they have sex. Outside sexual encounters of this nature, or "flings," are perceived to be purely for the purpose of sexual variety and excitement:

> It [sexual non-exclusivity] didn't detract us from the fact that I love [his partner]. It wasn't a case that I stopped loving him for somebody else. I think that is an important difference. I think being unfaithful is when you sort of want to love somebody else. I think that's what unfaithful means. I think having sex with somebody else isn't quite the same as being unfaithful. *(Resp. No. 11A)*

The search for sexual variety and excitement appears to be the main reason for sexual non-exclusivity, also reported by other researchers (e.g. McWhirter and Mattison 1984; Blasband and Peplau 1985). This search for sexual excitement is facilitated by the male sex-role socialisation process which encourages men, both gay and heterosexual, to place great importance on frequent and varied sexual activity (Berzon 1979) and the high degree of availability of casual and anonymous sex in the gay scene (Jay and Young 1977; Harry and DeVall 1978; Spada 1979; Peplau 1981). Having sexual encounters outside the partnership is not perceived to be unfaithfulness or infidelity, since both partners have either explicitly agreed or implicitly assumed that the partnership is of this nature. This attitude is pervasive among the respondents in this category.

The second reason for sexual non-exclusivity is closely related to the absence of normative guidelines for same-sex partnerships. Certain respondents considered this one of the advantages of being in a gay partnership. The flip side of the absence of a social script is the abundance of freedom and flexibility to negotiate the constitutional arrangement of the sexual dimension of their partnership. Therefore, the partnership would be non-exclusive if both partners agree that this should be the arrangement. This sentiment is clearly articulated in the following comment:

> The reason for sexual non-exclusivity is that we don't conform to marriage where men and women generally [do] . . . well they have sex with other people but it is more difficult because they have got children, social conventions and so on. Whereas with us the Church, for instance, has never recognised the relationship. So what the hell? And so, on the whole, sexuality seems much more liberal for us. *(Resp. No. 3A)*

The third reason in support of sexual non-exclusivity pertains to the issue of egalitarianism. It is argued that if a gay partnership should maximise the in-built freedom and flexibility, the partners themselves should empower each other to utilise such freedom. It also helps prevent interpartner possessiveness (Silverstein 1981; Peplau and Gordon 1983; McWhirter and Mattison 1984; Blasband and Peplau 1985). This sentiment is strongly shared:

> It's part of trust. It's so important not to cling on to someone in a relationship. Not trying to control their thoughts nor their body. If I feel anyone is trying to cling on to me, my immediate reaction is to pull myself away. So, I think in a relationship you give the person the freedom

to go away or to come towards you. And that's the sort of freedom which I hope I am giving [his partner]. And this is the sort of freedom I expect from him. I don't want to be hung on to. *(Resp. No. 48B)*

The account above illustrates the argument that giving the partner the freedom to sexually experiment if he so wishes is a sign of trust and love for him. Partners should liberate each other at the outset before they can fully experience the freedom of being in a gay partnership. In this connection, McWhirter and Mattison (1984:256) reported that, "the single most important factor that keeps couples together past the 10-year mark is the lack of possessiveness they feel. Many couples learn very early in their relationship that ownership of each other sexually can become the greater internal threat to their staying together [compared to non-exclusivity]." It follows that sexual non-exclusivity actually facilitates the development of a long-term partnership in this case (Warren 1974; Harry and DeVall 1978; Blasband and Peplau 1985).

Rejection of Conventional Christian Sexual Ethics Unlike some respondents in Category A who upheld the ethic of sexual monogamy, the majority of the respondents in this category firmly rejected the conventional Christian sexual ethics as the basis on which they should organise their sexual lifestyles. This rejection stems from their perception of the lack of recognition from the institutionalised Church and its silence on the sexual ethics for gay Christian partnerships:

I think one of the things that I constantly re-evaluate is, is our relationship just aping a heterosexual married couple? Because I think the Church has kind of excluded us on paper. There is no official marriage ceremony or anything like that. I think therefore we've got a license to do whatever we feel. I think we should make our own rule. I think what we actually do is what we are happy with, what we want to do. It's what we consider to be appropriate, or certainly what I consider to be appropriate. *(Resp. No. 1A)*

The lack of recognition and acceptance from the Church results in the distancing of many respondents from the Church and its official stance on homosexuality. It is therefore not surprising to observe that many respondents were more inclined to organise their sexual lifestyles on the basis of their own sense of morality which is broadly derived from what Ellison (1994:236–441) called "the ethic of common decency," with its emphasis on Christian principles such as love, justice and responsibility. However, the specific emphasis of the conventional sexual ethics on, for instance, exclusivity and monogamy, is rejected. This appears to be a manifestation of their rejection of the Church which discredits their gay identity and the individualised spirituality they have developed which is broadly based on Christian principles rather than specifically dependent on the Church. The following respondent offers a glimpse of such an attitude:

I suppose Paul talked about the body being a temple of God and actually treating your body like that in that sort of way really. But I don't think that necessarily means that you are exclusive. I guess it has to do with what is going on between that kind of encounter and how it relates to others and whether there is abuse. . . . I can't see what the Church can say . . . because it doesn't say anything, does it? But I think it does in the sense that I follow it from the point of view of responsibility. . . . I respect even people I meet in the cottage [public toilet], you know, that they are human beings. It [the Church] has got nothing to say. The relationship is something that it doesn't support. *(Resp. No. 3A)*

In rejecting the ethic of sexual monogamy, many respondents shared an affinity with the "theology of friendship" which encourages, *inter alia*, "inclusiveness" (Stuart 1992:2; see also Stuart 1995; Rudy 1996). However, their sexual attitudes still reflected the broad Christian norms of justice and love, although their rejection of the Church as their moral arbiter is indisputably strong.

Compared to couples in Category B, couples in this category relied less heavily on ground rules to manage their non-exclusive lifestyle. This is because both partners shared the expectation of a non-exclusive partnership. However, this does not preclude the existence of certain regulatory mechanisms to minimise the possible emergence of conflict. These mechanisms share great similarity with what I have discussed in the preceding section. I therefore will not repeat them here.

SEXUAL EXCLUSIVITY, RELATIONSHIP SATISFACTION AND COMMITMENT

Much research evidence suggests that there is no significant difference between sexually exclusive and non-exclusive couples in terms of relationship satisfaction, adjustment and commitment (e.g. Peplau 1981; Blumstein and Schwartz 1983; Blasband and Peplau 1985; Kurdek and Schmitt 1986; Kurdek 1988). This section assesses this claim by comparing all three categories of couples. Each respondent was asked to rate, on a scale of 1 to 10, his level of satisfaction with his sex life within the partnership and with the partnership in general.

Table 1 indicates that Category A has the *highest* mean rating for sex life within partnership and the partnership in general—7.6 and 9.0 respectively. Taking Category B and Category C together as behaviourally non-exclusive couples, the mean ratings for sex life within partnership and the partnership in general are 6.9 and 8.7, which are lower than those of Category A.

It can therefore be concluded that couples in Category A who are expectationally and behaviourally exclusive have a *higher* level of satisfaction with their sex life within partnership and the partnership in general, compared to their counterparts in Category B who are expectationally exclusive but behaviourally non-exclusive and those in Category C who are expectationally and behaviorally non-exclusive.

However, it must be emphasised that the difference in mean ratings in these cases is not significantly great. This seems to lend credence to the research evidence cited above. In terms of the partner's commitment to the longevity of the partnership, two couples were facing the prospect of dissolution due to non-sexual problems. The other twenty-eight couples (93.3 per cent), regard-

Table 1 Level of Satisfaction with Sex Life Within Partnership and Partnership in General

Category	Average rating for sex life with partnership	Average rating for partnership in general
A ($N = 9$)	7.6	9.0
B ($N = 8$)	7.2	8.4
C ($N = 13$)	6.7	8.8
B+ ($N = 21$)	6.9	8.7

less of their current sexual arrangements, reported that they were committed to a life-long partnership. Below is a typical demonstration of this commitment:

> Yes, that [to be together till death] is the intention and we have made wills with the intention. We have talked about the fact that we are totally committed to each other for the length of our shared life. . . . That is in a sense a duty on each other because it is a costly thing to promise to your partner that you will be faithful with the understanding that the other is doing the same thing. . . . So it is till death do us part. That is definitely our commitment. I mean we have formalised that in a ceremony. *(Resp. No. 52A)*

Thus, it can be concluded that sexual non-exclusivity need not be problematic to a gay partnership. Its impacts tend to depend on the current constitutional arrangement of the partnership and its ability to adapt to change. Sexually non-exclusive partnerships in terms of behaviour should not be misconstrued as being less committed, loving and stable than exclusive ones (Isay 1989).

REFERENCES

Bell, A. P., and M. S. Weinberg. 1978. *Homosexualities: A Study of Diversity Among Men and Women*. New York: Simon and Schuster.

Berzon, B. 1979. "Achieving Success as a Gay Couple." In B. Berzon and R. Leighton, eds., *Positively Gay*. Millbrae: Celestial Arts.

Blasband, D., and L. A. Peplau. 1985. "Sexual Exclusivity Versus Openness in Gay Male Couples." *Archives of Sexual Behaviour* 14 (5):395–412.

Blumstein, P., and P. Schwartz. 1983. *American Couples: Money, Work, Sex*. New York: Simon and Schuster.

Buunk, B. P., and B. Van Driel. 1989. *Variant Lifestyles and Relationships*. Newbury Park, CA: Sage.

Davies, P. M., F. C. I. Hickson, P. Weatherburn, and A. J. Hunt. 1993. *Sex, Gay Men and AIDS*. London: Falmer Press.

Ellison, M. M. 1994. "Common Decency: A New Christian Sexual Ethics." In J. B. Nelson and S. P. Longfellow, eds., *Sexuality and the Sacred: Sources for Theological Reflection*. Westminster: John Knox Press.

Goffman, E. 1963. *Stigma: Notes on the Management of Spoiled Identity*. Englewood Cliffs, NJ: Prentice-Hall.

Greeley, A. M. 1991. *Faithful Attraction: Discovering Intimacy, Love, and Fidelity in American Marriage*. New York: Doherty.

Harry, J. 1977. "Marriage Among Gay Males: The Separation of Intimacy and Sex." In S. G. McNall, ed., *The Sociological Perspective: Introductory Readings*, 4th ed. Boston: Little Brown.

Harry, J. and W. B. DeVall. 1978. *The Social Organization of Gay Males*. New York: Praeger.

Hickson, F. C. I. 1991. "Sexual Exclusivity, Non-Exclusivity and HIV." London: Project SIGMA Working Paper No. 13.

Humphreys, R. A. L. 1972. *Out of the Closet: The Sociology of Homosexual Liberation*. Englewood Cliffs, NJ: Prentice-Hall.

Isay, R. A. 1989. *Being Homosexual: Gay Men and Their Development*. Harmondsworth, England: Penguin.

Jay, K., and A. Young. 1977. *The Gay Report: Lesbian and Gay Men Speak Out About Sexual Experiences and Lifestyles*. New York: Summit Books.

John, J. 1993. *Permanent, Faithful, Stable: Christian Same-Sex Partnerships*. London: Affirming Catholicism.

Kurdek, L. A. 1988. "Relationship Quality of Gay Men in Closed or Open Relationships." *Journal of Homosexuality* 15, 93–118.

Kurdek, L. A. 1991. "The Dissolution of Gay and Lesbian Couples." *Journal of Social and Personal Relationships* 8, (2):265–78.

Kurdek, L. A., and J. P. Schmidtt. 1986. "Relationship Quality of Gay Men in Closed or Open Relationships." *Journal of Homosexuality* 12 (2):85–99.

Lawson, A., and L. Samson. 1988. "Age, Gender and Adultery." *British Journal of Sociology* 39:409–40.

Lyman, S. M., and M. B. Scott. 1970. *A Sociology of the Absurd*. Pacific Palisades, CA: Goodyear.

Mazur, R. 1977. "Beyond Jealousy and Possessiveness." In G. Clanton and L. Smith, eds. *Jealousy*. Englewood Cliffs, NJ: Prentice-Hall.

McWhirter, D. P., and A. M. Mattison. 1984. *The Male Couple: How Relationships Develop*. Englewood Cliffs, NJ: Prentice-Hall.

Mendola, M. 1980. *The Mendola Report: A New Look at Gay Couples*. New York: Crown.

Mills, C. W. 1940. "Situated Actions and Vocabularies of Motives." *American Sociological Review* 5:904–13.

Peplau, L. A. 1981. "What Homosexuals Want in Relationships." *Psychology Today* 15, March: 28–38.

Peplau, L. A. and S. D. Cochran. 1981. "Value Orientations in the Intimate Relationships of Gay Men." *Journal of Homosexuality* 6 (3):1–19.

Peplau, L. A., and S. L. Gordon. 1983. "The Intimate Relationships of Lesbians and Gay Men." In E. R. Allegeier and N. B. McCormick, eds., *Changing Boundaries: Gender Roles and Sexual Behavior*. Palo Alto, CA: Mayfield.

Reiss, I. L., Anderson, R. E., and G. C. Sponaugle. 1980. "A Multivariate Model of Determinants of Extramarital Permissiveness." *Journal of Marriage and the Family* 42:395–411.

Rudy, K. 1996. "Where Two or Three Are Gathered: Using Gay Communities as a Model for Christian Sexual Ethics." *Theology and Sexuality* 4:81–99.

Saghir, M., and E. Robins. 1973. *Male and Female Homosexuality: A Comprehensive Investigation*. Baltimore: Williams and Wilkins.

Silverstein, C. 1981. *Man to Man: Gay Couples in America*. New York: William Morrow.

Spada, K. J. 1979. *The Spada Report: The Newest Survey of Gay Male Sexuality*. New York: Signet.

Stuart, E. 1992. *Daring to Speak Love's Name: A Gay and Lesbian Prayer Book*. London: Hamish Hamilton.

Stuart, E. 1995. *Just Good Friends: Towards a Lesbian and Gay Theology of Relationships*. London: Mowbray.

Warren, C. A. B. 1974. *Identity and Community in the Gay World*. New York: Wiley.

Yip, A. K. T. 1995. "Gay Christian Partnerships." Unpublished doctoral dissertation, University of Surrey.

Yip, A. K. T. 1997a. "Attacking the Attacker: Gay Christians Talk Back." *British Journal of Sociology* 48 (1):113–27.

Yip, A. K. T. 1997b (forthcoming). "Gay Christians' Perceptions of the Christian Community in Relation to Their Sexuality." *Theology and Sexuality*.

Yip, A. K. T. 1997c (forthcoming). "The Politics of Counter-rejection: Gay Christians and the Church." *Journal of Homosexuality*.

LEAVING DEVIANCE

39

Shifts and Oscillations in Deviant Careers
The Case of Upper-Level Drug Dealers and Smugglers

PATRICIA A. ADLER AND PETER ADLER

Adler and Adler discuss the process by which people burn out of deviance in their selection on exiting drug trafficking. After spending several years in the upper echelons of the drug trade, many marijuana dealers and smugglers find that the drawbacks of the lifestyle exceed the rewards. Their initial excitement and thrills turn to paranoia, people whom they know get busted all around them, and their risk of arrest grows. Years of excessive drug use take its toll on them physically, and they come to re-evaluate the straight life they formerly rejected as boring. Yet they cannot easily quit dealing; they have developed a high-spending lifestyle that they are loathe to abandon. When they try to retire from trafficking, they quickly spend all their money and are drawn back into the business. Their patterns of exiting, thus, often resemble a series of quittings and re-startings, as they move out of deviance with great difficulty.

The upper echelons of the marijuana and cocaine trade constitute a world which has never before been researched and analyzed by sociologists. Importing and distributing tons of marijuana and kilos of cocaine at a time, successful operators can earn upwards of a half million dollars per year. Their traffic in these so-called "soft"[1] drugs constitutes a potentially lucrative occupation, yet few participants manage to accumulate any substantial sums of money, and most people envision their involvement in drug trafficking as only temporary. In this study we focus on the career paths followed by members of

From "Shifts and Oscillations in Deviant Careers: The Case of Upper-Level Drug Dealers and Smugglers," Patricia A. Adler and Peter Adler, *Social Problems*, Vol. 31, No. 2, 1983. © 1983 Society for the Study of Social Problems. Reprinted by permission of University of California Press Journals.

one upper-level drug dealing and smuggling community. We discuss the various modes of entry into trafficking at these upper levels, contrasting these with entry into middle- and low-level trafficking. We then describe the pattern of shifts and oscillations these dealers and smugglers experience. Once they reach the top rungs of their occupation, they begin periodically quitting and re-entering the field, often changing their degree and type of involvement upon their return. Their careers, therefore, offer insights into the problems involved in leaving deviance.

Previous research on soft drug trafficking has only addressed the low and middle levels of this occupation, portraying people who purchase no more than 100 kilos of marijuana or single ounces of cocaine at a time (Anonymous, 1969; Atkyns and Hanneman, 1974; Blum et al., 1972; Carey, 1968; Goode, 1970; Langer, 1977; Lieb and Olson, 1976; Mouledoux, 1972; Waldorf et al., 1977). Of these, only Lieb and Olson (1976) have examined dealing and/or smuggling as an occupation, investigating participants' career developments. But their work, like several of the others, focuses on a population of student dealers who may have been too young to strive for and attain the upper levels of drug trafficking. Our study fills this gap at the top by describing and analyzing an elite community of upper-level dealers and smugglers and their careers.

We begin by describing where our research took place, the people and activities we studied, and the methods we used. Second, we outline the process of becoming a drug trafficker, from initial recruitment through learning the trade. Third, we look at the different types of upward mobility displayed by dealers and smugglers. Fourth, we examine the career shifts and oscillations which veteran dealers and smugglers display, outlining the multiple, conflicting forces which lure them both into and out of drug trafficking. We conclude by suggesting a variety of paths which dealers and smugglers pursue out of drug trafficking and discuss the problems inherent in leaving this deviant world.

SETTING AND METHOD

We based our study in "Southwest County," one section of a large metropolitan area in southwestern California near the Mexican border. Southwest County consisted of a handful of beach towns dotting the Pacific Ocean, a location offering a strategic advantage for wholesale drug trafficking.

Southwest County smugglers obtained their marijuana in Mexico by the ton and their cocaine in Colombia, Bolivia, and Peru, purchasing between 10 and 40 kilos at a time. These drugs were imported into the United States along a variety of land, sea, and air routes by organized smuggling crews. Southwest County dealers then purchased these products and either "middled" them directly to another buyer for a small but immediate profit of approximately $2 to $5 per kilo of marijuana and $5,000 per kilo of cocaine, or engaged in "straight dealing." As opposed to middling, straight dealing usually entailed adulterating the cocaine with such "cuts" as manitol, procaine, or inositol, and then dividing the marijuana and cocaine into smaller quantities to sell them to the next-lower level of dealers. Although dealers frequently varied the amounts they bought and sold, a hierarchy of transacting levels could be roughly discerned. "Wholesale" marijuana dealers bought directly from the smugglers, purchasing anywhere from 300 to 1,000 "bricks" (averaging a kilo in weight) at a time and selling in lots of 100 to 300 bricks. "Multi-kilo" dealers, while not the smugglers' first connections, also engaged in upper-level trafficking, buying between 100 to 300 bricks and selling them in 25 to 100 brick quantities. These were then purchased by middle-level dealers who filtered the marijuana through low-level and "ounce" dealers before it reached the ultimate consumer. Each time the marijuana changed hands its price increase was dependent on a number of factors: purchase cost; the distance it was transported (including such transportation costs as packaging, transportation equipment, and payments to employees); the amount of risk assumed; the quality of the marijuana; and the prevailing prices in each local drug market. Prices in the cocaine trade were much more predictable. After purchasing kilos of cocaine in South America for $10,000 each, smugglers sold them to Southwest County "pound" dealers in quantities of one to 10 kilos for $60,000 per kilo. These pound dealers usually cut the cocaine and sold pounds ($30,000) and half-pounds ($15,000) to "ounce" dealers, who in turn cut it again and sold ounces for $2,000 each to middle-level cocaine dealers known as "cut-ounce" dealers. In this fashion the drug was middled, dealt, divided and cut—sometimes as many as five or six times—until it was finally purchased by consumers as grams or half-grams.

Unlike low-level operators, the upper-level dealers and smugglers we studied pursued drug trafficking as a full-time occupation. If they were involved in other businesses, these were usually maintained to provide them with a legitimate front for security purposes. The profits to be made at the upper levels depended on an individual's style of operation, reliability, security, and the amount of product he or she consumed. About half of the 65 smugglers and dealers we observed were successful, some earning up to three-quarters of a million dollars per year.[2] The other half continually struggled in the business, either breaking even or losing money.

Although dealers' and smugglers' business activities varied, they clustered together for business and social relations, forming a moderately well-integrated community whose members pursued a "fast" lifestyle, which emphasized intensive partying, casual sex, extensive travel, abundant drug consumption, and lavish spending on consumer goods. The exact size of Southwest County's upper-level dealing and smuggling community was impossible to estimate due to the secrecy of its members. At these levels, the drug world was quite homogeneous. Participants were predominantly white, came from middle-class backgrounds, and had little previous criminal involvement. While the dealers' and smugglers' social world contained both men and women, most of the serious business was conducted by the men, ranging in age from 25 to 40 years old.

We gained entry to Southwest County's upper-level drug community largely by accident. We had become friendly with a group of our neighbors who turned out be heavily involved in smuggling marijuana. Opportunistically (Riemer, 1977), we seized the chance to gather data on this unexplored activity. Using key informants who helped us gain the trust of other members of the community, we drew upon snowball sampling techniques (Biernacki and Waldorf, 1981) and a combination of overt and covert roles to widen our network of contacts. We supplemented intensive participant-observation, between 1974 and 1980,[3] with unstructured, taped interviews. Throughout, we employed extensive measures to cross-check the reliability of our data, whenever possible (Douglas, 1976). In all, we were able to closely observe 65 dealers and smugglers as well as numerous other drug world members, including dealers' "old ladies" (girlfriends or wives), friends, and family members. . . .

SHIFTS AND OSCILLATIONS

. . . Despite the gratifications which dealers and smugglers originally derived from the easy money, material comfort, freedom, prestige, and power associated with their careers, 90 percent of those we observed decided, at some point, to quit the business. This stemmed, in part, from their initial perceptions of the career as temporary ("Hell, nobody wants to be a drug dealer all their life"). Adding to these early intentions was a process of rapid aging in the career: dealers and smugglers became increasingly aware of the restrictions and sacrifices their occupations required and tired of living the fugitive life. They thought about, talked about, and in many cases took steps toward getting out of the drug business. But as with entering, disengaging from drug trafficking was rarely an abrupt act (Lieb and Olson, 1976: 364). Instead, it more often resembled a series of transitions, or oscillations,[4] out of and back into the business. For once out of the drug world, dealers and smugglers were rarely successful in making it in the legitimate world because they failed to cut down on their extravagant lifestyle and drug consumption. Many abandoned their efforts to reform and returned to deviance, sometimes picking up where they left off and other times shifting to a new mode of operating. For example, some shifted from dealing cocaine to dealing marijuana, some dropped to a lower level of dealing, and others shifted their role within the same group of traffickers. This series of phase-outs and re-entries, combined with career shifts, endured for years, dominating the pattern of their remaining involvement with the business. But it also represented the method by which many eventually broke away from drug trafficking, for each phase-out had the potential to be an individual's final departure.

Aging in the Career

Once recruited and established in the drug world, dealers and smugglers entered into a middle phase of aging in the career. This phase was characterized by a progressive loss of enchantment with their occupation. While novice dealers and smugglers found that participation in the drug world brought them thrills and status, the novelty gradually faded. Initial feelings of exhilaration and awe began to dull as individuals became increasingly jaded. This was the result of both an extended exposure to the mundane, everyday business aspects of drug trafficking and to an exorbitant consumption of drugs (especially cocaine). One smuggler described how he eventually came to feel:

> It was fun, those three or four years. I never worried about money or anything. But after awhile it got real boring. There was no feeling or emotion or anything about it. I wasn't even hardly relating to my old lady anymore. Everything was just one big rush.

This frenzy of overstimulation and resulting exhaustion hastened the process of "burnout" which nearly all individuals experienced. As dealers and smugglers aged in the career they became more sensitized to the extreme risks they faced. Cases of friends and associates who were arrested, imprisoned, or killed began to mount. Many individuals became convinced that continued drug trafficking would inevitably lead to arrest ("It's only a matter of time before you get caught"). While dealers and smugglers generally repressed their awareness of danger, treating it as a taken-for-granted part of their daily existence, periodic crises shattered their casual attitudes, evoking strong feelings of fear. They temporarily intensified security precautions and retreated into near-isolation until they felt the "heat" was off.

As a result of these accumulating "scares," dealers and smugglers increasingly integrated feelings of "paranoia"[5] into their everyday lives. One dealer talked about his feelings of paranoia:

> You're always on the line. You don't lead a normal life. You're always looking over your shoulder, wondering who's at the door, having to hide everything. You learn to look behind you so well you could probably bend over and look up your ass. That's paranoia. It's a really scary, hard feeling. That's what makes you get out.

Drug world members also grew progressively weary of their exclusion from the legitimate world and the deceptions they had to manage to sustain that separation. Initially, this separation was surrounded by an alluring mystique. But as they aged in the career, this mystique became replaced by the reality of everyday boundary maintenance and the feeling of being an "expatriated citizen within one's own country." One smuggler who was contemplating quitting described the effects of this separation:

> I'm so sick of looking over my shoulder, having to sit in my house and worry about one of my non-drug world friends stopping in when I'm doing business. Do you know how awful that is? It's like leading a double life. It's ridiculous. That's what makes it not worth it. It'll be a lot less money [to quit], but a lot less pressure.

Thus, while the drug world was somewhat restricted, it was not an encapsulated community, and dealers' and smugglers' continuous involvement with the straight world made the temptation to adhere to normative standards and "go straight" omnipresent. With the occupation's novelty worn off and the "fast life" taken-for-granted, most dealers and smugglers felt that the occupation no longer resembled their early impressions of it. Once they reached the upper levels of the occupation, their experience began to change. Eventually, the rewards of trafficking no longer seemed to justify the strain and risk involved. It was at this point that the straight world's formerly dull ambiance became transformed (at least in theory) into a potential haven.

Phasing Out

Three factors inhibited dealers and smugglers from leaving the drug world. Primary among these factors were the hedonistic and materialistic satisfactions the drug world provided. Once accustomed to earning vast quantities of money quickly and easily, individuals found it exceedingly difficult to return to the income scale of the straight world. They also were reluctant to abandon the pleasure of the "fast life" and its accompanying drugs, casual sex, and power. Second, dealers and smugglers identified with, and developed a commitment to, the occupation of drug trafficking (Adler and Adler, 1982). Their self-images were tied to that role and could not be easily disengaged. The years invested in their careers (learning the trade, forming connections, building reputations) strengthened their involvement with both the occupation and the drug community. And since their relationships were social as well as business, friendship ties bound individuals to dealing. As one dealer in the midst of struggling to phase-out explained:

> The biggest threat to me is to get caught up sitting around the house with friends that are into dealing. I'm trying to stay away from them, change my habits.

Third, dealers and smugglers hesitated to voluntarily quit the field because of the difficulty involved in finding another way to earn a living. Their years spent in illicit activity made it unlikely for any legitimate organizations to hire them. This narrowed their occupational choices considerably, leaving self-employment as one of the few remaining avenues open.

Dealers and smugglers who tried to leave the drug world generally fell into one of four patterns.[6] The first and most frequent pattern was to postpone quitting until after they could execute one last "big deal." While the intention was sincere, individuals who chose this route rarely succeeded; the "big deal" too often remained elusive. One marijuana smuggler offered a variation of this theme:

> My plan is to make a quarter of a million dollars in four months during the prime smuggling season and get the hell out of the business.

A second pattern we observed was individuals who planned to change immediately, but never did. They announced they were quitting, yet their outward actions never varied. One dealer described his involvement with this syndrome:

> When I wake up I'll say, "Hey, I'm going to quit this cycle and just run my other business." But when you're dealing you constantly have people

dropping by ounces and asking, "Can you move this?" What's your first response? Always, "Sure, for a toot."

In the third pattern of phasing-out, individuals actually suspended their dealing and smuggling activities, but did not replace them with an alternative source of income. Such withdrawals were usually spontaneous and prompted by exhaustion, the influence of a person from outside the drug world, or problems with the police or other associates. These kinds of phase-outs usually lasted only until the individual's money ran out, as one dealer explained:

> I got into legal trouble with the FBI a while back and I was forced to quit dealing. Everybody just cut me off completely, and I saw the danger in continuing, myself. But my high-class tastes never dwindled. Before I knew it I was in hock over $30,000. Even though I was hot, I was forced to get back into dealing to relieve some of my debts.

In the fourth pattern of phasing out, dealers and smugglers tried to move into another line of work. Alternative occupations included: (1) those they had previously pursued; (2) front businesses maintained on the side while dealing or smuggling; and (3) new occupations altogether. While some people accomplished this transition successfully, there were problems inherent in all three alternatives.

1. Most people who tried resuming their former occupations found that these had changed too much while they were away. In addition, they themselves had changed: they enjoyed the self-directed freedom and spontaneity associated with dealing and smuggling, and were unwilling to relinquish it.
2. Those who turned to their legitimate front business often found that these businesses were unable to support them. Designed to launder rather than earn money, most of these ventures were retail outlets with a heavy cash flow (restaurants, movie theaters, automobile dealerships, small stores) that had become accustomed to operating under a continuous subsidy from illegal funds. Once their drug funding was cut off they could not survive for long.
3. Many dealers and smugglers utilized the skills and connections they had developed in the drug business to create a new occupation. They exchanged their illegal commodity for a legal one and went into import/export, manufacturing, wholesaling, or retailing other merchandise. For some, the decision to prepare a legitimate career for their future retirement from the drug world followed an unsuccessful attempt to phase-out into a "front" business. One husband-and-wife dealing team explained how these legitimate side businesses differed from front businesses:

> We always had a little legitimate "scam" [scheme] going, like mail-order shirts, wallets, jewelry, and the kids were always involved in that. We made a little bit of money on them. Their main purpose was for a cover. But [this business] was different; right from the start this was going to be a legal thing to push us out of the drug business.

About 10 percent of the dealers and smugglers we observed began tapering off their drug world involvement gradually, transferring their time and money into a selected legitimate endeavor. They did not try to quit drug trafficking altogether until they felt confident that their legitimate business could support them. Like spontaneous phase-outs, many of these planned withdrawals into legitimate endeavors failed to generate enough money to keep individuals from being lured into the drug world.

In addition to voluntary phase-outs caused by burnout, about 40 percent of the Southwest County dealers and smugglers we observed experienced a "bustout" at some point in their careers.[7] Forced withdrawals from dealing or smuggling were usually sudden and motivated by external factors, either financial, legal, or reputational. Financial bustouts generally occurred when dealers or smugglers were either "burned" or "ripped-off" by others, leaving them in too much debt to rebuild their base of operation. Legal bustouts followed arrest and possibly incarceration: arrested individuals were so "hot" that few of their former associates would deal with them. Reputational bustouts occurred when individuals "burned" or "ripped-off" others (regardless of whether they intended to do so) and were banned from business by their former circle of associates. One smuggler gave his opinion on the pervasive nature of forced phase-outs:

> Some people are smart enough to get out of it because they realize, physically, they have to. Others realize, monetarily, that they want to get out of this world before this world gets them. Those are the lucky ones. Then there are the ones who have to get out because they're hot or someone else close to them is so hot that they'd better get out. But in the end when you get out of it, nobody gets out of it out of free choice; you do it because you have to.

Death, of course, was the ultimate bustout. Some pilots met this fate because of the dangerous routes they navigated (hugging mountains, treetops, other aircrafts) and the sometimes ill-maintained and overloaded planes they flew. However, despite much talk of violence, few Southwest County drug traffickers died at the hands of fellow dealers.

Re-entry

Phasing-out of the drug world was more often than not temporary. For many dealers and smugglers, it represented but another stage of their drug careers (although this may not have been their original intention), to be followed by a period of reinvolvement. Depending on the individual's perspective, re-entry into the drug world could be viewed as either a comeback (from a forced withdrawal) or a relapse (from a voluntary withdrawal).

Most people forced out of drug trafficking were anxious to return. The decision to phase-out was never theirs, and the desire to get back into dealing or smuggling was based on many of the same reasons which drew them into the field originally. Coming back from financial, legal, and reputational bustouts was possible but difficult and was not always successfully accomplished. They had to re-establish contacts, rebuild their organization and fronting arrangements, and raise the operating capital to resume dealing. More difficult was the problem of overcoming the circumstances surrounding their departure. Once smugglers and dealers resumed operating, they often found their former colleagues suspicious of them. One frustrated dealer described the effects of his prison experience:

> When I first got out of the joint [jail], none of my old friends would have anything to do with me. Finally, one guy who had been my partner told me it was because everyone was suspicious of my getting out early and thought I made a deal [with police to inform on his colleagues].

Dealers and smugglers who returned from bustouts were thus informally subjected to a trial period in which they had to re-establish their trustworthiness and reliability before they could once again move in the drug world with ease.

Re-entry from voluntary withdrawal involved a more difficult decision-making process, but was easier to implement. The factors enticing individuals to re-enter the drug world were not the same as those which motivated their original entry. As we noted above, experienced dealers and smugglers often privately weighed their reasons for wanting to quit and wanting to stay in. Once they left, their images of and hopes for the straight world failed to materialize. They could not make the shift to the norms, values, and lifestyle of the straight society and could not earn a living within it. Thus, dealers and smugglers decided to re-enter the drug business for basic reasons: the material perquisites, the hedonistic gratifications, the social ties, and the fact that they had nowhere else to go.

Once this decision was made, the actual process of re-entry was relatively easy. One dealer described how the door back into dealing remained open for those who left voluntarily:

> I still see my dealer friends, I can still buy grams from them when I want to. It's the respect they have for me because I stepped out of it without being busted or burning someone. I'm coming out with a good reputation, and even though the scene is a whirlwind—people moving up, moving down, in, out—if I didn't see anybody for a year I could call them up and get right back in that day.

People who relapsed thus had little problem obtaining fronts, reestablishing their reputations, or readjusting to the scene.

Career Shifts

Dealers and smugglers who re-entered the drug world, whether from a voluntary or forced phase-out, did not always return to the same level of transacting or commodity which characterized their previous style of operation. Many individuals underwent a "career shift" (Luckenbill and Best, 1981) and became involved in some new segment of the drug world. These shifts were sometimes lateral, as when a member of a smuggling crew took on a new specialization, switching from piloting to operating a stash house, for example. One dealer described how he utilized friendship networks upon his re-entry to shift from cocaine to marijuana trafficking:

> Before, when I was dealing cocaine, I was too caught up in using the drug and people around me were starting to go under from getting into "base" [another form of cocaine]. That's why I got out. But now I think I've got myself together and even though I'm dealing again I'm staying away from coke. I've switched over to dealing grass. It's a whole different circle of people. I got into it through a close friend I used to know before, but I never did business with him because he did grass and I did coke.

Vertical shifts moved operators to different levels. For example, one former smuggler returned and began dealing; another top-level marijuana dealer came back to find that the smugglers he knew had disappeared and he was forced to buy in smaller quantities from other dealers.

Another type of shift relocated drug traffickers in different styles of operation. One dealer described how, after being arrested, he tightened his security measures:

> I just had to cut back after I went through those changes. Hell, I'm not getting any younger and the idea of going to prison bothers me a lot more than it did 10 years ago. The risks are no longer worth it when I can have a comfortable income with less risk. So I only sell to four people now. I don't care if they buy a pound or a gram.

A former smuggler who sold his operation and lost all his money during phase-out returned as a consultant to the industry, selling his expertise to those with new money and fresh manpower:

> What I've been doing lately is setting up deals for people. I've got foolproof plans for smuggling cocaine up here from Colombia; I tell them how to modify their airplanes to add on extra fuel tanks and to fit in more weed, coke, or whatever they bring up. Then I set them up with refueling points all up and down Central America, tell them how to bring it up here, what points to come in at, and what kind of receiving unit to use. Then they do it all and I get 10 percent of what they make.

Re-entry did not always involve a shift to a new niche, however. Some dealers and smugglers returned to the same circle of associates, trafficking activity, and commodity they worked with prior to their departure. Thus, drug dealers' careers often peaked early and then displayed a variety of shifts, from lateral mobility, to decline, to holding fairly steady.

A final alternative involved neither completely leaving nor remaining within the deviant world. Many individuals straddled the deviant and respectable worlds forever by continuing to dabble in drug trafficking. As a result of their experiences in the drug world they developed a deviant self-identity and a deviant *modus operandi*. They might not have wanted to bear the social and legal burden of full-time deviant work but neither were they willing to assume the perceived confines and limitations of the straight world. They therefore moved into the entrepreneurial realm, where their daily activities involved some kind of hustling or "wheeling and dealing" in an assortment of legitimate, quasi-legitimate, and deviant ventures, and where they could be their own boss. This enabled them to retain certain elements of the deviant lifestyle, and to socialize on the fringes of the drug community. For these individuals, drug dealing shifted from a primary occupation to a sideline, though they never abandoned it altogether.

LEAVING DRUG TRAFFICKING

This career pattern of oscillation into and out of active drug trafficking makes it difficult to speak of leaving drug trafficking in the sense of final retirement. Clearly, some people succeeded in voluntarily retiring. Of these, a few managed to prepare a post-deviant career for themselves by transferring their drug money into a legitimate enterprise. A larger group was forced out of dealing and either didn't or couldn't return; the bustouts were sufficiently damaging that they never attempted re-entry, or they abandoned efforts after a series of unsuccessful attempts. But there was no way of structurally determining in advance whether an exit from the business would be temporary or permanent. The vacillations in dealers' intentions were compounded by the complexity of operating successfully in the drug world. For many, then, no phase-out could ever be definitely assessed as permanent. As long as individuals had skills, knowledge, and connections to deal they retained the potential to re-enter the occupation at any time. Leaving drug trafficking may thus be a relative phenomenon, characterized by a trailing-off process where spurts of involvement appear with decreasing frequency and intensity.

SUMMARY

Drug dealing and smuggling careers are temporary and fraught with multiple attempts at retirement. Veteran drug traffickers quit their occupation because of the ambivalent feelings they develop toward their deviant life. As they age in the career their experience changes, shifting from a work life that is exhilarating and free to one that becomes increasingly dangerous and confining. But just as their deviant careers are temporary, so too are their retirements. Potential recruits are lured into the drug business by materialism, hedonism, glamor, and excitement. Established dealers are lured away from the deviant life and back into the mainstream by the attractions of security and social ease. Retired dealers and smugglers are lured back in by their expertise, and by their ability to make money quickly and easily. People who have been exposed to the upper levels of drug trafficking therefore find it extremely difficult to quit their deviant occupation permanently. This stems, in part, from their difficulty in moving from the illegitimate to the legitimate business sector. Even more significant is the affinity they form for their deviant values and lifestyle. Thus few, if any, of our subjects were successful in leaving deviance entirely. What dealers and smugglers intend, at the time, to be a permanent withdrawal from drug trafficking can be seen in retrospect as a pervasive occupational pattern of mid-career shifts and oscillations. More research is needed into the complex process of how people get out of deviance and enter the world of legitimate work.

NOTES

1. The term "soft" drugs generally refers to marijuana, cocaine and such psychedelics as LSD and mescaline (Carey, 1968). In this paper we do not address trafficking in psychedelics because, since they are manufactured in the United States, they are neither imported nor distributed by the group we studied.

2. This is an idealized figure representing the profit a dealer or smuggler could potentially earn and does not include deductions for such miscellaneous and hard-to-calculate costs as: time or money spent in arranging deals (some of which never materialize); lost, stolen, or unrepaid money or drugs; and the personal drug consumption of a drug trafficker and his or her entourage. Of these, the single largest expense is the last one, accounting for the bulk of most Southwest County dealers' and smugglers' earnings.

3. We continued to conduct follow-up interviews with key informants through 1983.

4. While other studies of drug dealing have also noted that participants did not maintain an uninterrupted stream of career involvement (Blum et al., 1972; Carey, 1968; Lieb and Olson, 1976; Waldorf et al., 1977), none have isolated or described the oscillating nature of this pattern.

5. In the dealers' vernacular, this term is not used in the clinical sense of an individual psychopathology rooted in early childhood traumas. Instead, it resembles Lemert's (1962) more sociological definition which focuses on such behavioral dynamics as suspicion, hostility, aggressiveness, and even delusion. Not only Lemert, but also Waldorf et al. (1977) and Wedow (1979) assert that feelings of paranoia can have a sound basis in reality, and are therefore readily comprehended and even empathized with others.

6. At this point, a limitation to our data must be noted. Many of the dealers and smugglers we observed simply "disappeared" from the scene and were never heard from again. We therefore have no way of knowing if they phased-out (voluntarily or involuntarily), shifted to another scene, or were killed in some remote place. We cannot, therefore, estimate the numbers of people who left the Southwest County drug scene via each of the routes discussed here.

7. It is impossible to determine the exact percentage of people falling into the different phase-out categories: due to oscillation, people could experience several types and thus appear in multiple categories.

REFERENCES

Adler, Patricia A., and Peter Adler. 1982. "Criminal commitment among drug dealers." *Deviant Behavior* 3: 117–135.

Anonymous. 1969. "On selling marijuana." In Erich Goode (ed.), *Marijuana* (pp. 92–102). New York: Atherton.

Atkyns, Robert L., and Gerhard J. Hanneman. 1974. "Illicit drug distribution and dealer communication behavior." *Journal of Health and Social Behavior* 15(March): 36–43.

Biernacki, Patrick, and Dan Waldorf. 1981. "Snowball sampling." *Sociological Methods and Research* 10(2): 141–163.

Blum, Richard H., and Associates. 1972. *The Dream Sellers*. San Francisco: Jossey-Bass.

Carey, James T. 1968. *The College Drug Scene*. Englewood Cliffs, NJ: Prentice-Hall.

Douglas, Jack D. 1976. *Investigative Social Research*. Beverly Hills, CA: Sage.

Goode, Erich. 1970. *The Marijuana Smokers*. New York: Basic.

Langer, John. 1977. "Drug entrepreneurs and dealing culture." *Social Problems* 24(3): 377–385.

Lemert, Edwin. 1962. "Paranoia and the dynamics of exclusion." *Sociometry* 25(March): 2–20.

Lieb, John, and Sheldon Olson. 1976. "Prestige, paranoia, and profit: On becoming a dealer of illicit drugs in a university community." *Journal of Drug Issues* 6(Fall): 356–369.

Luckenbill, David F., and Joel Best. 1981. "Careers in deviance and respectability: The analogy's limitations." *Social Problems* 29(2): 197–206.

Mouledoux, James. 1972. "Ideological aspects of drug dealership." In Ken Westhues (ed.), *Society's Shadow: Studies in the Sociology of Countercultures* (pp. 110–122). Toronto: McGraw-Hill, Ryerson.

Riemer, Jeffrey W. 1977. "Varieties of opportunistic research." *Urban Life* 5(4): 467–477.

Waldorf, Dan, Sheila Murphy, Craig Reinarman, and Bridget Joyce. 1977. *Doing Coke: An Ethnography of Cocaine Users and Sellers*. Washington, DC: Drug Abuse Council.

Wedow, Suzanne. 1979. "Feeling paranoid: The organization of an ideology." *Urban Life* 8(1): 72–93.

POST-DEVIANCE

40

The Professional Ex-
An Alternative for Exiting the Deviant Career

J. DAVID BROWN

Brown highlights an alternate route out of deviance that has recently become more common in society: capitalizing on one's former deviance. Brown presents the case of former alcoholics, like himself, who are helped to sobriety through alcohol counseling and then go on to become counselors themselves. They are drawn to this practice for two main reasons: remaining involved helps them maintain their sobriety, and their status as former deviants gives them an experiential credential that has become respected in society. Like ex-alcoholics, we commonly see ex-cult members, ex-anorectics and bulimics, ex-addicts, and a host of other professional ex-s trading on their past to re-enter society. This article moves beyond our discussion of wanting to get out of deviance to actually making the move and sticking with it to find a post-deviant life.

This study explores the careers of professional ex-s, persons who have exited their deviant careers by replacing them with occupations in professional counseling. During their transformation professional ex-s utilize vestiges of their deviant identity to legitimate their past deviance and generate new careers as counselors.

Recent surveys document that approximately 72% of the professional counselors working in the over 10,000 U.S. substance abuse treatment centers are former substance abusers (NAADAC 1986; Sobell and Sobell 1987). This attests to the significance of the professional ex-phenomenon. Though not all ex-deviants become professional ex-s, such data clearly suggest that the majority of substance abuse counselors are professional ex-s.[1]

Since the inception of the notion of deviant career by Goffman (1961) and Becker (1963), research has identified, differentiated, and explicated the char-

---

From *The Sociological Quarterly*, Vol. 32, No. 2, pp 219–230. © 1991 by JAI Press.
Reprinted by permission of University of California Press.

acteristics of specific deviant career stages (e.g., Adler and Adler 1983; Luckenbill and Best 1981; Meisenhelder 1977; Miller 1986; Shover 1983). The literature devoted to exiting deviance primarily addresses the process whereby individuals abandon their deviant behaviors, ideologies, and identities and replace them with more conventional lifestyles and identities (Irwin 1970; Lofland 1969; Meisenhelder 1977; Shover 1983). While some studies emphasize the role of authorities or associations of ex-deviants in this change (e.g., Livingston 1974; Lofland 1969; Volkman and Cressey 1963), others suggest that exiting deviance is a natural process contingent upon age-related, structural, and social psychological variables (Frazier 1976; Inciardi 1975; Irwin 1970; Meisenhelder 1977; Petersilia 1980; Shover 1983).

Although exiting deviance has been variously conceptualized, to date no one has considered that it might include adoption of a legitimate career premised upon an identity that embraces one's deviant history. Professional ex-s exemplify this mode of exiting deviance.

Ebaugh's (1988) model of role exit provides an initial framework for examining this alternative mode of exiting the deviant career. Her model suggests that former roles are never abandoned but, instead, carry over into new roles. I elaborate her position and contend that one's deviant identity is not an obstacle that must be abandoned prior to exiting or adopting a more conventional lifestyle. To the contrary, one's lingering deviant identity facilitates rather than inhibits the exiting process.

How I gathered data pertinent to exiting, my relationship to these data, and how my personal experiences with exiting deviance organize this article, follow. I then present a four stage model that outlines the basic contours of the professional ex- phenomenon. Finally I suggest how the professional ex- phenomenon represents an alternative interpretation of exiting deviance that generalizes to other forms of deviance.

METHODS

Data for this research consist of introspective and qualitative material.

Introspective Data

My introspections distill 20 years of experience with substance abuse/alcoholism, social control agents/agencies, and professional counselor training. I spent 13 years becoming a deviant drinker and entered substance abuse treatment in 1979. For 5 years (1981–1986), I was a primary therapist and family interventionist for a local private residential treatment facility.

"Systematic sociological introspection" (Ellis 1987, 1990), "autoethnography" (Hayano 1979), and "opportunistic research" (Reimer 1977) accessed the introspective data. Each group status—abuser, patient, therapist—indicates the "complete membership role" (Adler and Adler 1987) that combines unique circumstances with personal expertise to enhance research. The four stage model of exiting described later is, in part, informed by reexamination of written artifacts of my therapeutic/recovery experiences (e.g., alcobiography, moral inventory, daily inventory journal) and professional counselor training (e.g., term paper, internship journal).

Qualitative Data

Qualitative data were collected over a six month period of intensive interviews with 35 counselor ex-s employed in a variety of community, state, and private institutions that treat individuals with drug, alcohol, and/or eating disorder problems.[2]

These professional ex-s worked in diverse occupations prior to becoming substance abuse counselors. A partial list includes employment as accountants, managers, salespersons, nurses, educators, and business owners. Although they claimed to enter the counseling profession within two years of discharge from therapy, their decision to become counselors usually came within one year. On the average they had been counselors for four and one half years. Except one professional ex- who previously counseled learning disabled children, all claimed they had not seriously considered a counseling career before entering therapy.

THE EXIT PROCESS

Ebaugh (1988) contends that the experience of being an "ex" of one kind or another is common to most people in modern society. Emphasizing the sociological and psychological continuity of the ex- phenomenon she states, "[I]t implies that interaction is based not only on current role definitions but, more important, past identities that somehow linger on and define how people see and present themselves in their present identities" (p. xiii). Ebaugh defines the role exit as the "process of disengagement from a role that is central to one's self-identity and the reestablishment of an identity in a new role that takes into account one's ex-role" (p. 1).

Becoming a professional ex- is the outcome of a four stage process through which ex-s capitalize on the experience and vestiges of their deviant career in order to establish a new identity and role in a respectable organization. This process comprises emulation of one's therapist, the call to a counseling career, status-set realignment, and credentialization.

Stage One: Emulation of One's Therapist

The emotional and symbolic identification of these ex-s with their therapists during treatment, combined with the deep personal meanings they imputed to these relationships, was a compelling factor in their decisions to become counselors. Denzin (1987, pp. 61–62) identifies the therapeutic relationship's significance thus: "Through a process of identification and surrender (which may be altruistic), the alcoholic may merge her ego and her self in the experiences and the identity of the counselor. The group leader . . . is the group ego ideal, for he or she is a successful recovering alcoholic. . . . An emotional bond is thus formed with the group counselor. . . .

Professional ex-s not only developed this emotional bond but additionally aspired to have the emotions and meanings once projected toward their therapists ascribed to them. An eating disorders counselor discussed her relationship with her therapist and her desire to be viewed in a similar way with these words:

> My counselor taught me the ability to care about myself and other people. Before I met her I was literally insane. She was the one who showed me that I wasn't crazy. Now, I want to be the person who says, "No, you're not crazy!" I am the one, now, who is helping them to get free from the ignorance that has shrouded eating disorders.

Counselors enacted a powerfully charismatic role in professional ex-s' therapeutic transformation. Their "laying on of verbal hands" provided initial comfort and relief from the ravaging symptoms of disease. They came to represent what ex-s must do both spiritually and professionally for themselves. Substance abuse therapy symbolized the "sacred" quest for divine grace rather than the mere pursuit of mundane, worldly, or "profane" outcomes like abstinence or modification of substance use/abuse behaviors; counselors embodied the sacred outcome.

Professional ex-s claimed that their therapists were the most significant change agent in their transformation. "I am here today because there was one very influential counselor in my life who helped me to get sober. I owe it all to God and to him," one alcoholism counselor expressed. A heroin addiction counselor stated, "The best thing that ever happened in my life was meeting Sally [her counselor]. She literally saved my life. If it wasn't for her I'd still probably be out there shootin' up or else be in prison, or dead."

Subjects' recognition and identification of a leader's charismatic authority, as Weber (1968) notes, is decisive in validating that charisma and developing absolute trust and devotion. The special virtues and powers professional ex-s perceived in their counselors subsequently shaped their loyalty and devotion to the career.

Within the therapeutic relationship, professional ex-s perform a priestly function through which a cultural tradition passes from one generation to the next. While knowledge and wisdom pass downward (from professional ex- to patient), careers build upward (from patient to professional ex-). As the bearers of the cultural legacy of therapy, professional ex-s teach patients definitions of the situation they learned as patients. Indeed, part of the professional ex-mystique resides in once having been a patient (Bissell 1982). In this regard,

> My counselor established her legitimacy with me the moment she disclosed the fact that she, too, was an alcoholic. She wasn't just telling me what to do, she was living her own advice. By the example she set, I felt hopeful that I could recover. As I reflect upon those experiences I cannot

think of one patient ever asking me about where I received my professional training. At the same time, I cannot begin to count the numerous times that my patients have asked my if I was "recovering."

Similar to religious converts' salvation through a profoundly redemptive religious experience, professional ex-s' deep career commitment derives from a transforming therapeutic resocialization. As the previous examples suggest, salvation not only relates to a changed universe of discourse; it is also identified "with one's personal therapist."[3]

At this stage, professional ex-s trust in and devote themselves to their counselors' proselytizations as a promissory note for the future. The promise is redemption and salvation from the ever-present potential for self-destruction or relapse that looms in their mental horizon. An eating disorders counselor shared her insights in this way:

> I wouldn't have gotten so involved in eating disorders counseling if I had felt certain that my eating disorder was taken care of. I see myself in constant recovery. If I was so self assured that I would never have the problem again there would probably be less of an emphasis on being involved in the field but I have found that helping others, as I was once helped, really helps me.

The substance abuse treatment center transforms from a mere "clinic" occupied by secularly credentialed professionals into a moral community of single believers. As Durkheim (1915) suggests, however, beliefs require rites and practices in order to sustain adherents' mental and emotional states.

Stage Two: The Call to a Counseling Career

At this juncture, professional ex-s begin to turn the moral corner on their deviance. Behaviors previously declared morally reprehensible are increasingly understood within a new universe of discourse as symptoms of a much larger disease complex. This recognition represents one preliminary step toward grace. In order to emulate their therapist, however, professional ex-s realize they must dedicate themselves to an identity and lifestyle that ensure their own symptoms' permanent remission. One alcoholism counselor illustrated this point by stating:

> I can't have my life, my health, my family, my job, my friends, or anything, unless I take daily necessary steps to ensure my continued recovery. My program of recovery has to come first. Before I can go out there and help my patients I need to always make sure that my own house is in order.

As this suggests, a new world-view premised upon accepting the contingencies of one's illness while maintaining a constant vigilance over potentially recurring symptoms replaces deviant moral and social meanings. Professional ex-s' recognition of the need for constant vigilance is internalized as their moral mission from which their spiritual duty (a counseling career) follows as a natural next step.

Although professional ex-s no longer engage in substance abuse behaviors, they do not totally abandon deviant beliefs or identity. "Lest we become complacent and forget from whence we came," as one alcoholism counselor indicated the significance of remembering and embracing the past.

Professional ex-s' identification with their deviant past undergirds their professional, experiential, and moral differentiation from other professional colleagues. A heroin addiction counselor recounted how he still identified himself as an addict and deviant:

> My perspective and my affinity to my clients, particularly the harder core criminals, is far better than the professor and other doctors that I deal with here in my job. We're different and we really don't see things the same way at all. Our acceptance and understanding of these people's diseases, if you will, is much different. They haven't experienced it. They don't know these people at all. It takes more than knowing about something to be effective. I've been there and, in many respects, I will always be there.

In this way, other counselors' medical, psychiatric, or therapeutic skills are construed as part of the ordinary mundane world. As the quotation indicates, professional ex-s intentionally use their experiential past and therapeutic transformations to legitimate their entrance into and authority in counseling careers.

Professional ex-s embrace their deviant history and identity as an invaluable, therapeutic resource and feel compelled to continually reaffirm its validity in an institutional environment. Certainly, participating in "12 Step Programs"[4] without becoming counselors could help others but professional ex-s' call requires greater immersion than they provide. An alcoholism counselor reflected upon this need thus:

> For me, it was no longer sufficient to only participate "anonymously" in A.A. I wanted to surround myself with other spiritual and professional pilgrims devoted to receiving and imparting wisdom.

Towards patients, professional ex-s project a saintly aura and exemplify an "ideal recovery." Internalization of self-images previously ascribed to their therapist and now reaffirmed through an emotional and moral commitment to the counseling profession facilitate this ideation. Invariably, professional ex-s' counseling careers are in institutions professing treatment ideologies identical to what they were taught as patients. Becoming a professional ex-symbolizes a value elevated to a directing goal, whose pursuit predisposes them to interpret all ensuing experience in terms of relevance to it.

Stage Three: Status-Set Realignment

Professional ex-s' deep personal identification with their therapist provides an ego ideal to be emulated with regard to both recovery and career. They immerse themselves in what literally constitutes a "professional recovery career" that provides an institutional location to reciprocate their counselors' gift, immerse themselves in a new universe of discourse, and effectively lead novitiates to salvation. "I wouldn't be here today if it wasn't for all of the help I received in therapy. This is my way of paying some of those people back by helping those still in need," one alcoholism counselor related this.

Professional ex-s' identities assume a "master status" (Hughes 1945) that differs in one fundamental respect from others' experiencing therapeutic resocialization. Specifically, their transformed identities not only become the "most salient" in their "role identity hierarchy" (Stryker and Serpe 1982), but affect all other roles in their "status-sets" (Merton 1938). One alcoholism counselor reflected upon it this way:

> Maintaining a continued program of recovery is the most important thing in my life. Everything else is secondary. I've stopped socializing with my old friends who drink and have developed new recovering friends. I interact differently with my family. I used to work a lot of overtime but I told my old boss that overtime jeopardized my program. I finally began to realize that the job just didn't have anything to do with what I was really about. I felt alienated. Although I had been thinking about becoming a counselor ever since I went through treatment, I finally decided to pursue it.

Role alignment is facilitated by an alternative identity that redefines obligations associated with other, less significant, role identities. In the previous example, the strains of expectations associated with a former occupation fostered a role alignment consistent with a new self-image. This phenomenon closely resembles what Snow and Machalek (1983, p. 276) refer to as "embracement of a master role" that "is not merely a mask that is taken off or put on according to the situation. . . . Rather, it is central to nearly all situations. . . . " An eating disorders counselor stated the need to align her career with her self-image, "I hid in my former profession, interacting little with people. As a counselor, I am personally maturing and taking responsibility rather than letting a company take care of me. I have a sense of purpose in this job that I never had before."

Financial remuneration is not a major consideration in the decision to become a professional ex-. The pure type of call, Weber (1968, p. 52) notes, "disdains and repudiates economic exploitation of the gifts of grace as a source of income. . . ." Most professional ex-s earned more money in their previous jobs. For instance, one heroin addiction counselor stated:

> When I first got out of treatment, my wife and I started an accounting business. In our first year we cleared nearly sixty thousand dollars. The money was great and the business showed promise but something was missing. I missed being around other addicts and I knew I wanted to do more with my life along the lines of helping out people like me.

An additional factor contributing to professional ex-s' abandonment of their previous occupation is their recognition that a counseling career could resolve lingering self-doubts about their ability to remain abstinent. In this respect becoming a professional ex allows "staying current" with their own recovery needs while continually reaffirming the severity of their illness. An eating disorder counselor explained:

> I'm constantly in the process of repeating insights that I've had to my patients. I hear myself saying, to them, what I need to believe for myself. Being a therapist helps me to keep current with my own recovery. I feel that I am much less vulnerable to my disease in this environment. It's a way that I can keep myself honest. Always being around others with similar issues prevents me from ignoring my own addiction clues.

This example illustrates professional ex-s' use of their profession to secure self-compliance during times of self-doubt. While parroting the virtues of the program facilitates recognition that they, too, suffer from a disease, the professional ex- role, unlike their previous occupations, enables them to continue therapy indirectly.

Finally, the status the broader community ascribes to the professional ex- role encourages professional ex-s' abandonment of previous roles. Association with an institutional environment and an occupational role gives the professional ex- a new sense of place in the surrounding community, within which form new self-concepts and self-esteem, both in the immediate situation and in a broader temporal framework.

The internal validation of professional ex-s' new identity resides in their ability to successfully anticipate the behaviors and actions of relevant alters. Additionally, they secure validation by other members of the professional ex- community in a manner atypical for other recovering individuals. Affirmation by this reference community symbolizes validation by one's personal therapist and the therapeutic institution, as a heroin addiction counselor succinctly stated:

> Becoming a counselor was a way to demonstrate my loyalty and devotion to helping others and myself. My successes in recovery, including being a counselor, would be seen by patients and those who helped me get sober. It was a return to treatment, for sure, but the major difference was that this time I returned victorious rather than defeated.

External validation, on the other hand, comes when others outside the therapeutic community accord legitimacy to the professional ex- role. In this regard, a heroin addiction counselor said:

> I remember talking to this guy while I was standing in line for a movie. He asked me what I did for a living and I told him that I was a drug abuse counselor. He started asking me all these questions about the drug problem and what I thought the answers were. When we finally got up to the door of the theater he patted me on the back and said, "You're doing a wonderful job. Keep up the good work. I really admire you for what you're trying to do." It really felt good to have a stranger praise me.

Professional ex-s' counseling role informs the performance of all other roles, compelling them to abandon previous work they increasingly view as mundane and polluting. The next section demonstrates how this master role organizes the meanings associated with their professional counselor training.

Stage Four: Credentialization

One characteristic typically distinguishing the professions from other occupations is specialized knowledge acquired at institutions of higher learning (Larson 1977; Parsons 1959; Ritzer and Walczak 1986, 1988). Although mastering esoteric knowledge and professional responsibilities in a therapeutic relationship serve as gatekeepers for entering the counseling profession, the moral and emotional essence of being a professional ex- involves much more.

Professional ex-s see themselves as their patients' champions. "Knowing what it's like" and the subsequent education and skills acquired in training legitimate claims to the "entitlements of their stigma" (Gusfield 1982), including professional status. Their monopoly of an abstruse body of knowledge and skill is realized through their emotionally lived history of shame and guilt as well as the hope and redemption secured through therapeutic transformation. Professional ex-s associate higher learning with their experiential history of deviance and the emotional context of therapy. Higher learning symbolizes rediscovery of a moral sense of worth and sacredness rather than credential acquisition. This distinction was clarified by an alcoholism counselor:

> Anymore, you need to have a degree before anybody will hire you. I entered counseling with a bachelors but I eventually received my MSW about two years ago. I think the greatest benefit in having the formal training is that I have been able to more effectively utilize my personal alcoholism experiences with my patients. I feel that I have a gift to offer my patients which doesn't come from the classroom. It comes from being an alcoholic myself.

These entitlements allow professional ex-s to capitalize on their deviant identity in two ways: the existential and phenomenological dimensions of their lived experience of "having made their way from the darkness into the light" provide their experiential and professional *legitimacy* among patients, the community, and other professionals, as well as occupational *income*. "Where else could I go and put bulimic and alcoholic on my resume and get hired?" one counselor put it.

Professional ex-s generally eschew meta-perspective interpretations of the system in which they work. They desire a counseling method congruent with their fundamental universe of discourse and seek, primarily, to perpetuate this system (Peele 1989; Room 1972, 1976). The words of one educator at a local counselor training institute are germane:

> These people [professional ex-s] . . . are very fragile when they get here. Usually, they have only been in recovery for about a year. Anyone who challenges what they learned in therapy, or in their program of recovery [i.e., A.A., Narcotics Anonymous, Overeaters Anonymous] . . . is viewed as a threat. Although we try to change some of that while they're here with us, I still see my role here as one of an extended therapist rather than an educator.

Information challenging their beliefs about how they, and their patients, should enact the rites associated with recovery is condemned (Davies 1963; Pattison 1987; Roizen 1977). They view intellectual challenges to the disease concept as attacks on their personal program of recovery. In a Durkheimian sense, such challenges "profane" that which they hold "sacred."

Within the walls of these monasteries professional ex-s emulate their predecessors as one generation of healers passes on to the next an age old message of salvation. Although each new generation presents the path to enlightenment in somewhat different, contemporary terms, it is already well lit for those "becoming a professional ex-."

DISCUSSION

Focusing on their lived experiences and accounts, this study sketches the central contours of professional ex-s' distinctive exit process. More generally, it also endeavors to contribute to the existing literature on deviant careers.

An identity that embraces their deviant history and identity undergirds the professional ex-s' careers. This exiting mode is the outcome of a four stage process enabling professional ex-s to capitalize on their deviant history. They do not "put it all behind them" in exchange for conventional lifestyles, values, beliefs, and identities. Rather, they use vestiges of their deviant biography as an explicit occupational strategy.

My research augments Ebaugh's (1988) outline of principles underlying role exit in three ways. First, her discussion suggests that people are unaware of these guiding principles. While this holds for many, professional ex-s' intentional rather than unintentional embracement of their deviant identity is the step by which they adopt a new role in the counseling profession. Second, Ebaugh states that significant others' negative reactions inhibit or interrupt exit. Among professional ex-s, however, such reactions are a crucial precursor to their exit mode. Finally, Ebaugh sees role exit as a voluntary, individually initiated process, enhanced by "seeking alternatives" through which to explore other roles. Professional ex-s, by contrast, are compelled into therapy. They do not look for this particular role. Rather, their alternatives are prescribed through their resocialization into a new identity.

Organizations in American society increasingly utilize professional ex-s in their social control efforts. For example, the state of Colorado uses prisoners to counsel delinquent youth. A preliminary, two year, follow-up study suggests that these prisoner-counselors show only 13% recidivism (Shiller 1988) and a substantial number want to return to college or enter careers as guidance counselors, probation officers, youth educators, or law enforcement consultants. Similarly, a local effort directed toward curbing gang violence, the Open Door Youth Gang Program, was developed by a professional ex- and uses former gang members as counselors, educators, and community relations personnel.

Further examination of the modes through which charismatic, albeit licensed and certified, groups generate professional ex- statuses is warranted. Although the examples just described differ from the professional ex-s examined earlier in this research in terms of therapeutic or "medicalized" resocialization, their similarities are even more striking. Central to them all is that a redemptive community provides a reference group whose moral and social standards are internalized. Professional ex- statuses are generated as individuals intentionally integrate and embrace rather than abandon their deviant biographies as a specific occupational strategy.

NOTES

1. Most individuals in substance abuse therapy do not become professional ex-s. Rather, they traverse a variety of paths not articulated here including (1) dropping out of treatment, (2) completing treatment but returning to substance use and/or abuse, and (3) remaining abstinent after treatment but feeling no compulsion to enter the counseling profession. Future research will explore the differences among persons by mode of exit. Here, however, analysis and description focus exclusively on individuals committed to the professional ex- role.

2. I conducted most interviews at the subject's work environment, face-to-face. One interview was with a focus group of 10 professional ex-s (Morgan 1988). Two interviews were in my office, one at my home, and one at a subject's home. I interviewed each individual one time for approximately one hour. Interviews were semi-structured, with open-end questions designed to elicit responses related to feelings, thoughts, perceptions, reflections, and meanings concerning subjects' past deviance, factors facilitating their exit from deviance, and their counseling career.

3. I contend that significantly more professional ex-s pursue their careers due to therapeutic resocialization than to achieving sobriety/recovery exclusively through the 12 Step Program (e.g., A.A.). It is too early, however, to preclude that some may enter substance abuse counseling careers lacking any personal therapy. My experiences and my interviews with other professional ex-s suggest that very few professional ex-s enter the profession directly through their contacts with the 12 Step Program. The program's moral precepts—that "sobriety is a gift from God" that must be "given freely to others in order to assure that one may keep the gift"—would appear to discourage rather than encourage substance abuse counseling careers. Financial remuneration for assisting fellow substance abusers directly violates these precepts. Further, professional ex-s are commonly disparaged in A.A. circles as "two hatters" (cf. Denzin 1987). They are, therefore, not a positive reference group for individuals recovering exclusively through the 12 Step Program. Sober 12 Step members are more inclined to emulate their "sponsors" than pursue careers with no experiential referents or direct relevance to their recovery. Further data collection and analysis will examine these differences. Extant data, however, strongly indicate that therapeutic resocialization and a professional role model provide the crucial link between deviant and substance abuse counseling careers.

4. "12 Step Program" refers to a variety of self-help groups (e.g., A.A., Narcotics Anonymous, Overeaters Anonymous) patterning their recovery model upon the original 12 Steps and 12 Traditions of A.A.

REFERENCES

Adler, Patricia, and Peter Adler. 1983. "Shifts and Oscillations in Deviant Careers: The Case of Upper-Level Drug Dealers and Smugglers." *Social Problems* 31: 195–207.

———, 1987. *Membership Roles in Field Research*. Newbury Park, CA: Sage.

Becker, Howard. 1963. *Outsiders: Studies in the Sociology of Deviance*. New York: Free Press.

Bissell, LeClair. 1982. "Recovered Alcoholism Counselors." Pp. 810–817 in *Encyclopedic Handbook of Alcoholism*, edited by E. Mansell Pattison and Edward Kaufman. New York: Gardner.

Davies, D. L. 1963. "Normal Drinking in Recovered Alcoholic Addicts" (comments by various correspondents). *Quarterly Journal of Studies on Alcohol* 24: 109–121, 321–332.

Denzin, Norman. 1987. *The Recovering Alcoholic*. Beverly Hills: Sage.

Durkheim, Emile. 1915. *The Elementary Forms of the Religious Life*. New York: Free Press.

Ebaugh, Helen Rose Fuchs. 1988. *Becoming an Ex: The Process of Role Exit*. Chicago: University of Chicago Press.

Ellis, Carolyn. 1987. "Systematic Sociological Introspection and the Study of Emotions." Paper presented to the annual meetings of the American Sociological Association. Chicago.

———. 1990. "Sociological Introspection and Emotional Experience." *Symbolic Interaction* 13(2): Forthcoming.

Frazier, Charles. 1976. *Theoretical Approaches to Deviance*. Columbus: Charles Merrill.

Goffman, Erving. 1961. *Asylums*. Garden City, NY: Anchor.

Gusfield, Joseph. 1982. "Deviance in the Welfare State: The Alcoholism Profession and the Entitlements of Stigma." *Research in Social Problems and Public Policy* 2: 1–20.

Hayano, David. 1979. "Auto-Ethnography: Paradigms, Problems and Prospects." *Human Organization* 38: 99–104.

Hughes, Everett. 1945. "Dilemmas and Contradictions of Status." *American Journal of Sociology* L: 353–359.

Inciardi, James. 1975. *Careers in Crime*. Chicago: Rand McNally.

Irwin, John. 1970. *The Felon*. Englewood Cliffs: Prentice-Hall.

Larson, Magali. 1977. *The Rise of Professionalism*. Berkeley: University of California Press.

Livingston, Jay. 1974. *Compulsive Gamblers*. New York: Harper and Row.

Lofland, John. 1969. *Deviance and Identity*. Englewood Cliffs: Prentice-Hall.

Luckenbill, David F., and Joel Best. 1981. "Careers in Deviance and Respectability: The Analogy's Limitations." *Social Problems* 29: 197–206.

Meisenhelder, Thomas. 1977. "An Exploratory Study of Exiting from Criminal Careers." *Criminology* 15: 319–334.

Merton, Robert. 1938. *Social Theory and Social Structure*. Glencoe: Free Press.

Morgan, David L. 1988. *Focus Groups as Qualitative Research*. Beverly Hills: Sage.

NAADAC. 1986. *Development of Model Professional Standards for Counselor Credentialing*. National Association of Alcoholism and Drug Abuse Counselors. Dubuque: Kendall-Hunt.

Parsons, Talcott. 1959. "Some Problems Confronting Sociology as a Profession." *American Sociological Review* 24: 547–559.

Pattison, E. Mansell. 1987. "Whither Goals in the Treatment of Alcoholism." *Drugs and Society* 2/3:153–171.

Peele, Stanton. 1989. *The Diseasing of America: Addiction Treatment Out of Control*. Toronto: Lexington.

Petersilia, Joan. 1980. "Criminal Career Research: A Review of Recent Evidence." Pp. 321–379 in *Crime and Justice: An Annual Review of Research,* vol. 2, edited by Norval Morris and Michael Tonry. Chicago: University of Chicago Press.

Reimer, Jeffrey. 1977. "Varieties of Opportunistic Research." *Urban Life* 5: 467–477.

Ritzer, George, and David Walczak. 1986. *Working: Conflict and Change.* 3rd ed. Englewood Cliffs: Prentice-Hall.

———. 1988. "Rationalization and the Deprofessionalization of Physicians." *Social Forces* 67: 1–22.

Roizen, Ron. 1977. "Comment on the Rand Report." *Quarterly Journal of Studies on Alcohol* 38: 170–178.

Room, Robin. 1972. "Drinking and Disease: Comment on the Alchologist's Addiction." *Quarterly Journal of Studies on Alcohol* 33: 1049–1059.

———. 1976. "Drunkenness and the Law: Comment on the Uniform Alcoholism Intoxication Treatment Act." *Quarterly Journal of Studies on Alcohol* 37: 113–114.

Shiller, Gene. 1988. "A Preliminary Report on SHAPE-UP." Paper presented to the Colorado District Attorneys Council, Denver.

Shover, Neil. 1983. "The Later Stages of Ordinary Property Offenders' Careers." *Social Problems* 31: 208–218.

Snow, David, and Richard Machalek. 1983. "The Convert as a Social Type." Pp. 259–289 in *Sociological Theory 1983,* edited by Randall Collins. San Francisco: Jossey-Bass.

Sobell, Mark B., and Linda C. Sobell. 1987. "Conceptual Issues Regarding Goals in the Treatment of Alcohol Problems." *Drugs and Alcohol* 2/3:1–37.

Stryker, Sheldon, and Richard Serpe. 1982. "Commitment, Identity Salience, and Role Behavior: Theory and Research Example." Pp. 199–218 in *Personality, Roles, and Social Behavior,* edited by William Ickes and Eric S. Knowles. New York: Springer-Verlag.

Volkman, Rita, and Donald Cressey. 1963. "Differential Association and the Rehabilitation of Drug Addicts." *American Journal of Sociology* 69: 129–142.

Weber, Max. 1968. *On Charisma and Institution Building.* Edited by S. N. Eisenstadt. Chicago: Unversity of Chicago Press.

References for the General and Part Introductions

Adler, Patricia A., and Peter Adler. 1987. *Membership Roles in Field Research,* Newbury Park, CA: Sage.

———. 1995. "The Demography of Ethnography." *Journal of Contemporary Ethnography* 24:3–29.

Becker, Howard S. 1963. *Outsiders: Studies in the Sociology of Deviance,* New York: Free Press.

———. 1973. "Labelling Theory Reconsidered." Pp. 177–212 in *Outsiders.* New York: Free Press.

Best, Joel, and David F. Luckenbill. 1980. "The Social Organization of Deviants." *Social Problems* 28(1): 14–31.

———. 1981. "The Social Organization of Deviance." *Deviant Behavior* 2: 231–59.

Cloward, Richard, and Lloyd Ohlin. 1960. *Delinquency and Opportunity.* Glencoe, IL: Free Press.

Cohen, Albert. 1995. *Delinquent Boys.* Glencoe, IL: Free Press.

Davis, Fred. 1961. "Deviance Disavowal: The Management of Strained Interaction by the Visibly Handicapped." *Social Problems* 9: 120–32.

Erikson, Kai T. 1966. *Wayward Puritans.* New York: Wiley.

Goffman, Erving. 1963. *Stigma.* Englewood Cliffs, NJ: Prentice-Hall.

Gusfield, Joseph. 1967. "Moral Passage: The Symbolic Process in Public Designations of Deviance." *Social Problems* 15: 175–88.

Henry, Jules. 1964. *Jungle People.* New York: Vintage.

Hilgartner, Stephen, and Charles L. Bosk. 1988. "The Rise and Fall of Social Problems: A Public Arenas Model." *American Journal of Sociology* 94: 53–78.

Hughes, Everett. 1945. "Dilemmas and Contradictions of Status." *American Journal of Sociology* (March): 353–59.

Kitsuse, John. 1962. "Societal Reactions to Deviant Behavior: Problems of Theory and Method." *Social Problems* 9: 247–56.

———. 1980. "Coming Out All Over: Deviants and the Politics of Social Problems." *Social Problems* 28: 1–13.

Krauthammer, Charles. 1993: "Defining Deviancy Up." *The New Republic*, November 22, pp. 20–25.

Lemert, Edwin. 1951. *Social Pathology*. New York: McGraw-Hill.

———. 1967. *Human Deviance, Social Problems, and Social Control*. New York: Prentice-Hall.

Luckenbill, David F., and Joel Best. 1981. "Careers in Deviance and Respectability: The Analogy's Limitation." *Social Problems* 296(2): 197–206.

Lyman, Stanford M. 1970. *The Asian in the West*. Reno/Las Vegas, Nevada: Western Studies Center, Desert Research Institute.

Matza, David. 1964. *Delinquency and Drift*. New York: Wiley.

Merton, Robert. 1938. "Social Structure and Anomie." *American Sociological Review* 3 (October): 672–82.

Miller, Walter. 1958. "Lower Class Culture as a Generating Milieu of Gang Delinquency." *Journal of Social Issues* 14(3): 5–19.

Moynihan, Daniel Patrick. 1993. "Defining Deviancy Down" *The American Scholar* 62(1): 17–30.

Polsky, Ned. 1967. *Hustlers Beats and Others*. Chicago: Aldine.

Schur, Edwin. 1979. *Interpreting Deviance*. New York: Harper and Row.

Scott, Marvin, and Stanford Lyman. 1968. "Accounts." *American Sociological Review* 33(1): 46–62.

Sellin, Thorsten. 1938. "Culture Conflict and Crime." A Report of the Subcommittee on Delinquency of the Committee on Personality and Culture. *Social Science Research Council Bulletin* 41, New York.

Sumner, William. 1906. *Folkways*. New York: Vintage.

Sutherland, Edwin. 1934. *Principles of Criminology*. Philadelphia: Lippincott.

Sykes, Gresham, and David Matza. 1957. "Techniques of Neutralization: A Theory of Delinquency." *American Sociological Review* 22:664–70.

Tannenbaum, Frank. 1938. *Crime and the Community*. Boston: Ginn.

Turner, Ralph H. 1972. "Deviance Avowal as Neutralization of Commitment." *Social Problems* 19 (Winter): 308–21.

Weatherford, Jack. 1986. *Porn Row*. New York: Arbor House.